Invasive Animals and Plants in Massachusetts Lakes and Rivers

There is a growing demand for appropriate management of aquatic invasive species in lakes and rivers worldwide. This book covers biology, invasion ecology, sightings, and control techniques of 4 invasive animals and 15 invasive plants in Massachusetts lakes and rivers. It provides valuable information on their biological characteristics and potential expansion pathways, as well as monitoring and management, including alternative management tools using updated biological, chemical, and mechanical methods. The book also includes the regulation of invasive species management to allow resource managers, biologists, practitioners, volunteers, and students a better understanding of compliance and enforcement with environmental law. An example of successful management is highlighted for each invasive species.

Invasive Animals and Plants in Massachusetts Lakes and Rivers is the first book to provide comprehensive, systematic coverage and illustrations for both freshwater invasive animals and plants. Although focusing on Massachusetts, it will provide readers with the bigger picture on multiple invasive species, from prevention, early detection, control, ecological restoration, to public education. Natural resource managers in Massachusetts and surrounding states will glean the detailed and valuable information they need to learn and use to prevent and manage freshwater invasive species in the field.

Invasive Animals and Plants in Massachusetts Lakes and Rivers

Lessons for International Aquatic Management

Wai Hing Wong

CRC Press
Taylor & Francis Group
Boca Raton London New York

CRC Press is an imprint of the
Taylor & Francis Group, an **informa** business

First edition published 2023
by CRC Press
6000 Broken Sound Parkway NW, Suite 300, Boca Raton, FL 33487–2742

and by CRC Press
4 Park Square, Milton Park, Abingdon, Oxon, OX14 4RN

CRC Press is an imprint of Taylor & Francis Group, LLC

© 2023 Wai Hing Wong

ISBN: 978-1-032-06186-3 (hbk)
ISBN: 978-1-032-06188-7 (pbk)
ISBN: 978-1-003-20110-6 (ebk)

DOI: 10.1201/9781003201106

Typeset in Times
by Apex CoVantage, LLC

Contents

Contents

Contents

Foreword

Invasive Animals and Plants in Massachusetts Lakes and Rivers: Lessons for International Aquatic Management by Dr. Wai Hing (David) Wong contains a unique mix of well-documented information useful to both regional handbooks addressing the diversity of problematic invasive species and the applied aquatic resource management literature. He is a certified lake manager (CLM) and currently the manager for two Massachusetts Department of Environmental Protection Programs: (1) WM04 Herbicide/Pesticide Application Program and (2) 401 Water Quality Certification Program. In his over 20 years of experience in studying aquatic invasive species in North America and Asia, he has been a principal investigator or co-investigator for over 30 projects funded by federal, state, and private entities. He has been a deputy editor and associate editor for BioInvasions Records, Management of Biological Invasions, and Aquatic Invasions.

This volume is a valuable introduction to those working in government agencies, natural resource venues, students, lake associations, concerned citizens and public education by taking advantage of the summaries and introductions to each chapter. Summaries are valuable to those requiring brief overview of the contents. Introductions provide easy access to information regarding the morphological characteristics, biology, and taxonomic status of the aquatic nuisance species found in Massachusetts.

Management professionals, academicians, and researchers will appreciate the details on the biology and autecology of the treated organisms and the recently documented distributions of each species in the state as included in maps and tabular formats. The native ranges and histories of invasions into Massachusetts, along with records accounting potential and actual ecological and economic impacts are provided. Details of agencies monitoring programs, relevant chemical, physical, and biological management strategies, products and case studies illustrating the success or lack of those strategies are presented. Extensive references listed after each chapter will be particularly valuable.

The contents are structured so that individuals in Massachusetts, other states in the United States, or other countries who are interested in one taxon can peruse all the information on that organism that is available in one chapter. Every chapter has an abstract and introduction in addition to the main narrative, therefore redundancies are unavoidable. This work volume will be especially valuable to readers in nearby states since most nuisance aquatic species in Massachusetts have been found in many nearby sates and a few that are not currently encountered or detected in their regions may soon appear there.

Dr. Willard N. Harman, CLM
Distinguished Service Professor
Rufus J. Thayer Chair, Otsego Lake Research
Director, SUNY Oneonta Biological Field Station
5838 State Highway 80
Cooperstown, NY 13326

Acknowledgments

The author appreciates all inspirations from Richard Chase, Laurie Kennedy, and Therese Beaudoin on the initiation and development of this book. This would not be possible without the bountiful efforts on data and other information collection by the Massachusetts Department of Environmental Protection (MassDEP) staff who have done biological surveys at more than 2,000 waterbodies in the Commonwealth of Massachusetts. Invasive species monitoring data are primarily from MassDEP's biannual assessment reports on Massachusetts Integrated List of Waters. Sighting data in this book are also collected by the following agencies and individuals: Massachusetts Department of Conservation and Recreation, Charles River Watershed Association, Mystic River Watershed Association, Nashua River Watershed Association, Wayland Surface Water Quality Committee, Massachusetts Water Watch Partnership at the University of Massachusetts, the US Geological Survey (USGS), US Fish and Wildlife Service (FWS), US Forest Service (USFS), Water Resource Services LLC, Northeast Aquatic Nuisance Species Panel, Illinois Natural History Survey, EBT Environmental Consultants Inc., Metcalf & Eddy, Inc., Baystate Environmental Consultants, Inc. (BEC), Environmental Science Services, Inc. (ESS), as well as residents (with confirmation from an expert in the field), Robert Bertin, Jamie Carr, Mike Lowery, Eileen McGourty, Bob Nuzzo, Matthew Reardon, Jim Straub, and Joy Trahan-Liptak.

The expertise and time from the following reviewers are gratefully acknowledged because their constructive comments and suggestions significantly improved the quality of this book: Dave Adams, Therese Beaudoin, Tobias O Bickel, Cynthia and Jeff Boettner, Carrie Brown-Lima, Greg Bugbee, Timothy Counihan, Joan Deely, Luc Denys, Tom Flannery, Willard Harman, Nathan Harms, Kimberly Jensen, Barbara Kickham, Susan Knight, Chuhua Liu, Paul Lord, Bob Maietta, Catherine McGlynn, Bob McMahon, John McPhedran, Meg Modley, Bob Nuzzo, Susan Pasko, Steven Pearson, Read Porter, Edward Rainer, Lisa Rhodes, Allison Roy, Kara Sliwoski, Amy Smagula, Rosali Smith, Michele Tremblay, Rebecca Urban, Ken Wagner, and Nathan Waltham.

The MassDEP's Michelle Waters-Ekanem Internship Program and the following interns are appreciated because they have assisted the author in field surveys, graphic-making, reference checking, and quality assurance and quality control: Aymane Akobi, Kira Becker, Alexander Carli-Dorsey, Hana Colwell, Rachel Holland, Peifeng Li (Derek), Salome Maldonado, Kristina McAvoy, Davis Miller, Niki Patel, Fionna Pennington, Jonathan Rickwood, Riani Suarez, Ursula Svoboda, Jiahui Wang, Mingkai Zhang, and Yichen Zhang.

In addition, I am grateful to CRC Press' Alice Oven and Shikha Garg whose expertise and advice make this book more attractive!

Finally, I am thankful to my wife, Ting, for her long-time support and understanding for my career development.

Disclaimer

Although findings and conclusions of this book and many activities such as data sources are associated with Masschusetts Department of Environmental Protection and the author himself is currently an employee of Massachusetts Department of Environmental Protection, the views in this book are solely of the author himself and do not reflect the policies of the Massachusetts Department of Environmental Protection.

Author Biography

Dr. Wai Hing Wong (David) is an Environmental Analyst at the Massachusetts Department of Environmental Protection, Rese-arch Associate Professor in State University of New York at Oneonta, and Adjunct Associate Professor in Worcester State University. David's interests and experience include environmental monitoring, biological invasions, risk assessment, eutrophication management, ecological restoration, and the Clean Water Act. He has over 20 years of experience in studying and managing aquatic invasive species in North America and Asia. He has a Ph.D. in Marine Ecology from City University of Hong Kong, and Master and Bachelor degrees in Aquaculture/Fisheries from Ocean University of China. He is a principal investigator or co-investigator for over 30 projects funded by federal, state, and private entities. Since 2011, he has been an associate editor for BioInvasions Records, Management of Biological Invasions, and Aquatic Invasions. He is a reviewer for 29 scientific journals and 12 grant agencies. He has been a member of the federal Aquatic Nuisance Species Task Force Research Subcommittee since 2012 and was the co-chair of Northeastern Aquatic Nuisance Species Panel from 2019 to 2020. Currently, he is managing the following two programs: (1) WM04 Chemical Application Program and (2) 401 Water Quality Certification Program.

1 Introduction to Freshwater Invasive Species in Massachusetts

ABSTRACT

Nineteen invasive species found in 701 Massachusetts lakes/ponds and 142 rivers/ streams are briefly described in this chapter. Among the 19 species, the most prevalent is purple loosestrife, found in 387 waterbodies, followed by variable milfoil, fanwort, curly-leaf pondweed, and Eurasian watermilfoil. The least widespread are water hyacinth and northern snakehead, which have only been detected in two waterbodies. Among these 843 waters with confirmed invasive species, the Charles River is home to the most invasive species (9). This book describes the basic biology and potential expansion pathways of these invasive species. It provides alternative management tools using updated biological, chemical, and physical/ mechanical methods. And it includes the regulation of invasive species management to allow resource managers, biologists, practitioners, volunteers, and students a better understanding of compliance and enforcement with environmental law. Although this book is about invasive species in Massachusetts lakes and rivers, the contents apply to other New England states, regions of the United States, and beyond where these invasive species have been detected.

INTRODUCTION

An invasive species is defined by a federal advisory committee as a species that is non-native to the ecosystem under consideration and whose introduction causes or is likely to cause economic or environmental harm or harm to human health (NISC 2007; Beck et al. 2008). Harmful invasive species can spread quickly, displacing native species, substantially affecting species richness and community composition, and attaining ecological dominance in many aquatic ecosystems.

Therefore, there is a growing global demand for effective management of aquatic invasive species.

Multiple invasive species have been detected in Massachusetts lakes and rivers. For this book, 19 species documented in Massachusetts lakes and rivers are described; species without official documentation, such as the sacred lotus (*Nelumbo nucifera*), Chinese soft-shelled marsh turtle (*Pelodiscus sinensis*), the freshwater jellyfish (*Craspedacusta sowerbii*), or some potential invasive species such as the spiny waterflea (*Bythotrephes longimanus*), are not included. Each subsequent chapter (Chapter 8 covers two milfoil species) is focused on a specific invasive species, in this format: (1) Introduction including basic biological characteristics (e.g., species identification, growth, reproduction, habitat requirements, etc.), native range, invasive pathways, ecological impacts, and economic impacts; (2) sightings where the specific invasive species has been documented; (3) management options, including prevention and control (i.e., chemical, physical, and biological methods); and (4) success story. Although the focus of this book is on invasive species found in Massachusetts lakes and rivers, it will serve other New England states, other regions of the United States, and other countries where similar invasive species occur in their waters.

THE 19 INVASIVE SPECIES IN MA LAKES AND RIVERS

This book covers the biology, invasion ecology, sightings, and control techniques of 4 invasive animals and 15 invasive plants documented in Massachusetts lakes and rivers, including Asian clam (*Corbicula fluminea*), zebra mussel (*Dreissena polymorpha*), Chinese mystery snail (*Bellamya chinensis*), northern snakehead (*Channa argus*), hydrilla (*Hydrilla verticillata*),

DOI: 10.1201/9781003201106-1

curly-leaf pondweed (*Potamogeton crispus*), Eurasian watermilfoil (*Myriophyllum spicatum*), variable milfoil (*Myriophyllum heterophyllum*), parrot-feather (*Myriophyllum aquaticum*), fanwort (*Cabomba caroliniana*), European naiad (*Najas minor*), South American waterweed (*Egeria densa*), swollen bladderwort (*Utricularia inflata*), water chestnut (*Trapa natans*), water hyacinth (*Eichhornia crassipes*), European waterclover (*Marsilea quadrifolia*), yellow floating heart (*Nymphoides peltata*), common reed grass (*Phragmites australis*), and purple loosestrife (*Lythrum salicaria*).

Unlike many other books that focus on one organism (e.g., zebra mussel), one group of organisms (e.g., bivalves), or one subject of a group (i.e., identification of aquatic plants), this book presents valuable information on biological characteristics (such as identification, reproduction, invasion pathways, etc.), monitoring, and management of both invasive animals and plants. Invasive species monitoring and sightings data are primarily documented in MassDEP's biennial Integrated

Lists of Waters, as well as reports by other state, federal, and individual entities (see the list in the Preface). In addition, an example of a successful management technique(s) or lesson(s) learned will be highlighted for each invasive species. Therefore, this book can provide readers a comprehensive view of multiple invasive animal/plant species, ranging from prevention, early detection, and control to ecological restoration and public education.

Among the 19 aquatic invasive species found in Massachusetts, the top 5 most widespread (in order of prevalence) are purple loosestrife (found in 379 lakes and rivers), variable milfoil, fanwort, curly-leaf pondweed, and Eurasian watermilfoil. The next most widespread are common reed grass, water chestnut, Chinese mystery snail, Asian clam, and European naiad (Table 1.1). The least "invasive" species are northern snakehead and water hyacinth, which have only been documented in two waterbodies (Table 1.1). This result is similar to a lake survey conducted from 1994 to 2003 that identified the top invasive

---

TABLE 1.1

Number of Infested Waterbodies of Freshwater Invasive Species in Massachusetts

Invasive Species	Sightings
Purple loosestrife	387
Variable milfoil	245
Fanwort	196
Curly-leaf pondweed	167
Eurasian watermilfoil	115
Water chestnut	114
Reed grass	112
Chinese mystery snail	37
Asian clam	33
European naiad	23
Swollen bladderwort	15
Hydrilla	11
European waterclover	9
South American waterweed	7
Parrot-feather	6
Yellow floating heart	6
Zebra mussel	3
Water hyacinth	2
Northern snakehead	2

TABLE 1.2

Massachusetts Watershed Codes and Names

Code	Watershed Name
11	Hudson: Hoosic River Basin
12	Hudson: Kinderhook River Basin
13	Hudson: Bashbish River Basin
21	Housatonic River Basin
31	Farmington River Basin
32	Westfield River Basin
33	Deerfield River Basin
34	Connecticut River Basin
35	Millers River Basin
36	Chicopee River Basin
41	Quinebaug River Basin
42	French River Basin
51	Blackstone River Basin
52	Ten Mile River Basin
53	Narragansett Bay (Shore) Drainage Area
61	Mount Hope Bay (Shore) Drainage Area
62	Taunton River Basin
70	Boston Harbor (Proper)
71	Boston Harbor: Mystic River Basin and Coastal Drainage Area
72	Charles River Basin
73	Boston Harbor: Neponset River Basin and Coastal Drainage Area
74	Boston Harbor: Weymouth and Weir River Basin and Coastal Drainage Area
81	Nashua River Basin
82	Concord River Basin
83	Shawsheen River Basin
84	Merrimack River Basin and Coastal Drainage Area
91	Parker River Basin and Coastal Drainage Area
92	Ipswich River Basin and Coastal Drainage Area
93	North Shore Coastal Drainage Area
94	South Shore Coastal Drainage Area
95	Buzzards Bay Coastal Drainage Area
96	Cape Cod Coastal Drainage Area
97	Islands Coastal Drainage Area

aquatic species as purple loosestrife, fanwort, the two non-native milfoil species, and common reed grass (Mattson et al. 2004a).

There are more than 4,230 miles of rivers and over 3,000 lakes and ponds located within 33 watersheds (Table 1.2 and Figure 1.1) in Massachusetts.[1] The largest of these waterbodies, Quabbin Reservoir (24,704 acres) and Wachusett Reservoir (4,160 acres) are manmade.

The largest natural lake is Assawompset Pond in southern Massachusetts (2,656 acres). At least 843 waterbodies (701 lakes and 142 rivers, Figure 1.1) are known to be infested by at least one invasive species. The Charles River from Hopkinton to Boston is the most affected, with nine invasive species: Asian clam, Chinese mystery snail, curly-leaf pondweed, Eurasian watermilfoil, fanwort, purple

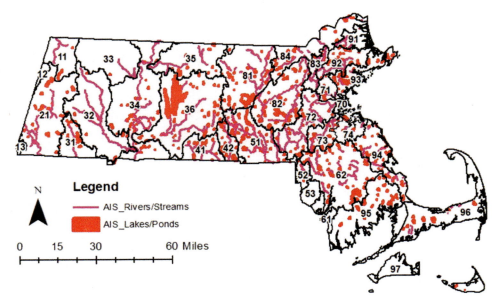

FIGURE 1.1   The 701 lakes/ponds and 142 rivers/streams infested by invasive species within 33 Massachusetts watersheds and drainage areas. (Table 1.1 shows the watershed codes and their corresponding names.)

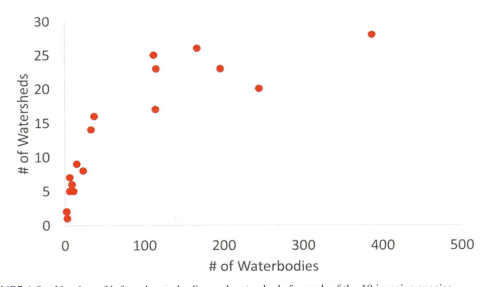

FIGURE 1.2   Number of infested waterbodies and watersheds for each of the 19 invasive species.

loosestrife, variable milfoil, water chestnut, and European waterclover (Table 1.3). Waterbodies with seven and six invasive species are listed in Table 1.3. The Blackstone River watershed has 82 waters infested by invasive species, followed by the Taunton (80 waters), Concord (59 waters), Nashua (58 waters), Connecticut (48 waters), and Chicopee (48 waters) watersheds. It is clear that

the more the waterbodies in which an invasive species is detected (Table 1.1), the more water-sheds this species has invaded (Figure 1.2).

Since the first invasive species was documented in Massachusetts, the rise of invasive species has occurred in tandem with the rise in population (Figure 1.3). The first documented case of an invasive species in Massachusetts

TABLE 1.3
Massachusetts Waterbodies Mostly Infested by Invasive Species

Waterbody	Asian clam	Chinese mystery snail	Curly-leaf pondweed	Eurasian water milfoil	European naiad	Fanwort	Purple loosestrife	Reed grass	South American waterweed	Swollen bladderwort	Variable milfoil	Water chestnut	European waterclover	Total # of species
Charles River	1	1	1	1	0	1	1	0	0	0	1	1	1	9
Monponsett Pond	0	0	1	1	0	1	1	1	0	1	1	0	0	7
Sudbury River	1	0	1	1	0	1	1	0	0	0	0	1	1	7
White Island Pond	0	0	1	1	1	1	1	1	0	1	0	0	0	7
Bartlett Pond	0	0	1	1	0	1	1	0	0	0	1	1	0	6
Billington Sea	0	0	1	1	0	1	1	1	0	0	1	0	0	6
Blackstone River	1	0	1	0	0	1	1	0	0	0	1	1	0	6
Flint Pond (Blackstone Watershed)	0	0	1	1	1	1	1	0	0	0	1	0	0	6
Fort Meadow Reservoir	1	0	0	1	0	1	1	1	0	0	1	0	0	6
Lake Cochituate	1	0	1	1	0	0	1	0	0	0	1	1	0	6
Lake Quinsigamond	0	1	1	1	0	1	0	0	0	0	1	1	0	6
Lake Rico	0	0	0	1	0	1	1	0	1	1	1	0	0	6
Mill River	0	0	1	0	0	1	1	1	0	0	1	1	0	6
North Pond	0	0	0	0	1	1	1	0	1	1	1	0	0	6
Webster Lake	1	0	1	1	0	1	1	0	0	0	1	0	0	6

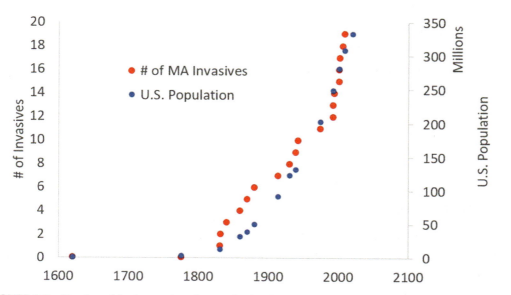

FIGURE 1.3 Number of freshwater invasive species in Massachusetts and US population size (millions).

was purple loosestrife in 1831 (see Chapter 19); freshwater invasive species have been recorded increasingly throughout the state since then (Figure 1.3). On average, from 1800 to 1900, the time period for a freshwater species to be established and detected was 17 years; from 1900 to 2000, that period decreased to 13 years and from 2000 to the present, it again decreased to 4 years. This is consistent with the strong relationship between the spread of invasive species and trade and transportation, i.e., the larger the human population within Massachusetts and the United States, the greater the trade and transportation activities (e.g., commercial and recreational) between waters within the Commonwealth and within the country (see "Spread of Freshwater Invasive Species" section). The latest documented invasive species was the zebra mussel in Laurel Lake in Lee and Lenox in 2009 (see Chapter 3).

SPREAD OF FRESHWATER INVASIVE SPECIES

Historically, trade and transportation have served as the major pathways for the invasion of new territory by invasive species for millennia; in recent decades, the world has entered the era of globalization, a new phase in terms of the magnitude and diversity of biological invasions (Hulme 2009). For coastal waters, the discharge of ballast water and biofouling from ocean-going ships is a major pathway for invasive species (Hewitt et al. 2009; Wan et al. 2021). Much of the ongoing spread of aquatic invasive species to inland waters throughout North America can be attributed to the overland movement of small watercraft (Bossenbroek et al. 2001; Johnson et al. 2001; Leung et al. 2006; Rothlisberger et al. 2010). These boats are usually less than 40 ft (12.2 m) in length, and include vessels such as powerboats, small commercial and recreational fishing boats, sailboats, personal watercraft, canoes and kayaks, and pontoon boats that can be towed overland on trailers (Rothlisberger et al. 2010). Translocation of invasive organisms by boaters can be intentional, but are most often unintentional (see Rothlisberger et al. 2010). These organisms can not only be carried in bilge water, live wells, and bait buckets, but also be entrained on boat exteriors, e.g., entangled

on propellers and trailers, or attached to other entangled organisms. The overland transport of small boats is reported to be responsible for the spread of spiny waterflea (MacIsaac et al. 2004; Muirhead and MacIsaac 2005), Eurasian watermilfoil (Buchan and Padilla 2000), and zebra mussels (Schneider et al. 1998; Leung et al. 2004; De Ventura et al. 2016). It has been postulated that the recent large-scale invasion of quagga mussels in the western United States resulting from their introduction to Lake Mead is from bilge water (i.e., bait or live wells) in a boat from the Great Lakes region (Stokstad 2007; Wong and Gerstenberger 2011). During an observational study of 85 boats, it was found that boats and trailers leaving the lakes were three times more likely to be carrying vegetation (as "hitchhikers") than those arriving; these hitchhikers included 13 species of plants and 51 small-bodied organisms (including 28 aquatic animals), and the invasive species Eurasian watermilfoil (*Myriophyllum spicatum*) was among them (Rothlisberger et al. 2010). Therefore, whenever a boat is transported overland from an infested waterway, the possibility exists that invasive species may reach uninfested waterways. The invasion pathway may vary among regions. For example, the major pathways of freshwater invasive species in Europe are aquaculture, the pet/aquarium trade, and stocking activities (Les 2002; Nunes et al. 2015). Aquatic invasive species can also spread by water currents, waterfowl, and other vectors. During invasion, an invasive species at different life stages can be transported such as plant seeds, tubers, buds, rhizomes, fragments, or animal adults and larvae. More information about the potential invasion pathways of each of the 19 species can be found in Chapters 2–19.

Recently, it was found that global online trade plays an increasingly influential role in the potential spread of invasive species, such as the trade of ornamental fish in Brazil (Borges et al. 2021). In addition, the broad-scale patterns of spread may mask considerable variation across regions, time periods, and even entities contributing to invasive species sampling. In some cases, observed temporal shifts in species discovery may be influenced by dramatic fluctuations in the number and spatial extent of individual observations, reflecting the possibility that shifts in sampling effort may obscure underlying rates of species introduction (Mangiante et al. 2018).

After individuals of an invasive species are transported from the source region (their native range) to the recipient region (non-native range), whether the transport is intentional (e.g., plants sold in the aquarium or rain garden trade) or unintentional (e.g., hitchhikers in commercial shipping vessels), establishment occurs if the transported individuals are able to survive and reproduce. The introduced species is considered established if it has formed a reproducing and self-sustaining population. The introduction and establishment transitions in different areas are usually independent of each other, and species that become widely established are successful overall because their introduction has been attempted in many locations, not because of a better establishment capability (García-Berthou et al. 2005). After that, secondary spread can occur in which the species extends its range within the non-native region. Recreational boats in tow between lakes are also a vector of secondary spread of invasive (and native) species among freshwater ecosystems (Muirhead and MacIsaac 2005; Vander Zanden and Olden 2008; Kelly et al. 2013).

CLIMATE CHANGE ON FRESHWATER INVASIVE SPECIES

Hundreds of invasive species are established in the United States (USGS 2020), impacting the United States both ecologically and economically, as shown subsequently (i.e., the section on ecological and economic damage). At the same time, over 11,000 scientists from around the world have declared that, according to clear and unequivocal indicators, the planet is facing a climate emergency (Ripple et al. 2020), given that any of the recent changes observed in the climate are unprecedented in thousands, if not hundreds of thousands, of years (IPCC 2021). For example, O'Reilly et al. (2015) revealed a statistically significant upward trend of $0.34°C$ per decade in the nighttime surface water temperature in some 300 lakes around the world from 1985 to 2009. As a consequence of the changing climate, invasive species could expand their ranges into the northern US states either through human-mediated pathways or through natural movement. Climate change will influence the likelihood of new species becoming established by eliminating cold temperatures or winter hypoxia that currently limit survival, and by increasing the reservoirs that serve as hot spots for invasive species (Rahel and Olden 2008). Important shifts in species ranges (depending on taxonomic group and scenario in Europe) toward the north and east of Europe are occurring at the unprecedented rate of 14–55 km/decade (Gallardo and Aldridge 2013). In Massachusetts, Asian clams have been detected in the Charles River since 2001 (Colwell et al. 2017). Long-term trend analysis shows that, from 1985 to 2016 (16 years before and after the first sighting of Asian clam in Massachusetts), the winter daily minimal temperature in Massachusetts increased. This higher winter temperature is more conducive to the successful establishment of Asian clams (Chapter 2). Further, the warming trend may be the key factor underlying the recent establishment of Asian clams in the United States north of the $40°$ latitude line, which is the clams' long-term incipient lower thermal limit (Mattice and Dye 1976). It has been documented that, while the distribution of Asian clams is controlled by both habitat variables (e.g., substrate and pH) and climate variables, climate variables are likely more important than the habitat variables (McDowell et al. 2014). The establishment of Asian clams in Massachusetts further confirms that climate variable is critical in shifting the range of invasive species. Therefore, resource agencies should be aware of and focus on protecting high-risk resource areas to prevent new invasive species invasion amid climate change.

ECOLOGICAL AND ECONOMICAL DAMAGE

Freshwater ecosystems are subject to a large range of anthropogenic threats such as habitat loss and fragmentation, hydrological alteration, climate change, overexploitation, pollution and introduction of alien species (Dudgeon et al. 2006). Biological invasions are responsible for high economic losses to society associated with the monetary expenditures from the management of these invasions. Invasive species have been attributed to a 42% decrease in populations of endangered species in the United States

(Stein and Flack 1997; Wilcove 1998). Many invasive species are also called ecosystem engineers because of the destruction to the waters they invade. Even specific impacts from individual species can be significantly different (see Chapters 2–19 for more information). Invasive species continue to drive major losses in biodiversity and ecosystem function across the globe. Among the most important hazards posed by invasive species, the introduction of such species outside their natural range is widely recognized to be one of the main threats to biodiversity and the second leading cause of animal extinctions (Millennium Ecosystem Assessment 2005). For example, 20% of the 680 species extinctions listed by the International Union for Conservation of Nature (IUCN) are due to invasive species (Clavero and García-Berthou 2005). Understanding the impact of invasive species on extinctions requires the development of more scientific and evidence-based approaches (Gurevitch and Padilla 2004).

Dealing with the effects of invasive species is particularly challenging in marine and freshwater habitats, because the pace at which invaders establish often greatly outstrips the resources available for their eradication (Green and Grosholz 2020). In the United States alone, economic costs are estimated to exceed $120 billion each year (Pimentel et al. 2005). The global cost of aquatic invasive species is conservatively $345 billion, with the majority attributed to invertebrates (62%), followed by vertebrates (28%) and plants (6%). The principal cost is from resource damages (74%) and only 6% of recorded costs were from management (Cuthbert et al. 2021). There is no doubt that the economic cost of invasive species is high and rising worldwide (Diagne et al. 2021).

Estimates of economic costs associated with zebra/quagga mussel infestations of water intake systems in North America range from $100 million to $1 billion per annum (Pimentel et al. 2005; Bidwell 2010). A study was conducted on the economic impacts of invasive quagga and zebra mussels on infrastructure and the economies of nine western states with a focus on the costs, which were mainly associated with mussel establishment prevention strategies and operations and maintenance (O&M) expenditures to control or mitigate mussel-related damages at US Bureau of Reclamation

hydropower facilities (Rumzie et al. 2021). Watercraft inspection and decontamination (WID) stations are the primary strategy used to prevent the spread and introduction of dreissenid mussels throughout the West and the 2019 average annual WID budget was approximately $1,605,900. Control cost data collected through a US Bureau of Reclamation survey showed that facilities have spent approximately $10 million in total on preventative control measures since mussel inception (in 2007). Facilities surveyed spend approximately $464,000 annually on increased maintenance, and the total recurring maintenance costs for facilities surveyed were $650,000 per invasive species occurrence; facilities also spend approximately $88,000 in total annually on monitoring (Rumzie et al. 2021).

MANAGEMENT

For the successful prevention/control of invasive species, a management plan must be developed, such as the Massachusetts Department of Conservation and Recreation (MA DCR) aquatic invasive species assessment and management plan (MA DCR 2010). The most efficient approach to controlling the spread of aquatic invasive species is to adopt an integrated management strategy (Henne et al. 2005; Lavergne and Molofsky 2006; Wong and Gerstenberger 2015) which usually includes (1) public education and awareness, (2) prevention, (3) monitoring/early detection and rapid response, (4) control (e.g., physical, chemical, or biological control) and restoration, and (5) regulation.

MA DCR developed decontamination protocols for boaters, to remove aquatic invasive species from boats and trailers before launching in any DCR and Massachusetts Water Resource Authority (MWRA) reservoirs (MA DCR 2010). Another example is invasive species monitoring and mapping which can raise public awareness of the presence of aquatic invasive species in Massachusetts lakes and rivers and keep water users from further spreading them to other waterbodies. The Housatonic River Watershed has been infested by zebra mussels since 2008; as a result, MassDEP developed equipment decontamination protocols for staff conducting water quality and biological monitoring in that watershed (Chase and Wong 2015). Equipment decontamination and gear cleanup are detailed in Appendix I.

PUBLIC EDUCATION AND AWARENESS

Efforts to increase public awareness about the negative impacts of invasive species might increase public support for management actions (Novoa et al. 2017). To increase public awareness of the threats posed by invasive species, appropriate signs and decontamination equipment should be provided at key locations around infested waterbodies to ensure that users are aware of the presence of invasive species, and the need to clean their boats thoroughly before leaving the waterbody. Mapping infested waterbodies in Massachusetts informs local residents and out-of-state travelers of infestations in lakes and rivers (see distribution maps of each invasive species from Chapters 2–19). It is critical to identify and engage stakeholders, such as anglers and other water users, in activities focused on invasive species in their communities. MA DCR has created brochures (Figure 1.4) and booklets to enhance public awareness of aquatic invasive species such as the guide to aquatic plants in Massachusetts (MA DCR 2016). Public awareness and outreach is key in educating all users to clean, drain, and dry their watercraft and equipment to prevent the spread of invasive species in our lakes and rivers. It can also enhance safeguarding invasives-free waters from the introduction of invasive species by facilitating and coordinating actions among federal, state, tribal, and other resource agencies. Finally, maximizing public support and minimizing regional spread are usually weighted much higher than other objectives to reflect decision-maker values; thus, they have greater influence on the overall performance of each alternative in addressing invasive species challenges

Remember:

Always remove all plant and animal fragments from your boat, trailer and gear. Dispose of livewell, bait bucket, and cooling water well away from the shore.

Anchors and rope or chain

Livewells and bait buckets

Transom wells and bilge water

Hitch

Raw water in engine cooling system

Trailer frame

Rollers/bunks

Lower Units and Propellor

FIGURE 1.4 Part of the brochure created by MA DCR for boaters on preventing the spread of invasive species.

for managers while using environmental DNA (eDNA) for early detections and containment alternatives (Sepulveda et al. 2022).

PREVENTION

Prevention refers to both protecting "clean" waterbodies from invasive species and containing invasive species within an infested waterbody and halting their further spread to other waterbodies. One example of a prevention strategy is the decontamination of watercraft and equipment to interrupt the transport pathway of an invasive species (Comeau et al. 2011; DiVittorio 2015; Zook and Phillips 2015). Prevention is an aggressive and proactive approach in minimizing the risk of aquatic invasive species introduction, and is generally regarded as the first line of defense against biological invasion. It is also the best management practice for dealing with any invasive species to avoid ecological impacts and reduce economic costs. More importantly, it has been shown to be far more effective and affordable than dealing with established populations of invasive animals and plants (Scholl 2006): an ounce of prevention is worth a pound of cure. This proactive approach usually includes collecting information about aquatic invasive species, the targeted waterbody ecosystem, early detection (discussed later), education and awareness (discussed earlier), developing a contingency plan and workload, engaging local government and lake community/associations, and funding allocations.

Biological invasions and natural disasters are similar phenomena as both generate enormous environmental damage, with the economic costs of invasions worldwide exceeding that of natural disasters. Although many nations invest in personnel training, disaster preparedness, and emergency response plans for extreme natural hazards (e.g., earthquakes), despite the rarity of such events (Ricciardi et al. 2011), few invest in precautions addressing invasive species (apart from infectious diseases), even though prevention is the most efficient and cost-effective approach to stopping or minimizing the spread of aquatic invasive species among all management options.

Border control and pathway management are the logical priority to prevent the introduction and spread of invasive species. Since the overland movement of small-craft boats is the major contributor for much of the ongoing spread of aquatic invasive species to inland waters throughout North America (see more in "Spread of Freshwater Invasive Species" section), among all management options, drain, clean, and dry watercraft and equipment is the prioritized strategy in prevention. Clean, drain, and dry is a simple three-step process that boaters can follow every time they move from one body of water to another. When boaters take preventive actions every time they leave the water, they can stop the spread of aquatic invasive species. Recently, CD3 Waterless Cleaning Systems are developed as a waterless, free, user-operated cleaning equipment that includes wet/dry vacuum, blower system, tethered hand tools, and lights for watercraft clean, drain, and dry.[2] Before the start of each water quality and biological monitoring season, the Massachusetts Department of Environmental Protection (MassDEP) develops standard operation procedures (SOPs) to prevent potential spread of aquatic invasive species between different waterbodies and/or watersheds (Chase and Wong 2015). MassDEP usually takes multiple approaches for boat and equipment decontamination, such as air-dry, high-pressured wash, or quarantine with disinfectants (Figure 1.5). Since the discovery of zebra mussels in Laurel Lake, MA, in 2009 (see Chapter 3), DCR has set up Watercraft Inspection Stations at Quabbin Reservoir for inspecting, cleaning, and sealing a boat before it enters the Quabbin Reservoir (Table 1.4). Mostly adopted watercraft quarantine methods include flushing (Davis et al. 2016), pressure-washing (Morse 2009; Comeau et al. 2011), rinsing with hot water (Comeau et al. 2015; Kappel et al. 2015), using cleaning agents (Chase and Wong 2015; Davis et al. 2015, 2016), or air-drying (Figure 1.5), even the extent of their efficacy is quite different (Mohit et al. 2021 and references therein). According to a review based on 37 studies to study the efficacy of different quarantine methods, the majority (70.3%) assessed air-drying, followed by hot water (32.4%), household chemicals (16.2%), and pressure-washing (2.7%) (Mohit et al. 2021). The recommended air-drying duration of up to one week produced high mortality (≥90%) among several invertebrate and macrophyte species, although survival was high for certain aquatic

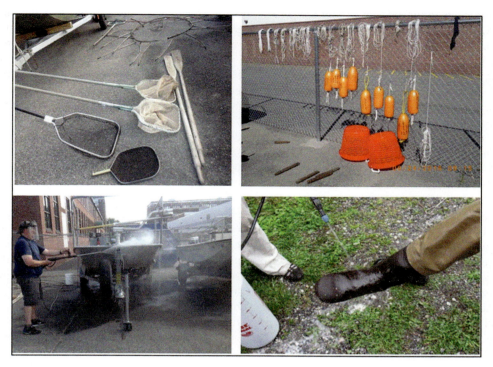

FIGURE 1.5 After sampling in a waterbody, equipment will be air-dried (upper panel) and boat will be cleaned with high-pressured spray (*bottom left*) by MassDEP staff. If more than one waterbody needs to be sampled, gears will be quarantined with chlorine or other disinfectants (*bottom right*). (Photos by W.H. Wong.)

TABLE 1.4

Quabbin Reservoir Boat Decontamination Program at Two Facilities from August 16 to September 26, 2009 (MA DCR 2010)

Facility	# 1 at Belchertown	#2 at Orange	Total
Number of inspections	280	382	662
Number of boats rejected	4	4	8
Total days of activity	12 days	19 days	N/A

snails. Aquatic plant survival and growth were inversely related to water loss (a function of drying time and relative humidity), and short or single fragments were less resistant to air-drying than larger or clustered fragments. Immersion in water ≥50°C for 15 minutes resulted in 100% mortality among mussels, small invertebrates, and some plant species. A higher temperature of 60°C was required for hot water spray applications lasting ≥5 seconds to achieve the same mortality rate among dreissenid mussels (Morse 2009; Comeau et al. 2011). High pressure-washing eliminated significantly more entangled plants and small organisms and seeds than low pressure. Household chemicals such as salt or bleach required specific doses and exposed durations to be lethal to targeted invasive species. The boat cleaning behavior could be significantly improved by changing how boat owners value the perceived costs and benefits of cleaning, as well as by increasing their awareness of the potential negative impacts on aquatic ecosystems caused by non-native species (De Ventura et al. 2017; Golebie et al. 2021). For a specific water district, science-based risk analysis needs to be performed to identify priority invasion pathways for introduction (i.e., watercraft, equipment, retail aquatic plant trade, aquarium, etc.) and species of concern. The most vulnerable waterbodies need to be prioritized for interventions in order for resources to be allocated in the most cost-effective manner.

As mentioned earlier, climate change is also facilitating the spread of invasive species by shifting their ranges. Massachusetts and other northeastern states have been identified as hot spots where up to 100 warm-adapted, range-shifting invasive plants could establish before 2050, it is suggested that only those high-impact species should be prioritized for practically monitoring and management (Rockwell-Postel et al. 2020). For freshwater invasive species such as Asian clams in Massachusetts and beyond (McDowell et al.2014; Colwell et al. 2017), as well as the global distribution and spread of hydrilla (Patrick and Florentine 2021), their ecological and socioeconomic impacts are substantial and should be monitored and prioritized for proactive management (see Chapters 2 and 6 in this book).

EARLY DETECTION AND RAPID RESPONSE (EDRR)

Eradication of the targeted invasive species is the primary goal of the EDRR. EDRR includes biological monitoring and assessment such as eDNA survey, and staff training workshops. A range of capabilities (e.g., risk assessments, monitoring programs, identification support, alert systems, eradication techniques, etc.) are necessary to support effective EDRR (Figure 1.6). EDRR also prioritizes high risk waters that may be invaded by potential invasive species and identifies obstacles and explores opportunities to establish an emergency rapid respond fund, etc. It can respond to new species detections in a timely manner to prevent their establishment and spread to other waterbodies. EDRR is generally necessary to achieve eradication which refers to actions that may be taken to eradicate populations of potentially invasive species that are new to the United States or contain the spread of known invasive species by eradicating satellite populations that could result in range expansions. A template for a decision-making process that incorporates the stages within the EDRR process is described in Figure 1.7.

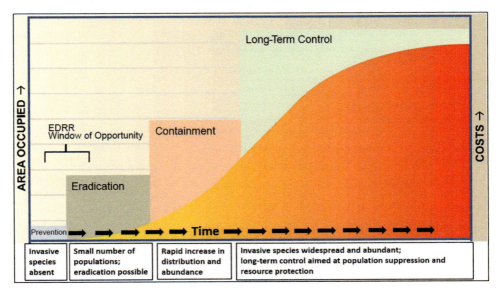

FIGURE 1.6 The invasion curve. (Modified from the U.S. Department of Interior (2016).)

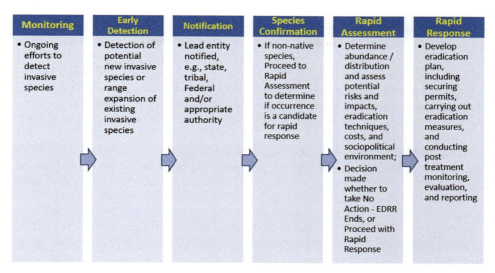

Monitoring	Early Detection	Notification	Species Confirmation	Rapid Assessment	Rapid Response
• Ongoing efforts to detect invasive species	• Detection of potential new invasive species or range expansion of existing invasive species	• Lead entity notified, e.g., state, tribal, Federal and/or appropriate authority	• If non-native species, Proceed to Rapid Assessment to determine if occurrence is a candidate for rapid response	• Determine abundance / distribution and assess potential risks and impacts, eradication techniques, costs, and sociopolitical environment; • Decision made whether to take No Action - EDRR Ends, or Proceed with Rapid Response	• Develop eradication plan, including securing permits, carrying out eradication measures, and conducting post treatment monitoring, evaluation, and reporting

FIGURE 1.7 Depiction of general early detection and rapid response decision-making process. (Modified from U.S. Department of Interior (2016, 2021).)

EARLY DETECTION AND MONITORING

Early detection is the process of surveying for, reporting, and verifying the presence of a non-native species, before the founding population becomes established or spreads so widely that eradication is no longer feasible (U.S. Department of Interior 2016). Early detection can allow for a swift, proactive management response, nipping an infestation in the bud to more successfully and cost-effectively prevent ecological damage (Goldberg et al. 2013, 2016; Trebitz et al. 2017; Sepulveda et al. 2020). In Massachusetts, usually traditional monitoring is applied for species detection and confirmation. For example, MassDEP will add a waterbody to the CWA 303d list if the presence of alive organisms are confirmed (MassDEP 2018). There is growing interest in using environmental DNA (eDNA) methods for monitoring aquatic invasive species. For example, it is found that the frequency of conducting molecular analyses for species identification published from 2012 to 2019 in *BioInvasions Records*, an international journal about invasive species records worldwide, has increased from 4.5% to 25.2% and is expected to further increase with the continuing development of molecular tools (Stranga and

Katsanevakis 2021). Therefore, eDNA is specifically described subsequently and should be a priority for future invasive species detection and monitoring.

eDNA is genetic material that is obtained directly from the environment after being shed or excreted by an organism (Thomsen and Willerslev 2014; Eichmiller et al. 2016). Using a variety of techniques, this material can be collected from the field, processed in a lab, and analyzed to uncover the identity of source organisms. The general method for finding the identity of a species based on a DNA sample is called "barcoding." As living things evolve over many generations, different regions of the genome change at different rates, leading to certain regions being more variable than others. Genes controlling respiration and other important cellular processes are more likely to be conserved between individuals of the same species than genes controlling eye color, for example. Some regions, such as the COI gene in animals and the rbcL in plants, are known through experimentation to vary from species to species, but not between individuals of the same species, and so genetic sequences from these regions are useful as "barcodes" to distinguish species based on genetic information (Hajibabaei et al. 2007; Bourlat et al. 2013). Using primers that are designed specially to latch onto these distinct barcode regions, biologists can detect the presence of a species' signature code within a mix of unknown DNA. Once primers have detected a matching segment in the sample, a polymerase chain reaction (PCR) can be used to "amplify" the segment, creating copies at an exponential rate. Conventionally, PCR can be used in this way to detect a target species (Thomsen and Willerslev 2014) by simply answering a yes-or-no question: if the target sequence (i.e., the barcode unique to the target species) is present in the sample, the primers will find it and PCR will proceed, producing a tube full of replicated target DNA; but if the target sequence is absent, PCR will not proceed, resulting in a tube of mostly water. Since PCR primers seek out their target region and make many copies of it, the technique is effective even if there is only a small amount of target DNA in the sample to begin with. Advanced techniques in analyzing eDNA include quantitative PCR (qPCR), high-throughput sequencing, and next-generation sequencing (Zaiko et al. 2018).

eDNA analysis has several advantages over traditional monitoring and sampling methods. Since eDNA samples can be collected from sediment or water, it is logistically simple to collect eDNA during routine water or sediment chemical sampling (Thomsen and Willerslev 2014; Zaiko et al. 2018). eDNA analysis does not require collection of or interaction with any live plants or animals, leading to a non-intrusive sampling process (Thomsen and Willerslev 2014). Identification and detection by eDNA do not require specific expertise of an organism's morphology. This is helpful in cases where it is a challenge to differentiate cryptic invertebrates, plants, or organisms in small, non distinct juvenile stages. (Trebitz et al. 2017; Thomsen and Willerslev 2014). As shown earlier, eDNA analysis can allow for early detection of invasive species, picking up on signs and traces of an organism before the invasive population has become fully established. eDNA analysis is becoming an increasingly powerful tool for ecosystem management, allowing for early detection of invasive species in waterways (Stranga and Katsanevakis 2021). For example, for invasive plants, different binding sites in water samples have been investigated to detect their sightings (Scriver et al. 2015) which include parrot-feather, water hyacinth, and Brazilian waterweed that are

found in Massachusetts. For aquatic invertebrates, Thomsen et al. (2011) detected 82% of insect larvae (compared to the results of conventional collection) and Goldberg et al. (2013) successfully detected the invasive New Zealand mud-snail from US field sites. Some others include invasive rusty crayfish (Dougherty et al. 2016), invasive soft-shelled clams (Ardura et al. 2015), and invasive pygmy mussels (Miralles et al. 2016). For fish and other vertebrates, eDNA has been documented in their detection and management (Darling and Mahon 2011; Adrian-Kalchhauser and Burkhardt-Holm 2016; Rey et al. 2018). It is thought that fish and amphibians are easier to detect from environmentally excreted DNA than arthropods or other shelled invertebrates because of the mucus on fish and amphibian skin and its tendency to slough off in the water (Thomsen et al. 2011; Goldberg et al. 2013).

However, the uncertainty associated with eDNA surveillance (Frischer et al. 2012), as well as the lack of robust guidelines and reporting standards for eDNA which has affected the confidence of natural resources managers to apply eDNA results in support of management actions (DFO 2020), makes it challenging to determine appropriate natural resource management responses. Gingera et al. (2017) provided a case study using eDNA as a tool for detecting dreissenid mussels in Lake Winnipeg, Canada, where zebra mussels were first recorded in 2013. They developed two species-specific and one Dreissena-specific qPCR-based eDNA assays designed as a single multiplexed reaction able to identify the presence of zebra mussel and infer the presence of quagga mussel in water samples from at-risk and invaded locations in 2014. This case study demonstrates that eDNA is an early indicator of the presence of zebra mussels. At the same time, there are still critical knowledge gaps for eDNA being recognized as a reliable management tool. For example, when three methods were used to detect the presence of quagga/zebra mussel veligers that cross-polarized light microscopy currently provides the most reliable means (96.3% accuracy) in plankton samples. Imaging flowcytometry was next with 91.7% accuracy, and eDNA was the least reliable (75.8% accuracy) (Frischer et al. 2012). The most prevalent type of error associated with all the methods was false negatives.

The decay of genetic materials might be one of other challenges for eDNA implementation due to uncontrollable environmental factors such as temperature, microbial loads, and pH that can affect eDNA degradation significantly (Lance et al. 2017). Multiple sources of error can give rise to positive detections of eDNA in a sample when that species is not present at the site and acting on an erroneous eDNA inference could result in needless costs (Sepulveda et al. 2022). A structured decision-making process to evaluate appropriate response actions following hypothetical eDNA detections of invasive dreissenid mussel (*Dreissena* spp.) eDNA in Jordanelle Reservoir, Utah, is therefore suggested. This approach needs to communicate and work with decision-makers and stakeholders to identify objectives and discrete management action alternatives, and assess alternative consequences and trade-offs. These alternatives ranged from no action to intensive and expensive control efforts. The alternative identified to most likely satisfy the objectives was delayed containment, where immediate attempts to confirm the eDNA detections using non-molecular sampling techniques were conducted and, regardless of those results, containment efforts were enacted to prevent dreissenid mussel spread to other waterbodies through mandatory watercraft exit inspections shortly thereafter (Sepulveda et al. 2022). Delayed containment had the lowest downside risk, and the highest upside gains relative to other alternative actions would be optimal, regardless of the decision-maker being risk-adverse, risk-neutral, or risk-tolerant. Sensitivity and elasticity analyses demonstrate that it is robust to parameter and outcome uncertainty (Sepulveda et al. 2022).

RAPID RESPONSE

Rapid response is the process that is employed to eradicate the founding population of a non-native species from a specific location (U.S. Department of Interior 2016). Rapid response for eradication of the founding invasive species population alleviates the need for expensive invasive species control programs that would have to be enacted over the long term. Analytics and decision tools will help determine what and when rapid response measures should be taken to find and eradicate

potential invasive species before they spread and become harmful ecologically and/or economically. When a potentially invasive species is detected, it is necessary for appropriate decision-makers to employ rapid response. Prevention is obviously the most desirable management option, but often this fails (Cheater 1992; Mattson et al. 2004a). Successful rapid response is contingent on preparedness: having the plans, tools, training, resources etc. in place to mount eradication efforts (U.S. Department of Interior 2016). Therefore, it needs support for response actions and preparedness activities, including intensive monitoring for many taxonomic groups, rapid response training, and tools for containment and eradication such as hand pulling or benthic barriers.

Among the 19 invasive species described in this book, rapid response for eradication is suggested for water hyacinth and northern snakehead because there are only two sightings for each of these two species in Massachusetts. For zebra mussel, even there are three waterbodies infested, but two of them are Housatonic River and Green River, which are not recommended for eradication. Rapid response for eradication can also be implemented to other species to an early infestation in a waterbody. In Massachusetts, MA Department of Conservation and Recreation Lakes and Ponds Program has developed a guide on two different types of manual control techniques rapid response which are suitable for volunteers who wish to implement a rapid response to a small infestation of aquatic invasive species, but they usually are not considered feasible techniques for large-scale infestations (MA DCR 2007). The guide is mainly for two techniques, hand pulling or/and benthic barriers, to potentially eradicate submerged, floating, or emergent aquatic invasive species. More detailed information on the rapid response standard operating procedure is provided in Appendix II.

CONTROL AND CONTAINMENT

While for species in their early invasion stage, rapid response and eradication is prioritized, for widely invaded species whose spread to an extent where eradication is no longer feasible, control and containment is the remaining management option. Habitat and ecosystem restoration such as planting or stocking organisms or improving predator--prey relationships to attain food webs similar to pre-invasion conditions should be conducted since rehabilitation is often necessary to restore ecological processes. Among the 19 invasive species described in this book, the following 17 species are well-established and are recommended for control and containment: purple loosestrife, variable milfoil, fanwort, curly-leaf pondweed, Eurasian watermilfoil, water chestnut, common reed grass, Chinese mystery snail, Asian clam, European naiad, swollen bladderwort, hydrilla, European waterclover, South American waterweed, parrot-feather, yellow floating heart, and zebra mussel (Table 1.1, in the order from most common to least common). Usually a control/contain plan should be developed to evaluate/justify the need for control and the projected consequences of no control and less than full control, as well as potential risk in harming non-target organisms, ecosystems, and the public health. The elimination or reduction in population of aquatic invasive species may promote an increase in indigenous plant and animal diversity and/or edge effect habitat and likely improve wildlife habitat and sever as restoration per Massachusetts Wetlands Protection Act 310 CMR 10.60 (Langley et al. 2004). The cost of the project is critical too in terms of technical feasibility and effectiveness of alternative control strategies and actions. It should be evaluated whether benefits of control, including costs avoided, exceed the costs of the program. Long-term control programs require substantial financial investments in perpetuity. Structuring the plant community to meet clear goals of invasive species management in an ecologically and ethically sound manner is more appropriate, although potentially still quite expensive. While most managers in North America now focus on ongoing containment and suppression interventions, they often lack quantitative guidance from which to set targets and evaluate success. Practical guidelines are recommended for identifying management targets for invasions for which eradication is unfeasible, based on achieving "functional" eradication—defined as suppressing invader populations below levels that cause unacceptable ecological effects—within high-priority locations. Identifying targets for suppression allows managers to estimate the degree of removal required to mitigate ecological impacts and the management resources needed to achieve sufficient control of an invasive species.

Depending on the site to be treated, usually invasive species control involves the use of variable methods which mainly consist of physical removal, chemical treatment, and biological control (Tables 1.5–1.7). Invasive plants should be managed aggressively. Due to the possibility of fragmentation spreading of some species, such as Eurasian watermilfoil, the best treatment is with a selective herbicide unless the growths are extremely localized and amendable to simple physical techniques (hand pulling at low density, benthic barriers at high density). All of the techniques to control plants have their unique advantages and disadvantages (Wagner 2001, 2004; Mattson et al. 2004a) which are summarized in Tables 1.5 (physical removal), 1.6 (chemical treatments), and 1.7 (biological control) in this chapter. Detailed information about these techniques can be found from the Massachusetts' Generic Environmental Impact Report (GEIR), Eutrophication and Aquatic Plant Management (Mattson et al. 2004a), and the accompanying The Practical Guide to Aquatic Plant Management in Massachusetts (Wagner 2004). Although these methods are mainly for aquatic plants, some of these techniques such as benthic barriers or copper treatment can also be used for some invasive animals such as Asian clams or zebra mussels. For more information about control and containment techniques of each specific invasive species, refer to Chapters 2–19 in this book.

PHYSICAL REMOVAL

Physical removal and control techniques are listed in Table 1.5 which includes (1) benthic barriers, (2) dredging, (3) dyes and surface covers, (4) mechanical removal ("harvesting"), and (5) water level control. If harvesting is proposed, the ultimate disposal location of the vegetative spoils should be identified and it should be outside of wetland resource areas or buffer zones (Langley et al. 2004). Plant material decomposition in the water column should be prevented that may result in lower dissolved oxygen.

CHEMICAL TREATMENT

Chemicals used for treating aquatic plants and animals in Massachusetts are listed in Table 1.6. Before each treatment, applicators must apply for a WM04 Chemical Application

TABLE 1.5

Physical Removal and Management Options for Control of Aquatic Invasive Plants

Technique Option	Technique Subdivisions	Mode of Action	Advantages	Disadvantages
Benthic barriers*	(a), (b), or (c) below	• Mat of variable composition laid on bottom of target area, preventing growth • Can cover area for as little as several months or permanently • Maintenance improves results • Usually applied around docks, in boating lanes, and in swimming areas	• Highly flexible control • Reduces turbidity from soft bottom sediments • Can cover undesirable substrate • Can improve fish habitat by creating edge effects	• May cause anoxia at sediment–water interface • May limit benthic invertebrates • Non-selective interference with plants in target area • May inhibit spawning/feeding
	(a) Porous or loose-weave synthetic materials	• Laid on bottom and usually anchored by weights or stakes • Removed and cleaned or flipped and repositioned at least once per year for maximum effect	• Allows some escape of gases which may be generated underneath • Panels may be flipped in place or removed for relatively easy cleaning or repositioning	• Allows some plant growth through pores • Gas may still build up underneath in some cases, lifting barrier from bottom
	(b) Non-porous or sheet synthetic materials	• Laid on bottom and anchored by many stakes, anchors or weights, or by layer of sand • Not typically removed, but may be swept or "blown" clean periodically	• Prevents all plant growth until buried by sediment • Minimizes interaction of sediment and water column	• Gas buildup may cause barrier to float upward • Strong anchoring makes removal difficult and can hinder maintenance
	(c) Improving sediment composition	• Sediments may be added on top of existing sediments or plants • Use of sand or clay can limit plant growths and alter sediment–water interactions • Sediments can be applied from the surface or suction dredged from below muck layer (reverse layering technique)	• Plant biomass can be buried • Seed banks can be buried deeper • Sediment can be made less hospitable to plant growths • Nutrient release from sediments may be reduced • Surface sediment can be made more appealing to human users • Reverse layering requires no addition or removal of sediment	• Lake depth may decline • Sediments may sink into or mix with underlying muck • Permitting for added sediment difficult • Addition of sediment may cause initial turbidity increase • New sediment may contain nutrients or other contaminants • Generally too expensive for large-scale application

(Continued)

TABLE 1.5
(Continued)

Technique Option	Technique Subdivisions	Mode of Action	Advantages	Disadvantages
Dredging	(a), (b), or (c) below	• Sediment is physically removed by wet or dry excavation, with deposition in a containment area for dewatering/disposal • Dredging can be applied on a limited basis, but is most often a major restructuring of a severely impacted system • Plants and seed beds are removed and regrowth can be limited by light and/or substrate limitation	• Plant removal with some flexibility • Increases water depth • Can reduce pollutant reserves • Can reduce sediment oxygen demand • Can improve spawning habitat for many fish species • Allows complete renovation of aquatic ecosystem • May allow for growth of desirable species	• Temporarily removes benthic invertebrates • May create turbidity • May eliminate fish community (complete dry dredging only) • Possible impacts from containment area discharge • Possible impacts from dredged material disposal • Interference with recreation or other uses during dredging • Usually very expensive
	(a) "Dry" excavation	• Lake drained or lowered to maximum extent practical • Target material dried to maximum extent possible • Conventional excavation equipment used to remove sediments	• Tends to facilitate a very thorough effort • May allow drying of sediments prior to removal • Allows use of less specialized equipment	• Eliminates most aquatic biota unless a portion left undrained • Eliminates lake use during dredging
	(b) "Wet" excavation	• Lake level may be lowered, but sediments not substantially dewatered • Draglines, bucket dredges, or long reach backhoes used to remove sediment	• Requires least preparation time or effort, tends to be least cost dredging approach • May allow use of easily acquired equipment • May preserve most aquatic biota	• Usually creates extreme turbidity • Tends to result in sediment deposition in surrounding area • Normally requires intermediate containment area to dry sediments prior to hauling • May cause severe disruption of ecological function • Impairs most lake uses during dredging

Method	Description	Advantages	Disadvantages
c) Hydraulic (or pneumatic) removal	• Lake level not reduced • Suction or cutterhead dredges create slurry which is hydraulically pumped to containment area • Slurry is dewatered; sediment retained, water discharged	• Creates minimal turbidity and limits impact on biota • Can allow some lake uses during dredging • Allows removal with limited access or shoreline disturbance	• Often leaves some sediment behind • Cannot handle extremely coarse or debris-laden materials • Requires advanced and more expensive containment area • Requires overflow discharge from containment area
Dyes and surface covers	• Water-soluble dye is mixed with lake water, thereby limiting light penetration and inhibiting plant growth • Dyes remain in solution until washed out of system • Opaque sheet material applied to water surface	• Light limit on plant growth without high turbidity or great depth • May achieve some control of algae as well • May achieve some selectivity for species tolerant of low light	• May not control peripheral or shallow water rooted plants • May cause thermal stratification in shallow ponds • May facilitate anoxia at sediment interface with water • Covers inhibit gas exchange with atmosphere and restrict recreation • Cannot be used in waterbodies with an active outlet
Mechanical removal ("harvesting") (a), (b), (c), (d), or (e) below	• Plants reduced by mechanical means, possibly with disturbance of soils • Collected plants may be placed on shore for composting or other disposal • Wide range of techniques employed, from manual to highly mechanized • Application once or twice per year usually needed	• Highly flexible control • May remove other debris • Can balance habitat and recreational needs	• Possible impacts on aquatic fauna • Non-selective removal of plants in treated area • Possible spread of undesirable species by fragmentation • Possible generation of turbidity
(a) Hand pulling*	• Plants uprooted by hand ("weeding") and preferably removed	• Highly selective technique	• Labor-intensive • Difficult to perform in dense stands • Can cause fragmentation
(b) Cutting (without collection)	• Plants cut in place above roots without being harvested	• Generally efficient and less expensive than complete harvesting	• Leaves root systems and part of plant for possible regrowth • Leaves cut vegetation to decay or to re-root • Not selective within applied area

(Continued)

TABLE 1.5
(Continued)

Technique Option	Technique Subdivisions	Mode of Action	Advantages	Disadvantages
	(c) Harvesting (with collection)	• Plants cut at depth of 2–10 ft and collected for removal from lake	• Allows plant removal on greater scale	• Limited depth of operation • Usually leaves fragments which may re-root and spread infestation • May impact lake fauna • Limited selectivity within applied area • More expensive than cutting
	(d) Rototilling	• Plants, root systems, and surrounding sediment disturbed with mechanical blades	• Can thoroughly disrupt entire plant	• Usually leaves fragments which may re-root and spread infestation • May impact lake fauna • Not selective within applied area • Creates substantial turbidity • More expensive than harvesting
	(e) Hydroraking	• Plants, root systems, and surrounding sediment and debris disturbed with mechanical rake, part of material usually collected and removed from lake	• Can thoroughly disrupt entire plant • Also allows removal of stumps or other obstructions	• Usually leaves fragments which may re-root and spread infestation • May impact lake fauna • Not selective within applied area • Creates substantial turbidity • More expensive than harvesting
Water level control	(a) or (b) below	• Lowering or raising the water level to create an inhospitable environment for some or all aquatic plants • Disrupts plant life cycle by desiccation, freezing, or light limitation	• Requires only outlet control to affect large area • Provides widespread control in increments of water depth • Complements certain other techniques (dredging, flushing)	• Potential issues with water supply • Potential issues with flooding • Potential impacts to non-target flora and fauna

Method		Possible impacts
(a) Drawdown	• Lowering of water over winter period allows desiccation, freezing, and physical disruption of plants, roots, and seed beds • Timing and duration of exposure and degree of dewatering are critical aspects • Variable species tolerance to drawdown; emergent species and seed bearers are less affected • Most effective on annual to once/three-year basis	• Control with some flexibility • Opportunity for shoreline cleanup/structure repair • Flood control utility • Impacts vegetative propagation species with limited impact to seed-producing populations	• Possible impacts on contiguous emergent wetlands • Possible effects on overwintering reptiles and amphibians • Possible impairment of well production • Reduction in potential water supply and fire fighting capacity • Alteration of downstream flows • Possible overwinter water level variation • Possible shoreline erosion and slumping • May result in greater nutrient availability for algae
(b) Flooding	• Higher water level in the spring can inhibit seed germination and plant growth • Higher flows which are normally associated with elevated water levels can flush seed and plant fragments from system	• Where water is available, this can be an inexpensive technique • Plant growth need not be eliminated, merely retarded or delayed • Timing of water level control can selectively favor certain desirable species	• Water for raising the level may not be available • Potential peripheral flooding • Possible downstream impacts • Many species may not be affected, and some may be benefitted • Algal nuisances may increase where nutrients are available

*Benthic barriers and hand pulling are also introduced in the "Early Detection and Rapid Response" section. See Appendix II for more detailed information.

Source: Modified from Wagner (2001, 2004) and Mattson et al. (2004a).

TABLE 1.6
Major Chemicals for Treating Aquatic Invasive Species in Massachusetts

Major Ingredient	Brand Name	Mode of Action	Advantages	Disadvantages
Fluridone	• AVAST! SC • SONAR SRP • SONAR Q • SONAR PR • SONAR GENESIS • SONAR ONE • SONAR AS • SONAR H4C	• Inhibition of carotenoid biosynthesis at the phytoene desaturase step	• Complete kill of susceptible vegetation • Can be used selectively on certain major invasive species at low doses • Slow death of plants minimizes oxygen demand and nutrient release • Minimal risk of any direct impacts on fauna	• Acts slowly in the aquatic environment; exposure time of up to 90 days needed • Highly diffusive; dilution will limit effectiveness in areas of high flushing activity
Diquat	• REWARD • TRIBUNE	• Interferes with photosynthesis by forming highly reactive and toxic free-radicals, such as peroxide and super oxide, in plant cells	• Effective against a wide variety of species • Relatively rapid kill of targeted vegetation • Can be used for spot treatments; limited drift or impact outside target area	Detriments • Not very selective; kills most species contacted • Does not damage portions of plants with which it does not contact; regrowth from roots is common • Potential for toxicity to fauna, but uncommon in practice
Glyphosate	• GLYPRO • RODEO • AQUAPRO • ACCORD • ROUNDUP CUSTOM	• Inhibition of 5-enolpyruvylshikimate-3-phosphate (EPSP) synthase; EPSP inhibition leads to depletion of the aromatic amino acids tryptophan, tyrosine, and phenylalanine, all needed for protein synthesis or for biosynthetic pathways leading to growth	• Effective on emergent vegetation • Kills entire plant for susceptible species • Selective by area and vegetation type (emergent/floating vs. submergent)	• Ineffective against submergent species • Precipitation (rain) interferes with uptake

Herbicide	Mode of action	Effects	Concerns
Triclopyr • RENOVATE • RENOVATE OTF • RENOVATE MAXG • RESTORATE • PATHFINDER II • GARLON 3A • GARLON 4 • VASTLAN	• As selective plant growth regulator to induce growth (auxin mimic) while preventing synthesis of essential plant enzymes, resulting in disruption of growth processes	• Complete kill of susceptible vegetation • Can be used selectively on certain major invasive species • Lower necessary exposure time allows for treatment in areas of greater water exchange • Low risk of any direct impacts on fauna	• Lowered oxygen levels are possible as a function of vegetation decay after treatment. • No experience yet with application approaches and control success in Massachusetts
2,4-D • AQUA-KLEEN • NAVIGATE • PLATOON • DMA 4 • CLEAN AMINE • SCULPIN G	• Absorbed by roots, leaves, and shoots and disrupts cell division throughout the plant	• Complete kill of susceptible vegetation, typically provides multiple years of control for target species • Acts relatively quickly in the aquatic environment; plant death may be delayed, but sufficient uptake occurs within three days • Can be used selectively on certain major invasive species at low doses, and for partial (especially shoreline) lake treatments	• Potential for toxicity to fauna, but a rare occurrence in practice • Use restrictions in or near drinking water supplies (surface or wells) limits application
Endothall • AQUATHOL-K • AQUATHOL SUPER K • HYDROTHOL • HYDROTHOL 191	• As plant protein phosphatase inhibitor	• Effective against a wide variety of species • Relatively rapid kill of targeted vegetation • Selective; limited drift or impact outside target area	• Not very selective; kills most species contacted • Does not damage portions of plants with which it does not contact; regrowth from roots is common • Potential for toxicity to fauna, but uncommon in practice
Imazapyr • HABITAT • POLARIS	• It prevents the production of acetolactate synthase and disrupts synthesis of amino acids	• Practically non-toxic (the EPA's lowest toxicity category) to fish, invertebrates, birds and mammals	• Not for use on submersed species • Unlike imazamox, imazapyr does not demonstrate any in-water herbicidal activity

(Continued)

TABLE 1.6
(Continued)

Major Ingredient	Brand Name	Mode of Action	Advantages	Disadvantages
Imazamox	• CLEARCAST	• Inhibition of acetolactate synthase and disrupts synthesis of amino acids	• Low toxicity to animals such as fish, birds, invertebrates, and amphibians, as well as to humans	• Usually limited to monocultural stands of target plants or in areas where selectivity is not a concern
Florpyrauxifen-benzyl	• PROCELLACOR EC	• Action like indole acetic acid (synthetic auxins)	• Non-toxic to fish and birds and has no drinking, fishing or swimming restrictions	• Registered by EPA in 2018 and lots of ongoing studies on it to collect more information
Flumioxazin	• CLIPPER • SCHOONER • REDEAGLE • PROPELLER	• Inhibits photosynthesis, bleaching of the chloroplasts, and cell death	• With a broader spectrum of activity compared to carfentrazone	• The primary breakdown pathway in water is highly dependent on water pH. Under high pH values (> 9), half-life in water is 15–20 minutes. Under more neutral pH values (7–8), half-life in water is around 24 hours
Carfentrazone	• STINGRAY	• Inhibition of protoporphyrinogen oxidase enzyme that is important in chlorophyll synthesis	• Rapid degradation in the aquatic environment, no accumulation in the sediment, low use rates; does not pose a hazard to mammals or other non-target organisms • Provides good selectivity and it will not control comingled non-target plants like pickerelweed or grasses	• Currently very few published literature on the efficacy of carfentrazone against submersed plants
Metsulfuron	• ESCORT XP	• Inhibition of acetolactate synthase (acetohydroxyacid synthase AHAS)	• A potent herbicide active ingredient with a very low use rate	• Do not apply directly to water, or to areas where surface water is present, or to intertidal areas below the mean high water mark

Sodium percarbonate + hydrogen peroxide hydrogen peroxide + peracetic acid sodium percarbonate	• GREEN CLEAN PRO • GREEN CLEAN LIQUID • PHYCOMYCIN • PAK 27	• Damaging cells by oxidation and disruption of cell membranes via the hydroxyl radical • Sodium carbonate peroxyhydrate is transformed into hydrogen peroxide and sodium carbonate in the presence of water	• Hydrogen peroxide residue levels dissipate rapidly and are mostly undetectable at 24 hours	• Undiluted granules should not be allowed to remain in an area where humans or animals may be exposed
Copper	• CUTRINE PLUS • NAUTIQUE • CAPTAIN • CAPTAIN XTR • K-TEA • KOMEEN • SECLEAR • POND BOSS PRO • EARTHTEC • SECLEAR G • HARPOON G • COPPER SULFATE	• The copper ion (copper II) is responsible for the toxicity of all of the formulations • Inhibition of photosynthesis leads to plant death • It may affect nitrogen metabolism	• Rapid kill of susceptible algae • Rapidly eliminated from water column, minimizing prolonged adverse impacts	• Toxic to many non-target organisms • Releases contents of most killed algal cells back into the water column; this may include nutrients, taste and odor compounds, and toxins • Ineffective on some algae; repeated treatments may favor those resistant algae, some of which are major nuisance species • Accumulates in sediments, although long-term impacts may not be severe
Aluminum	• POLYALUMINUM CHLORIDE • SODIUM ALUMINATE • ALUMINUM SULFATE • PHOSCLEAR	• Form a floc that can remove particulates, including algae, from the water column within minutes to hours and precipitate reactive phosphates • Reactions continue at the surface–water interface, binding phosphorus that could otherwise be released from the sediment	• Rapid removal of available phosphorus from the water column • Minimized internal loading of phosphorus • Potential removal of a variety of other contaminants and algae	• Risk of toxicity from aluminum comes from issues with pH due to either poorly buffered water or water with an already lower pH (<6.5). Alkalinization (when pH >8) can also lead to toxicity from aluminum • Limited longevity of effects if external loading is significant

Source: Modified mainly from Wagner (2001, 2004) and Mattson et al. (2004a, 2004b); it also includes information from these references: Durkin (2008, 2012); Glomski et al. (2006); MA DAR and MassDEP 2010, 2014, 2019; McComas (2011); NALMS (2004); Richardson et al. (2008); Sesin et al. (2018); Strid et al. (2018); U.S. EPA (1998); WSSA.

TABLE 1.7

Biological Control Options for Control of Aquatic Invasive Species in Massachusetts

Technique Option	Technique Subdivisions	Mode of Action	Advantages	Disadvantages
Biological introductions	(a), (b), (c), or (d) below	• Fish, insects or pathogens which feed on or parasitize plants are added to system to affect control • The most commonly used organism is the grass carp, but the larvae of several insects have been used more recently, and viruses are being tested	• Provides potentially continuing control with one treatment • Harnesses biological interactions to produce desired conditions • May produce potentially useful fish biomass as an end product	• Typically involves introduction of non-native species • Effects may not be controllable • Plant selectivity may not match desired target species • May adversely affect indigenous species
	(a) Herbivorous fish	• Sterile juveniles stocked at density which allows control over multiple years • Growth of individuals offsets losses or may increase herbivorous pressure. Grass carp are illegal in Massachusetts	• May greatly reduce plant biomass in single season • May provide multiple years of control from single stocking • Sterility intended to prevent population perpetuation and allow later adjustments	• May eliminate all plant biomass, or impact non-target species • Funnels energy into algae • Alters habitat • May escape upstream or downstream • Population control issues
	(b) Herbivorous insects	• Larvae or adults stocked at density intended to allow control with limited growth • Intended to selectively control target species • Milfoil weevil is best known, but still experimental	• Involves species native to region, or even targeted lake • Expected to have no negative effect on non-target species • May facilitate longer term control with limited management	• Population ecology suggests incomplete control likely • Oscillating cycle of control and regrowth • Predation by fish may complicate control • Other lake management actions may interfere with success
	(c) Fungal/ bacterial/viral pathogens	• Inoculum used to seed lake or target plant patch • Growth of pathogen population expected to achieve control over target species	• May be highly species specific • May provide substantial control after minimal inoculation effort	• Effectiveness and longevity of control not well-known • Infection ecology suggests incomplete control likely
	(d) Selective plantings	• Establishment of plant assemblage resistant to undesirable species • Plants introduced as seeds, cuttings, or whole plants	• Can restore native assemblage • Can encourage assemblage most suitable to lake uses • Supplements targeted species removal effort	• Largely experimental • Nuisance species may eventually return assemblage • Introduced species may become nuisances

Source: Modified from Wagner (2001, 2004) and Mattson et al. (2004a).

License from Massachusetts Department of Environmental Protection. At the same time, since most of these chemicals are not selective, non-target native species will be impacted, more or less. For example, glyphosate-based herbicides are now the most commonly used herbicides in the world and often promoted as "safe," despite scientific evidence of their harm to all facets of an ecosystem, including the soil, water, and non-target plants, amphibians, reptiles, invertebrates, animals, and humans (Rushton et al. 2016). It is reported that there was a huge invasion of bladderwort to Pierpont Meadow Pond in Dudley and Charlton, MA, about 20 years ago. The floating weed wrapped around docks impeded swimming and boating and was a terrible eyesore. So, yearly chemicals are applied to this lake since then. Lately, residents start to question if there may be a better way than to continue to add chemicals to the lake because they could not see native snails or crawfish in the water anymore which may or may not be a result of the yearly chemical application.[3] Although it is unlikely that snails or crawfish might not be eradicated from this lake because the chemical application permit only allows one-third of the entire lake and half of the littoral zone for chemical application, the negative impacts of accumulative chemical applications should not be overlooked. Therefore, it is important to reiterate that only herbicides products registered for use in Massachusetts through the Department of Agricultural Resources (DFA) may be used in Massachusetts, and then only by licensed applicators with proper permits (except in some water supply cases and ponds with no outlets). An updated list of registered herbicides can be obtained from the Massachusetts Department of Agricultural Resources. Reported chemical control methods on the 19 aquatic invasive species are summarized in Appendix III.

BIOLOGICAL CONTROL

Biological control has the objective of achieving control of plants without introducing toxic chemicals or using machinery. However, any introduction of organisms may have imperceptible impacts on the aquatic community structure and food web. Although still experimental at most times, biological controls may offer effective and low cost controls once properly developed. At the same time, because biological control is one of the most powerful techniques, it must be approached with some caution and respect (Mattson et al. 2004a). For example, water quality can be indirectly impacted by biological control. The most dramatic impacts are from induced grass carp, although not on a consistent basis. Increases in alkalinity, turbidity, and potassium concentrations and significant reductions in dissolved oxygen levels and chlorophyll have all been documented in waters stocked with grass carp (Mitzner 1978; Lembi et al. 1978; Leslie et al. 1987; Mattson et al. 2004a). Since biological controls may provide plant control benefits and represent another tool in addition to physical and chemical controls, interest has grown in biological control methods over the last decades. Various biological control options are listed in Table 1.7. Generally, a full biological survey and study should be conducted to determine what type of manipulation is best suited to achieve the desired goals while minimizing possible adverse impacts. Because of the experimental nature of most biological methods, results are not assured to the same degree as for other methods such as harvesting or herbicide treatment (Mattson et al. 2004a). Therefore, some biological agents for invasive species control, such as the introduction of grass carp to control invasive weeds, are not allowed in Massachusetts (Mattson et al. 2004a).

REGULATION AND PERMITTING

REGULATION

According to the National Sea Grant Law Center, primary responsibility for preventing the spread of invasive species within the United States lies with states and it is the state authority to implement desired management strategies for species and pathways of concern. In Massachusetts, monitoring, control, and management of invasive or native nuisance aquatic species are within the jurisdictions of Massachusetts Department of Conservation and Recreation (DCR), Massachusetts Department of Environmental Protection (DEP), and Massachusetts Department of Fish and Game (DFG). Chemicals used for invasive species treatments must be approved for use by the Massachusetts Department of Agricultural Resource (MA DAR) Pesticide Bureau. The MA laws statutes pertaining aquatic invasive/nuisance species are briefly summarized subsequently in Appendix IV. It is a collection of laws on introduction and possession of fish and other animals, introduction and possession of plants, and prevention and control of invasive plants. More information on statues and laws on aquatic invasive/nuisance species from Massachusetts as well as other New England States and New York can also be found in Appendix IV.

MassDEP provides a guidance for the issuing authority (the Conservation Commission or MassDEP) in the review of aquatic plant management projects proposed to control abundance and distribution of aquatic vegetation under the Wetlands Protection Act Regulations (310 CMR 10) (Langley et al. 2004). Aquatic plant management projects in lakes and ponds subject to the jurisdiction must comply with the general performance standards established for each applicable resource area in the regulations (i.e., 310 CMR 10.54, 10.55, 10.56, 10.57) unless the project is "limited" (310 CMR 10.53(4) or 310 CMR 10.53(3)(l)). If wildlife habitat thresholds described in the general performance standards are exceeded, the requirements of 310 CMR 10.60 for wildlife habitat must be met (e.g., a wildlife habitat evaluation shall be performed). No project may be permitted which will have any adverse effect on specified habitat sites of rare vertebrate or invertebrate species (310 CMR 10.59) (see "Permitting" section).

In Massachusetts, waterbodies infested with an invasive species will be added to the Clean Water Act (CWA) 303d list of impaired or threatened waters during water quality assessments to fulfill the requirements of the Clean Water Act. If an invasive species management project footprint is within an Estimated Habitat of Rare Wildlife or a Priority Habitat of Rare Species area, it will be subject to the Massachusetts Endangered Species Act (MESA) that prevents the "taking" of any rare plant or animal species listed as Endangered, Threatened, or of Special Concern unless specifically permitted for scientific, educational, propagation, or conservative purposes (321 CMR 10.00).

PERMITTING

Order of Conditions

Whether physical removal/mechanical harvesting, chemical treatment, or biological control, it should fulfill the requirements of Massachusetts Wetlands Protection Act (310 CMR10). Usually, an approval from the local conservation commission is needed. A negative determination of applicability (NDA) may be appropriate if the area to be treated affects less than 5,000 ft². A Notice of Intent must be sent to the Conservation Commission with a copy to the Department of Environmental Protection Regional Office. An Order of Conditions (OOC) issued by the local conservation commission may be required for areas greater than 5,000 ft² of native vegetation. Non-native vegetation identified on the rapid response usually can be removed without restriction on areal extent (MA DCR 2007).

401 Water Quality Certification (Dredge)

For projects within infested waters may be subject to the jurisdiction of the 401 Water Quality Certification (WQC), conditions on invasive species prevention and decontamination will be provided to avoid the spread of invasive species during construction. For projects involving dredging or disposal of sediment including detritus or biomass of aquatic plants with volumes 100 cubic yards or greater, a 401 WQC must be obtained from MassDEP to ensure that the proposed project is in compliance with existing water quality standards.

There are three categories of water quality certification for dredging: major project, minor project certifications, and Amendment of Certification for Dredging. Major project certification includes the dredging of 5,000 cubic yards or greater, minor dredge project certification includes projects not listed in the major project certification category; and those projects involving dredging less than 5,000 cubic yards but more than 100 cubic yards. If a water quality certification has been issued for a dredging project and the project is subsequently revised, the certification is not valid. An amended certification may be requested for project modifications, such as changes in the dredging area, dredged material volume, or construction methods for dredging or disposal. In addition, a change in one or more of the special conditions of certification may be allowed through an amended certification. The 401 Water Quality Certification applications must be submitted through ePLACE, located at https://eplace.eea.mass.gov/citizenaccess.

WM 04 Chemical Application License

The MassDEP WM04 Chemical Application License grants approval to apply chemicals for the control of nuisance aquatic vegetation in accordance with authority granted to the MassDEP by Massachusetts General Laws c. 111, s. 5E.

Anyone may apply chemicals to bodies of water except under the following conditions: (a) When treatment is undertaken by employees and agents of the Departments of Environmental Protection, Conservation and Recreation, and Fisheries, Wildlife and Environmental Law Enforcement, or of the Reclamation Board, or of related Federal agencies, while in the conduct of their official duties. (b) When treatment is undertaken with algaecide containing copper by a public water system. However, the public water system shall notify the department in writing prior to the application of such algaecides [310 CMR 22.20B(8), the Drinking Water Regulations]. (c) When treatment is undertaken in privately owned (single owner) ponds from which there are no flowing outlets.

What prerequisites should be considered before applying for this permit? (1) Chemical treatments to water using general-use pesticides shall only be performed by an applicator currently licensed by the Massachusetts Department of Agricultural Resources Pesticide Program in the aquatics category. Chemical treatments to Bordering Vegetated Wetlands (310 CMR 10.55(2)(a)) and Salt Marsh (310 CMR 10.32(2)) using general-use pesticides and techniques that ensure chemicals are not applied to water shall only be performed by an applicator currently licensed by the Massachusetts Department of Agricultural Resources Pesticide Program. Chemical treatments using restricted-use pesticides shall only be performed by an applicator currently certified by the Massachusetts Department of Agricultural Resources (MA DAR) Pesticide Program. (2) Chemicals used for treatments must be currently approved for use in the state by the MA DAR Pesticide Bureau. The website listing the approved chemicals is found here: www.mass.gov/herbicides-for-aquaticvegetation-management. (3) A valid Order of Conditions, Negative Determination of Applicability (Wetlands Protection Act), Administrative Consent Order or similar authorization must be obtained prior to the treatment.

What concurrent applications are related to this permit? Since chemical treatments constitute the alteration of wetland resources, a Notice of Intent must be filed in accordance with the Wetlands Protection Act (MGL c. 131, s. 40) and Wetlands Protection Regulations (310 CMR 10.00). The following website can be used to file an application—www.mass.gov/how-to/wm-04-herbicide-application.

MESA-associated Permits

If the proposed project occurs within an Estimated Habitat of Rare Wildlife in the most recent version of the Natural Heritage Atlas, a copy of the Notice of Intent must be submitted to the Natural Heritage and Endangered Species Program (NHESP) within the Massachusetts Department of Fish and Game for review. If the proposed project occurs within a Priority Habitat of Rare Species, the project proponent must submit project plans to the NHESP for an impact determination. An Order of Conditions must be obtained prior to work. In the case of lakes and ponds with rare species, a Conservation and Management Permit likely will be required, or an Alteration Permit if the project is within a formally designated "significant habitat." A Massachusetts Environmental Policy Act (MEPA) review may also be required, depending upon interpretation of the impact thresholds. For biological control, in addition to approval of the Conservation Commission, the importation or liberation of fish or wildlife requires permits from the Division of Fisheries and Wildlife. Generally, native species (native to Massachusetts) do not require extensive testing and quarantine as non-native species do. Introduction of foreign species would require additional testing, approval of Division of Fisheries and Wildlife and Massachusetts Department of Agricultural Resources. It may require quarantine and approval of the federal US Department of Agriculture, Animal Plant Health Inspection Service (USDA-APHIS).

WEAKNESS AND STRENGTH

The sightings data in this book may not be complete due to lack of monitoring efforts (e.g., funding and time). Each chapter just covers sightings using data collected by MassDEP and others before each chapter is drafted. Due to continuous monitoring efforts from MassDEP, MA DCR, and other agencies and institutions, there should be new detection sightings since then but are not included in this book. I hope updated information and new data can be added to future editions of this book. In addition, for some waterbodies that have been treated chemically, physically, or biologically, I am not certain if the targeted invasive species has been eradicated or not due to lack of monitoring data after management activities. Therefore, it is possible that there are no invasive species anymore in a few reported sightings because of management activities, even it is extremely difficult to eradicate an established invasive species population.

Usually, a book either focuses on one organism (e.g., zebra mussel), or one group of organisms (e.g., submerged aquatic plants or bivalves), or just one subject of a group (i.e., identification of aquatic plants). This book provides valuable information on biological characteristics, monitoring, and management of both invasive animals and invasive plants in freshwater systems in Massachusetts. Furthermore, an example of success management is highlighted for each invasive species. Therefore, this book provides readers a bigger picture on multiple invasive animal/plant species from basic biology, prevention, early detection, control, ecological restoration, to public education. This book provides a unique comprehensive, systematic coverage with illustrations for both freshwater invasive animals and plants. Natural resource managers and practitioners usually must handle multiple invasive species in a system. This book will provide detailed and valuable information natural resource managers need to understand and use to prevent and manage freshwater invasive species in the field. Therefore, this book can serve as a field guide for how to identify and respond to invasions for resource managers and practitioners. It can also be a textbook for researchers and college students majoring in ecology, environmental biology, fisheries and wildlife, lake management, environmental science, and environmental study. Although the focus of this book is on Massachusetts lakes and rivers, the method for species identification, data collection, and mapping sightings of aquatic invasive species, as well as control techniques and management strategies can certainly be applied to waterbodies beyond Massachusetts. That is, it can provide valuable lessons for those who are concerned about same invasive animals and plants in a freshwater system (lake or river). Indeed, some success story sites on invasive species management are located in other states/regions/countries. Invasive species know no boundaries and they can come from, spread to, and reestablish in surrounding areas. Therefore, an integrated, strategic approach needs to be taken (Henne et al. 2005; Lavergne and Molofsky 2006; Wong and Gerstenberger 2015) and individuals from local towns, watersheds, regions, states, and even international communities dealing with invasive species need to work together and learn from each other via exchange of information and development of technology (Ding et al. 2006; Wong et al. 2017; Abrahams et al. 2019). Doing so would avoid and/ or minimize ecological and economic impacts from aquatic invasive species in lakes and rivers.

NOTES

1 www.city-data.com/states/Massachusetts-Topography.html
2 See more from www.cd3systems.com/
3 According to an email from Ann Rose, resident at Charlton, MA, to Diane Manganaro, Massachusetts Department of Environmental Protection, dated September 21, 2021.

REFERENCES

Abrahams, B., N. Sitasa, and K.J. Esler. 2019. Exploring the dynamics of research collaborations by mapping social networks in invasion science. *Journal of Environmental Management* 229: 27–37.

Adrian-Kalchhauser, I., and P. Burkhardt-Holm. 2016. An eDNA assay to monitor a globally invasive fish species from flowing freshwater. *PLoS ONE* 11(1): e0147558.

Ardura, A., A. Zaiko, J.L. Martinez, A. Samulioviene, A. Semenova, and E. Garcia-Vazquez. 2015. eDNA and specific primers for early detection of invasive species—A case study on the bivalve rangia cuneata, Currently spreading in Europe. *Marine Environmental Research* 112: 48–55.

Beck, K.G., Z. Kenneth, J.D. Schardt, J. Stone, R.R. Lukens, S. Reichard, J. Randall, A.A. Cangelosi, D. Cooper, and J.P. Thompson. 2008. Invasive species defined in a policy context: Recommendations from the federal invasive species advisory committee. *Invasive Plant Science and Management* 1: 414–421.

Bidwell, J.R. 2010. Range expansion of dreissena polymorpha: A Review of Major Dispersal Vectors in Europe and North America. Backhuys. In: G. Van der Velde, S. Rajagopal, and A. Bij de Vaate (eds). *The Zebra Mussel in Europe*. Leiden: Backhuys, pp. 69–78.

Borges, A.K.M., T.P.R. Oliveira, I.L. Rosa, F. Braga-Pereira, H.A.C. Ramos, L.A. Rocha, and R.R.N. Alves. 2021. Caught in the (Inter)net: Online trade of ornamental fish in Brazil. *Biological Conservation*. DOI: 10.1016/j.biocon.2021.109344.

Bossenbroek, J.M., C.E. Kraft, and J.C. Nekola. 2001. Prediction of long-distance dispersal using gravity models: Zebra mussel invasion of inland lakes. *Ecological Applications* 11: 1778–1788.

Bourlat, S.J., A. Borja, J. Gilbert, M.I. Taylor, N. Davies, S.B. Weisberg, J.F. Griffith, T. Lettieri, D. Field, J. Benzie, F.O. Glöckner, N. Rodríguez-Ezpeleta, D.P. Faith, T.P. Bean, and M. Obst. 2013. Genomics in marine monitoring: New opportunities for assessing marine health status. *Marine Pollution Bulletin* 74(1): 19–31.

Buchan, L.A.J., and D.K. Padilla. 2000. Predicting the likelihood of Eurasian watermilfoil presence in Lakes, A macrophyte monitoring tool. *Ecological Applications* 10: 1442–1455.

Chase, R.F., and W.H. Wong. 2015. *Standard Operating Procedure: Field Equipment Decontamination to Prevent the Spread of Invasive Aquatic Organisms*. Cambridge: Massachusetts Department of Environmental Protection (File # CN 59.6.29 pp).

Cheater, M. 1992. Alien invasion. *Nature Conservancy* 92: 24–29.

Clavero, M., and E. García-Berthou. 2005. Invasive species are a leading cause of animal extinctions. *Trends in Ecology and Evolution* 20(3): 110.

Colwell, H., J. Ryder, R. Nuzzo, M. Reardon, R. Holland, and W.H. Wong. 2017. Invasive Asian clams (*Corbicula fluminea*) recorded from 2001 to 2016 in Massachusetts, USA. *Management of Biological Invasions* 8(4): 507–515.

Comeau, S., R.S. Ianniello, W.H. Wong, and S.L. Gerstenberger. 2015. Boat decontamination with hot water spray: Field validation. In: W.H. Wong, and S.L Gerstenberger (eds). *Biology and Management of Invasive Quagga and Zebra Mussels in the Western United States*. Florida: CRC Press, pp. 161–173.

Comeau, S., S. Rainville, W. Baldwin, E. Austin, S.L. Gerstenberger, C. Cross, and W.H. Wong. 2011. Susceptibility of quagga mussels (*Dreissena rostriformis bugensis*) to hot-water sprays as a means of watercraft decontamination. *Biofouling* 27(3): 267–274.

Cuthbert, R.N., Z. Pattison, N.G. Taylor, L. Verbrugge, C. Diagne, D.A. Ahmed, B. Leroy, E. Angulo, E. Briski, C. Capinha, J.A. Catford, T. Dalu, F. Essl, R.E. Gozlan, P.J. Haubrock, M. Kourantidou, A.M. Kramer, D. Renault, and R.J. Wasserman. 2021. Global economic costs of aquatic invasive alien species. *Science of the Total Environment*. DOI: 10.1016/j.scitotenv.2021.145238.

Darling, J.A., and A.R. Mahon. 2011. From molecules to management: adopting DNA-based methods for monitoring biological invasions in aquatic environments. *Environmental Research* 111: 978–988.

Davis, E.A., W.H. Wong, and W.N. Harman. 2015. Distilled white vinegar (5% acetic acid) as a potential decontamination method for adult zebra mussels. *Management of Biological Invasions* 6(4): 423–428.

Davis, E.A., W.H. Wong, and W.N. Harman. 2016. Livewell flushing to remove zebra mussel (driessena polymorpha) veligers. *Management of Biological Invasions* 7(4): 399–403.

De Ventura, L., N. Weissert, R. Tobias, K. Kopp, and J. Jokela. 2016. Overland transport of recreational boats as a spreading vector of zebra mussel dreissena polymorpha. *Biological Invasions* 18: 1451–1466.

De Ventura, L., N. Weissert, R. Tobias, K. Kopp, and J. Jokela. 2017. Identifying target factors for interventions to increase boat cleaning in order to prevent spread of invasive species. *Management of Biological Invasions* 8(1): 71–84.

DFO. 2020. Advice on the Use of Targeted Environmental DNA (eDNA) Analysis for the Management of Aquatic Invasive Species and Species at Risk. DFO Can. Science Advisory Secretariat. Science Advisory Report. 2020/058.

Diagne, C., B. Leroy, A.C. Vaissière, R.E. Gozlan, D. Roiz, I. Jarić, J.M. Salles, C.J.A. Bradshaw, and F. Courchamp. 2021. High and rising economic costs of biological invasions worldwide. *Nature* DOI: 10.1038/s41586-021-03405-6.

Ding, J.Q., R. Reardon, W.D. Fu, Y. Wu, H. Zheng. 2006. Biological control of invasive plants through collaboration between China and the United States of America: A perspective. *Biological Invasions* 8(7): 1439–145.

DiVittorio J. 2015. Equipment Inspection and Cleaning: The First Step in an Integrated Approach to Prevent the Spread of Aquatic Invasive Species and Pests. In W.H. Wong, and S.L. Gerstenberger (eds.). *Biology and Management of Invasive Quagga and Zebra Mussels in the Western United States*. London: CRC Press, pp. 225–246.

Dougherty, M.M., E.R. Larson, M.A. Renshaw, C.A. Gantz, S.P. Egan, D.M. Erickson, and D.M. Lodge. 2016. Environmental DNA (eDNA) detects the invasive rusty crayfish *Orconectes rusticus* at low abundances. *Journal of Applied Ecology* 53(3): 722–732.

Dudgeon, D., A.H. Arthington, M.O. Gessner, Z.-I. Kawabata, D.J. Knowler, C. Leveque, R.J. Naiman, R.A.H. Prieur, D. Soto, M.L.J. Stiassny, and C.A. Sullivan. 2006. Freshwater biodiversity: Importance, threats, status and conservation challenges. *Biological Reviews* 81: 163–182.

Durkin, P.R. 2008. *Fluridone: Human Health and Ecological Risk Assessment Final Report. Submitted to Paul Mistretta, USDA/Forest Service*. Atlanta, GA: Southern Region, p. 115.

Durkin, P.R. 2012. Aminocyclopyrachlor: Human Health and Ecological Risk Assessment Final Report. Submitted to Dr. Harold Thistle, USDA Forest Service Forest Health Technology Enterprise Team. Morgantown, WV, 26505.212 p.

Eichmiller, J.J., S.E. Best, and P.W. Sorensen. 2016. Effects of temperature and trophic state on degradation of environmental DNA in lake water. *Environmental Science & Technology* 50: 1859–1867.

Frischer, M.E., K.L. Kelly, and S.A. Nierzwicki-Bauer. 2012. Accuracy and reliability of dreissena spp. Larvae detection by cross-polarized light microscopy, Imaging flow cytometry, and polymerase chain reaction assays. *Lake and Reservoir Management* 28(3):265–276.

Gallardo B., and D.C. Aldridge. 2013. Evaluating the combined threat of climate change and biological invasions on endangered species. *Biological Conservation* 160: 225–233.

García-Berthou, E., C. Alcaraz, Q. Pou-Rovira, L. Zamora, G. Coenders, and C. Feo. 2005. Introduction pathways and establishment rates of invasive aquatic species in Europe. *Canadian Journal of Fisheries and Aquatic Sciences* 62: 453–463.

Gingera, T.D., R. Bajno, M.F. Docker, and J.D. Reist. 2017. Environmental DNA as a detection tool for zebra mussels *Dreissena polymorpha* (Pallas, 1771) at the forefront of an invasion event in Lake Winnipeg, Manitoba, Canada. *Management of Biological Invasions* 8: 287–300.

Glomski, L.A.M., A.G. Poovey, and K.D. Getsinger. 2006. Effect of carfentrazone-ethyl on three aquatic macrophytes. *Journal of Aquatic Plant Management* 44(1): 67–69.

Goldberg, C.S., A. Sepulveda, A. Ray, J. Baumgardt, and L.P. Waits. 2013. Environmental DNA as a new method for early detection of New Zealand mudsnails (*Potamopyrgus antipodarum*). *Freshwater Science* 32(3): 792–800.

Goldberg, C.S., C.R. Turner, K. Deiner, K.E. Klymus, P.F. Thomsen, M.A. Murphy, S.F. Spear, A. McKee, S.J. Oyler-McCance, R.S. Cornman, M.B. Laramie, A.R. Mahon, R.F. Lance, D.S. Pilliod, K.M. Strickler, L.P. Waits, A.K. Fremier, T. Takahara, J.E. Herder, and P. Taberlet. 2016. Critical considerations for the application of environmental DNA methods to detect aquatic species. *Methods in Ecology and Evolution* 7(11):1299–1307.

Golebie, E., C.J. van Riper, C. Suski, and R. Stedman. 2021 Reducing invasive species transport among recreational anglers: The importance of values and risk perceptions. *North American Journal of Fisheries Management*. DOI:10.1002/nafm.10696.

Green, S.J., and E.D. Grosholz. 2020. Functional eradication as a framework for invasive species control. *Frontiers in Ecology and the Environment*. DOI:10.1002/fee.2277.

Gurevitch, J., and D.K. Padilla. 2004. Are invasive species a major cause of extinctions? *Trends in Ecology and Evolution* 19(9): 470–474.

Hajibabaei, M., G.A.C. Singer, P.D.N. Herbert, and D.A. Hickey. 2007. DNA barcoding: How it complements taxonomy, Molecular phylogenetics and population genetics. *Trends in Genetics* 32(4): 167–172.

Henne, D.C., C.J. Lindgren, T.S. Gabor, H.R. Murkin, and R.E. Roughley. 2005. An integrated management strategy for the control of purple loosestrife *Lythrum salicaria* L. (Lythraceae) in the netley-libau marsh, Southern Manitoba. *Biological Control* 32: 319–325.

Hewitt, C.L., S. Gollasch, and D. Minchin. 2009. The Vessel as a Vector—Biofouling, Ballastwater and Sediments. In: G. Rilov and J.A. Crooks (eds). *Biological Invasions in Marine Ecosystems*. Berlin: Springer, p. 117–131.

Hulme, P.E. 2009. Trade, Transport and trouble: Managing invasive species pathways in an era of globalization. *Journal of Applied Ecology* 46: 10–18.

IPCC (The Intergovernmental Panel on Climate Change). 2021. www.ipcc.ch/2021/08/09/ar6-wg1-20210809-pr/ (Accessed October 14, 2021).

Johnson, L.E., A. Ricciardi, and J.T. Carlton. 2001. Overland dispersal of aquatic invasive species: A risk assessment of transient recreational boating. *Ecological Applications* 11: 1789–1799.

Kappel, M., S.L. Gerstenberger, R.F. McMahon, and W.H. Wong. 2015. Thermal Tolerance of Invasive Quagga Mussels in Lake Mead National Recreation Area. In: W.H. Wong and S.L (eds). *Gerstenberger Biology and Management of Invasive Quagga and Zebra Mussels in the Western United States*. London: CRC Press, pp. 83–93.

Kelly, N.E., K. Wantola, E. Weisz, and N.D. Yan. 2013. Recreational boats as a vector of secondary spread for aquatic invasive species and native crustacean zooplankton. *Biological Invasions* 15: 509–519.

Lance1, R.F., K.E. Klymus, C.A. Richter, X. Guan, H.L. Farrington, M.R. Carr, N. Thompson, D.C. Chapman, and K.L. Baerwaldt. 2017. Experimental observations on the decay of environmental DNA from bighead and silver carps. *Management of Biological Invasions* 8(3): 343–359.

Langley, L., L. Rhodes, and M. Stroman. 2004. *Guidance for Aquatic Plant Management in Lakes and Ponds As It Relates to the Wetlands Protection Act*. Cambridge, MA: Massachusetts Department of Environmental Protection, Wetlands/Waterways Program. Boston, p. 12.

Lavergne, S., and J. Molofsky. 2006. Control strategies for the invasive reed canary grass (*Phalaris arundinacea* L.) in north American wetlands: The need for an integrated management plan. *Natural Areas Journal* 26(2): 208–214.

Lembi, C.A., B.G. Ritenour, E.M. Iverson, and E.C. Forss. 1978. The effects of vegetation removal by grass carp on water chemistry and phytoplankton in Indiana ponds. *Transactions of the American Fisheries Society* 107(1): 161–171.

Les, D. 2002. Non-indigenous aquatic plants: A garden of earthly delight? *LakeLine* 22(2):20–24.

Leslie, A.J. Jr., J.M. Van Dyke, R.S. Hestand, and B.Z. Thompson. 1987. Management of aquatic plants in multi-use lakes with grass carp (*Ctenopharyngodon Idella*). *Lake and Reservoir Management* 3(1): 266–276.

Leung, B., J.M. Bossenbroek, and D.M. Lodge. 2006. Boats, pathways, and aquatic biological invasions: estimating dispersal potential with gravity models. *Biological Invasions* 8: 241–254.

Leung, B., J.M. Drake, and D.M. Lodge. 2004. Predicting invasions: Propagule pressure and the gravity of allee effects. *Ecology* 85: 1651–1660.

MA DAR (Massachusetts Department of Agricultural Resources) and MassDEP (Massachusetts Department of Environmental Protection). 2010. Hydrogen Peroxide, Peracetic Acid and Sodium Percarbonate. www.mass.gov/doc/sodium-carbonate-peroxyhydrate-and-hydrogen-peroxide/download (Accessed August 2, 2021).

MA DAR (Massachusetts Department of Agricultural Resources) and MassDEP (Massachusetts Department of Environmental Protection). 2014. Imazamox. www.mass.gov/doc/imazamox/download (Accessed July 10, 2021).

MA DAR (Massachusetts Department of Agricultural Resources) and MassDEP (Massachusetts Department of Environmental Protection). 2019. Florpyrauxifen-benzyl. www.mass.gov/doc/florpyrauxifen-benzyl/download (Accessed July 22, 2021).

MA DCR (Massachusetts Department of Conservation and Recreation). 2007. *Standard Operating Procedures: Using Hand Pulling and Benthic Barriers to Control Pioneer Populations of Non-Native Aquatic Species A Guide for Volunteers*. Boston, MA: MA DCR, p. 27.

MA DCR (Massachusetts Department of Conservation and Recreation). 2010. *Aquatic Invasive Species Assessment and Management Plan*. Boston, MA: MA DCR, p. 39.

MA DCR (Massachusetts Department of Conservation and Recreation). 2016. A Guide to Aquatic Plants in Massachusetts. Lakes and Ponds Program. P. 32.

MacIsaac, H.J., J.V.M. Borbely, J.R. Muirhead, and P.A. Graniero. 2004. Backcasting and forecasting biological invasions of inland lakes. *Ecological Applications* 14: 773–783.

Mangiante, M.J., A.J.S. Davis, S. Panlasigui, M.E. Neilson, I. Pfingsten, P.L. Fuller, and J.A. Darling. 2018. Trends in nonindigenous aquatic species richness in the united states reveal shifting spatial and temporal patterns of species introductions. *Aquatic Invasions* 13(3): 323–338.

MassDEP (Massachusetts Department of Environmental Protection). 2018. Consolidated Assessment and Listing Methodology (CALM) Guidance Manual for the 2018 Reporting Cycle. Massachusetts Division of Watershed Management Watershed Planning Program. CN455.0.74 p.

Mattice, J.S., and L.L. Dye. 1976. Thermal tolerance of adult Asiatic clam. *Thermal Ecology* 2: 130–135.

Mattson, M.D., P.L. Godfrey, R.A. Barletta, and A. Aiello. 2004a. *Eutrophication and Aquatic Plant Management in Massachusetts: Final Generic Environmental Impact Report*. Cambridge, MA: Commonwealth of Massachusetts, Executive Office of Environmental Affairs Commonwealth of Massachusetts, p. 514.

Mattson, M.D., P.L. Godfrey, R.A. Barletta, and A. Aiello. 2004b. *Eutrophication and Aquatic Plant Management in Massachusetts: Final Generic Environmental Impact Report—Appendices*. Cambridge, MA: Commonwealth of Massachusetts, Executive Office of Environmental Affairs Commonwealth of Massachusetts, p. 181.

McComas, S. 2011. Literature Review on Controlling Aquatic Invasive Vegetation with Aquatic Herbicides Compared to Other Control Methods: Effectiveness, Impacts, and Costs. Blue Water Science, Prepared for: Minnehaha Creek Watershed District. www.clflwd.org/documents/HerbicideLitReviewReport-2011.pdf (Accessed August 16, 2021).

McDowell, W.G., A.J. Benson, and J.E. Byers. 2014. Climate controls the distribution of a widespread invasive species: Implications for future range expansion. *Freshwater Biology* 59: 847–857.

Millennium Ecosystem Assessment. 2005. *Ecosystems and Human Wellbeing: Biodiversity Synthesis.* Washington, DC: World Resources Institute.

Miralles, L., E. Dopico, F. Devlo-Delva, and E. Garcia-Vazquez. 2016. Controlling populations of invasive pygmy mussel (*Xenostrobus secures*) through citizen science and environmental DNA. *Marine Pollution Bulletin* 110(1): 127–132.

Mitzner, L. 1978. Evaluation of biological control of nuisance aquatic vegetation by grass carp. *Transactions of American Fisheries Society* 107(1): 135–145.

Mohit, S., T.B. Johnson, and S.E. Arnott. 2021. Recreational watercraft decontamination: Can current recommendations reduce aquatic invasive species spread? *Management of Biological Invasions* 12(1):148–164.

Morse, J.T. 2009. Assessing the effects of application time and temperature on the efficacy of hot-water sprays to mitigate fouling by dreissena polymorpha (zebra mussels Pallas). *Biofouling* 25: 605–610.

Muirhead, J.R., and H.J. MacIsaac. 2005. Development of inland lakes as hubs in an invasion network. *Journal of Applied Ecology* 42: 80–90.

NALMS (North American Lake Management Society) 2004. *The Use of Alum for Lake Management.* Washington, DC: NALMS, p. 3.

National Invasive Species Council (NISC). 2007. *Training and Implementation Guide for Pathway Definition, Risk Analysis and Risk Prioritization.* National Invasive Species Council Materials, p. 27. https://digitalcommons.unl.edu/natlinvasive/27.

Novoa, A., K. Dehnen-Schmutz, J. Fried, and G. Vimercati. 2017. Does public awareness increase support for invasive species management? Promising evidence across taxa and landscape types. *Biological Invasions* 19(12): 3691–3705.

Nunes, A.L., E. Tricarico, V.E. Panov, A.C. Cardoso, and S. Katsanevakis. 2015. Pathways and gateways of freshwater invasions in Europe. *Aquatic Invasions* 10(4): 359–370.

O'Reilly, C.M., et al. 2015. Rapid and Highly Variable Warming of Lake Surface Waters Around the Globe. *Geophysical Research Letters.* DOI: 10.1002/2015GL066235.

Patrick, A.E.S., and S. Florentine. 2021. Factors affecting the global distribution of hydrilla Verticillata (L. fil.) Royle: A Review. *Weed Research* 61(4): 253–271.

Pimentel, D., R. Zuniga, and D. Morrison. 2005. Update on the environmental and economic costs associated with alien-invasive species in the united states. *Ecological Economics* 52: 273–288.

Rahel, F.J., and J.D. Olden. 2008. Assessing the effects of climate change on aquatic invasive species. *Conservation Biology* 22(3): 521–533.

Rey, A., O.C. Basurko, and N. Rodríguez-Ezpeleta. 2018. The challenges and promises of genetic approaches for ballast water management. *Journal of Sea Research* 133: 134–145.

Ricciardi, A., M.E. Palmer, and N.D. Yan. 2011. Should biological invasions be managed as natural disasters? *BioScience* 61(4): 312–317.

Richardson, R.J., R.L. Roten, A.M. West, S.L. True, and A.P. Gardner. 2008. Response of selected aquatic invasive weeds to flumioxazin and carfentrazone-ethyl. *Journal of Aquatic Plant Management* 46: 154–158.

Ripple, W.J., et al. 2020. World scientists' warning of a climate emergency. *BioScience* 70(1): 8–12.

Rockwell-Postel, M., B.B. Laginhas, and B.A. Bradley. 2020. Supporting proactive management in the context of climate change: prioritizing range-shifting invasive plants based on impact. *Biological Invasions* 22: 2371–2383.

Rothlisberger, J.D., W.L. Chadderton, J. McNulty, and D.M. Lodge. 2010. Aquatic invasive species transport via trailered boats: What Is being moved, Who is moving it, and What can be done. *Fisheries* 35(3): 121–132.

Rumzie, N., R. Christopherson, S. O'Meara, Y. Passamaneck, A. Murphy, S. Pucherelli, and J. Trujillo. 2021. Costs Associated with Invasive Mussels Impacts and Management. Science and Technology Program, Research and Development Office, Final Report No. ST-2021–8142–01. Bureau of Reclamation. Denver, CO, p. 35.

Rushton, S., A. Spake, and L. Chariton. 2016. *The Unintended Consequences of Using Glyphosate (The Main Ingredient in the Herbicide Roundup).* London: Sierra Club, p. 27.

Schneider, D.W., C.D. Ellis, and K.S. Cummings. 1998. A transportation model assessment of the risk to native mussel communities from zebra mussel spread. *Conservation Biology* 12: 788–800.

Scholl, C. 2006. *Aquatic Invasive Species: A Guide for Proactive & Reactive Management.* Wisconsin: Vilas County Land and Water Conservation Department, p. 58.

Scriver, M., A. Marinich, C. Wilson, and J. Freeland. 2015. Development of Species-specific Environmental DNA (eDNA) Markers for invasive Aquatic Plants. *Aquatic Botany* 122: 27–31.

Sepulveda, A.J., N.M. Nelson, C.L. Jerde, and G. Luikart. 2020. Are environmental DNA methods ready for aquatic invasive species management? *Trends in Ecology and Evolution* 35(8): 668–678.

Sepulveda, A.J., D.R. Smith, K.M. O'Donnell, N. Owens, B. White, C.A. Richter, C.M. Merkes, S.L. Wolf, M. Rau, M.E. Neilson, W.M. Daniel, and M.E. Hunter. 2022. Using structured decision making to evaluate potential management responses to detection of dreissenid mussel (Dreissena spp.) environmental DNA. *Management of Biological Invasions* 13.

Sesin, V., R.L. Dalton, C. Boutin, S.A. Robinson, A.J. Bartlett, and F.R. Pick. 2018. Macrophytes are highly sensitive to the herbicide diquat dibromide in test systems of varying complexity. *Ecotoxicology and Environmental Safety* 165: 325–333.

Stein, B.A., and S.R. Flack. 1997. *Species Report Card: The State of U.S. Plants and Animals.* Arlington, VA: The Nature Conservancy, p. 26.

Stokstad, E. 2007. Feared quagga mussel turns up in western United States. *Science* 315: 453–453.

Stranga, Y., and S. Katsanevakis. 2021. Eight years of bioinvasions records: Patterns and trends in alien and cryptogenic species records. *Management of Biological Invasions* 12(2): 221–239.

Strid, A., W. Hanson, A. Cross, and J. Jenkins. 2018. Triclopyr General Fact Sheet; National Pesticide Information Center, Oregon State University Extension Services. npic.orst.edu/factsheets/triclopyrgen.html. (Accessed September 12, 2021).

Thomsen, P.F., J. Kielgast, L.L. Iversen, C. Wiuf, M. Rasmussen, M.T.P. Gilbert, L. Orlando, and E. Willerslev. 2011. Monitoring endangered freshwater biodiversity using environmental DNA. *Molecular Ecology* 21(11): 2565–2573.

Thomsen, P.F., and E. Willerslev. 2014. Environmental DNA—An emerging tool in conservation for monitoring past and present biodiversity. *Biological Conservation* 183: 4–18.

Trebitz, A.S., J.C. Hoffman, J.A. Darling, E.M. Pilgrim, J.R. Kelly, E.A. Brown, W.L. Chadderton, S.P. Egan, E.K. Grey, S.A. Hashsham, K.E. Klymus, A.R. Mahon, J.L. Ram, M.T. Schultz, C.A. Stepien, and J.C. Schardt. 2017. Early detection monitoring for aquatic non-indigenous species: Optimizing surveillance, incorporating advanced technologies, and identifying research needs. *Journal Environmental Management* 202: 299–310.

U.S. Department of the Interior. 2016. *Safeguarding America's Lands and Waters from Invasive Species: A National Framework for Early Detection and Rapid Response*. Washington, DC: U.S. Department of the Interior, p. 55.

U.S. Department of the Interior. 2021. *U.S. Department of the Interior Invasive Species Strategic Plan, Fiscal Years 2021–2025*. Washington, DC: U.S. Department of the Interior, p. 54.

U.S. EPA (Environmental Protection Agency). 1998. EPA Red Facts Triclopyr. https://www3.epa.gov/pesticides/chem_search/reg_actions/reregistration/fs_G-82_1-Oct-98.pdf (Accessed October 5, 2021).

U.S. Geological Survey (USGS). 2020. *Nonindigenous Aquatic Species Database*. Gainesville, FL. https://nas.er.usgs.gov (Accessed October 10, 2020).

Vander Zanden, M.J., and JD. Olden. 2008. A management framework for preventing the secondary spread of aquatic invasive species. *Canadian Journal of Fisheries and Aquatic Sciences* 65: 1512–1522.

Wagner, K. 2001. Chapter 7 Management Techniques Within the Lake or Reservoir. In: C. Holdren, W. Jones, and J. Taggart (eds). *Managing Lake and Reservoirs*. Madison, WI: USEPA/NALMS, pp. 215–306.

Wagner, K. 2004. *The Practical Guide to Lake Management in Massachusetts A Companion to the Final Generic Environmental Impact Report on Eutrophication and Aquatic Plant Management in Massachusetts*. Cambridge, MA: Commonwealth of Massachusetts, Executive Office of Environmental Affairs Commonwealth of Massachusetts. 160 p.

Wan, Z., Z. Shi, A. Nie, J. Chen, and Z. Wang. 2021. Risk assessment of marine invasive species in Chinese ports introduced by the global shipping network. *Marine Pollution Bulletin* 173: 112950.

Wilcove DS, Rothstein D, Bulow J, Phillips Al, Losos E. 1998. Quantifying threats to imperiled species in the United States. *Bioscience* 48: 607–615.

Wong, W.H., and S.L. Gerstenberger. 2011. Quagga mussels in the western United States: Monitoring and management. *Aquatic Invasions* 6(2): 125–129.

Wong, W.H., and S.L. Gerstenberger. 2015. *Biology and Management of Invasive Quagga and Zebra Mussels in the Western United States*. Boca Raton, FL: CRC Press, p. 566.

Wong, W.H., M. Piria, F.P.L. Collas, P. Simonovic, and E. Tricarico. 2017. Management of invasive species in inland waters: Technology development and international cooperation. *Management of Biological Invasions* 8(3): 267–272.

WSSA (Weed Science Society of America). Summary of herbicide mechanism of action according to the Weed Science Society of America (WSSA). https://wssa.net/wp-content/uploads/WSSA-Mechanism-of-Action.pdf (Accessed November 5, 2021).

Zaiko, A., X. Pochon, E. Garcia-Vazquez, S. Olenin, and S.A. Wood. 2018. Advantages and limitations of environmental DNA/RNA tools for marine biosecurity: Management and surveillance of non-indigenous species. *Frontiers in Marine Science* 5: 322 (17 pages).

Zook, B., and S. Phillips. 2015. Uniform Minimum Protocols and Standards for Watercraft Interception Programs for Dreissenid Mussels in the Western United States. In W.H. Wong, and S.L. Gerstenberger (eds). *Biology and Management of Invasive Quagga and Zebra Mussels in the Western United States*. London: CRC Press, pp. 176–204.

2 Asian Clam (*Corbicula fluminea*)

ABSTRACT

Invasive Asian clams (*Corbicula fluminea*) were first recorded in the state of Massachusetts in 2001, in a section of the Charles River running through Watertown. As of 2019, the species has been found in 32 different bodies of water throughout state. All clams have been found in the warmer parts of the state, where water temperatures are above the clams' long-term incipient lower thermal limit; therefore, the increasing abundance of the Asian clam may be a sign and a result of rising temperatures in this region. The map of infested waterbodies created as a result of this project will serve to increase the general public's awareness of the issue as well as aid water resource managers in tracking the extent of infestations in Massachusetts, measures which will be helpful in preventing the further spread of the clams to new waterbodies. This report also describes the biological characteristics and potential invasive pathways of these invasive clams and review alternative management tools to control the species chemically, mechanically, and biologically.

INTRODUCTION

Taxonomic Hierarchy (ITIS 2021)

Kingdom	Animalia
Subkingdom	Bilateria
Infrakingdom	Protostomia
Superphylum	Lophozoa
Phylum	Mollusca
Class	Bivalvia
Subclass	Heterodonta
Order	Veneroida
Superfamily	Corbiculoidea
Family	Corbiculidae
Genus	*Corbicula*

The Asian clam is a small bivalve mollusk. Its light-colored shell often has concentric sulcations, with anterior and posterior lateral teeth that have many serrations (Foster et al. 2019). While light shells are the predominant shell type in the northeast United States, darker morphs can be found in the southwestern United States. This clam is generally less than 50 mm in shell length. In the Columbia River (CR), the zooplankton community of the lower CR is dominated by recently released juvenile *C. fluminea* suspended on turbulent water currents in late summer and early autumn (Dexter et al. 2015; 2020). The Asian clam is also a self-fertilizing hermaphroditic species which can produce up to 570 larvae per day, resulting in more than 68,000 larvae produced per individual per year (McMahon 1999). This enables a single individual to initiate reproduction in isolation and has been shown to lead to rapid reproduction (Gomes et al. 2016). The unique androgenetic reproductive biology of Asian clams where maternal chromosomes are extruded from the fertilized egg as two polar bodies during first meiosis leaves only the male pronucleus present in the egg. However, there is no second meiosis. The male pronucleus becomes the metaphase chromosomes during the first mitosis. Thus, only male genetic material is left in the developing embryo. More detailed androgenetic reproduction in *C. fluminea* can be found from Ishibashi et al. (2003). Suspended newly released juveniles usually settle by the time they reach shell lengths of ≤5 mm using a single byssal thread to attach to hard substratum. However, adults can leave the substratum to be carried long distances downstream over the substratum by water currents to settle in areas of slower water flow. Food availability has been implicated as an important factor to the growth and reproductive capacity of Asian clams in the lower Rhine River, Germany (Viergutz et al. 2012). Temperature

DOI: 10.1201/9781003201106-2

can affect individual growth rate (Cataldo and Boltovskoy 1998) as well as clam shell size and body mass (Denton et al. 2012; Rosa et al. 2012). The species has demonstrated a high degree of both physiological and ecological plasticity, making the Asian clam hard to control and manage (Coughlan et al. 2018).

ECOLOGICAL IMPACTS

The Asian clam, *Corbicula fluminea*, is considered to be one of the most ecologically and economically impactful aquatic invasive species in global aquatic ecosystems (Sousa et al. 2008). Over time, *C. fluminea* has spread from its native range in Eastern Asia and Africa to Europe, North America, South America, and other areas of the world. The initial introduction of Asian clams in North America in 1938 is thought to have been due to transoceanic ballast water exchange and use by Chinese immigrants as a food source (Counts 1981; 1986; Sousa et al. 2008). By the 1960s, the Asian clam had spread through the US waterways and reached as far as the Atlantic coast (McMahon and Bogan 2001). Asian clams have now established themselves in the United States above the 40° latitude line, where waters meet their temperature

tolerance. The species' invasive success and wide dispersion are mainly owed to its natural characteristics, including rapid growth, early sexual maturity, short life span, high fecundity, extensive dispersal capacities, tolerance to variable environmental variables, and their association with human activities (McMahon 2002; Karatayev et al. 2005; Sousa et al. 2008; Lucy et al. 2012). As shown in Figure 2.1, clams are able to attach to drifting plant materials, shells, or pebbles (Minchin and Boelens 2018). In a field experiment to study survival and growth of captively reared juveniles of four native mussel species in situ exposures at 17 sites in the Rockcastle River system, Kentucky, it is found that growth of these mussels was negatively related with the abundance of Asian clams, but survival rates were not affected (Haag et al. 2021). In recent years, Asian clams continue to infest new waters as a result of human activity, including ballast water transport and discharge, fish bait, release from aquariums, tourism boats (Sousa et al. 2008), and subsequently through natural water dispersal.

Like many other bivalve mollusks such as zebra or quagga mussels (Baker et al. 1998; Wong et al. 2003), Asian clams are suspension feeders that can drastically change the ecosystem.

 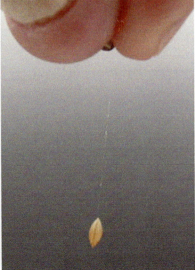

FIGURE 2.1   An 8-mm clam attached by a byssal thread to drifting plant material retrieved using a dredge (*left*) and a byssal thread of a clam removed from the mesh of a dredge (Minchin and Boelens 2018).

Feeding behavior studies of Asian clams demonstrated that adult clams in the Columbia River at the Pacific Northwest selectively feed on flagellates but avoid feeding on cyanobacteria (Bolam et al. 2019). It is also found that *C. fluminea* veligers are associated with cyanobacteria and higher water temperatures in the Columbia River (Hassett et al. 2017). In the Meuse River in Northwest Europe, the filter feeding of Asian clams caused a 70% decrease in phytoplankton abundance, thereby impacting the entire food web (Pigneur et al. 2014). In addition, Asian clams can also perform pedal feeding on benthic microbes such as flagellates, bacteria, and diatoms, even in the presence of abundant prey in the water column (Hakenkamp et al. 2001).

Although Asian clams have many negative impacts, there are also a few positive effects that stem from their presence. Empty shells can be used by other species as shelters, and the clams are a food source for some species at higher trophic levels, such as cyprinid species (e.g., *Barbus* spp. and *Luciobarbus* spp., *Cyprinus carpio* Linnaeus, 1758) and pumpkinseed sunfish *Lepomis gibbosus* (Linnaeus, 1758) (Pereira et al. 2016 and references therein).

ECONOMIC IMPACTS

Asian clams have significantly reduced ecosystem services and caused economic damages. It has been reported to have cost the United States approximately $1 billion per year to control Asian clams since 1980 (Pimentel et al. 2005). This is due to the damage Asian clams cause to equipment, such as water intake pipes, electric power plant cooling systems, and sewage treatment plants (McMahon 2002; Muller and Baur 2011). The ecological damage Asian clams cause is also significant. Asian clams are filter feeders that consume large amounts of microscopic plants and animals, creating more competition for food sources among native, aquatic life (Robinson 2004). Asian clams can reach densities greater than 6,000 clams/m² in rivers (Robinson 2004; Sheehan et al. 2014). Such expansive coverage alters the benthic community and has a large impact on macroinvertebrate communities (Coughlan et al. 2018). The clams also sequester a large portion of the available carbon in the water (Muller and Baur 2011). Native unionid mussel

species may be threatened with displacement by the Asian clam (Robinson 2004), although this is not always considered to be the case (McMahon and Bogan 2001).

MONITORING AND SIGHTINGS

The Massachusetts Department of Environmental Protection (MassDEP) is committed to protect, enhance, and restore the quality and value of the waters of the Commonwealth with guidance by the Federal Clean Water Act. It works to secure the environmental, recreational, and public health benefits of clean water for the citizens of Massachusetts. As a result, a long-term statewide benthic macroinvertebrate biomonitoring program has been in place since 1973. Benthic community diversity data is one of the indicators used by WPP staff to assess the health of a waterbody. Benthic samples have been collected from 410 different waterbodies statewide. The standard operating procedures, including sampling methodology, species identification, quality control and assurance, can be found in Nuzzo (2003). Sites with Asian clams were mapped as a data layer with ArcGIS® ArcMap 10.1 (ESRI, Redlands, California). Based on MassDEP's own records, Asian clams were only detected from 2 out of the 410 waterbodies statewide (Table 2.1). This ratio (2/410) is low because not every waterbody in the state is part of the benthic biomonitoring program.

To gain a more complete record of Asian clams in Massachusetts, sightings reported by the Massachusetts Department of Conservation and Recreation (Jim Straub, personal communication) and US Geological Survey (USGS) (Foster et al. 2019) staff were also included. The first documented sighting of live Asian clams in Massachusetts was in a section of the Charles River in Watertown, Massachusetts, in 2001 (Foster et al. 2019). In Massachusetts, as of 2016, there were a total of 25 sampling sites from 24 waterbodies with Asian clams known to MassDEP (Colwell et al. 2017). From 2001 to 2018, Asian clams have been found in 33 waterbodies, including 7 rivers and 26 ponds/lakes (Figure 2.2 and Table 2.1). These sightings are within 14 Massachusetts watersheds (Table 2.2 and Figure 2.2). The Asian clam has also been

TABLE 2.1

Latitude and Longitude for Each Waterbody Shown in Figure 2.1 and Source of Report

Site Number	First Year of Presence Report	Waterbody	Longitude	Latitude	Reporting Agency	Source of the Report
1	2001	Charles River	−71.18164	42.36057	MassDEP, USGS	DEP: Charles River Watershed 2002–2006 Water Quality Assessment Report 72-AC-4 CN136.5 April 2008, DEP: 2007018 Benthic Sample ID, USGS 8/22/2016 http://nas.er.usgs.gov/default.aspx
2	2005	Congamond Lakes (Middle Basin)	−72.75593101	42.02495588	MA DCR, USGS	DCR Data from Lakes and Ponds, 2005, and USGS 8/22/2016 http://nas.er.usgs.gov/default.aspx
3	2005	Congamond Lakes (North Basin)	NA	NA	MA DCR	DCR Data from Lakes and Ponds, 2005
4	2005	Congamond Lakes (South Basin)	NA	NA	MA DCR	DCR Data from Lakes and Ponds, 2005
5	2005	Fort Meadow Reservoir	−71.54811073	42.36864846	MA DCR, USGS	DCR: Lakes and Ponds, 2007, USGS 8/22/2016 http://nas.er.usgs.gov/default.aspx
6	2005	Tispaquin Pond	−70.85731719	41.86342594	MA DCR, USGS	DCR: Lakes and Ponds, 2006, USGS 8/22/2016 http://nas.er.usgs.gov/default.aspx
7	2005	Webster Lake	−71.84417968	42.04063346	MA DCR, USGS	DCR: Lakes and Ponds, 2005, USGS 8/22/2016 http://nas.er.usgs.gov/default.aspx
8	2006	East Brimfield Reservoir	−72.13218404	42.11613205	MA DCR	DCR: Lakes and Ponds, 2007, DCR: National Heritage 2006 DFW T. French, DCR: Lakes and Ponds, 2007
9	2006	Long Pond	−70.94455336	41.80145728	MA DCR	DCR: Lakes and Ponds, 2006
10	2007	Fivemile Pond	−72.51066308	42.14168287	MA DCR, USGS	DRC: Lakes and Ponds, 2007, USGS 8/22/2016 http://nas.er.usgs.gov/default.aspx
11	2007	Lake Sabbatia	−71.10816736	41.94456101	MA DCR, USGS	DCR: Lakes and Ponds, 2007, USGS 8/22/2016 http://nas.er.usgs.gov/default.aspx

#	Year	Location	Longitude	Latitude	Source	Reference
12	2007	Mashpee Pond	−70.48697269	41.66106628	MA DCR, USGS	DCR: National Heritage 2007 E. Nedeau, USGS 8/22/2016 http://nas.er.usgs.gov/default.aspx
13	2007	Oldham Pond	−70.83644943	42.0641901	MA DCR	DCR: J. Cordeiro 2007
14	2007	Sampson Pond	−70.75012337	41.85070814	MA DCR	DCR: Lakes and Ponds, 2007
15	2007	Winnecunnet Pond	−71.13126679	41.97093302	MA DCR	DCR: National Heritage 2007 E. Nedeau
16	2010	Middle Pond	−70.41564	41.66864	USGS	USGS 8/22/2016 http://nas.er.usgs.gov/default.aspx
17	2010	Norton Reservoir	−71.19718046	41.98368621	USGS	USGS 8/22/2016 http://nas.er.usgs.gov/default.aspx
18	2010	Taunton River	−71.07509	41.9031	USGS	USGS 8/22/2016 http://nas.er.usgs.gov/default.asp
19	2010	Taunton River	−70.95663889	41.93381794	USGS	USGS 8/22/2016 http://nas.er.usgs.gov/default.aspx
20	2013	Bennett Pond	−71.87779605	42.11796353	Individual	Personal communication 2/1/2017, Joy Traha
21	2013	Massapoag Lake	−71.177528	42.103486	Individual	Personal communication 2/1/2017, Joy Traha
22	2014	Nemasket River	−70.94243923	41.93623946	Individual	Personal communication 2/1/2017, Joy Traha
23	2014	Unnamed Forest Park Tributary	−72.56556396	42.07983422	MassDEP	14-C001–05, 14-C008–05
24	2015	Blackstone River	−71.61710088	42.09218317	Individual	Personal communication 2/1/2017, Joy Traha
25	2015	Hamblin Pond	−70.410532	41.663433	Individual	Personal communication 8/31/2016, Ken Wagner
26	2015	Spy Pond	−71.15275255	42.40706673	USGS	USGS 8/22/2016 http://nas.er.usgs.gov/default.aspx
27	2016	Concord River	−71.336	42.479	FWS	Personal communication 11/8/2016 Eileen McGourty, FWS
28	2016	Dudley Pond	−71.374	42.33	Wayland Surface Water Quality Committee	Personal communication, Mike Lowery, Wayland Surface Water Quality Committee, 10/24/2016
29	2016	Lake Cochituate-Carling Basin	−71.370429	42.30029	DCR	Personal communication, Tom Flannery 10/21/2016 DCR

(Continued)

TABLE 2.1
(Continued)

Site Number	First Year of Presence Report	Waterbody	Longitude	Latitude	Reporting Agency	Source of the Report
30	2016	Lake Cochituate-Middle Basin	−71.374753	42.310286	DCR	Personal communication, Tom Flannery 10/21/2016 DCR
31	2016	Sudbury River	−71.394404	42.335319	FWS	Personal communication, 9/6/2016, Eileen McGourty
32	2017	Heart Pond	−71.387734	42.565745	MassDEP	Fish Servey 5/23/2017
33	2018	Whitins Pond	NA	NA	MassDEP	Therese Beaudoin

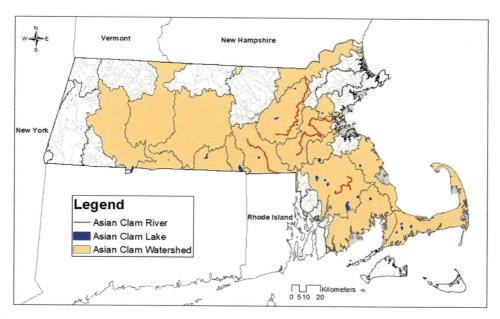

FIGURE 2.2 Distribution map depicting *Corbicula fluminea* in Massachusetts from 2001 to 2018.

recently confirmed in Onota Lake within the Housatonic River Watershed at Pittsfield[1] and Quinsigamond Lake within the Blackstone River Watershed between the City of Worcester and the Town of Shrewsbury.[2] The distribution of Asian clams throughout the state in rivers and ponds and the affected watersheds is also highlighted in Figure 2.2. Most of these waterbodies are in the southern and eastern parts of the state (Figure 2.2). Asian clams were found in both brackish and freshwater sections of the Taunton River (Colwell et al. 2017), signaling dispersion across various types of aquatic habitat. Table 2.1 provides the name of the waterbodies, as well as the latitude and longitude of a sampling site, the agency(s) that reported the sighting, and the more specific report source(s). The figure and table only represent sighting data from during or before the year 2018. Seen as the most widely distributed aquatic nuisance species in the contiguous United States (McMahon and Bogan 2001), Asian clams are an emerging issue for the Northeast states (Lake Champlain Basin Program 2016). The species' presence has recently been detected in other New England states as well. The first infestation of Asian clam in the State of Vermont was in the southwestern part of Lake Bomoseen in the summer

of 2016 (Ann Bove, personal communication). In the State of New Hampshire, Asian clams are already present in the lower Merrimack River and in several ponds (New Hampshire Department of Environmental Service 2016). Seen as the most widely distributed aquatic nuisance species in the contiguous United States (McMahon and Bogan 2001), Asian clams are an emerging issue for the Northeast states (Lake Champlain Basin Program 2016).

It is not surprising that the first Asian clam sighting in Massachusetts was in the Charles River. The Charles River is a central feature of Eastern Massachusetts' metropolitan landscape, frequently used recreationally by boaters, swimmers, and anglers. The river is about 129 km (80 miles) long, beginning at an outlet at Echo Lake, Hopkinton, and running through the City of Boston, where it flows into the Atlantic Ocean. The first Asian clam sighting was located in the segment between Watertown Dam and the Boston University Bridge, where boaters and canoers can be found year-round (Table 2.1). Infestations of non-native aquatic species often occur in recreational waterbodies. For example, the first invasion of quagga mussels (*Dreissena rostriformis bugensis*) in the western United States was in the Las Vegas Boat Harbor Marina

TABLE 2.2

Asian Clams Infested Watersheds and Waterbodies in Each Watershed

Watershed	Waterbody
Blackstone	Blackstone River
	Whitins Pond
Boston Harbor: Mystic	Spy Pond
Boston Harbor: Neponset	Massapoag Lake
Buzzards Bay	Sampson Pond
Cape Cod	Ashumet Pond
	Hamblin Pond
	Mashpee Pond
	Middle Pond
	Peters Pond
Charles	Charles River
Chicopee	Fivemile Pond
Concord (SuAsCo)	Concord River
	Dudley Pond
	Fort Meadow Reservoir
	Heart Pond
	Lake Cochituate
	Sudbury River
Connecticut	Unnamed Pond
	Unnamed Tributary
French	Webster Lake
Quinebaug	East Brimfield Reservoir
South Coastal	Oldham Pond
Taunton	Lake Sabbatia
	Long Pond
	Nemasket River
	Norton Reservoir
	Taunton River
	Tispaquin Pond
	Winnecunnet Pond
Westfield	Congamond Lakes, Middle Basin
	Congamond Lakes, North Basin
	Congamond Lakes, South Basin

of Lake Mead (Moore et al. 2015). Lake Mead is ranked as one of the most visited national parks in the United States. The invasive quagga mussel was introduced first into Lake Mead and rapidly spread to other areas of the western United States (McMahon 2011; Wong and Gerstenberger 2011). Likewise, the spread of the Asian clam in Massachusetts may be correlated with the recreational activity of boaters, swimmers, and anglers.

To specifically examine the abundance of the Asian clam in a tributary to Porter Lake (inside Forest Park) in Springfield, MA, samples were collected from four stations along the stream (Colwell et al. 2017). Invasive purple loosestrife (*Lythrum salicaria*) has been found in this lake

FIGURE 2.3 An unnamed tributary to Porter Lake in Springfield, Massachusetts. The white spots instream are empty *C. fluminea* shells. (Photos by Wai Hing Wong.)

FIGURE 2.4 Live clams (*left*) and dead empty shells (*right*) collected from sediment (the two photos are not in the same scale). (Photos by W.H. Wong.)

(Kennedy and Weinstein 2000) and there is also an unconfirmed report of invasive curly-leaved pondweed (*Potamogeton cripus*) (Carr and Kennedy 2008). Historically, this lake has very dense growth of aquatic macrophytes (primarily *Ceratophyllum demersum*) and there are floating algal and duckweed mats at the western end (Kennedy and Weinstein 2000). This unnamed tributary drains into Porter Lake, which is a 28-acre freshwater lake. The tributary is small and wadable, with vegetated banks, and black coarse sands being major components of the sediment (Figure 2.3). Both living Asian clams and dead clam shells were found at the sites along the stream (Figure 2.4). The mean density of Asian clams was 6,124 individuals/m^2 (Table 2.3) and the average shell length of clams from the four sites was 6.92, 6.88, 6.44,

TABLE 2.3

Density of Live Asian Clams and Dry Weight of Dead/Broken Shells in an Unnamed Tributary to Porter Lake in Springfield, Massachusetts

Station	Station 1	Station 2	Station 3	Station 4
Density (individuals/m^2)	7,104	6,932	6,416	4,047
Dry dead shell weight (g/m^2)	512.3	159.3	51.7	4.3

Source: From Colwell et al. (2017).

TABLE 2.4

The Size of Asian Clams at Different Stations in an Unnamed Tributary to Porter Lake in Springfield, Massachusetts (Colwell et al. 2017)

Station		Station 1	Station 2	Station 3	Station 4
Shell length (mm)	Mean	6.92	6.88	6.44	6.11
	Standard deviation	2.82	2.72	2.11	1.40
	Maximum	22.64	18.72	15.46	12.69
	Minimum	2.56	3.43	2.49	3.04
Shell height (mm)	Mean	5.83	5.82	5.46	5.16
	Standard deviation	2.41	2.39	1.80	1.20
	Maximum	19.05	15.88	13.64	11.06
	Minimum	2.13	2.89	2.10	2.56

and 6.11 mm, respectively (Table 2.4). The maximum shell length from the four sites was 22.64, 18.72, 15.46, and 12.69 mm, respectively.

This study demonstrates that the infestation of Asian clams in Massachusetts is not as severe as infestations in the southeast states (Foster et al. 2019), but still considerable compared with other locations in the United states. The density from the unnamed tributary to Porter Lake in Springfield, MA, at 6,124 clams/m^2 is moderate compared with densities reported at other sites such as 55 clams/m^2 in Lake Seminole, GA (Patrick et al. 2017), 722 clams/m^2 in Columbia River, OR/WA (McCabe et al. 1997), and more than 10,000 clams/m^2 in New River, VA (Graney et al. 1980). Asian clams that were found upstream were larger and more abundant than those found downstream. Based on the size structure, it is estimated that most Asian clams found at upstream stations are up to 2 years old, while the age of most individuals from the downstream station is about 1 year old (Mouthon 2001). The weight and maximum size of dead clam shells from upstream stations

were greater than that from downstream stations. It appears that a few clams reached up to 2 years old in the upstream stations and few, if any, appear to survive beyond the first year of life in the downstream station. This could possibly be attributed to the differences in habitats along this tributary. Asian clams may prefer upstream stations because they are composed of soft mud, unlike downstream sites where the sediments contain hard clays. It also may be that larger individuals are more likely to be displaced downstream during flood events. This may explain the loss of the 2-year-old cohort in the downstream station where the hard clay and coarse sand substrate suggest higher flow rates that could have exposed larger, older clams to being hydrologically transported downstream during flood events (Williams and McMahon 1986). By contrast, the deeper, coarse sand and soft mud substrata of the upstream sites may be indicative of a lower flow regime suggesting that the clams there are less hydrologically disturbed during flood events than the downstream station. To understand the density and size frequency of

Asian clams in other infested waterbodies, and better determine their population dynamics in the Commonwealth of Massachusetts, successive monthly samples from multiple locations should be taken over a period of at least one year (McMahon and Williams 1986).

The distribution of Asian clams is controlled by both habitat variables (e.g., substrate and pH) and climate variables (e.g., minimum temperature $\leq 2°C$ in the coldest month) (Mattice and Dye 1976). Invasive clams have typically been found to occur at higher densities in warmer habitats. Almost all Asian clam sightings are in the southern or eastern parts of the state: the warmer the water, the more likely Asian clams can be found. MassDEP has monitored water temperature in five rivers since 2012 where continuous water temperature data were collected every 30 min for each at the five monitoring stations (T1, T2, T3, T4, and T5). Based on data collected from November 1, 2013, to April 24, 2014, Station T1 had significantly lower temperature than Station T2 and T4; while the temperatures of the latter two stations were significantly lower than that of Station T3 and T5 (Analysis of Variance, $P < 0.0001$) (Nuzzo, unpublished data). Stations in the northern or western parts of the state had a lower temperature due to their higher latitude and/or altitude. Overall, the range of Asian clams appears to be expanding northward, given Asian clam sightings in the southern and eastern parts of Massachusetts from 2001 to 2016, as well as recent discoveries of Asian clams in Vermont and New Hampshire. This shift in range corresponds with the general warming trend observed in the Eastern United States.

Warming temperatures have been recorded recently in this region; ice cover has been found to freeze later in the winter with earlier ice breakup in the spring in northeastern lakes (USEPA 2012). In addition, the lowest daily winter temperatures (Blue Hill Observatory, unpublished data) from 1985 to 2000 (specifically, 16 years before Asian clams were found in Massachusetts) were significantly lower than that from 2001 to 2016 (16 years after the first-year Asian clam was confirmed in Massachusetts) which demonstrates that Massachusetts is becoming warmer from 1985 to 2016 (Wilcoxon Two-Sample Test, $N = 2,888$, $P = 0.02$). If trend analysis is used for analyzing the winter daily minimal temperature

in Massachusetts from 1985 to 2016, the temperature is increasing (Kendall's Tau Correlation Coefficient, $N = 2,888$, $P = 0.0002$). Therefore, the increasing winter temperature in Massachusetts is highly likely to help Asian clams establish their populations successfully at different waterbodies (Figure 2.2). Climate change is most likely the key reason Asian clams are now able to establish themselves in the United States above the $40°$ latitude line, as waters now remain above the clams' long-term incipient lower thermal limit (Mattice and Dye 1976). Therefore, the impact of climate change on the distribution of this species in Massachusetts and the northeast region should also be monitored because Asian clams are likely to expand well beyond their present distribution in Massachusetts and the United States (McDowell et al. 2014).

Apart from temperature, calcium concentration in the water is another limiting factor for Asian clam. Bivalve mollusks require calcium for the formation of their shells and calcium is often an important environmental predictor of the presence of freshwater bivalves (McMahon 1996; Lucy et al. 2012). Ferreira-Rodríguez et al. (2017) reported that the growth rates of *C. fluminea* are reduced significantly at lower calcium concentrations. Smith (1993) reported two watersheds, the Hoosic and Housatonic Rivers in Western Massachusetts with favorable conditions for zebra mussels to succeed, while the rest of the Connecticut River Watershed in Central Massachusetts and Eastern Massachusetts were marginally susceptible. So far, zebra mussels are only found in the Housatonic River watershed. However, many Asian clam sightings are found in Central and Southeast Part of Massachusetts. Therefore, Asian clams may be more tolerant than zebra mussels to lower calcium concentrations. This is also in agreement with report from McDowell et al. (2014) that climate variables are likely more important than the habitat variables, although Asian clams can be controlled by both their habitat and climate variables.

MANAGING ACTIONS

The best approach to preventing the spread of aquatic invasive species is an integrated management strategy (Wong and Gerstenberger 2015). An urgent biosecurity requirement is needed to stop or significantly limit the spread of this invasive species. For example, appropriate signage and decontamination equipment should be provided at infested waterbodies for users to be aware of the infestation and the need to clean their boats before leaving the waterbody. Mapping infested waterbodies in Massachusetts, such as the one (Figure 2.2) created in this study, is helpful in keeping local residents and out-of-state travelers aware of the infestations before and after visiting their targeted lakes and rivers, as well as helping resource managers track the expansion or contraction of the infestation. There should be at least four parts to this plan:

1. The primary means by which the species is being spread, whether it is human activity or a natural cause, needs to be identified. The most vulnerable waterbodies also need to be identified in order for resources to be allocated in the most cost-effective manner.
2. A means to slow or stop the spread needs to be created. This would include educational programs to raise public awareness, stricter boating codes, and new legislation (Sousa et al. 2013) to prevent the further spread of Asian clams such as building washing stations at the entrance to susceptible waters.
3. Analysis needs to be completed to determine the ecological and economic impacts of Asian clam infestations of new waterbodies.
4. Finally, a plan for emergency response to an invasion needs to be created (Sousa et al. 2013). This would proactively limit the invasive potential of the species.

The manner of dealing with Asian clams can be both reactive and proactive. Reactive methods of Asian clam control are often more costly than proactive approaches: an example of this would be using a benthic barrier to control their populations (Sousa et al. 2013). Proactive methods target Asian clams in the larval stage, such as with chemical treatments. Proactive methods tend to be more effective and versatile (Sousa et al. 2013). It is more difficult to stop the spread of Asian clams in open waters, where treatment methods are more likely to negatively impact non-target species.

CHEMICAL TREATMENT

Carefully regulated potassium treatments have the potential of eradicating nuisance molluscs while avoiding harm to non-molluscan wildlife and human health (Colwell et al. 2017, Sousa et al. 2013). Chemically innovative means of control such as BioBullets® have also been developed for specific use as molluscicides. These substances are designed to selectively target filter feeders by introducing poison to the water in the form of food particles preferred by molluscans. The poisons are then removed from the water when the molluscs feed, killing the molluscs while theoretically limiting harm to the rest of the aquatic ecosystem (Aldridge et al. 2006). While molluscicides such as these have shown promise in selectively controlling other biofouling organisms, such as the zebra mussel (Aldridge et al. 2006), more testing is required to determine an optimum protocol and dosage for treating Asian clam infestations. Potassium chloride seemed to be more lethal around 100–250 mg/l under hypoxic conditions and the clam mortality rate increased significantly at around 150–900 mg/l with the highest mortality rate resulting from around 450 mg/l under normoxic conditions (Rosa et al. 2015; Densmore et al. 2018). Several other chemicals have been tested which include ammonium nitrate, niclosamide, dimethoate, polyDADMAC, and chlorpyrifos (see more information in Appendix III Literature research of chemical control methods on 19 invasive species). In addition, preliminary experimentation has shown that thyme oil and extracts used for *Eichornia* spp. (water hyacinths) treatment could be employed in controlling Asian clam infestations, but definitive results have not yet been published in support of this theory. There are many other methods and molluscicides that are used to control Asian clam macrofouling of raw water using facilities. Therefore, treatment in raw water facilities may not be suitable for application to open waters (i.e., natural lakes, ponds, rivers, and streams).

PHYSICAL TREATMENT

Depletion of dissolved oxygen induced by the placement of gas-impermeable benthic barriers has been found to be effective on Asian clams in Lake Tahoe (Wittmann et al. 2012) and Lake

George (NY) (Colwell 2017; Sousa et al. 2013), although the strategy is relatively expensive and harmful to non-target benthic organisms. Thermal shock also may be effective in eradicating Asian clams. Treatment of Asian clam infestation using dry ice, hot steam, and open flame (on dry sand substrate) has also been attempted in a controlled setting, and in some cases found to be effective depending on the time and intensity of exposure (Coughlan et al. 2018; Coughlan et al. 2019). Winter lake drawdowns might also be a possible control measure to reduce the population and/or contain them in the deeper area of a lake. However, these methods could also be damaging to non-invasive species and the surrounding ecosystem, and the cost and benefit of their applications should be carefully weighed on a case-by-case basis. Though it is time-consuming, hand removal remains the most safe and effective strategy to combat Asian clam infestations in the interest of avoiding negative effects to other aquatic organisms, especially in the early stage of the infestation. Once established, eradication of Asian clams is often difficult, if not impossible, without destructive impacts to native species and their natural habitats. Therefore, prevention of spread and early detection should be prioritized.

BIOLOGICAL TREATMENT

Biological control agents for *C. fluminea* may be difficult to find or develop, because a predator must be large enough to consume a whole clam and not be limited by clam valve closure or it must have an ability to open the clam's shell valves. It must also not be able to feed extensively on native bivalves and other native species. A study that examined the stomach contents of various fish in Portugal found that only those of *Luciobarbus bocagei*, a species of barbel, contained shells of Asian clams (Pereira et al. 2016). The shells only represented a small portion of the overall diet of the fish, despite high clam abundance in the area. The research team noted that only barbel that were greater than 30 cm in size fed on the clams, perhaps meaning smaller fish avoided the clams as they had a gape limitation. The crayfish *Procambarus clarkia* also was recorded to feed on smaller Asian clams (Pereira et al. 2016). The best biological controls may be parasites or disease organisms that have specifically evolved to infest Asian clams.

SUCCESS STORIES

Due to the Asian clam's physiological and ecological plasticity, many forms of control can achieve a short-term effect. Long-term management can be more difficult to achieve and has led some researchers to investigate other experimental solutions to their management and control. Two case studies for Asian clam control, one conducted in Lake Tahoe, and a second, in the laboratory, are described in the following sections.

BENTHIC BARRIER CONTROL OF ASIAN CLAMS

Gas impermeable benthic barriers made of polyvinyl chloride have typically been used in lakes and ponds for the control of invasive macrophytes, either by smothering them or preventing their access to light. For this reason, the barriers are designed to be gas impermeable, in order to contain nutrient outgassing that results from the large-scale dying and decay of benthic plant life, which would be comparable to the effects of eutrophication if not controlled in this manner (Wittmann et al. 2012). At lake Tahoe, Whittmann et al. (2012) used similar gas-impermeable benthic barriers to create hypoxic to anoxic conditions on the lake floor in an effort to suffocate invasive populations of *C. fluminea*. When the barriers were applied, DO concentrations beneath the barrier dropped to nearly 0 mg/l within 72 hours. Though Asian clams have the documented ability to reduce their metabolism and withstand hypoxic conditions for long periods of time, in particular being more resistant to extended periods of hypoxia than the zebra mussel (*Dreissena polymorpha*) (Ortmann and Grieshaber 2003, Matthews and McMahon 1999), the hypoxia induced by the barrier treatment was sufficient to eliminate the majority of affected Asian clams, significantly reducing their numbers in the area where it was applied. After 120 days under the barrier, the density of *C. fluminea* in the treated area was reduced by nearly 100%, and for one year following the removal of the barrier, the invasive population remained over 90% reduced relative to the control region (Wittmann et al. 2012). While this treatment was, thus, demonstrated to be effective in eradicating molluscs, the effect of the induced hypoxia was not limited to *C. fluminea*: populations of the native pea clam (*Pisidium* spp.) and other benthic organisms were also drastically reduced within the treated area. The long-term effectiveness of the treatment on *C. fluminea* is also not currently known. However, this trial in Lake Tahoe has shown that the use of benthic barriers can be successful in the short-term control of Asian clams if care is taken to select treatment locations with especially dense invasive populations.

Dry Ice Study

Dry ice was used to kill and manage populations of Asian clams in both a laboratory and a simulated setting (Coughlan et al. 2018). Asian clam specimens were gathered from the River Barrow in Ireland and transported to a laboratory. The species was first detected in Ireland in 2010 in the River Barrow, where it was found to have densities of around 9,000 individuals/m^2 in the upper tidal portions of the river (Minchin 2014). It has since spread to additional waterbodies in Ireland. In the laboratory, four different quantities of dry ice were used to test their lethality on Asian clams: 50, 100, 200, and 300 g of 9 mm dry ice pellets. A control group was also created in which no dry ice was applied. Clams were assigned into groups based on their size, with small clams ranging from 8 to 14.9 mm, medium clams from 15 to 20.9 mm, and large clams from 21 to 32 mm. Each dry ice application lasted for a duration of either 1 or 5 minutes. Additionally, several other experimentational variants were performed, such as testing the efficacy of dry ice as a control agent when clams were covered in sediments such as fine gravel or mud. In control groups, survival rates ranged from 87% to 100%, while in groups that were exposed to dry ice, survival ranged from 0% to 100%. Longer exposure times, smaller clam size, and greater amounts of dry ice all resulted in greater clam mortality. The most lethal application was of 300 g of dry ice at a 5-minute exposure, resulting in complete clam mortality across all size ranges (small to large). In the experiments that simulated coverage by various types of sediment, results revealed that dry ice was still effective at

inducing mortality, even with increased coverage. The dry ice would bind itself to the clam in the mud or fine gravel and create a frozen layer in which the clam became incapacitated. These results showed the promising potential of dry ice as a control agent for Asian clams, especially at greater quantities of dry ice (300 g) and with longer exposures. It is possible that some non-target species could be affected by dry ice applications, specifically native molluscs and clams. However, if applied correctly in conjunction with a survey of the invaded area, the benefits of dry ice could outweigh the potential negative results. The promising results of this study also may be a window into future control methods of dry ice for other nuisance aquatic species, such as the zebra mussel and quagga mussels (Coughlan et al. 2018). Further testing and experimentation are required to validate to efficacy of dry ice application for controlling Asian clams and dreissenid mussels as well as its potential negative impacts on native, non-target species.

NOTES

1. According to an email from James McGrath, City of Pittsfield, to Tom Flannery and Jim Straub, Massachusetts Department of Conservation and Recreation, dated September 22, 2021.
2. According to an email from Barbara Kickham, Town of Shrewsbury, to David W.H. Wong, Massachusetts Department of Environmental Protection, dated December 21, 2021.

REFERENCES

Aldridge, D.C., P. Elliott, and G.D. Moggridge. 2006. Microencapsulated biobullets for the control of biofouling zebra mussels. *Environmental Science & Technology* 40: 975–979.

Baker, S.M., J.S. Levinton, J.P. Kurdziel, and S.E. Shumway. 1998. Selective feeding and biodeposition by zebra mussels and their relation to changes in phytoplankton composition and seston load. *Journal of Shellfish Research* 17(4): 1207–1213.

Bolam, B.A., G. Rollwagen-Bollens, and S.M. Bollens. 2019. Feeding rates and prey selection of the invasive Asian clam, *Corbicula fluminea*, on microplankton in the Columbia River, USA. *Hydrobiologia* 833(1): 107–123.

Carr, J.W., and L.E. Kennedy. 2008. Connecticut River Basin 2003 Water Quality Assessment Report. Massachusetts Department of Environmental Protection, Division of Watershed Management. Report # 34-AC-2; DWM Control # CN 105.5, 136 pp.

Cataldo, D., and D. Boltovskoy. 1998. Population dynamics of *Corbicula fluminea* (Bivalvia) in the Paraná River Delta (Argentina). *Hydrobiologia* 380(1): 153–163.

Colwell, H., J. Ryder, R. Nuzzo, M. Reardon, R. Holland, and W.H. Wong. 2017. Invasive Asian clams (*Corbicula fluminea*) recorded from 2001 to 2016 in Massachusetts, USA. *Management of Biological Invasions* 8(4): 507–515.

Coughlan, N.E., R.N. Cuthbert, S. Potts, E.M. Cunningham, K. Crane, J.M. Caffrey, F.E. Lucy, E. Davis, and J.T.A. Dick. 2019. Beds are burning: Eradication and control of invasive Asian clam, *Corbicula fluminea*, with rapid open-flame burn treatments. *Management of Biological Invasions* 10(3): 486–499.

Coughlan, N.E., D.A. Walsh, J.M. Caffrey, E. Davis, F.E. Lucy, R.N. Cuthbert, and J.T. Dick. 2018. Cold as Ice: A novel eradication and control method for invasive Asian clam, *Corbicula fluminea*, using pelleted dry ice. *Management of Biological Invasions* 9(4): 463–474.

Counts, C.L. 1981. *Corbicula fluminea* (Bivalvia: Sphaeriacea) in British Columbia. *Nautilus* 95(1): 12–13.

Counts, C.L. 1986. The zoogeography and history of the invasion of the United States by *Corbicula fluminea* (Bivalvia: Corbiculidae). *American Malacological Bulletin* (Special Edition) 2: 7–39.

Densmore, C.L., L.R. Iwanowicz, A.P. Henderson, V.S. Blazer, B.M. Reed-Grimmett, and L.R. Sanders. 2018. An Evaluation of the toxicity of potassium chloride, active compound in the molluscicide potash, on salmonid fish and their forage base. *USGS Science for a Changing World* 1–46.

Denton, M.E., S. Chandra, M.E. Wittmann, J. Reuter, and J.G. Baguley 2012. Reproduction and population structure of *Corbicula fluminea* in an oligotrophic subalpine lake. *Journal of Shellfish Research* 31(1): 145–152.

Dexter, E., S.M. Bollens, and G. Rollwagen-Bollens. 2020. Native and invasive zooplankton show differing responses to decadal-scale increases in maximum temperatures in a large temperate river. *Limnology and Oceanography Letters* 5(6): 403–409.

Dexter, E., S.M. Bollens, G. Rollwagen-Bollens, J. Emerson, and J. Zimmerman. 2015. Persistent vs. ephemeral invasions: 8.5 years of zooplankton community dynamics in the Columbia River. *Limnology and Oceanography* 60(2): 527–539.

Ferreira-Rodríguez, N., I. Fernandez, S. Varandas, R. Cortes, M. Cancela, and I. Pardo. 2017. The role of calcium concentration in the invasive capacity of *Corbicula fluminea* in crystalline basins. *Science of The Total Environment* 580: 1363–1370.

Foster, A.M., P. Fuller, A. Benson, S. Constant, D. Raikow, J. Larson, and A. Fusaro. 2019. *Corbicula fluminea*. USGS Nonindigenous Aquatic Species Database, Gainesville, FL. http://nas.er.usgs.gov/queries/factsheet.aspx?speciesid=92 (Accessed January 7, 2021).

Gomes, C., R. Sousa, T. Mendes, R. Borges, P. Vilares, V. Vasconcelos, L. Guilhermino, and A. Antunes. 2016. Low genetic diversity and high invasion success of *Corbicula fluminea* (Bivalvia, Corbiculidae) (Müller, 1774) in Portugal. *PLoS One* 11(7): e0158108.

Graney, R.L., D.S. Cherry, J.H. Rodgers, and J. Cairns. 1980. The influence of thermal discharges and substrate composition on the population-structure and distribution of the Asiatic clam, *Corbicula fluminea*, in the New River, Virginia. *Nautilus* 94: 130–135.

Haag, W.D., J. Culp, A.N. Drayer, M.A. McGregor, D.E.J. White, and S.J. Price. 2021. Abundance of an invasive bivalve, *Corbicula fluminea*, is negatively related to growth of freshwater mussels in the wild. *Freshwater Biology* 66(3): 447–457.

Hakenkamp, C.C., S.G. Ribblett, M.A. Palmer, C.M. Swan, J.W. Reid, and M.R. Goodison. 2001. The impact of an introduced bivalve (*Corbicula fluminea*) on the benthos of a sandy stream. *Freshwater Biology* 46(4): 491–501.

Hassett, W., S.M. Bollens, T.D. Counihan, G. Rollwagen-Bollens, J. Zimmerman, S. Katz, and J. Emerson. 2017. Veligers of

the invasive Asian clam *Corbicula fluminea* in the Columbia River Basin: Broadscale distribution, abundance, and ecological associations. *Lake and Reservoir Management* 33(3): 234–248.

Ishibashi, R., K. Ookubo, M. Aoki, M. Utaki, A. Komaru, and K. Kawamura. 2003. Androgenetic reproduction in a freshwater diploid clam *Corbicula fluminea* (Bivalvia: Corbiculidae). *Zoological Science* 20: 727–732.

ITIS (Integrated Taxonomic Information System). 2021. www.itis.gov/servlet/SingleRpt/SingleRpt#null (Accessed January 28, 2021).

Karataev, A.Y., L.E. Burlakova, and D.K. Padilla. 2005. *Contrasting Distribution and Impacts of Two Freshwater Exotic Suspension Feeders, Dreissena Polymorpha and Corbicula Fluminea. The Comparative Roles of Suspension-Feeders in Ecosystems, NATO Science Series IV: Earth and Environmental Series.* Dordrecht: Springer, pp. 239–262.

Kennedy, L.E., and M.J. Weinstein. 2000. Connecticut river Basin 1998 Water quality assessment report. Massachusetts Department of Environmental Protection, Division of Watershed Management. Report # 34-AC-1; DWM Control # CN 45, 110 pp.

Lake Champlain Basin Program. 2016. Invasive species task force recommends actions to control invasive Asian Clam in Lake Bomoseen, Vermont. www.lcbp.org/2016/09/invasive-species-task-force-recommends-actions-control-invasive-asian-clam-lake-bomoseen-vt/. (Accessed October 11, 2016).

Lucy, F., A. Karataev, and L. Burlakova. 2012. Predictions for the spread, population density, and impacts of Corbicula fluminea. *Aquatic Invasions* 7(4): 465–474.

Matthews, M.A., and R.F. McMahon. 1999. Effects of temperature and temperature acclimation in survival of zebra mussels (*Dreissena polymorpha*) and Asian clams (*Corbicula fluminea*) under extreme hypoxia. *Journal of Molluscan Studies* 65: 317–325.

Mattice, J.S., and L.L. Dye. 1976. Thermal tolerance of adult Asiatic clam. *Thermal Ecology* 2: 130–135.

McCabe, G., S.A. Hinton, R.L. Emmett, and B.P. Sandford. 1997. Benthic invertebrates and sediment characteristics in main channel habitats in the lower Columbia River. *Northwest Science* 71: 45–55.

McDowell, W.G., A.J. Benson, and J.E. Byers. 2014. Climate controls the distribution of a widespread invasive species: Implications for future range expansion. *Freshwater Biology* 59: 847–857.

McMahon, R.F. 1996. The Physiological Ecology of the Zebra Mussel, Dreissena polymorpha, in North America and Europe. *American Zoologist* 36(3): 339–363.

McMahon, R.F. 1999. Invasive characteristics of the freshwater bivalve *Corbicula fluminea*. In: R. Claudi, and J.H. Leach (eds). *Nonindigenous Freshwater Organisms: Vectors, Biology, and Impacts*. Boca Raton, FL: Lewis Publishers, pp. 315–343.

McMahon, R.F. 2002. Evolutionary and physiological adaptations of aquatic invasive animals: r selection versus resistance. *Canadian Journal of Fisheries and Aquatic Sciences* 59: 1235–1244.

McMahon, R.F. 2011. Quagga mussel (*Dreissena rostriformis bugensis*) population structure during the early invasion of Lakes Mead and Mohave January-March 2007. *Aquatic Invasions* 6: 131–140.

McMahon, R.F., and A.E. Bogan. 2001. Mollusca: Bivalvia. In: H. Thorp and A.P. Covich (eds). *Ecology and Classification of North American Freshwater Invertebrates* (2nd edition). New York: Academic Press, pp. 331–428.

McMahon, R.F., and C.J. Williams. 1986. A reassessment of growth rate, life span, life cycles and population dynamics in a natural population and field caged individuals of *Corbicula fluminea* (Müller) (Bivalvia:Corbiculacea). *American Mala-cological Bulletin* (Special Edition) 2: 151–186.

Minchin, D. 2014. The distribution of the Asian clam *Corbicula fluminea* and its potential to spread in Ireland. *Management of Biological Invasions* 5(2): 165–177.

Minchin, D., and R. Boelens. 2018. Natural dispersal of the introduced Asian clam *Corbicula fluminea* (Müller, 1774) (Cyrenidae) within two temperate lakes. *BioInvasions Records* 7(3): 259–268.

Moore, B., G.C. Holdren, S. Gerstenberger, K. Turner, and W.H. Wong. 2015. Invasion by Quagga Mussels (*Dreissena rostriformis bugensis* Andrusov 1897) into Lake Mead, Nevada-Arizona: The first occurrence of dreissenid species in the Western United States. In: W.H. Wong and S.L. Gerstenberger (eds). *Biology and Management of Invasive Quagga and Zebra Mussels in the Western United States*. Boca Raton, FL: CRC Press, pp. 17–31.

Mouthon, J. 2001. Life cycle and population dynamics of the Asian clam *Corbicula fluminea* (Bivalvia: Corbiculidae) in the Saone River at Lyon (France). *Hydrobiologia* 452: 109–119.

Muller, O., and B. Baur. 2011. Survival of the invasive clam *Corbicula fluminea* (Müller) in response to winter water temperature. *Malacologia* 53: 367–371.

New Hampshire Department of Environmental Service. 2016. *Stop Aquatic Hitchhikers! Boating on the Lakes* (2016 ed.). New Hampshire: New Hampshire Department of Environmental Service, pp. 54–55.

Nuzzo, R.M. 2003. *Standard Operating Procedures: Water Quality Monitoring in Streams Using Aquatic Macroinvertebrates*. Worcester, MA: Massachusetts Department of Environmental Protection. Division of Watershed Management, p. 39.

Ortmann, C., and M.K. Grieshaber. 2003. Energy metabolism and valve closure behaviour in the Asian clam *Corbicula fluminea*. *Journal of Experimental Biology* 206: 4167–4178.

Patrick, C.H., M.N. Waters, and S.W. Golladay. 2017. The distribution and ecological role of *Corbicula fluminea* (Müller, 1774) in a large and shallow reservoir. *BioInvasions Records* 6(1): 39–48.

Pereira, J.L., S. Pinho, A. Ré, P.A. Costa, R. Costa, F. Gonçalves, and B.B. Castro. 2016. Biological control of the invasive Asian clam, *Corbicula fluminea*: Can predators tame the beast? *Hydrobiologia* 779: 209–226.

Pigneur, L.M., E. Falisse, K. Roland, E. Everbecq, J.F. Deliège, J.S. Smitz, K.V. Doninck, and J.P. Descy. 2014. Impact of invasive Asian clams, *Corbicula spp.*, on a large river ecosystem. *Freshwater Biology* 59(3): 573–583.

Pimentel, D., R. Zuniga, and D. Morrison. 2005. Update on the environmental and economic costs associated with alien-invasive species in the United States. *Ecological Economics* 52: 273–288.

Robinson, M. 2004. Asian clam: An exotic aquatic species. www.mass.gov/eea/docs/dcr/watersupply/lakepond/factsheet/asian-clam.pdf (Accessed November 1, 2016).

Rosa, I.C., R. Garrido, A. Ré, J. Gomes, J. Peirera, F. Gonçalves, and R. Costa. 2015. Sensitivity of the invasive bivalve *Corbicula fluminea* to candidate control chemicals: The role of dissolved oxygen conditions. *Science of Total Environment* 536: 825–830.

Rosa, I.C., J.L. Pereira, R. Costa, F. Gonçalves, and R. Prezant 2012. Effects of upper-limit water temperatures on the dispersal of the Asian clam *Corbicula fluminea*. *PLoS ONE* 7(10).

Sheehan, R., J.M. Caffrey, M. Millane, P. McLoone, H. Moran, and F. Lucy. 2014. An investigation into the effectiveness of mechanical dredging to remove *Corbicula fluminea* (Müller, 1774) from test plots in an Irish river system. *Management of Biological Invasions* 5(4): 407–418.

Smith, D.G. 1993. The Potential for Spread of the Exotic Zebra Mussel (*Dreissena polymorpha*) in Massachusetts. Report MS-Q-11. Department of Biology, University of Massachusetts Amherst, Massachusetts and Museum of Comparative Zoology, Harvard University, Cambridge, MA. www.mass.gov/eea/docs/dcr/watersupply/lakepond/downloads/zebra-mussel-massdep-ms-q-11.pdf (Accessed June 27, 2018).

Sousa, R., C. Antunes, and L. Guilhermino. 2008. Ecology of the invasive Asian *clam Corbicula fluminea* (Müller, 1774) in aquatic ecosystems: An overview. *Annales de Limnologie—International Journal of Limnology* 44: 85–94.

Sousa, R., A. Novais, R. Costa, and D. Stayer. 2013. Invasive bivalves in fresh water: Impacts from individuals to ecosystems and possible control strategies. *Hydrobiologia* 735: 233–251.

USEPA. 2012. *Climate Change Indicators* (2nd ed.). New York: United States Environmental Protection Agency (USEPA), p. 3.

Viergutz, C., C. Linn, and M. Weitere. 2012. Intra- and interannual variability surpasses direct temperature effects on the clearance rates of the invasive clam *Corbicula fluminea*. *Marine Biology* 159: 2379–2387.

Williams, C.J., and R.F. McMahon. 1986. Power station entrainment of *Corbicula fluminea* (Müller) in relation to population dynamics, reproductive cycle and biotic and abiotic variables. *American Malacological Bulletin* (Special Edition) 2: 99–111.

Wittmann, M.E., S. Chandra, J.E. Reuter, S.G. Schladow, B.C. Allen, and K.J. Webb. 2012. The control of an invasive bivalve, *Corbicula fluminea*, using gas impermeable benthic barriers in a large natural lake. *Environmental Management* 49: 1163–1173.

Wong, W.H., and S.L. Gerstenberger. 2011. Quagga mussels in the western United States: monitoring and management. *Aquatic Invasions* 6: 125–129.

Wong, W.H., and S.L. Gerstenberger. 2015. *Biology and Management of Invasive Quagga and Zebra Mussels in the Western United States*. Boca Raton, FL: CRC Press, p. 566.

Wong, W.H., J.S. Levintion, B.S. Twining, and N. Fisher. 2003. Assimilation of micro- and mesozooplankton by zebra mussels: A demonstration of the food web link between zooplankton and benthic suspension feeders. *Limnology & Oceanography* 48(1): 308–312.

3 Zebra Mussel (*Dreissena polymorpha*)

ABSTRACT

The zebra mussel (*Dreissena polymorpha*), a freshwater bivalve mollusk that was introduced to the Great Lakes in the 1980s, is now thriving in the United States due to its vigorous fecundity, posing threats to native freshwater species. In Massachusetts, the species was first found in 2009 at Laurel Lake and currently is only found in the waters of the Housatonic River basin. This report covers monitoring and sightings of the zebra mussel, as well as alternative management tools to raise awareness to the general public for prevention and to monitor and control (chemical, physical, or biological treatments) of zebra mussels in lakes and rivers.

INTRODUCTION

Taxonomic Hierarchy (ITIS 2021)

Kingdom	Animalia
Subkingdom	Bilateria
Infrakingdom	Protostomia
Superphylum	Lophozoa
Phylum	Mollusca
Class	Bivalvia
Subclass	Heterodonta
Order	Veneroida
Superfamily	Dreissenoidea
Family	Dreissenidae
Genus	*Dreissena*
Species	*Dreissena polymorpha* (Pallas, 1771)

The zebra mussel (*Dreissena polymorpha*) is an invasive freshwater bivalve mollusk in North America which is native to Europe. It has an epifaunal mode of life (Mackie and Schloesser 1996), which usually grows 3.5–5.0 cm in length (Maxwell 1992). They have elongated D-shaped shells with light- and dark-colored bands (Figure 3.1). Due to their genetic and morphological plasticity (as the species name *polymorpha* implies), sometimes the species will have a yellow and brownish color (Feinberg 1979). According to studies from early Soviets, there were five dreissenid species at minimum, based on the discretionary morphological and physiological existence (O'Neill and MacNeill 1991). Like many members of the Class Bivalvia, zebra mussels have byssal threads to attach to nearby surfaces. According to Ekroat et al. (1993), there are two different types of byssal threads: temporary and permanent. Usually, temporary threads range from 1 to 6 threads in a tripod shape, and they are separated from major byssal threads. Permanent threads are clustered together and arranged in rows. The two types of byssal threads differ in numbers, length, thickness, arrangement, and morphology. To calculate the approximate total number of threads, a formula can be applied: number of threads = -8.59 + 19.26* (shell length in mm) (Claudi and Mackie 1994). The dominant constituent of byssal threads is a type of amino acid called DOPA (3, 4-dihydroxyphenyl-alanine) (Rzepecki and Waite 1993). Based on their proclivity to water currents, zebra mussels prefer living in water pipelines and water canals, which can result in decreased water flow and even produce disagreeable flavors in drinking water (Ellis 1978). Lakeshores and riverbanks are preferable sites of colonization for zebra mussels. Due to their high reproductive ability, the species sometimes appears in reef-like mats and can tether themselves to shells of other clams and mussels (Maxwell 1992). Generally, the densities of zebra mussel populations are between 7,000 and 114,000/m^2 with a standing crop biomass between 0.05 and 15 kg/m^2 (Thorp and Covich 1991). Colonization of zebra mussels is dependent on light intensity, water temperature, food sources, and many other factors. They usually live in the range of 2–7 meters below the water surface. They sometimes can be found as deep as 50 m (Maxwell 1992). Zebra mussels have a high tolerance to harsh environmental conditions. They prefer living in a temperature range of

DOI: 10.1201/9781003201106-3

FIGURE 3.1   An adult zebra mussel and a zebra mussel bed in Lake Erie. (Left photo by W.H. Wong; right photo by Berkman etal. (1998).)

20–25°C and in water currents ranging from 0.15 to 0.5 m/second. If water temperatures fall below 7°C or above 32°C, with a water current faster than 2 m/second, it will be considered as unsuitable conditions for their growth (Maxwell 1992). A survival threshold value of 12–20 mg/l of calcium is required for the growth of zebra mussels (Neary and Leach 1991). Based on parameters used by Burlakova (1998), a suitable waterbody should have a pH ≥7.5. For marginally suitable conditions, a pH between 7.2 and 7.5 is needed. A pH <7.2 will be considered as unsuitable. Major environmental factors that can limit the survival and/or facilitate the establishment of zebra mussels are summarized in Table 3.1 (Counihan et al. 2018). According to alkalinity or hardness, as well as pH, Smith (1993) reported two watersheds in Western Massachusetts with favorable conditions for zebra mussels to survive and establish sustainable populations: the Hoosic and Housatonic River Basins. The Connecticut River watershed (with the exception of the Millers and Chicopee watersheds) and most of Eastern Massachusetts were moderately to marginally susceptible. The Millers and Chicopee watersheds, the coastal plain of Southeastern Massachusetts, and Cape Cod could not support zebra mussels (Smith 1993).

Usually, a mature female mussel at age 1 or older will produce 30,000–40,000 eggs a year (Maxwell 1992) and up to 1 million eggs during one spawning event (Walz 1978; Sprung 1990). The reproductive cycle highly depends on environmental conditions (see Table 3.1), such as water temperature and plenitude of phytoplankton (Maxwell 1992). Based on O'Neill and MacNeill's (1991) research on Lake Erie,

supporting evidence shows that the reproductive rate of zebra mussels can be inhibited by cool water and high turbidity. Zebra mussels fertilize externally, by ejecting sperm and eggs into the surrounding area. After hatching from the fertilized egg, a planktonic veliger larva will float in the water and forage, remaining suspended for about eight to ten days before descending to the substratum (Thorp and Covich 1991). During the larva stage, the veligers can swim and follow the direction of the current to disperse themselves to other waterbodies causing new infestations (Thorp and Covich 1991). After about three weeks, zebra mussels will start to tether themselves to substrate such as cobble, bedrock, debris, boat hulls, water pipes, and breakwaters. Although zebra mussels have a high reproductive rate and are highly active, mortality rates can be as high as 99% or more. Reasons for mortality can be hypoxia, fluctuation of temperature, or failing to find an appropriate substrate (O'Neill and MacNeill 1991; Thorp and Covich 1991). Eventually, they will find a better location to grow and tether themselves via their provisional tiny byssal fibers (Maxwell 1992) and adapt to a sessile lifestyle. Younger zebra mussels are usually between 5 and 10 mm in length (O'Neill and MacNeill 1991; Thorp and Covich 1991). The life span of zebra mussels can vary, with most living up to five to six years (Thorp and Covich 1991). However, O'Neill and MacNeill (1991) argue that the average life span of a zebra mussel is 3.5 years, but occasionally can reach up to 8–10 years if production is less intense. It is also reported that the longevity of zebra mussels varies from 2 to 19 years and it is not clear to what

TABLE 3.1

Risk Categories of Zebra Mussel Invasions under Different Water Quality Parameters

Physical Risk Factor	Minimal Risk	Moderate Risk	High Risk
Calcium concentration	<8 mg/l	8–20 mg/l	>20 mg/l
pH	<7.0 or >9.6	7.0–7.3 or 9.4–9.6	>7.3–<9.4
Mean long-term (≥29 days) summer water temperature	>32°C	31–32°C	<31°C
Minimum temperature for spawning	<14°C	14–16°C	>16°C
Salinity	>5‰	4–5‰	<4‰
Percent full air oxygen saturation at 25°C	<20%	20–30%	>30%
Annual water level variation	No minimal risk	Extensive variation exposing mussels to air and/or submergence in a hypoxic hypolimnion	Relatively constant water levels

extent this variation is caused by biological variability and environmental conditions and what amount of the variation is caused by the methods used to assess age and longevity (Karatayev et al. 2006). Generally, in non-native ranges, the species can sharply increase in population at the early stage of an invasion. However, different long-term trajectories for populations of zebra mussels and other alien species have been illustrated by Strayer and Malcom (2006). In conclusion, rapid growth rate, robust fecundity, relatively short life spans, and the mobility during the larval stage contribute to the aggressiveness of zebra mussels in non-native areas (Maxwell 1992).

The species originated from drainage basins in the Black Sea, Caspian Sea, and Aral Sea in the latter part of the 18th century (Maxwell 1992). After that, zebra mussels settled down firmly around inland waterways of Europe. The zebra mussel was first documented in North America in 1988 in Lake St. Clair (Hebert et al. 1989). However, a recent paper (Carlton 2008) provided convincing evidence that it was present as early as 1986 in Lake Erie (Nalepa and Schloesser 2013). Up to 30,000–40,000 individuals/m² were found in the west basin of Lake Erie in 1988. By September 1991, zebra mussels had been reported in most of the Great Lakes. In the same year, the invasion of zebra mussels reached the Hudson River as far south as Catskill, New York (MRB Group 1991). In 1992, the species was reported in Vicksburg, Mississippi, west of Minneapolis, Minnesota, as well as north

and east of Quebec. By 1994, the species was found in upper Mississippi, Tennessee, Ohio, and Arkansas (Mackie and Schloesser 1996). In 2008, it was found in the San Justo Reservoir, near Monterrey Bay, California, the first sighting in the western United States (Britton 2015). Recently, another dreissenid mussel, the quagga mussel *Dreissena rostriformis bugensis*, is displacing zebra mussels in the Great Lakes (Mills et al. 1999; Nalepa et al. 2010). It has been documented that Laurentian Great Lakes have been dramatically reengineered by invasive bottom-dwelling dreissenid mussels, especially the quagga mussel which is currently regulating the phosphorus cycling in the invaded Great Lakes (Li et al. 2021). Although quagga mussels and zebra mussels are relatives and share a common native habitat, morphology, lifestyle, life history, and dispersal potential, the former was found to be a better invader than the latter at most spatial scales throughout their invasion history worldwide (Karatayev et al. 2011). Therefore, early detection on quagga mussels and corresponding rapid response actions in Massachusetts are needed to prevent its invasion and establishment.

The invasion pathway of zebra mussels can vary. Among all mechanisms, the planktonic veliger stage should be considered the most effective method of natural spreading. Artificial canals exacerbated the spreading in the former USSR and Britain (Kerney and Morton 1970). Settled mussels that attached to debris may also contribute to the spreading. Moreover, juvenile and adult individuals can be distributed via drifting by water currents (Martel 1993). Some animal-mediated dispersal mechanisms are possible (e.g., turtles, crayfish, and birds, see Carlton 1993); however, anthropogenic mechanisms are thought to be the most common methods of translocating zebra mussels, such as in ballast water used by cargo ships (Mackie and Schloesser 1996), and anchors and anchor lines (McMahon et al. 1991). Not all mechanisms have been verified, especially during the planktonic veliger stage (Nalepa and Schloesser 1993). Recent research shows that over 90% of samples taken from boat live wells/bilges contain five or fewer veligers, which suggests residual water may present a low risk for the spread of this invasive species (Montz and Hirsch 2016); however, Choi et al. (2013) demonstrated a greater likelihood of veliger transport in the residual water of trailered watercraft in favorable conditions when veliger survival times increased with increased level of larval development. Finally, it has been reported that wind-driven currents have an influential effect on the vertical distribution of zebra mussel veligers (Fraleigh et al. 1993). If wind velocities are below 8 km/hour, there will be a manifest depth effect, where 5% of populations are found at depths of 0–2 m. At 2–4 m, 30% of populations have been reported. Finally, 64% have been reported at the depths of 4–6 m (Fraleigh et al. 1993).

Zebra mussels are ecosystem engineers (Nalepa and Schloesser 1993, 2013; Strayer et al. 1999; Karatayev et al. 2002) because they have the potential to change the entire lake or river where they invade (Figure 3.2). Although these invasive mussels may impact ecosystems in different ways, they generally increase water clarity by removing suspended particles (e.g., phytoplankton, debris, silt, and microzooplankton) in the water column (Griffiths 1993, Leach 1993, MacIsaac 1996, Wong et al. 2003). Top-down control on primary production by invasive mussels is mainly due to their efficient filtering behavior (Table 3.2). Since zebra mussels were first discovered in Lake Erie in 1988, chlorophyll-a concentration decreased by 43% in the western basin and 27% in the west-central basin between 1988 and 1989 (Leach 1993). A 20-year (1983–2002) data set shows a downward trend of

−0.07 µg/l/year in the central basin of Lake Erie, but no trend was found before the arrival of zebra mussel (Rockwell et al. 2005). In the Hudson River, New York, zebra mussels were found to reduce chlorophyll concentrations by 90% of the pre-zebra mussel level (Caraco et al. 2006). In some areas, such as in Green Bay of Lake Michigan, chlorophyll-a decreased immediately following dreissenid invasion but increased afterward (Qualls et al. 2007, De Stasio et al. 2008). A model of Lake Erie suggests that zebra mussels decreased algal biomass when they first invaded and their population was small (6,000 individuals/m^2). However, when the population grew larger (120,000 individuals/m^2), dreissenid mussels increased non-diatom inedible algae by excreting a large amount of ammonia and phosphate (Zhang et al. 2008). The increased water clarity can then affect other ecosystem components; decreased densities of microalgae result in lower uptake rates for dissolved nutrients from the water column (Nicholls et al. 1999), and benthic algae and plants benefit from increased water clarity (Hecky et al. 2004; Mayer et al. 2013). It is estimated that within the first four to eight years of contact with invasive mussels, Unionidae populations could decrease by 90% (Ricciardi et al. 1998). Generally, there is a strong positive linear relationship between numbers of zebra mussels per unionid and zebra mussel density and there is an overall trend for increased weight of zebra mussels per unionid with increased unionid size during the first ten years subsequent to a zebra mussel invasion; however, the trend decreased ten years after the initial invasion (Lucy et al. 2013). Within the fish communities, planktivorous fish that mainly rely on zooplankton, especially on microzooplankton, may suffer due to food shortage as a result of energy transfer from the pelagic community to the benthic community while benthic feeding fishes may benefit from the establishment of mussels. The influence of invasive mussels on fish vary widely across ecosystems as a function of system morphology, factors that limit primary production, and diets of these fish species (Strayer et al. 2004). In the first decade after zebra mussels arrived at the Hudson River, the biomass of zooplankton and deepwater macrobenthos fell by 50%, while the biomass of littoral macrobenthos rose by 10%. These changes in the forage base were associated

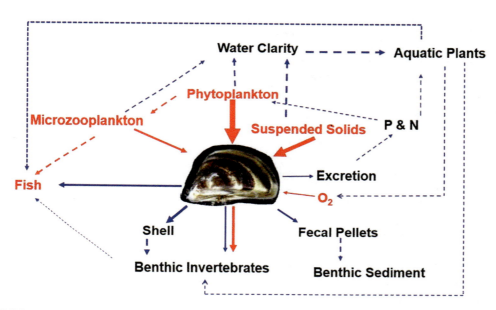

FIGURE 3.2 Potential ecological impacts of zebra mussels on an aquatic ecosystem. Positive (blue font) and negative (red font) effects are diagrammed. Solid and dashed lines represent direct and indirect impacts, respectively. The wider the line, the more profound the impact. Briefly, zebra mussels have potential to reduce the biomass and change the species composition of phytoplankton and zooplankton communities. Decreases in suspended solids and oxygen would increase water clarity. An increase in dissolved inorganic phosphorus and nitrogen would facilitate aquatic plant growth. Benthic production is expected to increase, which would positively impact some fish species. (Modified from Wong etal. (2011).)

with large, differential changes in the abundance, geographic distribution, and growth rates of open water and littoral fish (Strayer et al. 2014). In recent years, populations of zooplankton and deepwater macrobenthos have risen toward pre-invasion levels, while littoral macrobenthos remained unchanged. The analysis of large data sets for young-of-year fishes found no systematic change in the abundance or geographic distribution of either group of fish in the Hudson River, which shows that the ecological effects of a biological invasion may change over time (Strayer et al. 2014). Zebra mussels can also alter the morphological and physical properties of their habitat areas, thereby affecting the availability of resources such as competition on food and space in the food web (Karatayev et al. 1997, Vanderploeg et al. 2002, Hecky et al. 2004). Overall, they are efficient ecosystem engineers that can alter the ecology of a system directly or indirectly (Figure 3.2). The system-wide effect

of zebra and/or quagga mussels depends on water mixing rates, lake morphology, and turnover rates. Shortly after initial invasion, as populations increase, both dreissenids will have their largest effects on communities, and most of them will be direct effects. After the initial stage of invasion, impacts are less predictable, and more likely to be caused by indirect effects through changes in the ecosystem (Karatayev et al. 2015).

Economic damage caused by zebra mussels can be serious. Due to their byssal threads, zebra mussels can attach itself to hulls, increasing the weight of the vessel and consumption of fuel, and decreasing speeds. If the veligers or larval mussels are siphoned into engine cooling water intakes, it can overheat the engine and cause damage (Maxwell 1992). Dense mats of zebra mussels can be found on docks, piers, pilings, and ladders, increasing the weight by up to 20–30 pounds/m^2 and eventually sinking the structure due to the heavy colonization (O'Neill and

TABLE 3.2

Filtration Rate of a 15-mm (or 22-mg Total Dry Weight) Zebra Mussel

Mussel	Filtration Rate (ml/h/individual)	Suspended Particle	T (°C)	Reference
Zebra	7.9	*Cryptomonas* sp.	20	(Bunt et al. 1993)
	122.2	Hudson River water	20	(Roditi et al. 1996)
	38.0	*Chlamydomonas reinhardtii*	22	(Berg et al. 1996)
	99.7	*Chlamydomonas reinhardtii*	22	(Berg et al. 1996)
	93.0	*Chlamydomonas* sp.	23	(Horgan and Mills 1997)
	125.0	*Mallomonas* sp.	23	(Horgan and Mills 1997)
	159.5	*Cryptomonas pusilla*	23	(Horgan and Mills 1997)
	167.5	*Cryptomonas ovata*	23	(Horgan and Mills 1997)
	114.0	*Melosira* sp.	23	(Horgan and Mills 1997)
	78.8	*Chroococcus* sp. (globular colony)	23	(Horgan and Mills 1997)
	80.9	*Chroococcus* sp. filaments	23	(Horgan and Mills 1997)
	93.7	Natural assemblage + mixed aglal cultures	20	(Schneider et al. 1998)
	32.3	Clay	16	(Baker et al. 1998)
	63.8	Clay + *Microcystis*	16	(Baker et al. 1998)
	56.3	Clay + *Crucigenia*	16	(Baker et al. 1998)
	47.3	Clay + *Cyclotella*	16	(Baker et al. 1998)
	74.3	*Typha* detritus	16	(Baker et al. 1998)
	90.0	*Typha* detritus + *Microcystis*	16	(Baker et al. 1998)
	61.5	*Typha* detritus + *Crucigenia*	16	(Baker et al. 1998)
	29.6	*Typha* detritus + *Cyclotella*	16	(Baker et al. 1998)
	99.0	*Microcystis*	16	(Baker et al. 1998)
	32.3	*Microcystis* + *Micractinium*	16	(Baker et al. 1998)
	52.5	*Microcystis* + *Crucigenia*	16	(Baker et al. 1998)
	72.0	*Microcystis* + *Thalassiosira*	16	(Baker et al. 1998)
	30.8	*Microcystis* + *Cyclotella*	16	(Baker et al. 1998)
	9.0	*Microcystis* + *Scenedesmus*	16	(Baker et al. 1998)
	66.8	*Crucigenia*	16	(Baker et al. 1998)
	23.3	*Crucigenia* + *Micractinium*	16	(Baker et al. 1998)
	15.4	*Crucigenia* + *Thalassiosir*	16	(Baker et al. 1998)
	5.3	*Thalassiosira*	16	(Baker et al. 1998)
	21.0	*Thalassiosira* + *Cyclotella*	16	(Baker et al. 1998)
	14.3	*Cyclotella*	16	(Baker et al. 1998)
	37.5	*Cyclotella* + *Scenedesmus*	16	(Baker et al. 1998)
	6.0	*Cyclotella* + *Micractinium*	16	(Baker et al. 1998)
	71.6	Lillinonah Reservoir	16–20	(Bastviken et al. 1998)
	74.3	Hudson River + Lillinonah Reservoir (1:1)	16–20	(Bastviken et al. 1998)
	79.4	Hudson River + Lillinonah Reservoir (3:1)	16–20	(Bastviken et al. 1998)
	55.4	Hudson River + Lillinonah Reservoir (1:3)	16–20	(Bastviken et al. 1998)
	41.6	Hudson River + *Microcystis* + *Anabaena*	16–20	(Bastviken et al. 1998)
	42.2	Hudson River + *Microcystis* + *Anabaena* + Seston	16–20	(Bastviken et al. 1998)

Mussel	Filtration Rate (ml/h/individual)	Suspended Particle	T (°C)	Reference
	22–87	*Chlamydomonas*	20	(Baldwin et al. 2002)
	89.0	Suspended sediment	14	(Diggins 2001)
	55.0	Suspended sediment	8	(Diggins 2001)
	155.0	Suspended sediment	14	(Diggins 2001)
	110.0	Suspended sediment	22	(Diggins 2001)
	47.9	*Scenedesmus*	20	(Dionisio Pires et al. 2004)
	89.5	*Scenedesmus + Microcystis*	20	(Dionisio Pires et al. 2004)
	209.5	Cyanobacteria	20	(Dionisio Pires et al. 2004)
	80.0	Detritus	20	(Dionisio Pires et al. 2004)
	112.6	Non-Cyanobacteria phytoplankton	20	(Dionisio Pires et al. 2004)

MacNeill 1991). The shell of the zebra mussel can be sharp enough to cut bare feet. Beach and bathing footgear are used on some of the Great Lakes beaches to protect individuals from the sharp shells (Maxwell 1992). Deceased zebra mussels can generate foul odors. If they die in drinking water pipelines, it can produce unpleasant flavors and cause adverse health effects. Several billion dollars have been spent on facility and infrastructure repair and replacements (Roberts 1990). According to O'Neill (1997), $83 million has been spent on zebra mussel control from 339 facilities in 35 states and 3 provinces, with a mean expenditure of $248,000 per facility (Lovell et al. 2006). Among the expenses, $35,274,020 (51%) has been spent on managing Electric Power Generation and $21,435,610 (31%) has been spent on Water Treatment.

The species is considered as a good candidate for laboratory experiments. Zebra mussels are very easy to collect and they have both epifaunal and sessile life stages (Nichols 1996). For instance, the species has become very valuable due to its contribution to research on ciliary function (Silverman et al. 1996), ion transportation (Horohov et al. 1992), physiological resistance, and capacity adaptations (McMahon et al. 1993; McMahon 1996). It is also a good experimental species for toxicity studies (Roditi et al. 2000a; 2000b).

MONITORING AND SIGHTINGS

Massachusetts Department of Environmental Protection employees have done biological surveys at more than 2,000 waterbodies in Massachusetts since the 1980s. Field sheet quality control, database development, and map construction are under the guidance of the Standard Operating Procedures (SOP). The US Geological Survey (USGS) and other agencies and individuals also contributed to the process of data collection of *D. polymorpha* in Massachusetts (Benson et al. 2019; Wagner 2017). Tools for drafting maps and marking out data layers of sites with *D. polymorpha* were created via ArcGIS® ArcMap 10.1 (ESRI, Redlands, California).

In Massachusetts, adult zebra mussels were first discovered by a Town of Lee employee, Dmitri Consolati, and a lake abutter in Laurel Lake in July of 2009, and confirmed by state officials (Daley 2009). Currently, the Housatonic River Basin is the only watershed infested by zebra mussels (Figure 3.3 and Table 3.3). These waterbodies are located in the western part of the state. Since there are a total of 33 major watersheds in Massachusetts, the percentage of infested watersheds is about 3%.

Laurel Lake has favorable conditions for the growth of zebra mussels with a coarse bottom near the shore and ranging from sand to finer silt further out in deeper waters (Wagner 2017). There is plenty of gravel that is a good habitat for dense younger mussels to attach onto. It is believed that the invasion of Laurel Lake began in 2008 or even earlier. The Lake is the first documented waterbody that is infested by zebra mussels in Massachusetts. There are also other bivalves in Laurel Lake, such as Unionidae mussels and Sphaeriidae (tiny fingernail clams). They are not very common in the lake due to

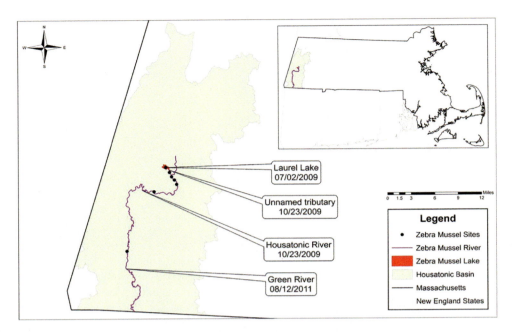

FIGURE 3.3   Distribution map depicting zebra mussel in Massachusetts.

the domination of the zebra mussel (Figure 3.4). The introduction of zebra mussels has placed great stress on native Unionidae populations around the country (see Introduction). The clarity of Laurel Lake ranges from medium to high (4–6.8 m or 13.1–22.3 ft) and the clarity may be improved by zebra mussels' filtering activities. Laurel Lake has been subject to a 3-ft drawdown each winter since 2010–2011 for the control of zebra mussels (and Eurasian watermilfoil) in shallow water. Although greater control is needed and desired, the drawdown has been limited to 3 ft by state regulation. Zebra mussel monitoring at different water depths over six years revealed that populations were denser in shallow water and became sparser as the depth increased (Figure 3.4) (Wagner 2017). Winter drawdowns in this lake have reduced the colonized zone of zebra mussels to depths of lower than 5 ft. This is the result of the desiccation or drying of mussels in the area that is dewatered and/or ice scour and exposure to freezing temperatures for animals exposed above and within an ice zone up to 2 ft below the water line. Deeper drawdowns have been proposed as a potential tool to further manage, and possibly eradicate zebra mussels from Laurel Lake. In

Laurel Lake, zebra mussels have been found to depths of 25 ft. It is unlikely that a drawdown could be achieved and maintained to eradicate zebra mussels and the environmental impact to other species and the lake itself would be significant and therefore cannot be ignored. For example, Laurel Lake is home to one of only two populations of the State Endangered Boreal Marstonia snail (*Marstonia lustrica*). Increase of winter drawdown depth will decrease the abundance of Boreal Marstonia snail because highest abundances have been found at depths of 4–6 ft in Laurel Lake.

Amid the arrival of zebra mussels at Laurel Lake, MA DCR developed protocols for watercraft users to fill a Clean Boat Certification Form before launching as well as clean, drain, and dry their boats (MA DCR 2010) to prevent spread of zebra mussels to other waterbodies. Working together with Massachusetts Department of Fish and Game (MA DFG), MA DCR also activated the Massachusetts Interim Zebra Mussel Action Plan to take actions in dealing with invasive mussels mainly in the Housatonic and Hoosic watershed which are highly susceptible to colonization by zebra mussels (Table 3.1). These actions include the following: (1)

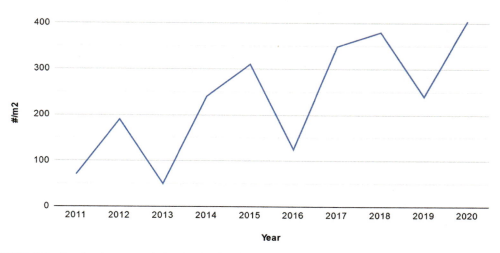

FIGURE 3.4 Dead zebra mussel density in the drawdown zone. For the years 2011–2016, densities of deceased zebra mussels were taken from Wagner (2017). For the years 2017–2020, estimations were provided following the same trend observed between 2011 and 2016. Estimated densities were designed to follow a logistic growth pattern. This was chosen as Wagner states that the substrate in Laurel Lake is a potential limiting factor in the spread of zebra mussels, meaning as the population increases and then decreases cyclically, it may eventually run out of substrate and the population will balance out (i.e., hitting some form of carrying capacity). This theory can be either confirmed or denied through further surveys of deceased zebra mussels during December drawdowns.

Self-certification by users that their boats have not been on waterbodies likely affected by zebra or quagga mussels within the last 30 days, or if they have, that the boats have been properly decontaminated in accordance with the procedures in the plan. Violations of these mandatory requirements are subject to fines and other penalties under OFBA regulations. (2) Extension of the temporary closure of the state boat ramp at Laurel Lake, the one Massachusetts waterbody that has a documented occurrence of zebra mussels to date, until October 15, 2009 (the updated sightings are listed in Table 3.1). (3) Extending the hours of three existing state boat ramp monitors and the addition of a fourth monitor to regularly monitor nine boat ramps in the Hoosic and Housatonic watersheds. (4) Implementation of a robust educational and technical assistance outreach effort designed to facilitate parallel prevention measures by other key stakeholders such as municipalities, lake associations, and recreational users of Massachusetts waterbodies

(MA DCR and MA DFG 2009). More systematic zebra mussel management is described subsequently.

MANAGING ACTIONS

Zebra mussels can spread rapidly even though there are many methods to inhibit their colonization of available invasion. Some evidence of biological control methods has been found in Europe. Crayfish are one of the groups that forages on veliger and post-veliger larvae, consuming about 100 mussels per day. However, areas located in deeper lakes with colder water temperatures are not a favorable habitat for crayfish to hunt in. Based on estimations, the loss of zebra mussels due to crayfish predation is only around 5% (Maxwell 1992). There are also some other Eurasian fish species that feed on *D. polymorpha*, such as the European roach or common roach (*Rutilus rutilus*). Reports show that about 90% of the European roach's diet consists of *D.*

TABLE 3.3

List of Town Names and Waterbodies that Contain Zebra Mussels

Town	Waterbody
Great Barrington	Green River
Lee	Housatonic River
	Laurel Lake
	Unnamed tributary
Stockbridge	Housatonic River

polymorpha. Also, *D. polymorpha* consists of on average 97% of the diets of certain species of waterfowl, specifically the common pochard (*Aythya ferina*), tufted duck (*Aythya fuligula*), and Eurasian coot, three species of overwintering diving ducks. Counts of zebra mussel larvae show fluctuations but no upsurge trends have been found, which provides evidence that the equilibrium in those areas is maintained by local predators (Walz 1991). In the Great Lakes, species like carp (*Cyprinidae*), eel (*Anguilliformes*), and sheepshead (*Archosargus*) have found to be promising biological control agents (O'Neill and MacNeill 1991). However, Asian carp itself is an invasive nuisance species and a collaboration of 28 US and Canadian agencies are currently working together to prevent the introduction and establishment of Asian carp in the Great Lakes (Asian Carp Regional Coordinating Committee 2020). Therefore, it is not suggested to introduce any carp species for zebra mussel control in Massachusetts. In an infested reservoir in California, Sweetwater Reservoir, stocking high-density (0.42 or 1.90 fish/m^2) redear sunfish (*Lepomis microlophus*) effectively removed another dreissenid mussel, the quagga mussel in an enclosure (Figure 3.5) and appeared to suppress their growth and recruitment (Wong et al. 2013). Redear sunfish is a species native to the Southeastern United States that has become a popular stocked sportfish in the Southwest. Based upon the experimental results in the Sweetwater Reservoir, biological control of dreissenid mussels with redear sunfish could be an effective approach (Hatcher and McClelland 2015).

Several experiments have been conducted with *Bucephalus polymorphus* to examine effects on *D. polymorpha* (Lajtner et al. 2008) because it is a type of parasite that treats *D.*

polymorpha as its first intermediate host (Molloy et al. 1997). However, considering the main impact to mussel gonads is less harmful (Molloy et al. 1997), using *B. polymorphus* as a biological control agent may not be effective. Additionally, a research project by Molloy and Griffin (1991) tested the potential lethality of over 260 different microorganisms on *D. polymorpha* in a laboratory setting over a two-year period and the bacterium *Pseudomonas fluorescens* (*Pf* CL 145A) was found to be an effective biological control agent. Both living and dead cells of *P. fluorescens* are effective in controlling *D. polymorpha*, as well as *D. bugensis* (the quagga mussel). The bacteria can be used within a wide temperature range (5–27°C), which means the agent is far more effective than artificial chemical molluscicides (normally, those copper-based chemical molluscicides can mainly be effective at temperatures above 18°C) (Molloy 2002). Nearly two decades of research on *Pf* CL 145A has led to the development of the first pesticide for the control of invasive zebra (and quagga) mussels, Zequanox® (Rackl and Link 2015). Five case studies from Rackl and Link (2015) (adult mussel control within a hydropower facility, adult mussel control within intake bays at a coal-powered plant, a comparison to chlorine at a drinking treatment facility, a mussel settlement management case within a hydropower facility, and open water treatments in a lake) revealed that Zequanox® can provide greater than 90% control in both flowing and static infrastructure in high volumes of water. Moreover, pilot research at Christmas Lake, Minnesota, has shown some eradication of zebra mussels in that area. By applying several treatments of Zequanox®, no evidence of zebra mussels in an experimental area was found (Smith 2015).

FIGURE 3.5   Quagga mussel study design at Sweetwater Reservoir, California. (Photo by Mark Hatcher, Hatcher, and McClelland (2015).)

Due to its tiny size, the value of *D. polymorpha* as a food source is unlikely. The species is also very hard to harvest because of their robust byssal fibers. Toxin bioaccumulation via their filter feeding can be harmful to humans as well (O'Neill and MacNeill 1991). According to many projects in Europe, the former USSR, and North America, it is better to extinguish the species during veliger stage when found in water pipelines, watercrafts, or other infrastructure (Choi et al. 2013; Dalton and Cottrell 2013; Davis et al. 2016; Delrose et al. 2015; DeShon et al. 2016; Moffitt et al. 2016; MRB Group 1991; Watters et al. 2015).

Several physical control methods can be potentially used to help lessen infestations. Strainers and filters can be applied to block the entry of larval, juvenile, and especially adult mussels (Maxwell 1992). The effectiveness is highly dependent on the size of the strainers and filters used. These tools are suggested to be installed at intakes and distribution systems. Scraping is considered very effective in larger conduits, only if the size is wide enough to allow for the use of equipment. However, this method requires high manpower and can be very expensive (O'Neill and MacNeill

1991). Some other methods have also been suggested such as the use of a type of pipe-crawling vehicle (Martin et al. 1991), centrifugal separators (Smythe et al. 1991b), low-voltage electric fields (Smythe et al. 1991a), the use of ultraviolet radiation (Chalker-Scott et al. 1991; Delrose et al. 2015; Stewart-Malone et al. 2015), acoustic energy (Menezes 1991), and coatings (Smithee and Kovalak 1991; Skaja et al. 2015). Periodically flushing water systems with heated water and tap water can be effective and less expensive. For hot water, the temperature should be higher than 37°C (99°F) and flushed through the area for 1 hour to achieve 100% eradication of mussels (O'Neill and MacNeill 1991) and, if it is for water craft decontamination, it is recommended to use 60°C (140°F) for 10 seconds by spraying a target area (Comeau et al. 2011; Comeau et al. 2015; Morse 2009). According to Wagner (2017), partial drawdown is a method that can be applied to reduce biomass of zebra mussels in shallow areas in a waterbody. However, partial drawdowns and/ or complete drawdowns will impact all species within the waterbody. The loss of *D. polymorpha* will also cut off the food web of the predatory fish *A. ferina* and *A. fuligula* (Leuven et al. 2014). It is possible that the discharged water may contain some overwinter veligers and larvae that can cause secondary infestations to downstream waterbodies.

There are two main categories of chemical control methods: one is the use of oxidizing agents (chlorine, chlorine dioxide, hydrogen peroxide, potassium permanganate, etc.), the other is the use of toxic chemicals (molluscicides, metallic ions, copper sulfate) to control zebra mussels (Maxwell 1992). Oxidizing agents should not remain in local environments for long periods; therefore, an immediate discharge may be needed (Maxwell 1992). Electrolytically dissolved metal ions can be considered a good method for small conduits that are not easily accessed. The use of chlorine has proved to be an effective control method. However, dechlorination is needed after the application. Concentrations of solutions from 0.25 mg/l to 1.0 mg/l, after two to three weeks of application, results in a 95–100% mortality rate of zebra mussels (O'Neill and MacNeill 1991). One should take proper precautions when applying chlorine treatments to public drinking waters due to the generation of trihalomethanes (THMs) which are potential carcinogens (Cotruvo and Regelski 1989). Acetic acid products (Claudi et al. 2012; Davis et al. 2015a) and potash such as NaCl and/or KCl (Davis et al. 2015b; 2017; Moffitt et al. 2016) are also effective alternative chemicals for zebra mussel control with relatively less side effects under appropriate environmental conditions or when used for watercraft decontamination. Specifically, KCl-based BioBullets® has been developed for specific use in zebra mussel control (Aldridge et al. 2006), which is designed to selectively target filter feeders by introducing poison to the water in the form of food particles preferred by molluscans. Therefore, the poisons are removed from the water when mussels feed and pose minimal impacts to the rest of the aquatic ecosystem (Aldridge et al. 2006). More information about chemical control on zebra mussel and other 18 invasive species can be found in Appendix III. Any chemical used for invasive species control requires a permit from the Massachusetts Department of Environmental Protection.

Among management options, prevention has proved to be the most efficient and cost-effective approach in stopping the spread of aquatic invasive species. The most efficient way to protect the environment from invasive species is integrated management, including prevention and education, early detection, control, ecosystem restoration, and the regulation and enforcement of environmental law (Wong and Gerstenberger 2015). To minimize the spreading of *D. polymorpha*, a combination of treatment methods should be used, which include increasing public awareness of invasive species as part of the integrated management approach. For example, the Massachusetts Department of Environmental Protection has developed and implemented standard operating procedures for internal field equipment decontamination during water quality monitoring in the field to prevent the spread of invasive aquatic organisms (Chase and Wong 2015). Specifically, citizens are strongly encouraged to inspect the hull of boats thoroughly after boating in infested waterbodies, flush hulls with heated water, drawdown wells, and expose the hulls, anchors, and chains under sunlight for drying. Clean, drain, and drying watercraft and experimental equipment and gear is critical to prevent the spread of invasive species from one waterbody to another.

SUCCESS STORY

One form of chemical control that has shown promising results is the application of copper-based pesticides. EarthTec QZ® is a copper-based (5% copper by weight) molluscicide that has been found to kill mollusks at certain concentrations. It is NSF-certified for safe use in drinking water and is approved by the EPA for use to control invasive species of mussels. EarthTec QZ® is a registered herbicide/pesticide in Massachusetts. It has been confirmed that EarthTec QZ® could prevent the settlement of dreissenid mussels (Watters et al. 2015) and it has been applied for eradication of either zebra or quagga mussels in various waterbodies (Hammond and Ferris 2019; Lund et al. 2018). This success story involves the use of EarthTec QZ® to eradicate an invasive population of mussels from a lake in Pennsylvania in 2017.

Billmeyer Quarry is a lake located in Conoy Township, Pennsylvania, with a surface area of 12 ha and a maximum depth of about 35 m in the west basin and 27 m in the east basin (Hammond and Ferris 2019). Billmeyer had become infested with quagga mussels (*Dreissena rostriformis bugensis*), with the first observations being recorded in 2005. Quagga mussels, a relative of the zebra mussel, are an aquatic bivalve mollusk that have also been listed as an invasive species in the United States, causing many of the same issues that the zebra mussel has. The molluscicide was applied to 50% of the lakes surface during each treatment and was allowed for mixing with the remaining waters of the lake naturally. Several cages filled with mussels were placed at various locations and depths around the lake. Treatments were to persist until every preplaced caged mussel was deceased. A total of three treatments occurred, with additional treatments being applied in order to keep the molluscicide at a target concentration of the 0.2 mg/l in the lake. On day 1 of the first treatment occurring in the fall of 2017, 6,245 l of EarthTec QZ® were dispersed at depths of 2, 6, and 9 m. The second treatment, occurring on day 7, dispersed 4,164 l at depths of 2, 6, 18, 21, and 26 m. The third and final treatment, occurring on day 37, dispersed 3,123 l at depths of 9–12 m. Based on historical measurements of water temperature and dissolved oxygen for Billmeyer Quarry, it was estimated that mussels could inhabit depths up to at least 24 m. Treatment methods therefore ensured that the EarthTec QZ® would reach a lethal concentration at that depth.

The first mussels began to die 3 days after the first treatment, with the final mussels dying after 40 days. Samples of plankton were taken several times throughout the study and then examined for the presence of mussel DNA. A sample taken in December of 2017 found very low levels of DNA below the detection limits of the tests used, meaning mussel DNA may have been present in the lake. However, in a sample taken in July 2018, over half a year since the application of the final treatment, no evidence of mussel DNA was found. This suggests that quagga mussels had been completely eradicated in the lake at that time. Future mussel surveys of the lake will be able to verify if this was true. The total cost of the project came out to $109,400, or about $0.06/m³ treated.

One potential risk that always comes with the use of chemical control agents is spillover deaths of non-target species. In Billmeyer Quarry, large numbers of various fish species were observed prior to the first treatment. The only incident of spillover from the effects of the molluscicide occurred after the second application when ten dead gizzard shad were found floating on the water. Because only one species was affected, it is possible that this species is overly sensitive to copper-based pesticides. It is also possible that, due to the fact that other members of the species were observed to be healthy during other periods of the project, the ten individuals that were killed were exposed to a pocket of a higher concentration of EarthTec QZ®. Another risk that comes specifically with molluscicides is undesired harm to native mollusks. Many native species of mollusks, such as several members of the family Uniondae, are already under great stress due to competition from the zebra and quagga mussels. Therefore, application of harmful molluscicides should be used sparingly in waterbodies inhabited by these native mollusks. In this specific project, the targeted lake had no populations of native mollusks.

The results of this study show that the use of copper-based pesticides such as EarthTec QZ® may be an effective and cost-efficient control agent for eradicating, or at least decreasing, populations of invasive mussels. However, chemical control becomes less effective on larger waterbodies, where the overall cost and needed manpower increase. The amount of chemicals needed to maintain target lethal concentrations would also increase. This makes chemical treatment with copper-based pesticides an effective reactive control agent. Coupled with good preventative measures, such as boat inspections and public education, the spread of invasive mussels in Massachusetts can hopefully be stopped.

REFERENCES

Aldridge, D.C., P. Elliott, and G.D. Moggridge. 2006. Microencapsulated biobullets for the control of biofouling zebra mussels. *Environmental Science & Technology* 40: 975–979.

Asian Carp Regional Coordinating Committee. 2020. *Asian Carp Action Plan for Fiscal Year 2020.* Asian Carp Regional Coordinating Committee, p. 167.

Baker, S.M., J.S. Levinton, J.P. Kurdziel, and S.E. Shumway. 1998. Selective feeding and biodeposition by zebra mussels and their relation to changes in phytoplankton composition and seston load. *Journal of Shellfish Research* 17: 1207–1213.

Baldwin, B.S., M.S. Mayer, J. Dayton, N. Pau, J. Mendilla, M. Sullivan, A. Moore, A. Ma, and E.L. Mills. 2002. Comparative growth and feeding in zebra and quagga mussels (*Dreissena polymorpha* and *Dreissena bugensis*): Implications for North American lakes. *Canadian Journal of Fisheries and Aquatic Sciences* 59: 680–694.

Bastviken, D.T.E., N.F. Caraco, and J.J. Cole. 1998. Experimental measurements of zebra mussel (*Dreissena polymorpha*) impacts on phytoplankton community composition. *Freshwater Biology* 39: 375–386.

Benson, A., D. Raikow, J. Larson, A. Fusaro, A.K. Bogdanoff, and A. Elgin. 2019. *Dreissena polymorpha (zebra mussel)*. https://nas.er.usgs.gov/queries/FactSheet.aspx?speciesID=5 (Accessed November 1, 2019).

Berg, D.J., S.W. Fisher, and P.F. Landrum. 1996. Clearance and processing of algal particle by zebra mussels (*Dreissena polymorpha*). *Journal of Great Lakes Research* 22: 779–788.

Berkman, P.A., M.A. Haltuch, E. Tichich, D.W. Garton, G.W. Kennedy, J.E. Gannon, S.D. Mackey, J.A. Fuller, and D.L. Liebenthal. 1998. Zebra mussels invade Lake Erie muds. *Nature* 393: 27–28.

Britton, D.K. 2015. History of western management actions on invasive mussels. In W.H. Wong and S.L. Gerstenberger (eds). *Biology and Management of Invasive Quagga and Zebra Mussels in the Western United States*. London: CRC Press, pp. 3–14.

Bunt, C.M., H.J. MacIsaac, and W.G. Sprules. 1993. Pumping rates and projected filtering impacts of juvenile zebra mussels (*Dreissena polymorpha*) in Western Lake Erie. *Canadian Journal of Fisheries and Aquatic Sciences* 50: 1017–1022.

Burlakova, L.E. 1998. Ecology of *Dreissena polymorpha* (Pallas) and its role in the structure and function of aquatic ecosystems. [In Russian.] Dissertation. Zoology Institute of the Academy of Science, Minsk, Belarus.

Caraco, N.F., J.J. Cole, and D.J. Strayer. 2006. Top down control from the bottom: Regulation of eutrophication in a large river by benthic grazing. *Limnology and Oceanography* 51: 664–670.

Carlton, J.T. 1993. Dispersal mechanisms of the zebra mussel. In T.F. Nalepa and D.W. Schoesser (eds). *Zebra mussels: Biology, impacts, and control*. London: Lewis/CRC Press, pp. 677–698.

Carlton, J.T. 2008. The zebra Mussel *Dreissena polymorpha* found in North America in 1986 and 1987. *Journal of the Great Lakes Research* 34: 770–773.

Chalker-Scott, L., J.D. Scott, C. Dunning, and K. Smith. 1991. Effect of ultraviolet-B radiation (280–320nm) on survivorship of zebra Mussel Larvae (*Dreissena polymorpha*): A potential control strategy. In Second international Zebra Mussel Research Conference, November 19–22. Rochester, NY, p. 35.

Chase, R.F., and W.H. Wong. 2015. Standard Operating Procedure: Field Equipment Decontamination to Prevent the Spread of Invasive Aquatic Organisms. Massachusetts Department of Environmental Protection. File # CN 59.6.29 pp.

Choi, W.J., S.L. Gerstenberger, R.F. McMahon, and W.H. Wong. 2013. Estimating survival rates of quagga mussel (*Dreissena rostriformis bugensis*) veliger larvae under summer and autumn temperature regimes in residual water of trailered watercraft at Lake Mead, USA. *Management of Biological Invasions* 4(1): 61–69.

Claudi, R., A. Graves, A.C. Taraborelli, R.J. Prescott, and S. Mastitsky. 2012. Impact of pH on survival and settlement of dreissenid mussels. *Aquatic Invasions* 7(1): 21–28.

Claudi, R., and G.L. Mackie. 1994. *Practical Manual for Zebra Mussel Monitoring and Control*. London: Lewis Publishers, p. 227.

Comeau, S., R.S. Ianniello, W.H. Wong, and S.L. Gerstenberger. 2015. Boat decontamination with hot water spray: Field validation. In W.H. Wong and S.L. Gerstenberger (eds) *Biology and Management of Invasive Quagga and Zebra Mussels in the Western United States*. London: CRC Press, pp. 161–173.

Comeau, S., S. Rainville, W. Baldwin, E. Austin, S.L. Gerstenberger, C. Cross, and W.H. Wong. 2011. Susceptibility of quagga mussels (Dreissena rostriformis bugensis) to hot-water sprays as a means of watercraft decontamination. *Biofouling* 27(3): 267–274.

Cotruvo J.A., and M. Regelski. 1989. Issues in developing national primary drinking water regulations for disinfection and disinfection-by-products. In: E.J. Calabrese, C.E., Gilbert, and H. Pastids (eds.), *Safe Drinking Water Act: Amendments, Regulations, and Standards*. Chelsea: Lewis Publishers, pp. 57–69.

Counihan, T., W.H. Wong, and R.F. McMahon. 2018. Early detection monitoring for invasive mussels in the Columbia River Basin. Columbia River Basin Dreissenid Mussel Monitoring Forum. October 31, 2018. Boise, ID.

Daley, B. 2009. Invasive mussel found in Bay State lake. Boston Globe. Retrieved from http://archive.boston.com/news/local/massachusetts/articles/2009/07/09/invasive_mussel_found_in_mass_lake/ (Accessed November 1, 2020).

Dalton, L.B., and S. Cottrell. 2013. Quagga and zebra mussel risk via veliger transfer by overland hauled boats. *Management of Biological Invasions* 4(2): 129–133.

Davis, E.A., W.H. Wong, and W.N. Harman. 2015a. Distilled white vinegar (5% acetic acid) as a potential decontamination method for adult zebra mussels. *Management of Biological Invasions* 6(4): 423–428.

Davis, E.A., W.H. Wong, and W.N. Harman. 2015b. Comparison of three sodium chloride chemical treatments for adult zebra mussel decontamination. *Journal of Shellfish Research* 34(3): 1029–1036.

Davis, E.A., W.H. Wong, and W.N. Harman. 2016. Livewell flushing to remove zebra mussel (*Driessena polymorpha*) veligers. *Management of Biological Invasions* 7(4): 399–403.

Davis, E.A., W.H. Wong, and W.N. Harman. 2017. Toxicity of potassium chloride compared to sodium chloride for zebra mussel decontamination. *Journal of Aquatic Animal Health* 30: 3–12.

De Stasio, B.T., M.B. Schrimpf, A.E. Beranek, and W.C. Daniels. 2008. Increased chlorophyll a, phytoplankton abundance, and cyanobacteria occurrence following invasion of Green Bay, Lake Michigan by dreissenid mussels. *Aquatic Invasions* 3: 21–27.

Delrose, P., S.L. Gerstenberger, and W.H. Wong. 2015. Effectiveness of the SafeGUARD ultraviolet radiation system as a system to control quagga mussel veligers (*Dreissena rostriformis bugensis*). In: W.H. Wong and S.L. Gerstenberger (eds). *Biology and Management of Invasive Quagga and Zebra Mussels in the Western United States*. London: CRC Press, pp. 479–485.

DeShon, D.L., W.H. Wong, D. Farmer, and A.J. Jensen. 2016. The ability of scent detection canines to detect the presence of quagga mussel (*Dreissena rostriformis bugensis*) veligers. *Management of Biological Invasions* 7(4): 419–428.

Diggins, T.P. 2001. A seasonal comparison of suspended sediment filtration by quagga (*Dreissena bugensis*) and zebra (*D. polymorpha*) mussels. *Journal of Great Lakes Research* 27: 457–466.

Dionisio Pires, L.M., R.R. Jonker, E. Van Donk, and H.J. Laanbroek. 2004. Selective grazing by adults and larvae of the zebra mussel (*Dreissena polymorpha*): Application of flow cytometry to natural seston. *Freshwater Biology* 49: 116–126.

Ekroat, L.R., E.C. Masteller, J.C. Shaffer, and L.M. Steele. 1993. The byssus of the zebra mussel (Dreissena polymorpha): Morphology, byssal thread formation, and detachment. In: T.F. Nalepa and D.W. Schoesser (eds). *Zebra Mussels: Biology, Impacts, and Control*. London: Lewis/CRC Press, pp. 239–263.

Ellis, A.E. 1978. *British Freshwater Bivalve Mollusca: Keys and Notes for the Identification of Species*. London: Academic Press, p. 109.

Feinberg, H.S. (ed.) 1979. *Simon and Schuster's Guide to Shells.* London: Simon and Schuster, p. 512.

Fraleigh, P.C., P.L. Klerks, G. Gubanich, G. Matisoff, and R.C. Stevenson. 1993. Abundance and settling of zebra mussel (*Dreissena polymorpha*) veligers in western and central Lake Erie. In: T.F. Nalepa, and D.W. Schoesser (eds). *Zebra Mussels: Biology, Impacts, and Control.* London: Lewis/CRC Press, pp. 129–142.

Griffiths, R.W. 1993. Effects of zebra mussels (Dreissena polymorpha) on the benthic fauna of Lake St. Clair. Lewis Publishers. In: T.F. Nalepa, and D.W. Schoesser (eds). *Zebra Mussels: Biology, Impacts, and Control.* London: Lewis/CRC Press, pp. 439–451.

Hammond, H., and G. Ferris. 2019. Low doses of EarthTec QZ ionic copper used in effort to eradicate quagga mussels from an entire Pennsylvania Lake. *Management of Biological Invasions* 10(3): 500–516.

Hatcher, M., and S. McClelland. 2015. Monitoring and control of quagga mussels in sweetwater reservoir. In W.H. Wong and S.L. Gerstenberger (eds). *Biology and Management of Invasive Quagga and Zebra Mussels in the Western United States.* London: CRC Press, pp. 401–424.

Hebert, P.D.N., B.W. Muncaster, and G.L. Mackie. 1989. Ecological and genetic studies on Dreissena polymorpha (Pallas)—A new mollusk in the Great Lakes. *Canadian Journal of Fisheries and Aquatic Sciences* 46: 1587–1591.

Hecky, R.E., R.E.H. Smith, D.R. Barton, S.J. Guildford, W.D. Taylor, M.N. Charlton, and T. Howell. 2004. The nearshore phosphorus shunt: A consequence of ecosystem engineering by dreissenids in the Laurentian Great Lakes. *Canadian Journal of Fisheries and Aquatic Sciences* 61: 1285–1293.

Horgan, M.J., and E.L. Mills. 1997. Clearance rates and filtering activity of zebra mussel (*Dreissena polymorpha*): Implications for freshwater lakes. *Canadian Journal of Fisheries and Aquatic Sciences* 54: 249–255.

Horohov, J., H. Silverman, J.W. Lynn, and T.H. Dietz. 1992. Ion transport in the freshwater zebra mussel, *Dreissena polymorpha. Biological Bulletin* 183: 297–303.

ITIS (Integrated Taxonomic Information System). 2021. www.itis. gov/servlet/SingleRpt/SingleRpt#null (Accessed February 8, 2021).

Karatayev, A.Y., L. Burlakova, S. Mastitsky, and D.K. Padilla. 2011. Rates of spread of two congeners, *Dreissena polymorpha* and *Dreissena rostriformis bugensis*, at different spatial scales. *Journal of Shellfish Research* 30(3): 923–931.

Karatayev, A.Y., L.E. Burlakova, and D.K. Padilla. 1997. The effects of *Dreissena polymorpha* (Pallas) invasion on aquatic communities in Eastern Europe. *Journal of Shellfish Research* 16: 187–203.

Karatayev, A.Y., L.E. Burlakova, and D.K. Padilla. 2002. Impacts of zebra mussels on aquatic communities and their role as ecosystem engineers. In E. Leppakoski, S. Gollach, and S. Olenin (eds). *Invasive Aquatic Species of Europe: Distribution, Impacts and Management.* Dordrecht: Kluwer Academic Publishers, pp. 433–446.

Karatayev, A.Y., L.E. Burlakova, and D.K. Padilla. 2015. Zebra versus quagga mussels: A review of their spread, population dynamics, and ecosystem impacts. *Hydrobiologia* 746: 97–112.

Karatayev, A.Y., D.K. Padilla, and L.E. Burlakova. 2006. Growth rate and longevity of *Dkareissena polymorpha* (Pallas): A review and recommendations for future study. *Journal of Shellfish Research* 25: 23–32.

Kerney, M.P., and B.S. Morton. 1970. The distribution of *Dreissena polymorpha.* (Pallas) in Britain. *Journal of Conchology* 27: 97–100.

Lajtner, J., A. Lucić, M. Marušić, and R. Erben. 2008. The effects of the trematode Bucephalus polymorphus on the reproductive cycle of the zebra mussel *Dreissena polymorpha* in the Drava River. *Acta Parasitologica* 53(1): 85–92.

Leach, J.H. 1993. Impacts of zebra mussel (*Dreissena polymorpha*) on water quality and fish spawning reefs in Western Lake Erie. In: T.F. Nalepa and D.W. Schoesser (eds). *Zebra Mussels: Biology, Impacts, and Control.* London: Lewis/CRC Press, pp. 381–397.

Leuven, R., F. Collas, K.R. Koopman, J. Matthews, and G.V. Velde. 2014. Mass mortality of invasive zebra and quagga mussels by desiccation during severe winter conditions. *Aquatic Invasions* 9(3): 243–252.

Li, J., V. Ianaiev, A. Huff, J. Zalusky, T. Ozersky, and S. Katsev. 2021. Benthic invaders control the phosphorus cycle in the world's largest freshwater ecosystem. *Proceedings of the National Academy of Sciences of the United States of America* 118(6): e2008223118. https://doi.org/10.1073/pnas.2008223118.

Lovell, S.J., S.F. Stone, and L. Fernandez. 2006. The economic impacts of aquatic invasive species: A review of the literature. *Agricultural and Resource Economics Review* 35(1): 195–208.

Lucy, F.E., L.E. Burlakova, A.Y. Karatayev, S.E. Mastitsky, and D.T. Zanatta. 2013. Zebra mussel impacts on unionids: A synthesis of trends in North America and Europe. In: T.F. Nalepa and D.W. Schoesser (eds). *Zebra Mussels: Biology, Impacts, and Control.* London: Lewis/CRC Press, pp. 623–646.

Lund K., K.B. Cattoor, E. Fieldseth, J. Sweet, and M.A. McCartney. 2018. Zebra mussel (*Dreissena polymorpha*) eradication efforts in Christmas Lake, Minnesota. *Lake and Reservoir Management* 34(1): 7–20.

MA DCR (Massachusetts Department of Conservation and Recreation). 2010. *Aquatic Invasive Species Assessment and Management Plan.* Boston, MA: MA DCR, p. 39.

MA DCR (Massachusetts Department of Conservation and Recreation) and MA DFG (Massachusetts Department of Fish and Game). 2009. *Massachusetts Interim Zebra Mussel Action Plan.* Cambridge, MA: MA DCR, p. 27.

MacIsaac, H.J. 1996. Potential abiotic and biotic impacts of zebra mussels on the inland waters of North America. *American Zoologist* 36: 287–299.

Mackie, G.L., and D.W. Schloesser. 1996. Comparative biology of zebra mussels in Europe and North America: An overview. *American Zoologist* 36(3): 244–258.

Martel, A. 1993. Dispersal and recruitment of zebra mussels (*Dreissena polymorpha*) in a nearshore area in west-central Lake Erie: The significance of postmetamorphic drifting. *Canadian Journal of Fisheries and Aquatic Sciences* 50: 3–12.

Martin, B., and S.E. Landsberger. 1991. Design of pipe-crawling vehicles for zebra mussel control. In Second international zebra mussel research conference, November 19–22, Rochester, NY, p. 32.

Maxwell, L. 1992. *The Biology, Invasion and Control of the Zebra Mussel (Dreissena polymorpha) in North America.* New York: State University of New York at Oneonta, Biological Field Station, pp. 1–26.

Mayer, C.M., L.E. Burlakova, P. Eklöv, D. Fitzgerald, A.Y. Karatayev, S.A Ludsin, S. Millard, E.L. Mills, A.P. Ostapenya, L.G. Rudstam, B. Zhu, and T.V. Zhukova. 2013. Benthification of freshwater lakes: Exotic mussels turning ecosystems upside down. In: T.F. Nalepa and D.W. Schoesser (eds). *Zebra Mussels: Biology, Impacts, and Control.* London: Lewis/CRC Press, pp. 575–585.

McMahon, R.F. 1996. The physiological ecology of the zebra mussel, *Dreissena polymorpha*, in North America and Europe. *American Zoologist* 36: 339–363.

McMahon, R.F., B.S. Payne, and T. Ussery. 1991. Effects of temperature and relative humidity on desiccation resistance in Zebra Mussels (*Dreissena polymorpha*): Is aerial exposure a viable control option? In Second International Zebra Mussel Research Conference, November 19–22, Rochester, NY.

McMahon, R.F., T.A. Ussery, A.C. Miller, and B.S. Payne. 1993. Thermal tolerance in zebra mussels (*Dreissena polymorpha*) relative to rate of temperature increase and acclimation temperature. In L. Tsou and Y. G. Mussalli (eds.). *Proceedings: Third International Zebra Mussel Conference* (EPRI TR-102077). Palo Alto, CA: Electric Power Research Institute, pp. 4-97–4-118.

Menezes, J. 1991. Zebra Mussel control using acoustic energy. In Second International Zebra Mussel Research Conference, November 19–22. Rochester, NY, p. 36.

Mills, E.L., J.R. Chrisman, B. Baldwin, R.W. Owens, R. O'Gorman, T. Howell, E.F. Roseman, and M.K. Raths. 1999. Changes in the Dreissenid Community in the Lower Great Lakes with Emphasis on Southern Lake Ontario. *Journal of Great Lakes research* 25(1): 187–197.

Moffitt, C.M., K.A. Stockton-Fiti, and R. Claudi. 2016. Toxicity of potassium chloride to veliger and byssal stage dreissenid mussels related to water quality. *Management of Biological Invasions* 7(3): 257–268.

Molloy, D.P. 2002. Biological control of Zebra Mussels. In Proceedings of the Third California Conference on Biological Control. University of California, Davis. Division of Research and Collections, New York State Museum, The State Education Department, Albany, NY, pp. 86–94. www.researchgate.net/profile/Daniel_Molloy2/publication/288844393_Biological_control_of_zebra_mussels/links/573d289608ae-a45ee841a752.pdf (Accessed June 28, 2018).

Molloy, D.P., and B. Griffin. 1991. Biological Control of Zebra Mussels: Screening for the Lethal Microorganisms. In Second International Zebra Mussel Research Conference, Nov. 19–22, 1991, Rochester, NY, p. 30.

Molloy D.P., A.Y. Karatayev, L.E. Burlakova, D.P. Kurandina, and F. Laruelle. 1997. Natural enemies of zebra mussels: Predators, parasites, and ecological competitors. *Reviews in Fisheries Science* 5: 27–97.

Montz, G., and J. Hirsch. 2016. Veliger presence in residual water— Assessing this pathway risk for Minnesota watercraft. *Management of Biological Invasions* 7(3): 235–240.

Morse, J.T. 2009. Assessing the effects of application time and temperature on the efficacy of hot-water sprays to mitigate fouling by *Dreissena polymorpha* (zebra mussels Pallas). *Biofouling* 25(7): 605–610.

MRB Group. 1991. *The Impact of Zebra Mussel Infestations on Public Water Supplies: A Summary Report*. Rochester, NY: MRB Group Engineering, Architecture and Surveying, p. 5.

Nalepa, T.F., D.L. Fanslow, and S.A. Pothoven. 2010. Recent changes in density, biomass, recruitment, size structure, and nutritional state of Dreissena populations in southern Lake Michigan. *Journal of Great Lakes Research* 36(Supplement 3): 19–25.

Nalepa, T.F., and D.W. Schloesser (eds.). 1993. *Zebra Mussels: Biology, Impacts, and Control*. London: Lewis/CRC Press, p. 810.

Nalepa, T.F., and D.W. Schloesser (eds.). 2013. *Quagga and Zebra Mussels: Biology, Impacts, and Control*. London: CRC Press, p. 815.

Neary, B.P., and J.H. Leach. 1991. Mapping the potential spread of the zebra mussel (*Dreissena polymorpha*) in Ontario. *Canadian Journal of Fisheries and Aquatic Sciences* 49: 406–415.

Nicholls, K.H., S.J. Standke, and G.J. Hopkins. 1999. Effects of dreissenid mussels on nitrogen and phosphorus in north shore waters of Lake Erie. In M. Munawar, T. Edsall, and I.F. Munawar (eds.), *State of Lake Erie (SOLE)-Past, Present, and Future*. Leiden: Backhuys Publishers, pp. 323–336.

Nichols, S.J. 1996. Variations in the reproductive cycle of *Dreissena polymorpha*, in North America and Europe. *American Zoologist* 36: 311–325.

O'Neill, C.R., Jr. 1997. Economic impact of Zebra Mussels: Results of the 1995 Zebra Mussel information clearinghouse study. *Great Lakes Res. Review* 3(1): 35–42.

O'Neill, C.R., Jr., and D.B. MacNeill. 1991. The Zebra Mussel (*Dreissena polymorpha*): An unwelcome North American Invader. Sea Grant. Cornell cooperative extension, State New York: University of New York and Coastal Resources Fact Sheet.

Qualls, T.M., D.M. Dolan, T. Reed, M.E. Zorn, and J. Kennedy. 2007. Analysis of the impacts of the zebra mussel, *Dreissena polymorpha*, on nutrients, water clarity, and the chlorophyll-phosphorus relationship in lower Green Bay. *Journal of Great Lakes Research* 33: 617–626.

Rackl, S., and C. Link. 2015. Zequanox®: Bio-based control of invasive Dreissena Mussels. In W.H. Wong and S.L. Gerstenberger (eds). *Biology and Management of Invasive Quagga and Zebra Mussels in the Western United States*. London: CRC Press, pp. 515–535.

Ricciardi, A., R.J. Neves, and J.B. Rasmussen. 1998. Impending extinctions of North American freshwater mussels (Unionoida) following the zebra mussel (*Dreissena polymorpha*) invasion. *Journal of Animal Ecology* 67(4): 613–619.

Roberts, L. 1990. Zebra Mussel invasion threatens U. S. waters. *Science* 249: 1370–1372.

Rockwell, D.C., G.J. Warren, P.E. Bertram, D.K. Salisbury, and N.M. Burns. 2005. The U.S. EPA Lake Erie indicators monitoring program 1983–2002: Trends in phosphorus, silica, and chlorophyll a in the central basin. *Journal of Great Lakes Research* 31(Supplement 2): 23–34.

Roditi, H.A., N.F. Caraco, J.J. Cole, and D.L. Strayer. 1996. Filtration of Hudson River water by the zebra mussel (*Dreissena polymorpha*). *Estuaries* 19: 824–832.

Roditi, H.A., N.S. Fisher, and S.A. Sanudo-Wilhelmy. 2000a. Field testing a metal bioaccumulation model for zebra mussels. *Environmental Science & Technology* 34: 2817–2825.

Roditi, H.A., N.S. Fisher, and S.A. Sanudo-Wilhelmy. 2000b. Uptake of dissolved organic carbon and trace elements by zebra mussels. *Nature* 407: 78–80.

Rzepecki, L.M., and J.H. Waite. 1993. Zebra mussel byssal glue: An unwelcome bond between man and mollusk? *Proceedings: Third International Zebra Mussel Conference*. Pleasant Hill, CA: Electric Power Research Institute, Inc, pp. 2-93–2-107.

Schneider, D.W., S.P. Madon, J.A. Stoeckel, and R.E. Sparks. 1998. Seston quality controls zebra mussel (*Dreissena polymorpha*) energetics in turbid rivers. *Oecologia* 17: 331–341.

Silverman, H., J.W. Lynn, E.C. Achberger, and T.H. Dietz. 1996. Gill structure in zebra mussels: Bacterial-sized particle filtration. *American Zoologist* 36: 373–384.

Skaja, A.D., D. Tordonato, and B. Jo Merten. 2015. Coatings for invasive Mussel control: Colorado river field study. In W.H. Wong and S.L. Gerstenberger (eds). *Biology and Management of Invasive Quagga and Zebra Mussels in the Western United States*. London: CRC Press, pp. 451–466.

Smith, D.G. 1993. The potential for spread of the exotic Zebra Mussel (*Dreissena polymorpha*) in Massachusetts. *Report MS-Q-11*. Department of Biology, University of Massachusetts Amherst, Massachusetts and Museum of Comparative Zoology, Harvard University, Cambridge, Massachusetts. www.mass.gov/eea/docs/dcr/watersupply/lakepond/downloads/zebra-mussel-massdep-ms-q-11.pdf (Accessed June 27, 2018).

Smith, K. 2015. Treatments on Christmas Lake kill off zebra mussels. www.startribune.com/treatments-on-christmas-lake-kill-off-zebra-mussels/300673491/ (Accessed June 28, 2019).

Smithe, R.D., and W.P. Kovalak. 1991. Control of zebra mussels fouling by coatings. In Second International Zebra Mussel Research Conference, November 19–22, Rochester, NY, p. 36.

Smythe, A.G., C.L. Lange, J.F. Doyle, and P.M. Sawyko. 1991b. Application of low voltage electric fields to deter attachment of zebra Mussel to structures. In Second International Zebra Mussel Research Conference, November 19–22. Rochester, NY, p. 34.

Smythe, A.G., C.L. Lange, T.M. Short, and R. Tuttle. 1991a. Application of centrifugal separators for control of zebra mussels in raw water systems. In Second International Zebra Mussel Research Conference, November 19–22, Rochester, NY, p. 33.

Sprung, M. 1990. Costs of reproduction: A study on metabolic requirements of the gonads and fecundity of the bivalve *Dreissena polymorpha*. *Malacologia* 32: 267–274.

Stewart-Malone, A., M. Misamore, S. Wilmoth, A.R. Reyes, W.H. Wong, and J. Gross. 2015. The effect of UV-C exposure on larval survival of the dreissenid Quagga Mussel. *PLoS ONE* 10(7): e0133039.

Strayer, D.L., N.F. Caraco, J.J. Cole, S. Findlay, and M.L. Pace. 1999. Transformation of freshwater ecosystems by bivalves: A case study of zebra mussels in the Hudson River. *Bioscience* 49: 19–27.

Strayer, D.L., K.A. Hattala, and A.W. Kahnle. 2004. Effects of an invasive bivalve (*Dreissena polymorpha*) on fish in the Hudson River estuary. *Canadian Journal of Fisheries and Aquatic Sciences* 61: 924–941.

Strayer, D.L., K.A. Hattala, A.W. Kahnle, and R.D. Adams. 2014. Has the Hudson River fish community recovered from the zebra mussel invasion along with its forage base? *Canadian Journal of Fisheries and Aquatic Sciences* 71: 1146–1157.

Strayer, D.L., and H.M. Malcom. 2006. Long term demography of a zebra mussel (*Dreissena polymorpha*) population. *Freshwater Biology* 51: 117–130.

Thorp, J.H., and A.P. Covich. 1991. *Ecology and Classification of North American Freshwater Invertebrates*. London: Academic Press, Inc, p. 911.

Vanderploeg, H.A., T.F. Nalepa, D.J. Jude, E.L. Mills, K.T. Holeck, J.R. Liebig, I.A. Grigorovich, and H. Ojaveer. 2002. Dispersal and emerging ecological impacts of Ponto-Caspian species in the Laurentian Great Lakes. *Canadian Journal of Fisheries and Aquatic Sciences* 59: 1209–1228.

Wagner, K. 2017. *Laurel Lake Drawdown Summary Report: 2010–2016*. Wilbraham, MA: Water Resource Services, p. 64.

Walz, N. 1978. The energy balance of the freshwater mussel *Dreissena polymorpha* in laboratory experiments and in Lake Constance. 2. Reproduction. *Archiv für Hydrobiologie—Supplement* 55: 106–119.

Walz, N. 1991. New invasions, increase, and ecological equilibrium of *Dreissena Polymorpha* Populations in Central and Southern Europe Lakes and Rivers. In Second International Zebra Mussel Research Conference. November 19–22, Rochester, NY.

Watters A., S.L. Gerestenberger, and W.H. Wong. 2015. Use of a molluscicide on preventing Quagga Mussel colonization. In W.H. Wong and S.L. Gerstenberger (eds). *Biology and Management of Invasive Quagga and Zebra Mussels in the Western United States*. London: CRC Press, pp. 507–513.

Wong, W.H., and S.L. Gerstenberger (eds.). 2015. *Biology and Management of Invasive Quagga and Zebra Mussels in the Western United States*. London: CRC Press, p. 566.

Wong, W.H., S.L. Gerstenberger, M.D. Hatcher, D.R. Thompson, and D. Schrimsher. 2013. Invasive quagga mussels can be attenuated by redear sunfish (*Lepomis microlophus*) in the Southwestern United States. *Biological Control* 64: 276–282.

Wong, W.H., S.L. Gerstenberger, J.M Miller, B. Moore, and C.J. Palmer. 2011. A standardized design for quagga mussel monitoring in Lake Mead, Nevada-Arizona. *Aquatic Invasions* 6(2): 205–215.

Wong, W.H., J.S. Levinton, B.S. Twining, and N. Fisher. 2003. Assimilation of micro- and mesozooplankton by zebra mussels: A demonstration of the food web link between zooplankton and benthic suspension feeders. *Limnology & Oceanography* 48: 308–312.

Zhang, H.Y., D.A. Culver, and L. Boegman. 2008. A two-dimensional ecological model of Lake Erie: Application to estimate dreissenid impacts on large lake plankton populations. *Ecological Modeling* 214: 219–241.

4 Chinese Mystery Snail (*Bellamya chinensis*)

ABSTRACT

Chinese mystery snail (*Bellamya chinensis*) is a freshwater snail native to east and southeast Asia, and invasive in North America. In Massachusetts, this species was first reported in 1914 in the Muddy River, which is part of the Charles River Watershed. Until 2018, it has been found in 37 waterbodies within 16 different watersheds in Massachusetts; however, there has been no formal assessment on the ecological and economic impacts of *B. chinensis*. A comprehensive monitoring plan for *B. chinensis* is needed to better understand their current status and to help curtail their further spread to other waterbodies in Massachusetts and beyond. Information on chemical, physical, and biological control methods are also provided in this report.

INTRODUCTION

Taxonomic Hierarchy (ITIS 2021)

Kingdom	Animalia—animal, animaux, animals
Subkingdom	Bilateria
Infrakingdom	Protostomia
Superphylum	Lophozoa
Phylum	Mollusca—mollusques, molusco, molluscs, mollusks
Class	Gastropoda Cuvier, 1797
Subclass	Prosobranchia Milne-Edwards, 1848
Order	Architaenioglossa
Family	Viviparidae Gray, 1847
Subfamily	Bellamyinae Rohrbach, 1937
Genus	*Cipangopaludina Hannibal*, 1912
Species	*Cipangopaludina chinensis* (Gray 1834)

The Chinese mystery snail (*Bellamya chinensis*)—also known as the black snail, trapdoor snail, or Chinese vivipara—is a species of freshwater snail within the family Viviparidae (Figure 4.1). This species has an inflated shell, and it is large compared to Massachusetts' other freshwater snails, reaching up to 63 mm (2.5 inches) in height, or slightly larger than a golf ball (Figure 4.2). The color of the shell is typically solid olive green to brown and the nacre (inner coloration) of the shell is white to pale blue (Clarke 1981). The aperture of the snail's shell is covered by a reddish operculum.

The specific nomenclatures within and surrounding Viviparidae are not well-determined, and morphological and molecular analyses to this date suggest that groupings within the family and the larger Caenogastropoda clade are paraphyletic, meaning some species have not been correctly placed on the evolutionary tree according to their true relatedness (Lu 2014; Wang et al. 2017). The species known by the common name Chinese mystery snail is referred to by many synonymous genus and species names (*Bellamya chinensis*, *Cipangopaludina chinensis*, *Viviparus malleatus*, and many more names including subspecies designations), as its taxonomic placement has been the subject of debate (USGS 2020). For the purposes of this report, the invasive snail in question will be scientifically referred to as *Bellamya chinensis*.

Despite complexities about the species' placement in Viviparidae, *B. chinensis* are distinct from South America "mystery snails" or "apple snails" in the genus *Pomacea* (family Ampullaridae), which are endemic to South America and have become invasive in China since their introduction in the 1980s, causing agricultural and environmental damage (Yang 2018). Snails in both the genus *Pomacea* and *Bellamya* are commonly sold and distributed as pets for home aquaria and both are often labeled as "mystery snails" in the aquarium trade. Individual snails from both species originating from the home aquarium trade can vary in color and appearance, as they are often selectively bred by hobbyists to achieve certain shell and body colorations. Chinese and

DOI: 10.1201/9781003201106-4

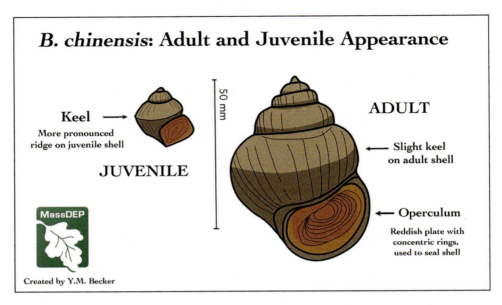

FIGURE 4.1   Adult and juvenile Chinese mystery snail. (Photo by Y.M. Becker.)

South American "mystery snails" are most easily distinguished by the shape of their shell: South American mystery snails have more inflated shells, with a proportionately large body whorl compared to their apical whorls, whereas Chinese mystery snails (*Bellamya*) have more conical shells with more evenly sized whorls (Figure 4.2). *Bellamya* snails are also much more tolerant of cold water than Pomacea apple snails and so they pose a greater threat of invasion to cold Massachusetts waterways.

It is also important to distinguish Chinese mystery snails and banded mystery snails (*Viviparus georgianus*), which are native to North America and can be found from Florida to the St. Lawrence river in Canada (Clench and Fuller 1965). The two species can be difficult to tell apart, since their shells are a very similar shape with only very minute anatomical differences between them. The easiest ways to distinguish these two species are by size and coloration. Banded mystery snails do not grow as large as Chinese mystery snails, the former reaching about 33 mm, only half the maximum size of a fully grown Chinese mystery snail (Figure 4.2). The shells of banded mystery snails also often carry a distinctive banded color pattern, with approximately three evenly spaced

dark stripes running along following the spiral of the shell (Clench and Fuller 1965). Chinese mystery snails sometimes possess a striped pattern on their own shells or variations in color, but the stripes on the invasive snails are often not evenly spaced and perpendicular to the shell's spiral, or they may appear as many small pinstripes following the spiral, in contrast to the distinctive three bands found on *V. georgianus*.

North American native *Campeloma* species also have a similar appearance to the Chinese mystery snail, and grow larger than *V. georgianus*, but *Campeloma spp.* tend to have shouldered flattened whorls with a longer, narrower shape to their shells than *Bellamya* (Clench and Fuller 1965), and *Campeloma* snails also have a very wide, flat foot compared to *Bellamya* (visible when alive and active). Juvenile Chinese mystery snails have notably sharper carinae (keel) on their shells than adult snails, which could be used to distinguish juvenile *B. chinensis* from visually similar snails that are smaller at the adult stage (Figure 4.1).

Chinese mystery snails can be found in a variety of freshwater environments, such as lakes, ponds, streams, canals, and roadside ditches (Jokinen 1982). Like other members of the Viviparidae, Chinese mystery snails are

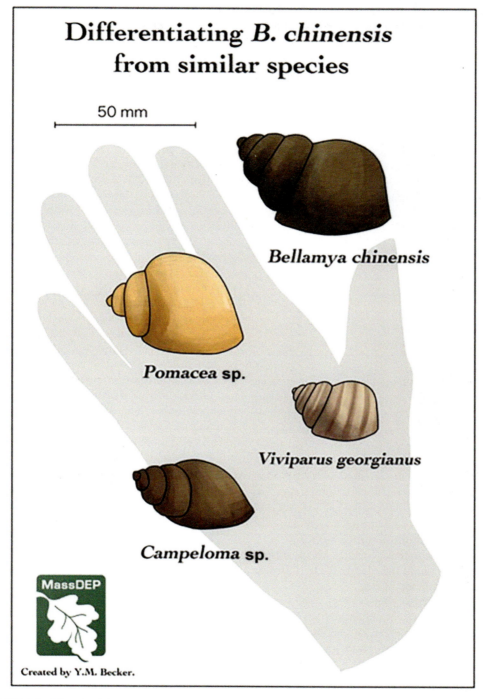

FIGURE 4.2 Visual guide to differentiating *B. chinensis* from visually similar species that may be found in North America (*Pomacea* sp., *Viviparus georgianus*, and *Campeloma* sp.). (Photo by Y.M. Becker.)

ovoviviparous, giving birth to live offspring (Waltz 2008; Martin 1999). Apart from being a viviparous animal (giving live birth) and reproducing sexually (Waltz 2008), it is also dioecious, meaning individual snails are reproductively either male or female. The females can live up to five years, while males can live for three to four years (Jokinen 1982). From the research of Jokinen (1982), it is believed that a female can produce more than 169 embryos over a lifetime. The embryos are usually released as live offspring between June and October, and the snail will spend the winter in the deeper water after reproduction (Jokinen 1982). Chinese mystery snails can tolerate very cold and even freezing water temperatures (Burnett et al. 2018). The snails can live on a variety of different substrate types. They primarily eat by grazing periphyton, but larger adult snails are also capable of filter-feeding (Haak 2015).

Native and Invasive Range, Mechanisms of Spread

Bellamya chinensis is native to Burma, Thailand, South Vietnam, China, Korea, eastern Russia, Japan, Philippines, and Java (Waltz 2008; Kipp 2020). The first documented sighting in the United States was in the Chinese markets in San Francisco in the late 1800s (Wood 1892). The species was most likely first imported as a food source, but were released into freshwater habitats. Since entering natural waterbodies, the invasive range of this species has expanded to many areas in the United States, especially the northeast and the midwest regions (Kipp 2020). In 1914, *B. chinensis* was found in Massachusetts for the first time, in the Muddy River in Boston (Johnson 1915; Kipp 2020).

The spread of *B. chinensis* in some regions of the country has been enhanced by boat access between lakes and rivers, suggesting boating as a vector of dispersion for this species. Experimentation has shown that these snails can survive upward of nine weeks exposed to air, theoretically allowing the snails to cling to the bottoms of boats for long periods of time and travel between bodies of water when the boats are transported by land (Unstad 2013; Havel 2011). Since Chinese mystery snails are also frequently sold as pets for home aquaria, it is possible that they are released outdoors at new locations either accidentally or intentionally by individuals who have purchased them from pet stores.

Ecological Impacts

The influence of Chinese mystery snail on native species has varied among studies. Competition between *B. chinensis* and native snails might depend on the population densities or spatial scale of analysis, and on whether other invasive species or environmental stressors are exerting additional pressure (Johnson et al. 2009; Solomon et al. 2010). A study by Johnson et al. (2009) simulated freshwater ecosystems by building mesocosms stocked with native and invasive species and found that a high density of *B. chinensis* stagnates or decreases the growth of native snail species *Physa gyrina* and *Lymnaea stagnalis*, even leading to the elimination of certain species in the experimental mesocosms when compounded with the presence of the invasive rusty crayfish *Orconectes rusticus*. However, a field survey of 42 lakes published the following year (Solomon et al. 2010) found no increased likelihood of native species loss where *B. chinensis* is present. Twardochleb and Olden (2016) found that when native snail populations are reduced in lakes as a result of human environmental degradation, lakes with introduced populations of *B. chinensis* may be more ecologically stable, as *B. chinensis* could provide molluscivorous predators, including several fish species, with an alternative food source. *B. chinensis* might also serve as food more favorably suited for native crayfish than for invasive crayfish (Olden et al. 2009).

Through examination of water chemistry in their mesocosms and the excretory habits of *B. chinensis*, Johnson et al. (2009) discovered that *B. chinensis* might have an innate ability to sequester or serve as a "sink" for phosphorus, thus raising the N:P ratio of a habitat where it has been introduced. It is generally understood that a high N:P ratio is associated with an oligotrophic ecosystem and a low N:P ratio with a eutrophic one (Downing 1992; Schindler 2008). Atomic nitrogen and phosphorus both are required in a certain proportion for plant and algal growth, and phosphorus is typically the limiting factor in exponential growth. Likewise, reduced phosphorus concentrations lead to reduced algal growth and bloom frequencies (Walker and Havens 1995).

Therefore, the phosphorus sequestering ability of *B. chinensis* may potentially be useful in reversing the effects of eutrophication in heavily polluted waterbodies, though not without the risk of posing as competition to native snail species.

B. chinensis might also serve as a vector for parasites. In the Boston area specifically, *B. chinensis* has been found to carry *Aspidogaster conchiola*, a parasite that begins its life inside mollusks and then moves on to fish (Michelson 1970). *A. conchiola* is not known to be dangerous to humans or to fish in most cases (Petty and Francis-Floyd 2015), but its transmission and spread could have ecological consequences. The role and effects of parasites in an ecosystem are often cryptic and difficult to identify, but worth consideration (Poulin 1999; Hechinger and Lafferty 2005). Certain native parasite species may be less capable of infecting invasive snails, leading to a reduction in their numbers along with the loss of native snail species (Lam 2016).

Human Impacts

Some of the parasites spread by Chinese mystery snails can impact human health. Echinostomate parasites (also called flukes or trematodes) have been found to use Chinese mystery snails as hosts (Sohn et al. 2013). These parasites can infect humans if ingested via raw or undercooked mollusks or fish and cause gastrointestinal sickness and damage to the intestines (CDC 2019). Many species of snails, including *B. chinensis*, can also carry *Angiostrongylus cantonensis*, also known as rat lungworm (Chen 1991). *A. cantonensis* affects the central nervous system, typically causing symptoms that last for two to eight weeks, including headaches and nausea. Though serious complications are uncommon, the illness occasionally leads to severe meningitis and even death (CDC 2015). Due to the severity of parasitic illnesses transmitted by gastropods, snails collected from the wild should not be eaten or used as bait (to avoid passing the parasite to fish or birds, which can serve as intermediate or primary hosts and perpetuate their life cycles).

The large number of dead snails or decaying shells on the wrack line at an infested waterbody may diminish the aesthetic value, as observed at heavily infested sites (Bury et al. 2007; Collas et al. 2017). However, these waste shells could have economic and medical value as well:

researchers have found that the shells of Chinese mystery snails can be recycled and used for the synthesis of hydroxyapatite, a bioceramic material with important biomedical applications, including bone repair and regeneration (Zhou et al. 2016). Chemical derivatives from the flesh of the Chinese mystery snail could also have biomedical benefits, namely, polysaccharides with anti-inflammatory and anti-angiogenic properties. These compounds could possibly have applications in treating damage caused by immunological disease, or even as a treatment for cancer, blocking the growth of tumors (Xiong et al. 2017; Jiang et al. 2013).

MONITORING AND SIGHTINGS

The invasive species sightings data were collected by the Massachusetts Department of Environmental Protection (MassDEP) and US Geological Survey's non-indigenous aquatic species database (USGS 2020). MassDEP implemented the standard operating procedures, including sampling methodology, species identification, quality control, and quality assurance, during the sample collections and species identification and quantification (Nuzzo 2003). Lake samples collected from 2016 to 2018 were composites of ten littoral zone samples taken at even intervals around each waterbody (so these records convey information about a whole waterbody but not a specific point within the waterbody). Information collected by USGS is from journal articles and the database *Exotic Aquatic Mollusk Collections in the Illinois Natural History Survey*. Sites with *B. chinensis* were mapped as a data layer with ArcGIS® ArcMap 10.1 (ESRI, Redlands, California). This tool is able to pinpoint the infestations and potential for further spreading areas, which can help to inform future decisions and regulations for controlling the spread of this invasive species (Figure 4.3).

B. chinensis was first documented in Massachusetts in Muddy River which is part of the Charles River Watershed in 1914 (Figure 4.3). Currently, there are 37 known waterbodies infested by *B. chinensis* (Tables 4.1 and 4.2) in 16 watersheds in Massachusetts (Table 4.3). The potential economic loss and ecological impact need to be estimated according to different invasive status at different waterbodies. An effective control method should be applied

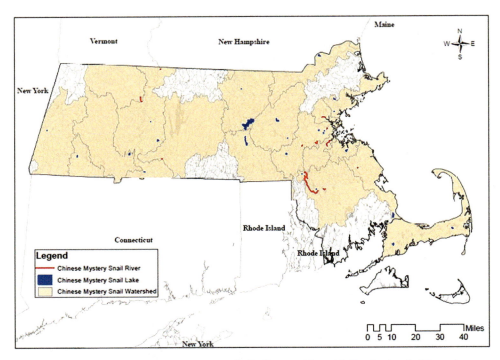

FIGURE 4.3 Map depicting Chinese mystery snail (*Bellamya chinensis*) distribution in Massachusetts.

for the infested waterbody. For waterbodies that have not been invaded or are under the risk of being invaded, prevention measures should be implemented to stop the invasion of the species. More comprehensive monitoring on *B. chinensis* should be developed for a better understanding of their infestation status. This can enhance the public awareness of the presence of invasive mystery snail in Massachusetts lakes and rivers and help to minimize their further spread to other waterbodies in Massachusetts and beyond.

MANAGING ACTIONS

The management of Chinese mystery snail usually includes two major components: prevention and control. Prevention measures usually include boat inspection and decontamination before and/or leaving a waterbody, regulation on trading aquarium pets, and general public education. Early detection and response measures need to be developed to monitor the establishment and spread of *B. chinensis* when it is introduced to new locations. Innovative eDNA (Environmental

DNA) analyses have been graded as an effective technique for the early detection of *B. chinensis* and other mollusk species, and this method can also provide an estimate of the quantity of snails (Peñarrubia et al. 2016; Matthews et al. 2017).

The control mainly involves chemical treatment and physical removal, as well as biological control. The best method for controlling Chinese mystery snail populations depends on the region, climate, management budget, regulation restrictions, and other factors.

CHEMICAL CONTROL

Rotenone and copper sulfate are considered by Haak et al. (2014) as two main chemical treatment options to consider for eradicating *B. chinensis*. Rotenone is typically used to eliminate fish, but it is also generally toxic to insects and most invertebrates. In an experiment by Haak et al. (2014), adult Chinese mystery snails were placed in buckets of lake water treated with rotenone (at twice the dosage recommended to kill fish) for 72 hours, after which 100% of the

TABLE 4.1

Sightings and Report Sources of Infested Waterbodies

Waterbody	Agency	Report_Source	Longitude	Latitude	First Year Reported
Boons Pond	MassDEP	Robert Nuzzo	42.39611	−71.494999	2017
Card Pond	MassDEP	Robert Nuzzo	42.3261883	−73.366693	2016
Charles River		USGS 2020	42.25301	−71.17261	1995
Chicopee River		USGS 2020	42.15583	−72.43295	1997
Coonamessett Pond	MassDEP	Robert Nuzzo	41.6192836	−70.566974	2018
Crystal Lake at 113 Crystal Lake Road, private residence		USGS 2020	42.79965	−71.15197	2017
Emmanuel Pond and the Muddy River		E.H. Michelson, 1970	42.33232	−71.15753	1968
Forest Park Ponds		USGS 2020	42.0736	−72.57345	1986
Forge Pond		USGS 2020	42.27004	−72.46934	2001
Fuller Swamp Brook		USGS 2020	42.53564	−72.61078	1977
Gleason Pond	MassDEP	Robert Nuzzo	42.2865424	−71.412718	2017
Great Herring Pond		USGS 2020	41.79984	−70.55985	2015
Hawes Pond		William James Clench and Samuel L.H. Fuller, 1965	42.063432	−71.297832	1965
Hinckleys Pond	MassDEP	Robert Nuzzo	41.7113503	−70.086067	2018
Indian Head River	MassDEP	Robert Nuzzo			1983
Jamaica Pond		William James Clench and Samuel L.H. Fuller, 1965	42.316764	−71.12033	1965
Lake Quinsigamond		William James Clench and Samuel L.H. Fuller, 1965	42.2709267	−71.75507	1918
Lake Winthrop		USGS 2020	42.1889	−71.42252	1992
Leverett Pond		William James Clench and Samuel L.H. Fuller, 1965	42.329264	−71.113386	1965
Lower Spectacle Pond	MassDEP	Robert Nuzzo	42.163092	−73.11985	2016
Muddy River		William James Clench and Samuel L.H. Fuller, 1965	42.351486	−71.092552	1914
Muddy River		D.S. Dundee, 1974	42.597355	−71.787114	1974
Pauchaug- Louisiana Brook		USGS 2020	42.7185	−72.44828	1978
Peacock Pond		William James Clench and Samuel L.H. Fuller, 1965	41.967404	−71.18278	1965
Pequot Pond	MassDEP	Robert Nuzzo	42.1813088	−72.69846	2016

(Continued)

TABLE 4.1
(Continued)

Waterbody	Agency	Report_Source	Longitude	Latitude	First Year Reported
Pillings Pond		William James Clench and Samuel L.H. Fuller, 1965	42.529872	−71.028671	1965
Pine Tree Brook		USGS 2020	42.25912	−71.08398	1982
Pine Tree Brook and pond		William James Clench and Samuel L.H. Fuller, 1965	42.245099	−71.089495	1965
Pond at Medford		USGS 2020	42.41866	−71.11248	1974
Pond at Mount Hood		William James Clench and Samuel L.H. Fuller, 1965	42.451763	−71.032551	1965
Snake River		USGS 2020	41.96715	−71.12613	1995
Trout Brook		William James Clench and Samuel L.H. Fuller, 1965	42.242137	−71.293625	1965
Turners Pond		William James Clench and Samuel L.H. Fuller, 1965	42.26121	−71.078106	1965
UMass pond		USGS 2020	42.38961	−72.52659	1973
Wachusett Reservoir		USGS 2020	42.40318	−71.68854	1988
Wading River	MassDEP	Robert Nuzzo			2006
Whitmans Pond		William James Clench and Samuel L.H. Fuller, 1965	42.206269	−70.934944	1965

snails survived, suggesting that *B. chinensis* has some resistance to rotenone. In experiments measuring rotenone toxicity on other species of aquatic snails, rotenone was found to cause mortality within hours of exposure at standard dosage (Dalu et al. 2015) and over the course of multiple days at a low dosage (Vehovsky et al. 2007). In Dalu et al.'s (2015) study, several different species of aquatic invertebrates were monitored in a system exposed to rotenone at the typical dosage and timescale recommended to achieve 100% fish mortality. In this experiment, some small aquatic snails such as *Physa acuta* were killed, but the snails did not see as high mortality as other aquatic invertebrates in the study, supporting Haak et al.'s (2014) conclusion that short-term exposure to rotenone is not an effective pesticide treatment for aquatic snails. The long-term effects of rotenone treatment on *B. chinensis* specifically have not been evaluated, but, considering rotenone's toxicity to

other forms of aquatic life, long-term application of the pesticide is not likely to be an ecologically sound option.

Copper sulfate, along with other copper compounds, is well-known as a molluscicide, specifically used to eradicate unwanted snails. Copper sulfate's effectiveness in killing Chinese mystery snails has been reported inconsistently in different cases, and its toxicity may depend on the method of exposure (i.e., whether snails ingest the toxic metal or are only exposed to dissolved copper in the water column). Parallel to the rotenone experiment, Haak et al. (2014) exposed adult Chinese mystery snails to copper sulfate solutions in buckets of water, observing the snails over 96 hours. At the end of this experiment, all but one of the snails survived, leading the team to conclude that *B. chinensis* are resistant to copper sulfate. However, there is evidence that copper sulfate is more effective as a molluscicide if it is ingested by snails,

TABLE 4.2

Towns and Infested Waterbodies

Number	Waterbody	Town
1	Boons Pond	Stow
2	Card Pond	West Stockbridge
3	Charles River	Dedham
4	Chicopee River	Wilbraham
5	Coonamessett Pond	Falmouth
6	Crystal Lake at 113 Crystal Lake Road private residence	Haverhill
7	Emmanuel Pond and the Muddy River	Brookline
8	Forest Park Ponds	Springfield
9	Forge Pond	Granby
10	Fuller Swamp Brook	Deerfield
11	Gleason Pond	Framingham
12	Great Herring Pond	Plymouth and Bourne
13	Hawes Pond	Dover
14	Hinckleys Pond	Harwich
15	Indian Head River	Hanover and Hanson
16	Jamaica Pond	Boston
17	Lake Quinsigamond	Worcester
18	Lake Winthrop	Holliston
19	Leverett Pond	Boston
20	Lower Spectacle Pond	Sandisfield
21	Muddy River	Boston
22	Muddy River	Fitchburg
23	Pauchaug-Louisiana Brook	Northfield
24	Peacock Pond	Lexington
25	Pequot Pond	Westfield
26	Pillings Pond	Lynnfield
27	Pine Tree Brook	Milton
28	Pine Tree Brook and pond	Milton
29	pond at Medford	Medford
30	pond at Mount Hood	Melrose
31	Snake River	Norton
32	Trout Brook	Dover
33	Turners Pond	Milton
34	Umass pond	Amherst
35	Wachusett Reservoir	Clinton
36	Wading River	Norton
37	Whitman's Pond	Weymouth

as several studies have found that toxic metal uptake in both land and aquatic snails is greater when the metals are concentrated in soil, sediment, or food rather than dissolved in water (Hoang and Rand 2009; Gomot-de Vaufleury and Pihan 2009; Pang et al. 2012). This would also explain a case in the fall of 2010, where copper sulfate crystals were experimentally applied

TABLE 4.3

Infested Watersheds

Number	Infested Watersheds
1	Blackstone
2	Boston Harbor (Neponset River Basin and Coastal Drainage Area)
3	Cape Cod
4	Charles
5	Chicopee
6	Concord (SuAsCo)
7	Deerfield
8	Farmington
9	Housatonic
10	Connecticut
11	Merrimack
12	Narragansett
13	Nashua
14	South Coastal
15	Taunton
16	Westfield

in two heavily snail-infested bodies of water, successfully killing large numbers of *B. chinensis*. Two ponds in Jackson County, Oregon, were treated with copper sulfate at a dosage of about 2 ppm. Reportedly, over 30,000 dead Chinese mystery snails were retrieved from the ponds within a month as a result of this action, but some live snails were still found, indicating that eradication was not complete. There has also not been any information reported on the broader environmental impact of the copper sulfate treatment, effects on soil or water chemistry, or whether the invasive snail populations have recovered in the decade following 2011, at which point their populations had recovered and required another round of pesticide dosage, according to a local news report (Freeman 2011). Regardless, this story is a sign that copper sulfate treatment is worth consideration as a tool for eradicating *B. chinensis*. Further studies and experimentation are required to learn definitively whether copper compounds can be used to kill Chinese mystery snails, as well as to find the most effective methods and dosages. More information about chemical control on Chinese mystery snail and other 18 invasive species can be found in Appendix III.

PHYSICAL CONTROL

Some physical strategies might be effective in controlling *B. chinensis*, including thermal treatment, manual removal, and hydrologic controls. Hot water treatment has shown potential to be useful against *B. chinensis*. Burnett et al. (2018) tested the thermal tolerance limits of the snail by exposing wild-caught *B. chinensis* to extreme high and low temperature conditions. While they found no lower temperature range, indicating that the snail is well-adapted to cold climates and capable of withstanding freezing cold water, *B. chinensis* was unable to tolerate hot water in the range of 40–45°C. Targeted hot water treatment might therefore be useful as a method of eradicating *B. chinensis*, if more study and experimentation are carried out to find safe and effective implementation. There is also at least one recorded instance of natural *B. chinensis* die-offs occurring following periods of especially hot weather (Haak et al. 2013).

Removal by hand could potentially reduce *B. chinensis* populations; larger adult Chinese mystery snails tend to produce more offspring each season, so selectively removing the largest snails from a waterbody could be a relatively easy task with an outcome of slowing population growth (Haak 2015).

Drawdown or treatments that involve exposing the snails to air for long periods of time seem to be effective against *B. chinensis* juveniles, but not adult snails. In a study by Unstad et al. (2013), adult and juvenile *B. chinensis* snails, along with native snail species, were subjected to desiccation for periods of five to nine weeks at a time. Most of the adult *B. chinensis* snails survived desiccation for as long as nine weeks, while none of the juvenile *B. chinensis* or native snails survived even as long as five weeks. Drawdown treatment therefore might be useful in eradicating juvenile snails (in addition to any native small snail present) but have little effect on adult *B. chinensis*. This study also illuminates the fact that since juveniles of *B. chinensis* cannot tolerate long periods of exposure, there is less evidence to say that juvenile snails are able to be transmitted between waterbodies by attaching to boats and withstanding long trips over land. Watercraft owners might take this into account and note that after storing a boat for a long period of time, adult snails are more likely to remain attached to the hull than juveniles. Physical prevention, targeting the vectors that *B. chinensis* may be using to reach new habitats, could be an effective response to control infestation. These strategies could include encouraging freshwater boaters to drain, dry, and clean their watercraft, checking carefully to ensure that no snails are attached while in transport (Matthews et al. 2017). Educating fish and aquarium owners about invasive species might also prevent introduction of Chinese mystery snails and other aquatic invaders. Pet stores could be asked to distribute informational material instructing against release with the purchase of a known invasive species.

It may also be possible to slow or stop the migration of snails between connected waterbodies by establishing high-velocity currents as boundaries. Some research has found that Chinese mystery snails' movement can be limited by strategic placements of culverts (Rivera 2008). Similarly to the studies on air exposure, larger snails seem more tolerant of high-velocity currents and more capable of moving in spite of them (Rivera 2008). Creating structures that induce high-velocity currents could help control invasion by Chinese mystery snails but could

affect other aquatic wildlife that should be considered and mitigated.

BIOLOGICAL (ECOLOGICAL) CONTROL

While *B. chinensis* is a good food source for native and invasive crayfish, possibly being more accessible as prey to native crayfish than to invasive species (Olden et al. 2009), there have been no documented attempts of introducing predators to control Chinese mystery snails and no research seeking to find the effectiveness of biological control on this species.

SUCCESS STORY

So far, no method is known to consistently be effective against *B. chinensis*, and there are no recorded instances of complete or near-total eradication of the species from an ecosystem that has been invaded. However, as already mentioned in the "Chemical Control" section, a period of high mortality was once caused by an applications of copper sulfate in Oregon ponds, according to local press and newsletters (Edwards 2010; Freeman 2010).

In 2009, the Chinese mystery snail was found in two ponds of the Jackson County Sports Park, Oregon, believed to have been released by someone who purchased them through the aquarium trade (Edwards 2010). In the fall of 2010, Oregon Department of Fish and Wildlife (ODFW) detected the new infestation in Klamath County's Lost River, Crane Prairie Reservoir, and Big Butte Pond near Bend, and decided to respond with an experimental chemical treatment in two infested ponds. The more heavily infested pond, with 166 snails/m^2, was dosed with a total of 47.5 pounds of copper sulfate. The pond with a lighter infestation was treated with only 13 pounds. Following the pesticide application, around 30,000 dead snails were collected from the ponds, but live snails were still caught in traps following the application. According to a *Mail Tribune* report, the snails returned the following year, leading ODFW to repeat the pesticide trial with a higher dosage than before (Freeman 2011). In 2011, the sale of Chinese mystery snails was also officially banned in the state of Oregon (ODFW 2011).

Altogether, the chemical treatment caused substantial, but not complete, mortality in the ponds where it was applied. Though this experiment was carried out over ten years ago and is sparsely documented, the circumstances and results of the trial being recorded only in a handful of local news sources, the experiment remains to date the most successful attempted eradication of Chinese mystery snails from a North American waterbody. These results suggest that copper sulfate may be useful in controlling Chinese mystery snail invasions, if applied in solid form at the right dosage, and perhaps combined with other control methods.

REFERENCES

Burnett, J.L., K.L. Pope, A. Wong, C.R. Allen, D.M. Haak, B.J. Stephen, and D.R. Uden. 2018. Thermal tolerance limits of the Chinese mystery snail (*Bellamya chinensis*): Implications for management. *American Malacological Bulletin* 36(1): 140–144.

Bury, J.A., B.E. Sietman, and B.N. Karns. 2007. Distribution of the Non-Native Viviparid Snails, *Bellamya chinensis* and *Viviparus georgianus*, in Minnesota and the First Record of *Bellamya japonica* from Wisconsin. *Journal of Freshwater Ecology* 22(4): 697–703. DOI: 10.1080/02705060.2007.9664830.

Center for Disease Control and Prevention. 2015. Angiostryliasis (also known as *Angiostrongylus* Infection). www.cdc.gov/parasites/angiostrongylus/disease.html (Accessed January 14, 2021).

Center for Disease Control and Prevention. 2019. Echinostomiasis. www.cdc.gov/dpdx/echinostomiasis/index.html (Accessed January 14, 2021).

Chen, E.R. 1991. Current status of food-borne parasitic zoonoses in Taiwan. *Southeast Asian Journal of Tropical Medicine and Public Health* 22: 62–64.

Clarke, A.H. 1981. The freshwater molluscs of Canada. In: *National Museum of Natural Sciences*. Ottawa: National Museums of Canada, p. 447.

Clench, W.J., and S.L.H. Fuller. 1965. The genus *Viviparus* (Viviparidae) in North America. *Harvard Univ. Mus. of Comp. Zool. Occ. Pap* 2(32): 385–412.

Collas, F.P.L., S.K.D. Breedvold, J. Matthews, G. van der Velde, and R.S.E.W. Leuven. 2017. Invasion biology and risk assessment of the recently introduced Chinese mystery snail, *Bellamya (Cipangopaludina) chinensis* (Gray, 1834), in the Rhine and Meuse River basins in Western Europe. *Environmental Sciences* 557: 1–75.

Dalu, T., R.J. Wasserman, M. Jordaan, W.P. Froneman, and O.L.F. Weyl. 2015. An Assessment of the effect of rotenone on selected non-target aquatic fauna. *PLoS One* 10(11): e0142140.

Downing, J.A., and E. McCauley. 1992. The nitrogen: Phosphorus relationship in lakes. *Limnology and Oceanography* 37(5): 936–945.

Dundee, D.S. 1974. Catalog of introduced mollusks of eastern North America (north of Mexico). *Sterkiana* 55: 1–37.

Edwards, R. 2010. Lake Wise: A voice for Quiet Waters. *The Oregon Lakes Association Newsletter*. www.oregonlakes.org/Resources/Documents/Lakewise/2010_Nov.pdf (Accessed May 15, 2020).

Freeman, M. 2010. This snail problem is hardly a mystery. *Mail Tribune*, August 12. https://mailtribune.com/oregon-outdoors/this-snail-problem-is-hardly-a-mystery (Accessed January 26, 2021).

Freeman, M. 2011. Invasive snails return to White City ponds. *Mail Tribune*, May 28. https://mailtribune.com/archive/invasive-snails-make-return-to-white-city-ponds (Accessed January 26, 2021).

Gomot-de Vaufleury, A., and F. Pihan. 2009. Methods for toxicity assessment of contaminated soil by oral or dermal uptake in land snails: Metal bioavailability and bioaccumulation. *Environmental Toxicology* 21(4): 820–827.

Haak, D.M. 2015. Bioenergetics and habitat suitability models for the Chinese mystery snail (*Bellamya chinensis*). University of Nebraska-Lincoln, Nebraska, p. 234.

Haak, D.M., N.M. Chaine, B.J. Stephen, A. Wong, and C.R. Allen. 2013. Mortality estimate of Chinese mystery snail, *Bellamya chinensis* (Reeve, 1863) in a Nebraska reservoir. *Biological Invasions* 2(2): 137–139.

Haak, D.M., B.J. Stephen, R.A. Kill, N.A. Smeenk, C.R. Allen, and K.L. Pope. 2014. Toxicity of copper sulfate and rotenone to Chinese mystery snail (*Bellamya chinensis*). *Management of Biological Invasions* 5: 371–375.

Havel, J.E. 2011. Survival of the Exotic Chinese Mystery Snail (*Cipangopaludina chinensis malleata*) during Air Exposure and Implications for Overland Dispersal by Boats. *Hydrobiologia* 668(1): 195–202.

Hechinger, R.F., and K.D. Lafferty KD. 2005. Host diversity begets parasite diversity: bird final hosts and trematodes in snail intermediate hosts. *Proc Biol Sci* 272(1567): 1059–1066.

Hoang, T.C., and G.M. Rand. 2009. Exposure routes of copper: Short term effects on survival, weight, and uptake in Florida apple snails (*Pomacea paludosa*). *Chemosphere* 76(3): 407–414.

ITIS (Integrated Taxonomic Information System). 2021. www.itis.gov/servlet/SingleRpt/SingleRpt?search_topic=TSN&search_value=18408#null (Accessed September 28, 2021).

Jiang, C., Y. Jiao, X. Chen, X. Li, W. Yan, B. Yu, and Q. Xiong. 2013. Preliminary characterization and potential hepatoprotective effect of polysaccharides from *Cipangopaludina chinensis*. *Food and Chemical Toxicology* 59: 18–25.

Johnson, C.W. 1915. *Viviparus malleatus* Reeve in Massachusetts. *The Nautilus* 29: 35.

Johnson, P.T.J., J.D. Olden, C.T. Solomon, and M.J. Vander Zanden. 2009. Interactions among invaders: community and ecosystem effects of multiple invasive species in an experimental aquatic system. *Oecologia* 159: 161–170.

Jokinen, E.H. 1982. *Cipangopaludina chinensis* (Gastropoda: Viviparidae) in North America, review and update. *Nautilus* 96(3): 89–95.

Jokinen, E.H. 1992. *The Freshwater Snails (Mollusca: Gastropoda) of New York State*. Albany, NY: The University of the State of New York, The State Education Department, The New York State Museum, p. 112.

Kipp, R.M., A.J. Benson, J. Larson, A. Fusaro, and C. Morningstar C. 2020. *Cipangopaludina chinensis* (Gray, 1834): U.S. Geological Survey, Nonindigenous Aquatic Species Database, Gainesville, FL. https://nas.er.usgs.gov/queries/FactSheet.aspx?SpeciesID=1044, Revision Date: 1/15/2020 (Accessed May 13, 2020).

Lake George Association (LGA). n.d. www.lakegeorgeassociation.org/educate/science/lake-george-invasive-species/chinese-mystery-snail/ (Accessed May 13, 2020).

Lam, J. 2016. *Cipangopaludina Chinensis: Effects of Temperatures and Parasite Prevalence*. Notre Dame: University of Notre Dame.

LaRoe, E.T., G.S. Farries, C.E. Puckett, P.D. Doran, and M.J. Mac. 1995. *Our Living Resources: A Report to the Nation on the Distribution, Abundance, and Health of US Plants, Animals, and Ecosystems. US Department of the Interior*. Washington, DC: National Biological Service, pp. 205–209.

Lu, H., L. Du, Z. Li, X. Chen, and J. Yang. 2014. Morphological analysis of the Chinese *Cipangopaludina* species (Gastropoda; Caenogastropoda: Viviparidae). *Zoological Research* 35(6): 510–527.

Martin, S.M. 1999. Freshwater snails (Mollusca: Gastropoda) of maine. *Northeastern Naturalist* 6(1): 39–88.

Matthews, J., F.P.L. Collas, L. Hoop, G. van der Velde, and R.S.E.W. Leuven. 2017. Management approaches for the Alien Chinese mystery snail (Bellamya Chinensis). *Environmental Sciences* 558: 1.

McCann, M.J. 2014. Population dynamics of the non-native freshwater gastropod, *Cipangopaludina chinensis* (Viviparidae): A capture-mark-recapture study. *Hydrobiologia* 730: 17–27.

Michelson, E.H. 1970. *Aspidogaster conchicola* from fresh water gastropods in the USA. *Journal of Parasitology* 56(4): 709–712.

Nuzzo, R.M. 2003. *Standard Operating Procedures: Water Quality Monitoring in Streams Using Aquatic Macroinvertebrates*. Worcester, MA: Massachusetts Department of Environmental Protection, Division of Watershed Management, p. 39.

Olden, J.D., E.R. Larson, and M.C. Mims. 2009. Home-field advantage: Native signal crayfish (*Pacifastacus leniusculus*) out consume newly introduced crayfishes for invasive Chinese mystery snail (*Bellamya chinensis*). *Aquatic Ecology* 43: 1073.

Oregon Department of Fish and Wildlife (ODFW). 2011. Oregon Fish and Wildlife Commission Minutes, February 4, p. 15. www.dfw.state.or.us/agency/commission/minutes/11/08_aug/Approved_Feb%204%202011_Commission%20Minutes_080511.pdf (Accessed January 25, 2021).

Pang, C., H. Selck, S.K. Misra, D. Berhanu, A. Dybowska, E. Valsami-Jones, and V.E. Forbes. 2012. Effects of sediment-associated copper to the deposit-feeding snail, *Potamopyrgus antipodarum*: A comparison of Cu added in aqueous form or as nano- and micro- CuO particles. *Aquatic Toxicology* 106–107: 114–122.

Peñarrubia, L., C. Alcaraz, A. bij de Vaate, N. Sanz, C. Pla, O. Vidal, and J. Viñas. 2016. Validated methodology for quantifying infestation levels of mussels in environmental DNA (eDNA) samples. *Scientific Reports* 6: 39067.

Petty, B.D., and R. Francis-Floyd. 2015. Parasitic Diseases of Fish. *MERCK Manual Veterinary Manual*. www.merckvetmanual.com/exotic-and-laboratory-animals/aquarium-fishes/parasitic-diseases-of-fish (Accessed January 14, 2021).

Poulin, R. 1999. The functional importance of parasites in animal communities: Many roles at many levels? *International Journal for Parasitology* 29(6): 903–914.

Rivera, C.J.R. 2008. Obstruction of upstream migration of the invasive snail Cipangopaludina chinensis by high water currents. In: *Summer UNDERC Project. South Bend*. Notre Dame: University of Notre Dame, Department of Biological Sciences, Galvin Life Sciences Center, p. 13.

Schindler, D.W., R.E. Hecky, D.L. Findlay, M.P. Stainton, B.R. Parker, M.J. Paterson, K.G. Beaty, M. Lying, and S.E.M. Kasian. 2008. Eutrophication of lakes cannot be controlled by reducing nitrogen input: Results of a 37-year whole-ecosystem experiment. *PNAS* 105(32): 11254–11258.

Sohn, W.M., J.Y. Chai, B.K. Na, T.S. Yong, K.S. Eom, H. Park, D.Y. Min, and H.J. Rim. 2013. *Echinostoma macrorchis* in Lao PDR: Metacercariae in *Cipangopaludina* snails and adults from experimentally infected animals. *Korean Journal Parasitol* 51(2): 191–196.

Solomon, C.T., J.D. Olden, P.T.J. Johnson, R.T. Dillon, and M.J. Vander Zanden. 2010. Distribution and Community-level Effects of the Chinese Mystery Snail (*Bellamya chinensis*) in Northern Wisconsin Lakes. *Biological Invasions* 12(6): 1591–605.

Twardochleb, L.A., and J.D. Olden. 2016. Non-native Chinese Mystery Snail (*Bellamya chinensis*) Supports Consumers in Urban Lake Food Webs. *Ecosphere* 7(5): E01293.

U.S. Geological Survey (USGS). 2020. Specimen observation data for *Cipangopaludina chinensis* (Gray, 1834), Nonindigenous Aquatic Species Database, Gainesville, FL, https://nas.er.usgs.gov/queries/CollectionInfo.aspx?SpeciesID=1044&State=MA, (Accessed May 13, 2020).

Unstad, K.M., D.R. Uden, C.R. Allen, N.M. Chaine, D.M. Haak, R.A. Kill, K.L. Pope, B.J. Stephen, and A. Wong. 2013. Survival and behavior of Chinese mystery snails (Bellamya chinensis) in response to simulated water body drawdowns and extended air exposure. *Nebraska Cooperative Fish & Wildlife Research Unit—Sta Publications* 123.

Vehovsky, A., H. Szabó, L. Hiripi, C.J.H. Elliott, and L. Hernádi. 2007. Behavioral and neural deficits induced by rotenone in the pond snail *Lymnaea stagnalis*. A possible model for Parkinson's disease in an invertebrate. *European Journal of Neuroscience* 25(7): 2123–2130.

Walker, W.W., and K.E. Havens. 1995. Relating Algal Bloom Frequencies to Phosphorus Concentrations in Lake Okeechobee. *Lake and Reservoir Management* 11(1): 77–83.

Waltz, J. 2008. *Chinese Mystery Snail (Bellamya chinensis) Review*. Washington, DC: Univ. of Washington.

Wang, J., D. Zhang, I. Jakovlić, and W. Wang. 2017. Sequencing the complete mitochondrial genomes of eight freshwater snail species exposes pervasive paraphyly within the Viviparidae family (Caenogastropoda). *PLoS One* 12(7): e0181699.

Wood, W.M. 1892. *Paludina japonica* Mart. for sale in the San Francisco Chinese markets. *Nautilus* 5: 114–115.

Xiong, Q., H. Hao, L. He, Y. Jing, T. Xu, J. Chen, H. Zhang, T. Hu, Q. Zhang, X. Yang, J. Yuan, and Y. Huang. 2017. Anti-inflammatory and Anti-angiogenic Activities of a Purified Polysaccharide from Flesh of *Cipangopaludina chinensis*. *Carbohydrate Polymers* 176: 152–59.

Yang, Q., S. Liu, C. He, and X. Yu. 2018. Distribution and the origin of invasive apple snails, *Pomacea canaliculata* and *P. maculata* (Gastropoda: Ampulliariidae) in China. *Scientific Reports* 8, 1185.

Zhou, H., M. Yang, M. Zhang, S. Hou, S. Kong, L. Yang, and L. Deng. 2016. Preparation of Chinese mystery snail shells derived hydroxyapatite with different morphology using condensed phosphate sources. *Ceramics International* 42(15): 16671–16676.

5 Northern Snakehead (*Channa argus*)

ABSTRACT

Northern Snakehead (*Channa argus*) is a popular food fish throughout its native distribution in Asia, but an aggressive pest in the United States, where its import has been federally prohibited since 2002. Hardy and prolific, this fish has become especially abundant in Maryland's Potomac River watershed. While the northern snakehead has not yet become established in the state of Massachusetts, fishermen, watershed managers, and citizens alike should remain on guard for appearances of northern snakehead, as an unchecked snakehead population in Massachusetts would be both ecologically damaging and profoundly difficult to reverse. This chapter discusses the basic biology of the northern snakehead, the impacts of its spread in North America, and a range of methods currently in use for monitoring and controlling its populations.

INTRODUCTION

Taxonomic Hierarchy (ITIS 2021)

Kingdom	Animalia
Subkingdom	Bilateria
Infrakingdom	Deuterostomia
Phylum	Chordata
Subphylum	Vertebrata
Infraphylum	Gnathostomata
Superclass	Actinopterygii
Class	Teleostei
Superorder	Acanthopterygii
Order	Perciformes
Suborder	Channoidei
Family	Channidae
Genus	*Channa* Scopoli, 1777
Species	*Channa argus* (Cantor, 1842)

Fish in the family Channidae are commonly called "snakeheads" (Figure 5.1) because of their elongated bodies and wide, flat heads, covered in large scales that are similar in appearance to the epidermal scales on the heads of snakes (Courtenay and Williams 2004). All snakeheads have elongated dorsal and anal fins, small tubular nostrils, and a large mouth full of sharp teeth. Depending on the species, snakeheads are either obligate or facultative airbreathers, meaning they either must regularly swim up to the surface and breathe air (obligate) or they have the ability to supplement water breathing with oxygen from the air (facultative) (Courtenay and Williams 2004). Due to this special adaptation, many snakehead species, including the northern snakehead, are capable of surviving out of the water for several days and can wiggle short distances to new waterbodies (Bressman et al. 2019).

Northern snakehead (*Channa argus*) has become established in several states, benefited by its broad temperature tolerance, ranging from 0°C to >30°C (Kramer et al. 2017). This fish is able to withstand freezing winters by lying dormant under the ice, forgoing eating and even respiration for long periods of time. The northern snakehead is an obligate air breather that thrives in shallow freshwater habitats with aquatic vegetation (Orrell and Weigt 2005; Kramer et al. 2017). It can tolerate slightly saline water, but it cannot live in water with a salinity higher than 10 parts per thousand (Courtenay and Williams 2004; Orrell and Weigt 2005).

Snakeheads, including the northern snakehead, generally reproduce in the summer. Northern snakeheads construct nests by clearing out circular areas of vegetation. The uprooted vegetation is often used above or beside the clearing to form cover for the nest, where females will lay bright yellow-orange eggs (Courtenay and Williams 2004; Landis and Lapointe 2010). Male and female northern snakeheads remain stationed together on or near the nest to guard their offspring. Eggs hatch after one to two days, but parents remain with the young for around four weeks (Qin et al. 2012). The young are reportedly

DOI: 10.1201/9781003201106-5

Invasive Snakeheads in North America

Illustrated by Y.M. Becker

Northern Snakehead (*Channa argus*)

Bullseye Snakehead (*Channa marulius*)

Giant Snakehead (*Channa maculata*)

FIGURE 5.1 Invasive snakeheads in the United States. *From top to bottom:* Northern snakehead (USGS 2003; Cermele 2019; GA DNR 2019; Trammel n.d.), Bullseye snakehead (Randall 2015; RAWWFishing 2017; Everglades CISMA 2018), and Giant snakehead (The Phuket News 2019; Raleigh n.d.).

seen schooling together for a prolonged time after leaving the nest, initially accompanied by a parent (Landis and Lapointe 2010). Northern snakeheads are capable of reproducing multiple times in a year, with female northern snakeheads reportedly bearing up to five separate broods per season (Courtenay and Williams 2004). At 2 years old, a female northern snakehead can lay up to 15,000 eggs (Chesapeake Bay Program n.d.). Larger or older individual fish have been found to produce larger numbers of eggs, and the observed lifespan of the species is 8 years (Qin et al. 2012) or up to 15 years (Susan Pasko, personal communication).

The northern snakehead has a fairly distinctive appearance. It is brown, with dark-colored fins, yellowish-white coloration on the lower part of its body, and large dark-brown blotches covering its sides in a repeating pattern. Like all snakehead species, the northern snakehead has elongated anal and dorsal fins on the bottom and top of its body, and it is covered in large glossy scales (Figure 5.1). Juvenile northern snakeheads appear light yellow-green with translucent fins, but begin to develop adult coloration at around four weeks, at which point they are typically 20 mm long (slightly under 1 inch) (Qin et al. 2012).

Northern snakehead are commonly confused with two fish species indigenous to North America: the burbot (*Lota lota*) and the bowfin (*Amia calva*) (U.S. Fish & Wildlife Service 2003). The burbot is a cod relative that inhabits cold freshwater in North America, Europe, and Asia. Burbot have a similar shape and general appearance to the northern snakehead, but can be easily distinguished from their invasive look-alikes by their slimy skin and lack of visible scales on its body. Burbot also tend to have a less defined, more mottled coloration on their body than snakehead, as well as a split dorsal fin and barbels (whiskers) on its mouth (ADFG n.d.). The bowfin is another visually similar species, common across the eastern United States (Fuller 2021). Bowfin can be distinguished from snakehead in several ways. The body of the bowfin is mostly solid-colored, with faint patterning toward the tail (Figure 5.2) (Fuller 2021). The head of a bowfin is smooth, appearing to lack scales, and male bowfin have a singular eyespot on the base of their tail (TPW n.d.). The most definitive feature that can be used to distinguish bowfin from northern snakehead is the shape and size of fins on the underside of its body: while snakeheads have a very long anal fin, mirroring the dorsal fins on their backs, bowfin have a small, short anal fin. The differences between northern snakehead and these two species are depicted and explained in the guide attached to this report (Figure 5.2).

While the northern snakehead is the highest snakehead species of concern in Massachusetts, on account of New England's climactic suitability for its establishment, there are two species of tropical snakehead that pose a risk of invasion to other areas of the United States. The bullseye snakehead (*Channa marulius*) prefers warmer climates, and so it poses a greater threat to southern states, including Florida, where the species has already established (SPDC 2014). Bullseye snakehead are the largest species of snakehead, capable of growing up to 70 inches (1,800 mm) in length (Courtenay and Williams 2004). They are similar in shape to northern snakehead, but have very different coloration, being a much darker, mostly solid brown or orange color, with some individuals having five or six dull blotches along their sides and sparse white speckles along the posterior edges of the blotches (Figure 5.1). Bullseye snakeheads are named for a distinctive "bullseye" mark on the base of their tails, which appears as a black oval rimmed in bright orange, when it is visible (Figure 5.1). The giant snakehead (*Channa maculata*) is another mostly tropical species of snakehead, which poses an invasion risk primarily to Florida and Hawaii (SPDC 2014). It has a very different appearance to that of the northern snakehead, with bright, almost iridescent-looking yellow-green scales arranged in irregular blotchy or striped patterns interspersed with black across the top of its body. It has a thick, very dark black stripe across the length of its body, and a white underside (Figure 5.1). Juvenile giant snakeheads are light yellow-green, with a red tail and two thin black stripes running the length of the body.

NATIVE AND INVASIVE RANGE, MECHANISMS OF SPREAD

The northern snakehead demonstrates hardiness to a variety of environmental conditions and an ability to spawn in large numbers, increasing its abundance at a rapid rate (Love and Newhard 2018; Kusek 2007). Such characteristics are common to species that travel outside their native range and become invasive. Native to China, Korea, and Russia, the fish was first legally imported to North America for sale in the live food fish market and became established in waterways after unauthorized intentional release (Fuller et al. 2019; SPDC

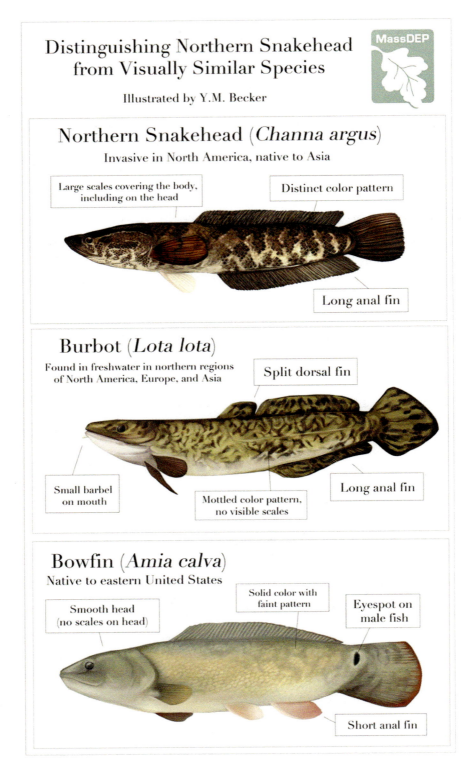

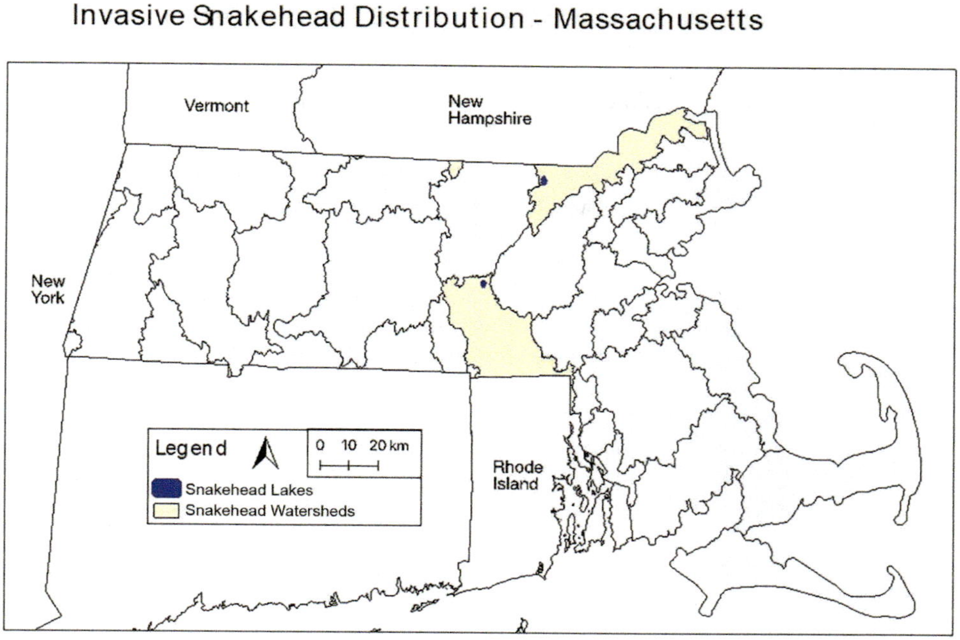

Invasive Snakehead Distribution - Massachusetts

FIGURE 5.3 Distribution map depicting the two waterbodies where northern snakehead (*Channa argus*) has been sighted in Massachusetts.

2014). Other minor contributions to the introduction of northern snakehead include unintentional release during commercial transport of live fish, release associated with the aquarium trade, and release of captive animals as part of religious ceremonies (SPDC 2014).

In 2002, the federal Lacey Act (18 U.S.C. 42) was amended to include all fish in the Channidae family on the list of injurious wildlife. This legislation now prohibits the import and of any snakehead species without a permit, imposing significant fines and a maximum of six months in prison as penalties on those found guilty of injurious wildlife violations (SPDC 2014). Regulation of possession and intrastate transport varies by state. In Massachusetts, it is illegal to own any unpermitted exotic (non-domesticated) wildlife from a species that is not exempt under the conditions of 321 CMR 9.00. Exemption criteria include that "accidental release of the fish . . . will not result in an adverse effect on the ecology of the state," and so possession of snakeheads is therefore illegal in the state of Massachusetts.

The passage and enforcement of state and federal regulations have resulted in fewer introductions of northern snakehead to the United States However, secondary spread from existing established snakehead populations still pose a great threat (SPDC 2014). Introduced northern snakeheads have been sighted in Massachusetts waterways on two different occasions (see "Monitoring and Sightings" section). Once established, a northern snakehead population is capable of rapidly increasing its numbers, as observed in the Potomac River system (Odenkirk and Owens 2011).

The life history of northern snakeheads in their non-indigenous range has in some cases been observed to vary from their typical life history, leading some to assert that snakeheads generally grow larger and faster outside of their native range (Landis et al. 2011). However, a 2014 metanalysis found no difference in northern snakehead growth rates between native and invasive populations, but that growth rates correlate with regional precipitation rates (Rypel 2014). Northern snakeheads may be able to take advantage of high-water or storm events to spread to new waterbodies, and even have the ability to travel overland on account of their air-breathing nature, having been observed to "crawl" slowly over land (Courtenay and Williams 2004; Odenkirk 2018; Bressman et al. 2019). When migrating to new territory through water, snakeheads have a tendency to swim upstream, typically expanding their range in an upstream direction (SPDC 2014). Elevation and precipitation levels may also be predictive factors for risk of northern snakehead invasion (Poulos et al. 2012; Rypel 2014). Middle latitudes (maximum air temperatures of 18–20°C) in the United States are susceptible to its invasion which matches the latitudinal range of its native distribution in China.

A tagging program carried out during 2009–2011 found that the majority (around 90%) of northern snakeheads did not leave the waterbody where they were initially tagged, but the few that did travel were able to travel very far, with one fish being recovered nearly 30 miles upstream (SPDC 2014; Lapointe et al. 2013). The team theorized that the fish could have traveled even further, if they had not been restricted by dams upstream. The majority of movement also seemed to occur in the spring prior to spawning or during high-flow events, and there was no relationship found between movement and fish size or age (SPDC 2014).

ECOLOGICAL IMPACTS

In the Eastern United States, largemouth bass is an ecologically and economically important species, serving as an apex predator in freshwater ecosystems and as a sport-fish commonly sought after by anglers (Ludsin and DeVries 1997; US FWS n.d.). Introduced northern snakeheads pose a potential threat to largemouth bass, as the invaders inhabit a similar ecological niche to native bass, with both species being large, fish-eating ambush predators. Prey availability is thought to be crucial to recruitment (survivability of young past their first year) of largemouth bass, so competition for prey is largely detrimental to young bass (Gutreuter and Anderson 1985). A modeling experiment evaluating the impact of northern snakehead on largemouth bass populations in the Potomac River found that if snakehead expansion continues at its current rate without control measures, largemouth bass abundance could be decreased by 35.5% (Love and Newhard 2012).

Largemouth bass also tend to be sensitive to low water temperatures or reduced dissolved oxygen in water (US FWS n.d.), conditions

which northern snakehead is notoriously capable of withstanding (SPDC 2014). Thus, northern snakeheads may have a competitive advantage over largemouth bass following especially harsh winters, or in waterbodies where oxygen is low due to eutrophication, caused by pollution or other environmental stressors. Massachusetts is at a particularly high risk for providing northern snakehead with conditions that favor these advantages: winters in Massachusetts are cold, and eutrophication is a severe issue. Long-term water quality monitoring data has shown that cultural eutrophication (low-oxygen conditions resulting from anthropogenic nutrient pollution) is directly or indirectly linked to 48% of water quality impairments in the state (Wong et al. 2018). However, the long-term data set shows that water chemistry in rivers and streams has been improving over the years, thanks to rigorous protections set in place by the federal Clean Water Act and enforcement by state agencies (Wong et al. 2018). With continued diligence and enforcement of water quality protections, conditions are likely to keep improving with lakes and ponds being prioritized for protection and restoration.

HUMAN RISKS AND BENEFITS

In many parts of Asia, within the fish's native range, northern snakehead is considered an economically important food fish (Wang et al. 2019). In China, specifically, captive cultivation of snakehead produces an annual volume of fish worth the equivalent of US$1.6 billion (Xu et al. 2017). Snakeheads that are collected by fishing from wild or feral populations are also eaten and enjoyed by many anglers. In fact, at the 2018 First International Snakehead Symposium, a large gathering of snakehead-interested researchers and watershed managers, fried snakehead appetizers were served to attendees to demonstrate the quality of the fish as food (Odenkirk 2018).

However, despite snakeheads' value as food in certain settings, they also pose some risks to human livelihood. As discussed in the previous section, northern snakehead has the potential to negatively impact recreational largemouth bass fisheries, which can be detrimental to tourism and to the overall recreational value of an area. Snakeheads also have sharp teeth, and while they are not typically aggressive toward humans, they

TABLE 5.1

Sightings and Report Sources of Infested Waterbodies

Waterbody	Agency	Report_Source	Longitude	Latitude	First Year Reported
Massapoag Pond	MassDEP	USGS NAS Database	−71.496	42.654	2004
Newton Pond	MassDEP	USGS NAS Database	−71.75	42.308	2011

TABLE 5.2

Infested Waterbodies, Towns, and Watersheds

Number	Waterbody	Town_1	Town_2	Watershed
2	Massapoag Pond	Dunstable	Groton/Tyingsborough	Merrimack
1	Newton Pond	Shrewsbury	Boylston	Blackstone

can be defensive of their nesting sites (USGS n.d.a). In Delaware, a young child was once bitten by a juvenile northern snakehead after reaching out to touch it (Howard 2016). Fortunately, the child's injury was not serious, but to avoid these kinds of unpleasant encounters, snakehead should be handled with caution.

MONITORING AND SIGHTINGS

So far, northern snakehead has only been sighted twice in the state of Massachusetts (Figure 5.1 and Table 5.1).[1] The first sighting was in 2004 in Massapoag Pond which is located in north central Massachusetts near the New Hampshire border. The second fish was seen in Newton Pond in Shrewsbury, in central Mass (USGS n.d.b) (Table 5.2). Each sighting was only of a single individual fish, and there is no evidence of northern snakehead reproducing or becoming established anywhere in the state. This is in contrast to nearby states, including Delaware, New York, and Maryland, where established northern snakehead populations are increasingly becoming an issue (SPDC 2014). In order to prevent sporadic individual introductions from going unnoticed and growing into statewide nuisances, it is important to maintain thorough monitoring efforts, paired with investment in public awareness and proactive plans for management in the case that northern snakehead is found.

Attentive monitoring is important to preventing the spread of northern snakehead, since smaller, newer populations are easier to control.

Northern snakehead monitoring can be carried out either by conventional, physical surveys or by using newer biological techniques such as environmental DNA (eDNA) analysis. Conventional surveys assessing snakehead populations have involved trapping, netting, and electrofishing to collect snakehead fish where they may be present (Orrell and Weigt 2005; SPDC 2014; Lapointe et al. 2019). Tagging experiments have also been carried out in parts of the United States to track the movement of northern snakehead and the expansion of its invasive range (Lapointe et al. 2013).

Detection by eDNA is another useful method for monitoring northern snakehead, especially useful for passive surveillance over larger areas of concern where conventional surveys are not feasible, or to supplement conventional surveys in measuring the success of control operations. Extensive work has been carried out on elucidating cytochrome oxidase subunit I (COI) gene sequences associated with different species in the Channidae family, so that snakeheads can be identified through DNA barcoding methods used in eDNA analysis (Serrao et al. 2014). eDNA monitoring has recently been used to detect the invasive bullseye snakehead (*C. marulius*) in southern Florida (Hunter et al. 2019). Another surveillance project in 2015 was able to detect northern snakehead by eDNA from the Muskingum River watershed in Ohio, demonstrating the usefulness of molecular screening for monitoring northern snakehead invasion (Simmons et al. 2015).

MANAGING ACTIONS

Demographic models have shown that removal of northern snakehead should occur during pre-spawn periods or prior to juvenile dispersal, since spawning events produce hundreds of new fish (SPDC 2014; Jiao et al. 2009). Eradication also appears to be more successful when it is attempted in confined ponds and lakes, as there are several recorded instances of northern snakehead being completely or almost completely eliminated from lakes (Boesch 2002; Mahon et al. 2016), while cases of treatment within extensive river systems, such as the White River system in Arkansas (SPDC 2014) and in the Potomac River (MDNR n.d.), have been reported to be more difficult and less effective. Since lake and pond infestations are more treatable, and populations within rivers are likely to become established when fish spread into river drainages from nearby ponds, effective preventative management can be achieved through diligent monitoring of lakes and ponds adjacent to large river systems. When northern snakehead is discovered in a pond or a lake, a rapid response is important to prevent the invasive population from growing to an unmanageable extent.

Massachusetts Division of Fisheries and Wildlife has a suite of regulations to prevent importation and liberation of anything vertebrate (321 CMR 2.15 and 4.09). Import permits not only protect natural resources from invasive organisms, but also help to protect them from diseases. All fish, except aquarium trade fish, need a permit and cannot be released to the wild. Even it is exempt from permits to import and possess, aquarium trade fish cannot be released into the wild. In addition, the Massachusetts Endangered Species Act prohibit authorized take; therefore, the general public should be aware that snakeheads cannot be possessed due to its harm to the environment; state agency will not issue permits for taking northern snakehead. Public awareness with regard to those who might eat snakehead and a concerted effort to educate fishermen and women to be on the lookout would be the most effective way of minimizing any intentional or unintentional release. There is currently no established northern snakehead population in Massachusetts, and so it is critical at this stage to invest in strong monitoring and prevention, as well as in raising public awareness to prevent introduction and spread by human means. Snakehead populations grow very quickly, and a population can be most effectively controlled if management is carried out imminently. In order not to be caught off guard by an emergent northern snakehead infestation rapidly propagating and spreading throughout a system, agencies should widely monitor for the presence of northern snakehead, and have a plan in place for a response to be carried out if the invasive fish are detected. Information on effective control methods can be found in the following sections, as well as two case studies describing successful responses to emergent northern snakehead infestations.

CHEMICAL CONTROL

The standard fish poison rotenone has been found to be highly successful in eradicating northern snakehead (Lazur et al. 2011; Nanda et al. 2009). However, rotenone is non-specific and toxic to native species of fish; thus, its use could do more ecological harm than good if applied without caution (Turner et al. 2007; Stinson 2018; Roop and Williams 2020). Ideally, applications of rotenone should be confined to high-density areas of snakehead infestation. Additional practicable precautions to be taken include preemptively capturing and removing native fish to holding tanks, and neutralizing rotenone using potassium permanganate at the end of treatment (Roop and Williams 2020).

Antimycin has not been tested on northern snakehead but is approved for use on other invasive fish species in the United States and could be effective (Abdel-Fattah 2011). Deltamethrin is also found to induce fish tissues damage, oxidative damage, and immunotoxicity (Kong et al. 2021). More information about chemical control on northern snakehead and other 18 invasive species can be found in Appendix III. Before any chemical application, it is required to obtain an aquatic nuisance species control license (BRP WM04 Form) from Massachusetts Department of Environmental Protection.

PHYSICAL CONTROL

Like many other fish, the northern snakehead can be enticed out of the water by a baited hook and a fishing rod. Organized efforts have been carried out in multiple states to collaborate with sport fishermen on the removal of northern snakehead, and to raise public awareness of the risks invasive snakeheads pose to the integrity of native freshwater fisheries. As recreational fish harvest has caused depression of fish biomass for many species in the past, employing sport fishermen to remove invasive fish has the potential to be a strong supplementary method of population control (SPDC 2014).

In the state of Maryland, where northern snakehead invasion is most severe, there have been several campaigns initiated to incentivize snakehead sport fishing. Maryland's Department of Natural Resources (MDNR) holds an annual Snakehead Contest—an event where local anglers compete to catch the biggest snakehead. In 2019, MDNR reported 107 participants in the Snakehead Derby, and 25 northern snakeheads caught in within 6 hours (MDNR 2019). As another incentive program, MDNR once held a photo raffle, where anglers entered photos of caught and killed northern snakehead for a chance to win prizes, such as Bass Pro Shops gift cards or free sport fishing licenses (SPDC 2014).

While these kinds of public outreach campaigns have been valuable in raising awareness of invasive fish and aiding removal efforts, there are risks to encouraging snakehead harvest with incentives. If bounties placed on snakeheads are significant enough to generate a large profit, or even if the entertainment value of snakehead capture is realized, well-intentioned control efforts could create pressure to sustain snakehead populations or introduce them into new areas, rather than to eliminate them (Pasko and Goldberg 2014). Fishermen might be incentivized to cultivate snakehead on their own to continue profiting from bounties or release them to the wild in order to populate a favorite fishing spot. On account of careful planning and consideration, no known case of intentional release thus far has been motivated by the desire to profit from incentives (SPDC 2014).

It is also possible to physically remove snakeheads from the water using electrofishing, netting, or trapping, but physical removal alone is unlikely to fully remove northern snakehead from an area (SPDC 2014). Electrofishing may also be less effective in collecting juvenile or smaller fish (Cohen and MacDonald 2016).

BIOLOGICAL (ECOLOGICAL) CONTROL

There is no known method of biological control that can be used on northern snakehead. As an apex predator in the fish food web, the northern snakehead has few underwater predators. However, piscivorous birds such as osprey (*Pandion haliaetus*) and great blue heron (*Ardea herodias*) have been observed to feed on northern snakehead in the Potomac River system (Lapointe et al. 2019). Northern snakeheads may be especially vulnerable to bird predation due to their biological obligation to rise to the surface of the water and sip air, placing them visibly within reach of birds (Lapointe et al. 2019). Additionally, eastern mosquitofish (*Gambusia holbrooki*) have been observed in the wild feeding on northern snakehead fish eggs and fry in large numbers, a single mosquitofish eating 60 eggs and fry within a short span of time (Landis and Lapointe 2010). Though there is currently no known means of biological control for northern snakehead, conserving and restoring habitat for native herons, osprey, and mosquitofish could be a useful, low-impact way to keep snakehead populations under control and mitigate the ecological effects caused by the invader.

SUCCESS STORIES

MACQUILLIAM POND, CROFTON, MARYLAND

In spring of 2002, an angler caught an 18-inch long northern snakehead from MacQuilliam Pond, a small pond in Crofton, Maryland, located on private property beside a strip-mall (MDNR 2019, n.d.; The Washington Times 2002). The identification of the fish was confirmed by an expert, and several more snakeheads, including juveniles, were found in the same pond. Maryland Department of Natural Resources (MDNR) began monitoring the population with electrofishing surveys. The first survey caught over 100 young-of-the-year northern snakeheads, confirming MacQuilliam Pond to be home to the first established, reproductive population of northern snakehead in the United States (Boesch 2002; SPDC 2014). Investigations by Maryland Natural Resources Police found the source of the infestation to have been a resident, who admitted to having released a pair of 12–14-inch snakehead fish into the pond a couple of years earlier, after the fish outgrew his home aquarium (Boesch 2002; Kobell 2002). The original purpose for this resident to buy the fish was to serve to a sick sister who believed to have medicinal properties (Susan Pasko, personal communication).

Amid urgent concerns that the invasive population would expand into the nearby Little Patuxent River, and somewhat tense negotiations with the pond's owners, eradication efforts finally began (The Washington Times 2002). In September 2002, MacQuilliam Pond was treated with rotenone, a fish poison. Herbicides were also applied to the pond, in an effort to reduce refuge area for the fish and to lower the oxygen content of the pond. Sandbags and silt fences were employed around the periphery of the pond to prevent any fish from escaping over land in the case of heavy rain (Boesch 2002).

During the eradication process, over 1,200 dead snakeheads were recovered. Though MDNR officials had expected the rotenone treatment to kill all of the native fish in the pond and had planned to restock the pond with native fish after the fact to mitigate the ecological damage, it was found that some indigenous species, including bluegills, eels, and pickerel, had survived the poisoning in small numbers (Kiehl 2002). Thanks to the MDNR's swift and aggressive treatment of the snakehead invasion, northern snakeheads have not been sighted in MacQuilliam Pond since 2002. However, separate introductions in the state of Maryland have resulted in widespread established populations of northern snakehead throughout the Potomac and Patuxent river watersheds today that are incredibly difficult to manage, highlighting the importance of early, targeted, preventative responses such as these to control emerging snakehead populations.

RIDGEBURY LAKE

In 2008, northern snakehead fish were collected from Ridgebury Lake, a lake in New York State which drains into a tributary of the Hudson River. The fish were discovered in the spring of 2008, and that summer, Ridgebury Lake and several adjacent ponds in the Catlin Creek

drainage area were treated with rotenone. Over 200 dead snakeheads were recovered following treatment, and most were juveniles, indicating that the population was established and reproducing. Treatment was repeated in select ponds the following year in 2009, and subsequent monitoring did not detect any northern snakehead in the area, leading managers to believe that the populations were successfully eradicated (SPDC 2014). To gain more information and to improve their certainty of the snakeheads' eradication, the New York State Department of Environmental Conservation (NYSDEC) worked in collaboration with a team from Central Michigan University (CMU) to begin environmental DNA (eDNA) monitoring in 2012 (SPDC 2014; Wegleitner and Mahon 2014). The sampling protocol was refined, updated, and repeated several times over the course of three years following the eradication efforts (Mahon et al. 2016).

Originally, 308 field samples were collected from the Catlin Creek drainage and 525 from the Erie Canal Corridor (ECC), the large artificial canal system connecting the Great Lakes, the Lake Champlain Basin, and the Mohawk–Hudson River Basin (Mahon et al. 2016). Using species-specific primers and standard PCR, the team from CMU found that none of the samples tested positive for the presence of northern snakehead eDNA. In 2014, after improving filtration and extraction efficiency and changing to a more sensitive method for screening results, digital droplet PCR (ddPCR), Mahon et al. (2016) collected another 294 and 725 samples from the Catlin Creek drainage and ECC, respectively. While once again there were zero positive detections from the ECC, three of the samples from the Catlin Creek drainage returned positive for snakehead eDNA. A memo was sent to the NYSDEC detailing the locations of each positive detections on maps with conclusions and recommendations. Sampling was repeated again in 2015 and this time returned no positive eDNA snakehead detection (Mahon et al. 2016).

This case provides an example for advanced monitoring techniques being used in conjunction with chemical control to ensure successful management of invasive species in vulnerable, highly connected waterways. The CMU team were able to monitor a large, high-risk area, pinpoint the precise location of any emergent evidence for the species' presence, and confirm that northern snakehead was prevented from actively expanding its range into the Hudson River (Mahon et al. 2016).

NOTE

1 According to a comment from Susan Pasko, a recent third sighting has been confirmed in the Canton Reservoir on August 27, 2021. This new infestation site is not mapped in this chapter.

REFERENCES

Abdel-Fattah, S. 2011. *Aquatic Invasive Species Early Detection and Rapid Response- Assessment of Chemical Response Tools.* International Joint Commission, Great Lakes Regional Office, p. 104.

Alaska Department of Fish and Game (ADFG). n.d. Burbot (*Lota lota*) species profile. www.adfg.alaska.gov/index.cfm?adfg=burbot.main (Accessed May 13, 2021).

Boesch, D. 2002. Snakehead Scientific Advisory Panel First Report to the Maryland Secretary of Natural Resources. https://msa.maryland.gov/megafile/msa/speccol/sc5300/sc5339/000113/004000/004189/unrestricted/20071038e.pdf. (Accessed May 7, 2021).

Bressman, N.R., J.W. Love, T.W. King, C.G. Horne, and M.A. Ashley-Ross. 2019. Emersion and Terrestrial Locomotion of the Northern Snakehead (*Channa argus*) on Multiple Substrates. *Integrative Organismal Biology* 1(1): obz026.

Cermele, J. 2019. Is the snakehead epidemic happening all over again? www.fieldandstream.com/is-snakehead-epidemic-happening-all-over-again/ (Accessed May 20, 2021).

Chesapeake Bay Program. n.d. Northern snakehead. www.chesapeakebay.net/discover/field-guide/entry/northern_snakehead#:~:text=Reproduction%20and%20Life%20Cycle,nest%2C%20which%20both%20parents%20guard.&text=After%20that%2C%20they%20feed%20mostly,small%20crustaceans%20and%20fish%20larvae (Accessed May 13, 2021).

Cohen, M.K., and J.A. MacDonald. 2016. Northern Snakeheads in New York City. *Northeastern Naturalist* 23(1): 11–24.

Courtenay, W.R., and J.D. Williams. 2004. *Snakehead (Pisces, Channidae)—A biological synopsis and risk assessment.* New York: U.S. Geological Survey Circular 1251, p. 143.

Everglades Center for Invasive Species and Ecosystem Health (CISMA). 2018. Bullseye Snakehead (Channa marulius). www.evergladescisma.org/the-dirty-dozen/bullseye-snakehead/ (Accessed May 20, 2021).

Fuller, P.L. 2021. *Amia calva* Linnaus, 1766. USGS Nonindigenous Aquatic Species Database, Gainesville, FL. https://nas.er.usgs.gov/queries/FactSheet.aspx?SpeciesID=305. (Accessed May 13, 2021).

Fuller, P.L., A.J. Benson, G. Nunez, A. Fusaro, and M. Neilson. 2019. *Channa argus* (Cantor, 1842). USGS Nonindigenous Aquatic Species Database, Gainesville, FL. https://nas.er.usgs.gov/queries/factsheet.aspx?speciesid=2265. (Accessed May 7, 2021).

Georgia Department of Natural Resources (GA DNR). 2019. Two juvenile northern snakehead fish that were caught in a pond in Gwinnett County, Ga. Via www.nytimes.com/2019/10/10/us/snakehead-fish-georgia.html. (Accessed May 20, 2021).

Gutreuter, S.J., and R.O. Anderson. 1985. Importance of Body Size to the Recruitment Process in Largemouth Bass Populations. *Transactions of the American Fisheries Society* 114(3): 314–327.

Howard, B.C. 2016. Fishermen Battle Invasive 'Frankenfish' Snakeheads. *National Geographic.* www.nationalgeographic.com/animals/article/160317-snakeheads-potomac-river-chesapeake-bay-invasive-species-fish?loggedin=true. (Accessed May 7, 2021).

Hunter, M., P. Schofield, G. Meigs-Friend, M. Brown, and J. Ferrante. 2019. Environmental DNA (eDNA) detection of nonnative bullseye snakehead in southern Florida. *American Fisheries Society.* p. 21.

ITIS (Integrated Taxonomic Information System). 2021. www.itis.gov/servlet/SingleRpt/SingleRpt?search_topic=TSN&search_value=166680#null (Accessed April 15, 2021).

Jiao, Y., N.W.R. Lapointe, P.L. Angermeier, and B.R. Murphy. 2009. Hierarchical demographic approaches for assessing invasion dynamics of non-indigenous species: An example using northern snakehead (*Channa argus*). *Ecological Modelling* 220(13–14): 1681–1689.

Kiehl, S. 2002. Hobbyist freed snakeheads in Md. Pond. *The Baltimore Sun.* www.baltimoresun.com/bal-te.md.snakehead-12jul12-story.html (Accessed May 7, 2021).

Knie, A. 2019. Jurassic Park: The Hunt for Jersey Bowfin. www.thefisherman.com/article/jurassic-park-the-hunt-for-jersey-bowfin/ (Accessed May 20, 2021).

Kobell, R. 2002. Snakehead extermination approved. *The Baltimore Sun.* www.baltimoresun.com/maryland/bal-md.snakehead16aug16-story.html (Accessed May 7, 2021).

Kong, Y, M. Li, X. Shan, G. Wang, and G Han. 2021. Effects of deltamethrin subacute exposure in snakehead fish, Channa argus: Biochemicals, antioxidants and immune responses. *Ecotoxicology and Environmental Safety* 209: 111821.

Kramer, A.M., G. Annis, M.E. Wittmann, W.L. Chadderton, E.S. Rutherford, D.M. Lodge, L. Mason, D. Beletsky, C. Riseng, J.M. Drake. 2017. Suitability of Laurentian Great Lakes for invasive species based on global species distribution models and local habitat. *Ecosphere* 8(7): e01883.

Kusek, J.L. 2007. Fishing for a solution: How to prevent the introduction of invasive species such as the snakehead fish. *Penn State Environmental Law Review* 15(2): 331–354.

Landis, A.M.G., and N.W.R. Lapointe. 2010. First record of a Northern Snakehead (*Channa argus* Cantor) Nest in North America. *Northeastern Naturalist* 17(2): 325–332.

Landis, A.M.G., N.W.R. Lapointe, and P.L. Angermeier. 2011. Individual growth and reproductive behavior in a newly established population of northern snakehead (*Channa argus*), Potomac River, USA. *Hydrobiologia* 661: 123–131.

Lapointe, N.W.R., J.S. Odenkirk, and P.L. Angermeier. 2013. Seasonal movement, dispersal, and home range of Northern Snakehead *Channa argus* (Actinopterygii, Perciforms) in the Potomac River catchment. *Hydrobiologia* 701: 73–87.

Lapointe, N.W.R., R.K. Saylor, and P.L. Angermeier. 2019. Diel Feeding and Movement Activity of Northern Snakehead. *American Fisheries Society Symposium* 89: 14.

Lazur, A., S. Early, and J.M. Jacobs. 2011. Acute toxicity of 5% rotenone to Northern Snakeheads. *North American Journal of Fisheries Management* 26(3): 628–630.

Love, J.W., and J.J. Newhard. 2012. Will the expansion of Northern Snakehead negatively affect the fishery for largemouth Bass in the Potomac River (Chesapeake Bay)? *North American Journal of Fisheries Management* 32: 859–868.

Love, J.W., and J.J. Newhard. 2018. Expansion of Northern Snakehead in the Chesapeake Bay Watershed. *Transactions of the American Fisheries Society* 147: 342–349.

Ludsin, S.A., and D.R. DeVries. 1997. First-year recruitment of largemouth bass: the interdependency of early life stages. *Ecological Applications* 7: 1024–1038.

Mahon, A.R., W.L. Chadderton, and A. Tucker. 2016. Final Report—Assessing Erie canal corridor invasion risk using environmental DNA. Institute for Great Lakes Research, Department of Biology, Central Michigan University, p. 43.

Maryland Department of Natural Resources (MDNR). 2019. Freedom and fishing Snakehead Derby. https://dnr.maryland.gov/fisheries/Pages/snakehead_derby.aspx (Accessed May 7, 2021).

Maryland Department of Natural Resources (MDNR). n.d. Northern Snakehead. https://dnr.maryland.gov/fisheries/pages/snakehead.aspx. (Accessed May 7, 2021).

Nanda, N.B.P., P.C. Das, and J. Jena. 2009. Use of rotenone as piscicide: Toxicity levels in a few common freshwater predatory and weed fishes. *Journal of Applied Aquaculture* 21(4): 241–249.

Odenkirk, J.S. 2018. The first international Snakehead symposium: News from the front(s). *Fisheries* 44: 123–128.

Odenkirk, J.S., and S. Owens. 2011. Expansion of a Northern Snakehead population in the Potomac River system. *Transactions of the American Fisheries Society* 136(6): 1633–1639.

Orrell, T.M., and L. Weigt. 2005. The Northern Snakehead *Channa argus* (Anabantomorpha: Channidae), a non-indigenous fish species in the Potomac River, U.S.A. *Proceedings of the Biological Society of Washington* 118(2): 407–415.

Pasko, S., and J. Goldberg. 2014. Review of harvest incentives to control invasive species. *Management of Biological Invasions* 5(3): 263–277.

Poulos, H.M., B. Chernoff, P.L. Fuller, and D. Butman. 2012. Ensemble forecasting of potential habitat for three invasive fishes. *Aquatic Invasions* 7(1): 59–72.

Qin, J., B. Cudmore, and B. Schroeder. 2012. *Channa argus* (northern snakehead). *CABI Invasive Species Compendium*. www.cabi.org/isc/datasheet/89026 (Accessed May 13, 2021).

Raleigh, S. n.d. Giant Snakehead (Channa micropeltes). https://dreamlake-fishing.com/fish-species-of-thailand/giant-snakehead-channa-micropeltes/ (Accessed May 20, 2021).

Randall, Z. 2015. Channa marulius (Hamilton 1822) family channidae. *Florida Museum Ichthyology Collection*. www.floridamuseum.ufl.edu/discover-fish/florida-fishes-gallery/bullseye-snakehead/ (Accessed May 20, 2021).

Raver, D. n.d. Arkansas Game and Fish Commission. www.uaex.edu/environment-nature/ar-invasives/invasive-animals/invasive-fish.aspx (Accessed May 20, 2021).

RAWWFishing. 2017. Fishing for big Bullseye Snakeheads in South Florida with topwater frogs. www.youtube.com/watch?v=afOjLc8ktoA (Accessed May 20, 2021).

Roop, J., and A. Williams. 2020. First record of occurrence and genetic characterization of a population of northern snakehead *Channa argus* (Cantor, 1842) in Georgia, USA. *BioInvasions Records* 9: 842–852.

Rypel, A.L. 2014. Do invasive freshwater fish species grow better when they are invasive? *Oikos* 123: 279–289.

Schloeffel, A.R. 2016. A Burbot at the ground of Lake Walchensee, Germany (Bavarian). https://commons.wikimedia.org/wiki/File:Tr%C3%BCsche_Walchensee.jpg (Accessed May 20, 2021).

Serrao, N.R., D. Steinke, and R.H. Hanner. 2014. Calibrating snakehead diversity with DNA barcodes: Expanding taxonomic coverage to enable identification of potential and established invasive species. *PLoS One* 9(6): e99546.

Siitari, K. 2018. New Kootenai River burbot fishery opens for anglers on Jan. 1. https://idfg.idaho.gov/blog/2018/12/kootenai-river-burbot-fishery-open-january-1 (Accessed May 20, 2021).

Simmons, M., A. Tucker, W.L. Chadderton, C.L. Jerde, and A.R. Mahon. 2015. Active and passive environmental DNA surveillance of aquatic invasive species. *Canadian Journal of Fisheries and Aquatic Sciences* 73(1): 76–83.

Snakehead Plan Development Committee (SPDC). 2014. National Control and Management Plan for Members of the Snakehead Family (Channidae), p. 80.

Stinson, H. 2018. The Northern Snakehead, *Channa argus*, as an invasive species. *Eukaryon* 14: 101–102.

Texas Parks & Wildlife (TPW). n.d. Bowfin vs. Snakeheads: Distinguishing features. https://tpwd.texas.gov/huntwild/wild/species/exotic/snakehead_bowfin.phtml (Accessed May 13, 2021).

The Phuket News. 2019. Fish the Exotic: Catching the Giant Snakehead. www.thephuketnews.com/fish-the-exotic-catching-the-giant-snakehead-69909.php. (Accessed May 20, 2021).

The Washington Times. 2002. Pond owner says no to snakehead plan. www.washingtontimes.com/news/2002/aug/8/20020808-035835-2594r/ (Accessed May 7, 2021).

Trammel, S. n.d. Northern Snakehead. Via https://nature.mdc.mo.gov/discover-nature/field-guide/snakeheads (Accessed May 20, 2021).

Turner, L., S. Jacobson, and L. Shoemaker. 2007. Risk Assessment for Piscicidal Formulations of Rotenone. Compliance Services International, p. 104.

U.S. Fish & Wildlife Service. 2003. Recognizing Northern Snakehead. www.fws.gov/fisheries/ans/pdf_files/Recognizing_Snakeheads.pdf (Accessed May 13, 2021).

U.S. Fish & Wildlife Service (US FWS). n.d. Largemouth Bass. *Fish and Aquatic Conservation*. www.fws.gov/fisheries/freshwater-fish-of-america/largemouth_bass.html (Accessed May 7, 2021).

USGS. 2003. Northern snakehead (Channa argus). Via https://en.wikipedia.org/wiki/Northern_snakehead#/media/File:Northern_snakehead.jpg. (Accessed May 20, 2021).

USGS. n.d.a. Can snakehead fish harm humans? *Biology and Ecosystems*. www.usgs.gov/faqs/can-snakehead-fish-harm-humans?qt-news_science_products=0#qt-news_science_products (Accessed May 7, 2021).

USGS. n.d.b. Nonindigenous Aquatic Species (NAS). *Channa argus*. https://nas.er.usgs.gov/queries/CollectionInfo.aspx?SpeciesID=2265&State=MA (Accessed May 10, 2021).

Wang, L., N. Xie, Y. Shen, B. Ye, G. Yue, and X. Feng. 2019. Constructing high-density genetic maps and developing sexing markers in Northern Snakehead (*Channa argus*). *Marine Biotechnology* 21: 348–358.

Wegleitner, B., and A.R. Mahon. 2014. Evaluation of eradication success of Northern Snakehead in the Catlin Creek Drainage, NY using environmental DNA. *American Fisheries Society 144th Annual Meeting*.

Wong, W., J.J. Dudula, T. Beaudoin, K. Groff, W. Kimball, and J. Swigor. 2018. Declining ambient water phosphorus concentrations in Massachusetts' rivers from 1999 to 2013: Environmental protection works. *Water Research* 139: 108–117.

Xu, J., C. Bian, K. Chen, G. Liu, Y. Jiang, Q. Luo, X. You, W. Peng, J. Li, Y. Huang, Y. Yi, C. Dong, H. Deng, S. Zhang, H. Zhang, Q. Shi, and P. Xu. 2017. Draft genome of the Northern snakehead, *Channa argus*. *GigaScience* 6(4): gix011.

6 Hydrilla (*Hydrilla verticillata*)

ABSTRACT

Hydrilla (*Hydrilla verticillata*) is among the most troublesome invasive aquatic plants in many southern states and has recently become established in Massachusetts. The plant crows out native vegetation, harms fisheries, limits recreation, impedes navigation, and reduces property values. Hydrilla is a relatively new introduction to Massachusetts with the first recorded sighting at Long Pond in Cape Cod in 2001. Currently, there are 11 waterbodies infested by hydrilla in Massachusetts. Hydrilla can reproduce and spread in different ways: through fragmentation, turions, tubers, and possibly seeds. Management practices are summarized in this chapter: manual (diver-assisted suction harvesting), mechanical (e.g., drawdowns), biological (e.g., triploid grass carp and some insects), and chemical (e.g., endothall and fluridone) treatments. The documented distribution of invasive hydrilla in Massachusetts' lakes and rivers will allow the general public to become aware of the presence of this invasive species and minimize unintentional transportation and spread of this invasive species to other waterbodies.

INTRODUCTION

Taxonomic Hierarchy (ITIS 2021)

Kingdom	Plantae
Subkingdom	Viridiplantae
Infrakingdom	Streptophyta
Superdivision	Embryophyta
Division	Tracheophyta
Subdivision	Spermatophytina
Class	Magnoliopsida
Superorder	Lilianae
Order	Alismatales
Family	Hydrocharitaceae
Genus	*Hydrilla*
Species	*Hydrilla verticillata* (L.f.) Royle

The invasive hydrilla, also known as Florida elodea and waterthyme, is native to tropical and subtropical areas of Eurasia and Africa and was introduced into the United States in Florida during the 1950s as an aquarium plant (Langeland 1996; Crow and Hellquist 2006; Jacono et al. 2020). It quickly spread after being improperly disposed of into native waterbodies and is now found in 28 states (Jacono et al. 2020). Two genetically and morphologically different biotypes of hydrilla exist: monoecious and dioecious (True-Meadows et al. 2016; Jacono et al. 2020; Tippery et al. 2020). The female and male reproductive organs of dioecious hydrilla are found on different plants, while monoecious hydrilla plants have both sets of reproductive organs (Cook and Lüönd 1982). In the United States, dioecious hydrilla is generally found in the south, while monoecious hydrilla is usually in the north but can be found north of South Carolina (Jacono et al. 2020). A new strain of hydrilla has been recently documented in the Connecticut River from Agawam, MA, to southern Connecticut. It appears more robust but lacks tubers.

Hydrilla is a submersed perennial herb that is adaptable, aggressive, and extremely competitive (Figure 6.1). It is found in freshwater lakes, ponds, rivers, impoundments, and canals. Hydrilla grows submerged, rooted to the bottom of waterbodies with long stems that grow toward the surface where it expands horizontally, forming dense mats (Langeland 1996; Jacono et al. 2020). It can flourish in any freshwater environment and can grow in both low- and high-nutrient conditions. Hydrilla can be found at depths of more than 20 ft, can withstand very low-light conditions. Its leaves are pointed and arranged in whorls of three to eight, ranging from 6–20 mm long and 2–4 mm wide (Figure 6.1). They have serrated margins with sharp teeth, and have spines or glands underneath the reddish midrib. Stems can grow up to 9 m in length. Female flowers can be identified by three whitish sepals and three translucent petals that are

DOI: 10.1201/9781003201106-6

FIGURE 6.1   Hydrilla. (Photos by Summer Stebbins.)

10–50 mm long, 4–8 mm wide, and float on the surface of the water. Male flowers can be identified by three whitish or reddish petals that are about 2 mm long and have three stamen. The female flowers extend via a long stalk to float on the water surface. The male flowers, on the other hand, detaches from the plant and unfold the petals on the surface, where the anthers will then dehisce explosively and release the pollen to fertilize the female flowers. Flowers are rarely observed in New England. Hydrilla reproduces in four ways: fragmentation, tubers, turions, and through seed dispersal. Fragmentation is the most common form of reproduction for hydrilla (Langeland 1996; Jacono et al. 2020). The biology of turion production is comprehensively reviewed by Netherland (1997). A fragment of hydrilla that contains only a single whorl has nearly a 50% chance of sprouting a new plant. Tubers of hydrilla are found at the end of the plant's rhizomes as deep as 30 cm in the sediment, are 5–10 mm long, and are often an off-white or yellowish color (Figure 6.2). They are highly resistant to various control methods, surviving for years in undisturbed sediment and leading to reintroductions years after treatment has stopped. Some genotypes do not appear to produce tubers—the hydrilla recently found in the Connecticut River is a new genotype (Tippery et al. 2020) which does not produce tubers (Greg Bugbee, personal communication). Turions of hydrilla are small buds that fall from the plant (Figure 6.2) and sit on the sediment, and they are 5–8 mm long, with a shiny dark green appearance. Hydrilla grows optimally at 20–27°C in the United States.

ECOLOGICAL EFFECTS

Hydrilla displaces native species in the plant community, such as native pondweed. The dense mats block sunlight and inhibit the growth of some native species such as eelgrass (Langeland 1996). Hydrilla can grow up to an inch per day and can absorb carbon from the water more effectively than can other aquatic plants (Cornell Cooperative Extension 2019). Hydrilla can also change the chemical and physical composition of lakes by decreasing the dissolved oxygen within waterbodies, reducing water flow, and causing flooding and damage to canal banks (Langeland 1996). Decomposing hydrilla releases nutrients that can accelerate eutrophication and degrade water quality. Stratification of the water column can also occur resulting in fish kills in lakes that have populations of hydrilla (Jacono et al. 2020).

Hydrilla may indirectly impact other species. A survey of three Brazilian reservoirs found the

FIGURE 6.2 Hydrilla tubers (white) and turions (green) and Connecticut River hydrilla aboveground vegetative turion development. (Left photo by Lyn Gettys; right photo by Mark (Heilman 2020).)

invasive bivalve *Limnoperna fortunei* became a problem because hydrilla provided better biomass for attachment compared to native macrophytes (Michelan et al. 2014). In addition, a toxic cyanobacterium, *Aetokthonos hydrillicola*, growing on hydrilla leaves causes a fatal neurological disease in birds that consume it, avian vacuolar myelinopathy (AVM) (Bidigare et al. 2009). It was originally hypothesized that a neurotoxin produced by the cyanobacterium and then ingested by birds was the underlying cause of illness (Williams et al. 2007). Recent research has confirmed that a neurotoxin, aetokthonotoxin (AETX), produced by the cyanobacterium (*A. hydrillicola*) not only affects birds, but also fish and invertebrates (Breinlinger et al. 2021). Fish and waterfowl are exposed to the toxin when they consume hydrilla. The toxin moves up the food chain and afflicts bald eagles. Deaths of bald eagles have been reported in areas of dense *H. verticillata* and *A. hyrillicola* populations (Breinlinger et al. 2021). The cyanobacterium produces the toxin from a bromide solution that likely enters the environment from power plants or other anthropogenic sources as well as diquat dibromide treatments to control hydrilla and other nuisance aquatic vegetation (Stokstad 2021). The toxin is found to be highly potent even at low levels (Pelletier et al. 2019) and mammals may also be susceptible.

Although hydrilla is an invasive species, it may have some beneficial aspects. The plant can provide habitat for fish and wildlife, especially for some animals' early life stages. It is very difficult to maintain the ideal abundance of hydrilla in the habitat. Maintenance of hydrilla could also have a positive effect in some other ways. The main management method of hydrilla has been chemical herbicides, but alternative methods, such as the use of harvester machines for biogas, benefit compost production (Evans and Wilkie 2010). However, use of harvesters is not recommended for hydrilla control because of fragmentation which may facilitate the spread of hydrilla to other waters.

ECONOMIC IMPACTS

Hydrilla has been responsible for many different economic impacts in multiple states. In the states of Massachusetts, Maine, New York, Connecticut, and Florida, hydrilla has inhibited water recreation and angling water quality, flood control, and tourism can also be heavily impacted (Langeland 1996; 2016). Maine has spent close to $200,000 on control in one inland lake and New York is currently spending more than $1 million each year to control several large-scale hydrilla infestations with the aid of US Army Corps of Engineers and many local stakeholders (NEANS 2021). Langeland (1990) reported that the annual control cost to manage 7,600 ha of hydrilla in Florida exceeds $5 million. Hydroelectric power generation is also hindered by fragmented plant material that builds up on trash racks and clogs intakes. For example, during 1991, hydrilla at Lake Moultrie, South Carolina, shut down the St. Stephen powerhouse operations for seven weeks resulting in $2,650,000 of expenses due to repairs, dredging, and fish loss. In addition, during this repair period, there was an estimated $2,000,000 loss in power generation for the plant (Balciunas et al. 2002).

NATIVE AND INVASIVE RANGE

Hydrilla is native to Asia, Africa, and Australia. Specifically, the monoecious strain of hydrilla is thought to originate in Korea, while the dioecious strain is most likely from the Indian subcontinent (Tippery et al. 2020). Hydrilla has been an emerging invasive species that has been slowly invading Massachusetts. It has also been found to be invasive in Maine, Connecticut, New York, New Jersey, Pennsylvania, Maryland, and Florida, as well as many other states (USDA National Invasive Species Information Center 2019). It is the most abundant aquatic plant in Florida public waters (Schardt 1994). A specimen of hydrilla collected at Mystic, Connecticut, in 1989 is the first authenticated record of hydrilla in New England, but it was unnoticed previously because of its misidentification as the South American waterweed (aka Brazillian waterweed, *Egeria densa*) (Les et al. 1997). A new strain of hydrilla has been recently documented in the Connecticut River from Agawam, MA, to southern Connecticut. It has spread rapidly with nearly 900 acres documented (Bugbee and Stebbins 2021).

VECTORS OF SPREAD

Because hydrilla can propagate from fragments, turions, tubers, and seeds, it can spread rapidly to other aquatic habitats. Turions are overwintering buds found where leaves attach to stems. They may break off and be carried by the current to new locations where they can settle, overwinter, and grow into new plants. Tubers are the p, otato-like reproductive structures that form on the roots of hydrilla plants each fall and allow the plants to store energy and regenerate the in following spring or in future seasons. The newly discovered hydrilla strain poses a severe risk to the entire northeast region. Already occurring in the river

in Massachusetts, the strain could easily spread into the portions of the river in Vermont and New Hampshire, as well as lakes, ponds, and other rivers throughout the Northeast. In Massachusetts, the Connecticut River has been surveyed at Agawam section and no turions or tubers have been found on any of the hydrilla plants (James Straub 2019, email communication). Nor have tubers been found in the Connecticut River within the boundaries of Connecticut (Greg Bugbee 2019, personal communication).

Hydrilla fragments can spread by attaching to kayaks or boats, trailers, or fishing equipment. The 2019 and 2020 surveys by the Connecticut Agricultural Experiment Station revealed 842 acres of hydrilla in the main channel of Connecticut River from Agawam, MA, downstream to its mouth at Long Island Sound (Bugbee and Stebbins 2021). Based on knowledge of boater movement, the Northeast Aquatic Nuisance Species (NEANS) Panel members have determined that the infestation poses a high risk of spread to other waterways throughout the northeastern United States (NEANS 2020). For example, before boating at Lake George, New York, the last stop for many boaters is the Connecticut River (Table 6.1). These data reinforce the urgency of preventing hydrilla from leaving the Connecticut River (Heilman 2020). In addition to boater movement, waterfowls are vectors with tubers that are regurgitated, not defecated, being viable (Joyce et al. 1980).

MONITORING AND SIGHTINGS

Hydrilla grows rapidly in many waterbodies throughout the United States (Poovey and Getsinger 2010) with multiple infestations confirmed in Massachusetts (Table 6.2). Location and distribution data have been collected from the Massachusetts Department of Environmental Protection (MassDEP) and the Department of Conservation and Recreation (DCR) and mapped using Geographic Information Systems (GIS) to aid in future decisions regarding waterbodies of Massachusetts (Figure 6.3).

GIS-produced maps will also inform the development of regulations that involve efforts to limit the spread of this aquatic invasive species. Hydrilla is a very aggressive invasive species and therefore can spread throughout Massachusetts in a relatively

TABLE 6.1

New York State Department of Environmental Conservation Boat Stewardship Inspection Program Locations/Events Citing Connecticut River as Last Waterbody Visited in 2020

Waterbody Name	Inspections	Waterbody Name	Inspections
Lake George	143	Cayuga Lake	1
Lake Champlain	42	Indian Lake	1
Hudson River	4	Lake Kushaqua	1
Fulton Chain Lakes	3	Lake Placid	1
Great Sacandaga Lake	3	Lake Pleasant	1
Second Pond	3	Mohawk River	1
Canada Lake	2	Oneida Lake	1
Long Lake	2	Otisco Lake	1
Saratoga Lake	2	Round Lake	1
Cayuga Lake	1	Salmon River	1
Indian Lake	1	Seneca Lake	1
Lake Kushaqua	1	Skaneateles Lake	1
		Others	23

FIGURE 6.3　Distribution of aquatic invasive Hydrilla in Massachusetts.

short period of time (Table 6.2). Hydrilla is a relatively new introduction to Massachusetts and was first identified in Long Pond in Cape Cod in 2001 but reported in the Cape Cod Water Quality Assessment report in 2002 (Table 6.2) (Mattson et al. 2004). Currently, there are 11 waterbodies in Massachusetts (Figure 6.3 and Table 6.2) infested with hydrilla. The sighting at Agwam is the first town north of the Connecticut–Massachusetts state line within the Connecticut River. These waterbodies are distributed in five major Massachusetts watersheds or drainage basins: Cape Cod, South Coast, Boston Harbor (Mystic River), Connecticut River, and Nashua River.

TABLE 6.2

List of Waterbodies with Invasive Hydrilla

Waterbody	Agency	Report Source	Longitude	Latitude	First Year Reported
Long Pond	MassDEP	Cape Cod Watershed Water Quality Assessment Report. CN 50.0			2002
Hobomock Pond	DCR	DCR Data from Lakes and Ponds, September 2008 MR			2008
Mystic Lake	MassDEP	Cape COD Coastal Drainage Areas 2004–2008 Surface Water Quality Assessment Report 96-AC-2 CN 171.0, May 2011			2011
Coachlace Pond	DCR	Jamie Carr	−71.697104	42.413367	2016
South Meadow Pond (East Basin)	DCR	Jamie Carr	−71.705732	42.415348	2016
South Meadow Pond (West Basin)	DCR	Jamie Carr	−71.713092	42.414889	2016
Mossy Pond	DCR	Jamie Carr	−71.701034	42.417156	2016
Connecticut River	DCR	*Hydrilla verticulata* and *Trapa natans* Survey on the Connecticut River, November 2018	−72.610896	42.038974	2018
Connecticut River	DCR	*Hydrilla verticulata* and *Trapa natans* Survey on the Connecticut River, November 2018	−72.610888	42.038902	2018
Connecticut River	DCR	*Hydrilla verticulata* and *Trapa natans* Survey on the Connecticut River November 2018	−72.610781	42.038236	2018
Connecticut River	DCR	*Hydrilla verticulata* and *Trapa natans* Survey on the Connecticut River, November 2018	−72.610799	42.038232	2018
Connecticut River	DCR	*Hydrilla verticulata* and *Trapa natans* Survey on the Connecticut River, November 2018	−72.610494	42.036364	2018
Connecticut River	DCR	*Hydrilla verticulata* and *Trapa natans* Survey on the Connecticut River, November 2018	−72.610769	42.036247	2018
Connecticut River	DCR	*Hydrilla verticulata* and *Trapa natans* Survey on the Connecticut River, November 2018	−72.610425	42.035436	2018
Connecticut River	DCR	*Hydrilla verticulata* and *Trapa natans* Survey on the Connecticut River, November 2018	−72.610353	42.034750	2018
Connecticut River	DCR	*Hydrilla verticulata* and *Trapa natans* Survey on the Connecticut River, November 2018	−72.610691	42.034287	2018
Connecticut River	DCR	*Hydrilla verticulata* and *Trapa natans* Survey on the Connecticut River, November 2018	−72.610805	42.034108	2018
Magoun Pond	DCR	Vanessa Curran			2019
Oakman Pond	DCR	Vanessa Curran			2019
Lower Woburn Street Pond	DCR	Vanessa Curran	−71.143992	42.522064	2019

FIGURE 6.4   Informational sign to prevent spread of Hydrilla (northeastans.org).

MANAGEMENT

In Massachusetts, waterbodies infested with hydrilla are, or will be, included in the 303d list of impaired or threatened waters during water quality assessments, fulfilling the requirements of the Clean Water Act. For projects within infested watersheds subject to the jurisdiction of the 401 Water Quality Certification, conditions on invasive species prevention and decontamination are required. In addition, water quality monitoring gear and watercraft used in these infested watersheds shall be decontaminated before being used in other watersheds to prevent the potential spread of invasive hydrilla.

Prevention and early detection are critical for minimizing the further spread of hydrilla. Voluntary prevention and detection efforts rely on education and outreach to change behavior and foster stewardship. For instance, the NEANS Panel has made hydrilla informational signs that are used by Massachusetts and other New England states to encourage recreationists to report potential hydrilla infestations (Figure 6.4). If reports are made while the infestations cover a small area, there is often a greater likelihood of control or even eradication.

Many different management and control methods have been tested in different waterbodies. Treatments to control monoecious hydrilla are generally the same as for dioecious hydrilla, with chemical control and grass carp generally being the most effective (True-Meadows et al. 2016). Research on management of the monoecious biotype is more limited than dioecious. Monoecious hydrilla active management should begin several weeks after tubers first sprout, or as soon as possible after discovery of new infestations. However, new infestations of monoecious hydrilla are often not detected until much later in the season, when shoot growth reaches the surface of the waterbody, which increases the level of difficulty to achieve control.

Triploid grass carp (*Ctenopharyngodon idella* Val.) are an effective biological control agent for hydrilla (Sutton 2019). However, biological control by grass carp is not being considered for the Connecticut River hydrilla infestations due to the large river size, connectivity to other waters, and the inability to prevent carp from escaping to tributaries (NEANS 2020). Moreover, it is recently demonstrated that management of hydrilla using grass carp may result in unintended consequences. For example, after stocking triploid grass carp, applying herbicide, and planting native macrophyte in experimental ponds infested with hydrilla, some native plants were impacted. In addition, a secondary expansion and infestation of invasive Asian clams was also observed because grass carp ate much of the aquatic vegetation resulting in approximately 60–90% bare, benthic substrate in the experimental ponds which was ideal habitat for the clams (Holbrook et al. 2020). More than two dozen non-US species of insects or pathogens have been considered as potential biological control agents for hydrilla (Balciunas et al. 2002). Laboratory and field data have been gathered to gain approval for release of two weevils and two leaf-mining flies. Although none of these four insects were strictly monophagous, hydrilla was greatly preferred, and the risk to the few other alternative hosts was considered minimal. For example, the tuber feeding weevil (*Bagous affinis*) and the Indian leaf-mining fly (*Hydrellia pakistanae*) have been used for biological control of hydrilla in Florida. The leaf-mining fly was first released in the United States during October 1987 as a potential agent for the biological control of hydrilla and as of 1997, nearly 3 million individuals have been released at more than 25 separate locations in Florida and other states in southeastern United States (Center et al. 1997). Permanent establishment at many locations is attributed to modifications of rearing and release procedures that considered the biology of the agent in relation to various habitat characteristics.

Herbicide such as endothall and fluridone have been successful for hydrilla management. Endothall has been used to control submersed aquatic plants since 1960, providing broad-spectrum control of aquatic weeds by disrupting plants' photosynthesis activities (Ortiz et al. 2019). Although endothall is considered a contact herbicide, many field observations suggest that it might have systemic activity (Ortiz et al. 2019). Because the efficacy to roots, turions, and tubers is poor, long-term control takes many annual applications. Fluridone has been used in New York and other parts of the United States to control hydrilla. It is a systemic herbicide that is absorbed through the roots of the plant and also disrupts photosynthesis. Netherland and Getsinger (1995) suggest the following treatment strategy for fluridone to adequately control hydrilla: an initial treatment of >10 mg/l of fluridone is applied, followed by long-term maintenance of doses lower than 10 mg/l. These herbicides need to be applied repeatedly for several consecutive seasons until the tuber bank in the sediment is exhausted. Diquat- and copper-based herbicides have also been used to control hydrilla (Mattson et al. 2004). Deciding which control method to use is based on the budget of the project and the waterbodies to be protected. Whichever method is chosen, it should be used in conjunction with other preventive measures such as watercraft inspections and proper signage at high-risk waterbodies (NEANS 2020). Additionally, an adaptive management plan, including prevention, early detection, monitoring, control, and restoration, is needed. Using the outcomes of each season's monitoring to assess effectiveness and make informed changes to future management plans is a key part of the adaptive management necessary to controlling hydrilla. Before chemical is applied in Massachusetts waters, it is required to obtain an aquatic nuisance species control license (BRP WM04 Form) from Massachusetts Department of Environmental Protection.

True-Meadows et al. (2016) summarized herbicide treatments on hydrilla and found that diquat and endothall (both dipotassium and monoamine salts) produced similar results on both US biotypes. Diquat and endothall are often more successful when applied early to mid-June, when there is less dense monoecious hydrilla biomass. A second application can be applied in mid-August if regrowth occurs. A concentration of 0.25 mg/l diquat for two days is lethal to both monoecious and dioecious hydrilla; endothall is effective on monoecious hydrilla, but appropriate exposure times and doses must be met (Hofstra and Clayton 2001; Poovey and Getsinger 2010). Fluridone controls monoecious hydrilla and reduces tuber density over time (Nawrocki 2011). However, after years of repetitive fluridone treatments within Florida, fluridone-resistant dioecious hydrilla has been documented (Michel et al. 2004; Giannotti et al. 2014). Mutations of the phytoene desaturase gene at codon 304 have been shown to confer fluridone resistance in dioecious hydrilla (Benoit and Les 2013). Utilizing different herbicides over a season with different modes of action is expected to reduce the possibility of resistance emerging within a population. To date, resistance has only been documented in dioecious hydrilla, but there is no evidence that monoecious hydrilla cannot develop resistance (Arias et al. 2005). As described previously, the occurrence of AVM has been linked to the cyanobacterium growing on hydrilla and the cause is the toxin AETX biosynthesis which relies on the availability of bromide (Breinlinger et al. 2021). As of 2022, no VM has been documented in any Massachusetts waterbody, even though bromide product such as Tribune or Reward has been permitted by MassDEP. It remains to be seen if there are any elevated bromide levels in waters of Massachusetts because of anthropogenic and geologic sources (Breinlinger et al. 2021).

More information about chemical control on hydrilla and other 18 invasive species can be found in Appendix III. Any chemical used for invasive species control requires the WM04 Chemical Application License from the Massachusetts Department of Environmental Protection.

Mechanically, hydrilla can be controlled by dredging, drawdowns, diver-assisted suction harvesting (DASH), or through benthic barriers for very small populations. For effective dredging management for hydrilla, it is critical to remove the entirety of the patch to prevent fragmentation and leaving behind tubers. Dredge spoils will need to be transported off-site and deposited at a safe location far from any waterbodies.

Hand harvesting with a vacuum hose is also a standard removal process for aquatic plants such as Eurasian watermilfoil (*Myriophyllum spicatum*) and is used for small infestations. Benthic barriers, while effective for smaller areas, are not useful for eradication and are not effective in flowing waters (NEANS 2020). Mechanical control is therefore usually not recommended for hydrilla management because of the fragmentation that occurs, regrowth form tubers and turions, the cost, and other negative impacts (True-Meadows et al. 2016). The Connecticut River hydrilla may be an exception because the unique strain does not produce tubers.

Based on molecular phylogenetics, a different strain of hydrilla has been found in a portion of the Connecticut River in Hartford County, Connecticut. It represents a novel introduction likely from northern Eurasia (Tippery et al. 2020). Therefore, genetic analysis of Massachusetts hydrilla is recommended as different management actions could be necessary. In addition, because certain strains of hydrilla have become resistant to fluridone (Michel et al. 2004), it is recommended that low concentrations of herbicides and/or a possible combination of herbicides be used to reduce the likelihood of resistance developing. Consistent long-term monitoring will enable managers to quickly detect and track any plants that have survived herbicide treatment. Intermediately and highly resistant phenotypes of hydrilla were found to need three to seven times more fluridone to control properly, which could impact project budgets and increase risk of non-target impacts. It is suggested that fragments of plants with the resistance mutation left over from previous fluridone treatments may produce new plants and spread into new waterbodies (Michel et al. 2004; Tippery et al. 2020). Because of this, it is critical to ensure no fragments of hydrilla remain when treating a waterbody.

SUCCESS STORY

PICKEREL POND IN MAINE

Maine confirmed the state's first known infestation of *Hydrilla verticillata* in 2002. When discovered in the 52-acre Pickerel Pond, Limerick,

Maine, the infestation was well-established, with an average depth of 16 ft and many areas exceeded 60–70% cover. The Maine Department of Environmental Protection (MDEP) estimated the population to be at least 4 years old. The initial objective of MDEP's response was to prevent infestation of other waters. The response measures included outlet screens to limit downstream migration, restrictions on boat access, public outreach to landowners, tuber bank and plant monitoring, herbicide treatments with fluoridone brand name Sonar, and annual dive surveys. After two years of herbicide treatment, MDEP determined to attempt eradication and continued annual treatments of fluoridone for a total of nine consecutive years (2003–2011). No hydrilla was found during the two years of herbicide treatment (2010 and 2011). In 2012, the first year without herbicide treatment, one plant was found during an annual diver survey. Continued annual diver surveys 2013–2017 resulted in no detection of hydrilla (personal communication with John McPhedran, March 17, 2022).

CAYUGA LAKE AND FALL CREEK IN NEW YORK

There have been some successful instances of controlling hydrilla in the State of New York. For example, hydrilla was first discovered in Cayuga Lake and Fall Creek in New York in 2011. A massive coordinated rapid response effort, carried out by local and state agencies, researchers, and community groups, revealed that 166 acres of the Cayuga Inlet was infested with hydrilla. The Local Cayuga Lake Hydrilla Task Force was formed to oversee the herbicide treatment process and annual monitoring. At the Cayuga Inlet, chemical treatments (fluridone and endothall) were part of the long-term plan to eradicate hydrilla from Cayuga Lake beginning in 2012. In 2018, there were no observations of hydrilla and therefore no chemical treatment needed to be continued. While invasive hydrilla is an ongoing problem, success stories like this help others learn how to eradicate these aquatic invasive species in order to maintain the quality of different waterbodies (Fighting Hydrilla in the Cayuga Lake Watershed, cce-tompkins.org/ accessed November 10, 2020).

Using the Cayuga Inlet as an example and led by the Northeast Aquatic Nuisance Species Panel (NEANS), several New England states and US Army Corps of Engineers are working together on monitoring and managing the hydrilla recently confirmed in the Connecticut River (NEANS 2020; Tippery et al. 2020). It is not too late to prevent the spread of hydrilla into additional states and control it in select locations in the Connecticut River. Most importantly, the Connecticut River Hydrilla Management Plan can be used as a template for control efforts for hydrilla in other states or for other aquatic invasive plants in Massachusetts.

REFERENCES

Arias, R.S., M.D. Netherland, B.E. Scheffler, A. Puri, and F.E. Dayan. 2005. Molecular evolution of herbicide resistance to phytoene desaturase inhibitors in *Hydrilla verticillata* and its potential use to generate herbicide-resistant crops. *Pest Management Science* 61(3): 258–268.

Balciunas, J.K., M.J. Grodowitz, A.F. Cofrancesco, and J.F. Shearer. 2002. Hydrilla. In F.V. Driesche, B. Blossey, M. Hoodle, S. Lyon, and R. Reardon (eds). *Biological Control of Invasive Plants in the Eastern United States*. United States Department of Agriculture Forest Service. Forest Health Technology Enterprise Team. Morgantown, West Virginia. FHTET-2002–04.91–114 pp.

Benoit, L.K., and D.H. Les. 2013. Rapid identification and molecular characterization of phytoene desaturase mutations in fluridone-resistant hydrilla (*Hydrilla verticillata*). *Weed Science* 61(1): 32–40.

Bidigare R.R., S.J. Christensen, S.B. Wilde, and S.A. Banack. 2009. Cyanobacteria and BMAA: Possible linkage with avian vacuolar myelinopathy (AVM) in the south-eastern United States. *Amyotrophic Lateral Sclerosis* 10: 71–73.

Breinlinger, S., T.J. Phillips, B.N. Haram, J. Mareš, J.A. Yerena, P. Hrouzek, R. Sobotka, W.M. Henderson, P. Schmieder, S.M. Williams, J.D. Lauderdale, H.D. Wilde, W. Gerrin, A. Kust, J.W. Washington, C. Wagner, B. Geier, M. Liebeke, H. Enke, T.H.J. Niedermeyer, and S.B Wilde. 2021. Hunting the eagle killer: A cyanobacterial neurotoxin causes vacuolar myelinopathy. *Science*, 371(6536): 1335.

Bugbee, G.J., and SE. Stebbins. 2021. *Connecticut River Middle and Upper Sections East Haddam, CT to Agawam, MA: Invasive Aquatic Vegetation Survey Hydrilla Overwintering and Spread Management Options*. New Haven, CT: The Connecticut Agricultural Experiment Station, p. 115.

Center, T.D., M.J. Grodowitz, A.F. Cofrancesco, G. Jubinsky, E. Snoddy, and J.E. Freedman. 1997. Establishment of *Hydrellia pakistanae* (Diptera: Ephydridae) for the biological control of the submersed aquatic plant *Hydrilla verticillate* (Hydrocharitaceae) in the Southeastern United States. *Biological Control* 8(1): 65–73.

Cook, C.D.K., and R. Lüönd. 1982. A revision of the genus Hydrilla (Hydrocharitaceae). *Aquatic Botany* 13: 485–504.

Cornell Cooperative Extension. Ecological impacts of hydrilla. http://erie.cce.cornell.edu/invasive-species/ecological-impacts-of-hydrilla. (Accessed October 11, 2019).

Crow, G.E., and C.B. Hellquist. 2006. *Aquatic and Wetland Plants of Northeastern North America, Volume II: A Revised and Enlarged Edition of Norman C. Fassett's A Manual of Aquatic Plants, Volume II: Angiosperms: Monocotyledons* (Vol. 2). Madison, WI: University of Wisconsin Press.

Evans, J.M., and A.C. Wilkie. 2010. Life cycle assessment of nutrient remediation and bioenergy production potential from the harvest of hydrilla (Hydrilla verticillata). *Journal of Environmental Management*, 91(12): 2626–2631.

Giannotti, A.L., T.J. Egan, M.D. Netherland, M.L. Williams, and A.K. Knecht. 2014. Hydrilla shows increased tolerance to fluridone and endothall in the winter park chain of lakes: considerations for resistance management and treatment options. https://conference.ifas.ufl.edu/aw14/Presentations/Grand/Thursday/Session%209A/0850%20Giannotti.pdf (Accessed January 29, 2021).

Heilman, M. 2020. *Technical Perspectives on Connecticut River Hydrilla Infestation and Its Potential Future Management*, p. 6.

Hofstra, D.E., and J.S. Clayton. 2001. Evaluation of selected herbicides for the control of exotic submerged weeds in New Zealand: I The use of endothall, triclopyr and dichlobenil. *Journal of Aquatic Plant Management* 39: 20–24.

Holbrook, D.L., A.N. Schad, G.O. Dick, L.L. Dodd, and J.H. Kennedy. 2020. Invasive bivalve establishment as a secondary effect of eradication-focused nuisance aquatic plant management. *Lake and Reservoir Management* 36(4): 423–431.

ITIS (Integrated Taxonomic Information System). 2021. www.itis.gov/servlet/SingleRpt/SingleRpt?search_topic=TSN&search_value=38974#null (Accessed January 28, 2021).

Jacono, C.C., M.M. Richerson, H. Morgan, and I.A. Pfingsten. 2020. *Hydrilla verticillata (L.f.) Royle: U.S. geological survey, nonindigenous aquatic species database*. Gainesville, FL. https://nas.er.usgs.gov/queries/FactSheet.aspx?speciesID=6.

Joyce, J.C., W.T. Haller, and D.E. Colle. 1980. Investigation of the presence and survivability of hydrilla propagules in waterfowl. *Aquatics* 2: 10–14.

Langeland, K.A. 1990. Hydrilla (*Hydrilla verticillata* (L.f.) Royle): A continuing problem in Florida waters. University of Florida Coop. Extension Service Circular No. 884. University of Florida, Gainesville, FL.

Langeland, K.A. 1996. *Hydrilla verticillata* (LF) Royle (Hydrocharitaceae), "the perfect aquatic weed". *Castanea* 61(3): 293–304.

Langeland, K.A. 2016. Hydrilla Management in Florida Lakes. https://edis.ifas.ufl.edu/ag370.

Les, D.H., L.J. Mehrhoff, M.A. Cleland, and J.D. Gebal. 1997. *Hydrilla verticillata* (Hydrocharitaceae) in Connecticut. *Journal of Aquatic Plant Management* 35: 10–14.

Mattson, M.D., P.J. Godfrey, R.A. Barletta, and A. Aiello. 2004. Eutrophication and aquatic plant management in Massachusetts. Final Generic Environmental Report. In: K.J. Wagner (ed.), *Department of Environmental Protection and Department of Conservation and Recreation, EOEA Commonwealth of Massachusetts*. Cambridge, MA: EOEA.

Michel, A., R.S. Arias, B.E. Scheffler, S.O. Duke, M. Netherland, and F.E. Dayan. 2004. Somatic mutation-mediated evolution of herbicide resistance in the nonindigenous invasive plant hydrilla (*Hydrilla verticillata*). *Molecular Ecology* 13(10): 3229–3237.

Michelan, T.S., M.J. Silveira, D.K. Petsch, G.D. Pinha, and S.M. Thomaz. 2014. The invasive aquatic macrophyte *Hydrilla verticillata* facilitates the establishment of the invasive mussel *Limnoperna fortunei* in Neotropical reservoirs. *Journal of Limnology* 73(3): 598–602.

Nawrocki, J.J. 2011. Environmental and Physiological Factors Affecting Submersed Aquatic Weed Management. Master's thesis. North Carolina State University. http://repository.lib.ncsu.edu. (Accessed January 29, 2021).

Netherland, M.D. 1997. Turion ecology of hydrilla. *Journal of Aquatic Plant Management* 35: 1–10.

Netherland, M.D., and K.D. Getsinger. 1995. Laboratory evaluation of threshold fluridone concentrations under static conditions for controlling hydrilla and Eurasian watermilfoil. *Journal of Aquatic Plant Management* 33: 33–36.

Northeast Aquatic Nuisance Species Panel (NEANS). 2020. *Connecticut River Hydrilla Control Project Five-Year Management Plan*. Penacook, NH: NEANS.

Northeast Aquatic Nuisance Species Panel (NEANS). 2021. *A Letter to Coalition of Northeast Governors, Northeast Committee on the Environment*. Penacook, NH: NEANS.

Ortiz, M.F., S.J. Nissen, and C.J. Gray. 2019. Endothall behavior in Myriophyllum spicatum and Hydrilla verticillate. *Pest Management Science* 75(11): 2942–2947.

Pelletier, A.R. 2019. Trophic Transfer of a Novel Cyanotoxin in Fishes. Masters of Science thesis, The University of Georgia, Athens.

Poovey, A.G., and K.D. Getsinger. 2010. Comparative Response of Monoecious and Dioecious Hydrilla to Endothall. *Journal of Aquatic Plant Management* 48: 15–20.

Schardt, J. 1994. 1994 Florida aquatic plant survey report. Florida Department of Environmental Protection, Bureau of Aquatic Plant Management, Tallahassee, Florida.

Stokstad, E. 2021. Mysterious eagle killer identified. *Science* 371(6536): 1298–1298.

Sutton, D.L., V.V. Vandiver, and J.E. Hill. 2019. *Grass Carp: A Fish for Biological Management of Hydrilla and Other Aquatic Weeds in Florida*. Gainesville: University of Florida.

Tippery, N.P., G.J. Bugbee, and S. Stebbins. 2020. Evidence for a genetically distinct strain of introduced *Hydrilla verticillata* (Hydrocharitaceae) in North America. *Journal of Aquatic Plant Management* 58: 1–6.

True-Meadows, S., E.J. Haug, and R.J. Richardson. 2016. Monoecious hydrilla—A review of the literature. *Journal of Aquatic Plant Management* 54: 1–11.

USDA National Invasive Species Information Center. 2019. Hydrilla. www.invasivespeciesinfo.gov/profile/hydrilla (Accessed September 9, 2019).

Williams, S.K., J. Kempton, S.B. Wilde, and A. Lewitus. 2007. A novel epiphytic cyanobacterium associated with reservoirs affected by avian vacuolar myelinopathy. *Harmful Algae* 6(3): 343–353.

7 Curly-leaf Pondweed (*Potamogeton crispus*)

ABSTRACT

In the early 1840s, the invasive species curly-leaf pondweed was first introduced into Philadelphia, Pennsylvania. Following this occurrence, in the 1880s, the curly-leaf pondweed was detected in Spy Pond, Arlington, Massachusetts. The sightings of invasive curly-leaf pondweed in Massachusetts' lakes and rivers have increased to include 167 waterbodies. These sightings were mapped to gain a spatial understanding of the status of this species in Massachusetts lakes and rivers with the goal of increasing public awareness of this invasive species to minimize its spread. In addition, this document provides an overview of the management methods such as mechanical and chemical treatments of curly-leaf pondweed.

INTRODUCTION

Taxonomy Hierarchy (ITIS 2021)

Kingdom	Plantae
Subkingdom	Viridiplantae
Infrakingdom	Streptophyta
Superdivision	Embryophyta
Division	Tracheophyta
Subdivision	Spermatophytina
Class	Magnoliopsida
Superorder	Lilianae
Order	Alismatales
Family	Potamogetonaceae
Genus	*Potamogeton* L.
Species	*Potamogeton crispus* L.

Curly-leaf pondweed (*Potamogeton crispus*) (Figure 7.1) has several common names such as curly-leaf pondweed, crispy-leaved pondweed, crisped pondweed, and curly pondweed (MA DCR 2002, 2015; Bugbee et al. 2015). Curly-leaf pondweed is a perennial submergent herbaceous aquatic plant that can grow longer than 4 m (GISD 2006). The sessile, serrated, oblong leaves can grow up to 7.6 cm (3 inches) long with clear veining on the surface (MA DCR 2002). The leave edges are serrated which helps distinguish them from other pondweeds (Acy 2019). The leaves are light to dark green, and sometimes have a reddish hue (Mikulyuk and Nault 2009). There are usually three to five veins on a prominent midvein. Stems of curly-leaf pondweed are flat and arranged in a spiral. Flowers of the plant are small and have four petal-shaped lobes that produce small reddish-brown fruits each year.

Curly-leaf pondweed can survive in a wide range of conditions, but grows well in waters that are alkaline, and brackish, and highly eutrophic conditions. Curly-leaf pondweed thrives in low-light and cooler temperature conditions (Mikulyuk and Nault 2009).

Yet another characteristic aids this invasive species, as it can reproduce quickly by rhizomatic spread and vegetative propagules called turions (Mikulyuk and Nault 2009). Turions are produced in the late spring. They will be dormant throughout the winter. The plant will begin to germinate and grow slowly until the following spring when rapid growth and turion production occurs (Figure 7.1, right panel). Curly-leaf pondweed dies back in early summer and when it is no longer a nuisance. Due to this uncommon growing cycle, curly-leaf pondweed avoids competition from other species. As summer ends, water temperatures cool down which creates prefect conditions for turion germination. Germination rates are very high according to lab experiments (100%) and field experiments (>60%) (Yeo 1966; Bolduan et al. 1994).

CURLY-LEAF PONDWEED AS AN INVASIVE SPECIES

Curly-leaf pondweed is a nonnative submersed macrophyte that inhabits temperate areas in North America, southern South America,

DOI: 10.1201/9781003201106-7

FIGURE 7.1 Curly-leaf pondweed (*left*) and turions (*right*). (Left photo by Greg Bugbee; right photo by Summer Stebbins.)

FIGURE 7.2 Abundant curly-leaf pondweed in Crystal Lake, Connecticut. (Photo by Greg Bugbee.)

and New Zealand (Kaplan and Fehrer 2004). It can be found in rivers, freshwater lakes, wetlands, estuaries, ponds, and even ditches (Figure 7.2). In North America, the first verified identification was in Philadelphia, PA, in 1841–1842 (Stuckey 1979). By the 1880s, it had spread to Massachusetts and New York (Bolduan et al. 1994). Though accidental introduction is likely to have occurred, curly-leaf pondweed may have been introduced intentionally as it can serve as habitat and provide food for native wildlife (Stuckey 1979). Because this plant spreads mainly through turion germination, even a small fragment of curly-leaf pondweed with attached turions could have been unintentionally transported by boat. Since this plant is highly competitive and aggressive, it may displace native species and disrupt the habitat of waterbodies. Curly-leave pondweed is listed as a prohibited plant species that is banned from the trade in Massachusetts.

ECOLOGICAL IMPACT

Curly-leaf pondweed can dominate aquatic plant communities and threaten native *Potamogeton* species. For example, the native *P. ogdenii* in New England, New York, and Canada was outcompeted by curly-leaf pondweed (Tomaino 2004). It has also been found that curly-leaf pondweed can accumulate metals from sediments via their roots and rhizomes. A decomposition experiment on curly-leaf pondweed fragments discovered an increase in the concentrations of Al, Cd, Cr, Fe, Mn, and Pb (Deng et al. 2015). It should be noted that the application of phytoremediation by submerged macrophytes could give a chance for subsidiary pollution during the decomposition of curly-leaf pondweed. Additionally, during July, curly-leaf pondweed becomes senescent. The decay of the large volume of organic matter can deplete oxygen in the water and addition of nutrients, which may result in the death of some native organisms and algal blooms (MA DCR 2002).

ECONOMIC IMPACT

Curly-leaf pondweed can decrease native plant diversity and deplete oxygen in the water causing the death of organisms. This can negatively impact local and regional economics as curly-leaf pondweed can degrade recreational activities such as fishing, boating, and swimming. For example, curly-leaf pondweed forms dense "mat like," monotypic stands that inhibit navigation, block boat docks, and clog water (Poovey 2008). In addition, high densities of curly-leaf pondweed and resulting algal blooms can damage the aesthetic value of waterbodies and reduce real estate values, tourism, and recreation

(IL DNR 2009; Jensen 2009; WI DNR 2012). Furthermore, the cost for controlling of curly-leaf pondweed can range from less than $100 to greater than $1,000 to treat a one-acre pond (Washington State Noxious Weed Control Board 2004; Swistock 2021).

POSITIVE ASPECTS

Despite the deleterious effect of curly-leaf pondweed, there are some beneficial aspects that it provides. First, curly-leaf pondweed is a good indicator of waterbody eutrophication (Cao and Wang 2007). In addition, the absorption process by the plant can help purify water (Wu et al. 2009). Because of this, it can be applied as a purifier for industrial aqueous waste which could reduce the need for chemical treatment (Hafez et al. 1992; Hafez et al. 1998). Several reports have shown that the impact on the native plant community could be overestimated as curly-leaf pondweed dominates the ecosystem during the winter and spring, when most plants are less prominent (Bolduan et al. 1994). It is also reported that curly-leaf pondweed is a good source of carotenoids (Ren and Zhang 2008) with medicinal and cosmetics properties, including antioxidation, tumor reduction, and immunity-regulation (Ren and Zhang 2008; Mikulyuk and Nault 2009; Du et al. 2015).

MONITORING AND SIGHTINGS

The data in this report were collected by MassDEP staff during field visits and recorded in field sheets. Field sheet quality control, database development, and map construction are under the guidance of the MassDEP Standard Operating Procedures (SOP). Massachusetts Department of Conservation and Recreation (MA DCR), the College of the Holy Cross, the United States Geological Survey (USGS), and individual researcher Joy Trahan-Liptak also contributed to the process of data collection of curly-leaf pondweed in Massachusetts. Tools for drafting the map and marking out data layers of sites with curly-leaf pondweed were created via ArcGIS® ArcMap 10.1 (ESRI, Redlands, California).

Curly-leaf pondweed was found in many lakes and rivers in Massachusetts (Tables 7.1 and 7.2).

TABLE 7.1

List of Town Names and Waterbodies that Contain *Potamogeton crispus*

Town	Waterbody	Town	Waterbody
Amesbury	Powwow River	Marlborough	Assabet River
Amherst	Fort River		Hager Pond
Andover	Gravel Pit Pond	Mashpee	Quashnet River
	Shawsheen River	Maynard	Assabet River
Arlington	Spy Pond	Medfield	Charles River
Ayer	Bowers Brook	Milford	Charles River
	Flannagan Pond	Millbury	Kettle Brook
	Grove Pond		Woolshop Pond
Barnstable	Middle Pond	Milton	Pine Tree Brook
Barre	Ware River		Russell Pond
Becket	Center Pond	Monterey	Lake Buel
Bedford	Concord River		Lake Garfield
	Shawsheen River		Stevens Pond
Belchertown	Swift River	Natick	Charles River
Belmont	Clay Pit Pond		Dug Pond
Berlin	North Brook		Fisk Pond
Blackstone	Blackstone River		Lake Cochituate
Bolton	Still River		Nonesuch Pond
Boston	Muddy River	New Bedford	Sassaquin Pond
	Neponset River	New Marlborough	York Lake
Boxford	Baldpate Pond	Norfolk	Mill River
Braintree	Monatiquot River	North Adams	Windsor Lake
Brimfield	Quinebaug River	North Attleborough	Ten Mile River
Brookfield	Dunn Brook	Northampton	Connecticut River
Cambridge	Charles River	Northborough	Assabet River
	Fresh Pond		Bartlett Pond
	Little Fresh Pond		Little Chauncy Pond
Canton	Massapoag Brook	Northbridge	Fish Pond
Chelmsford	Newfield Pond	Norton	Wading River
Cheshire	Cheshire Reservoir (Middle Basin)	Norwood	Neponset River
	Cheshire Reservoir (North Basin)	Orange	Millers River
Clinton	Mossy Pond	Otis	Benton Pond
	Coachlace Pond	Palmer	Chicopee River
	South Meadow Pond (East Basin)		Quaboag River
	South Meadow Pond (West Basin)		Ware River
Cohasset	Lily Pond	Pembroke	Stetson Pond
Concord	Assabet River	Pepperell	Nashua River
	Concord River	Petersham	East Branch Fever Brook
	Sudbury River	Pittsfield	Onota Lake
Deerfield	Connecticut River	Plymouth	Beaver Dam Brook
Dennis	Chase Garden Creek		Billington Sea

Town	Waterbody	Town	Waterbody
	Quivett Creek		Town Brook
Douglas	Manchaug Pond		Unnamed Tributary
Dover	Charles River		White Island Pond(East Basin)
Dracut	Beaver Brook	Plymouth	White Island Pond(West Basin)
	Long Pond	Richmond	Richmond Pond
Dudley	French River	Rochester	Leonards Pond
Dunstable	Massapoag Pond	Royalston	Priest Brook
Egremont	Mill Pond	Salisbury	Smallpox Brook
	Prospect Lake	Sheffield	Mill Pond
Erving	Millers River	Shirley	Nashua River
Essex	Chebacco Lake		Squannacook River
Falmouth	Childs River	Shrewsbury	Lake Quinsigamond
	Unnamed Tributary	Southborough	Sudbury Reservoir
Framingham	Cochituate Brook	Southwick	Congamond Lakes(Middle Basin)
	Eames Brook		Congamond Lakes(North Basin)
	Farm Pond		Congamond Lakes(South Basin)
	Sudbury River	Spencer	Sevenmile River
Gill	Barton Cove	Springfield	Porter Lake
Grafton	Flint Pond	Stockbridge	Lily Brook
	Quinsigamond River		Stockbridge Bowl
Great Barrington	Green River	Stow	Assabet River
	Mansfield Pond	Sturbridge	Breakneck Brook
Groton	Cow Pond Brook	Sudbury	Carding Mill Pond
	Nashua River		Grist Mill Pond
Halifax	Monponsett Pond		Hop Brook
	Stetson Brook		Sudbury River
Hanover	Forge Pond	Sutton	Welsh Pond
Hanson	Reservoir	Tyngsborough	Lake Mascuppic
Harvard	Mirror Lake	Uxbridge	West River
	Robbins Pond		Rice City Pond
Harwich	Hinckleys Pond	Wakefield	Lake Quannapowitt
Hingham	Weir River	Walpole	Neponset River
Holyoke	Connecticut River	Ware	Ware River
	Lake Bray	Watertown	Charles River
Hudson	Assabet River	Wayland	Dudley Pond
Ipswich	Turner Hill Country Club Pond		Lake Cochituate
Kingston	Jones River		Sudbury River
Lancaster	Cranberry Pond	Webster	Webster Lake
	Fort Pond	Wellesley	Charles River
	White Pond	Wendell	Whetstone Brook
Lanesborough	Cheshire Reservoir(South Basin)	West Brookfield	Quaboag River
	Pontoosuc Lake	West Stockbridge	Card Pond
Lee	Goose Pond Brook		Shaker Mill Pond
	Laurel Lake	Westfield	Pequot Pond

(Continued)

TABLE 7.1
(Continued)

Town	Waterbody	Town	Waterbody
Littleton	Long Pond	Westford	Forge Pond
	Mill Pond		Nabnasset Pond
	Spectacle Pond		Stony Brook
Lowell	Concord River	Winchendon	Millers River
Lunenburg	Catacoonamug Brook		Otter River
	Lake Whalom	Winchester	Upper Mystic Lake
Lynn	Flax Pond	Woburn	Horn Pond
Mansfield	Rumford River	Wrentham	Crocker Pond
	Wading River		Mirror Lake

Curly-leaf pondweed was first observed at Spy Pond, Arlington, Massachusetts in 1880 (see Stuckey 1979). From 1994 to 2016, MassDEP employees have surveyed 2,050 waterbodies in Massachusetts. During this period, curly-leaf pondweed was found in 132 rivers and lakes in Massachusetts (Figure 7.3) among the 2,050 waterbodies investigated. Out of the 132 waterbodies, 64 are rivers (3.0%), 67 are lakes (3.0%), and 1 is an estuary (<0.1%). These numbers do not represent curly-leaf pondweed in wetlands and the sampling is not randomly designed. In addition, the detection ratio (132 out of 2,050 waterbodies based on map information) likely underestimates the presence of curly-leaf pondweed in Massachusetts because many of the observations occurred during the summer after curly-leaf pondweed senesced. If data from other agencies and organizations (MA DCR, the College of the Holy Cross, USGS and biologist Joy Trahan-Liptak) are counted, there are a total of 163 lakes and rivers within 26 watersheds in Massachusetts that contain curly-leaf pondweed. Most of these waterbodies are in the northern and eastern parts of the state (Figure 7.3). Since there are 33 major watersheds in Massachusetts, the percentage of invasive watersheds is about 79%.

MANAGEMENT

MECHANICAL AND PHYSICAL METHODS

Mechanical and physical removal methods, such as cutting at the surface of the sediment, have proved to be efficient as well as overwinter drawdowns and shallow dredging are equivocal, but some research has shown no serious impact on this species (Mikulyuk and Nault 2009; Nichols and Shaw 1986). This may suggest that there are many limits to mechanical removal methods. For example, drawdowns are not species-selective and pose as threat to local native species by interrupting reproductive cycles. In addition, hand removal serves as an effective technique but is laborious and time-consuming.

Physical barriers using plastics and other materials can be applied to control *P. crispus*. Barrier blocks (anchored on the sediment) may be used to block sunlight and upward growth of the plant which can limit the development of curly-leaf pondweed. Materials used to construct physical barriers vary. Jute is a natural plant used to construct barriers. It is relatively inexpensive and can naturally decay. Due to its permeability, sunlight can transmit through jute, having a reduced impact on benthic fauna compared to other materials (Caffrey et al. 2010). Even so, jute is limited in controlling *P. crispus* because it does not target long-lived propagules (Caffrey et al. 2010). Similarly, fiberglass and polyethylene fabrics can also be used for management via shading sunlight. But due to its buoyancy, the *P. crispus* can move and billow the barrier from natural waves. As a result, sedimentation of *P. crispus* can occur above the barrier and allow for the regrowth of new individuals. Therefore, annual removal and deployment of barriers is needed to help reduce the infestation (Barr and Ditomaso 2014). Another method used to control involves rubber bottom barriers which is an impermeable material used to block sunlight. Due to its impermeability, gas

TABLE 7.2

Latitude and Longitude for Each Waterbody in Figure 7.3

Waterbody	Agency	Report Source	Longitude	Latitude	First Year Reported
Spy Pond	USGS	USGS 4/24/2017 http://nas.er.usgs.gov/default.aspx	−71.155600	42.407900	1943
Spy Pond	USGS	USGS 4/24/2017 http://nas.er.usgs.gov/default.aspx	−71.155600	42.407900	1943
Fresh Pond	USGS	USGS 4/24/2017 http://nas.er.usgs.gov/default.aspx	−71.149000	42.384300	1943
Fresh Pond	USGS	USGS 4/24/2017 http://nas.er.usgs.gov/default.aspx	−71.149000	42.384300	1943
Fresh Pond	USGS	USGS 4/24/2017 http://nas.er.usgs.gov/default.aspx	−71.149000	42.384300	1943
Fresh Pond	USGS	USGS 4/24/2017 http://nas.er.usgs.gov/default.aspx	−71.149000	42.384300	1943
Fresh Pond	USGS	USGS 4/24/2017 http://nas.er.usgs.gov/default.aspx	−71.149000	42.384300	1943
Fresh Pond	USGS	USGS 4/24/2017 http://nas.er.usgs.gov/default.aspx	−71.149000	42.384300	1943
Fresh Pond	USGS	USGS 4/24/2017 http://nas.er.usgs.gov/default.aspx	−71.149000	42.384300	1943
Fresh Pond	USGS	USGS 4/24/2017 http://nas.er.usgs.gov/default.aspx	−71.149000	42.384300	1943
Sudbury River	USGS	USGS 4/24/2017 http://nas.er.usgs.gov/default.aspx	−71.358400	42.465100	1943
Mill Pond	USGS	USGS 4/24/2017 http://nas.er.usgs.gov/default.aspx	−73.423300	42.157600	1972
Lake Buel	MassDEP	Housatonic River Watershed 2002 Water Quality Assessment Report 21-AC-4 CN141.5, September 2007	−73.275920	42.169491	1972
Lake Buel	USGS	USGS 4/24/2017 http://nas.er.usgs.gov/default.aspx	−73.274600	42.169500	1972
Fisk Pond	DCR	DCR Data from Lakes and Ponds, 2008	−71.368351	42.281982	1972
Fisk Pond	USGS	USGS 4/24/2017 http://nas.er.usgs.gov/default.aspx	−71.367800	42.283400	1972
Fisk Pond	USGS	USGS 4/24/2017 http://nas.er.usgs.gov/default.aspx	−71.367800	42.283400	1972
Mill Pond	USGS	USGS 4/24/2017 http://nas.er.usgs.gov/default.aspx	−73.370600	42.123300	1972
Stockbridge Bowl	USGS	USGS 4/24/2017 http://nas.er.usgs.gov/default.aspx	−73.318300	42.334800	1972
Flint Pond	College of the Holy Cross	From Robert Bertin	−71.725087	42.241352	1995
Lake Quinsigamond	College of the Holy Cross	From Robert Bertin	−71.749052	42.256062	1995
Lake Quinsigamond	College of the Holy Cross	From Robert Bertin	−71.749052	42.256062	1995

(Continued)

TABLE 7.2
(Continued)

Waterbody	Agency	Report Source	Longitude	Latitude	First Year Reported
Lake Quinsigamond	MassDEP	Blackstone River Basin 1998 Water Quality Assessment Report 51-AC-1 48.0, May 2011	−71.749052	42.256062	1995
Lake Quinsigamond	USGS	USGS 4/24/2017 http://nas.er.usgs.gov/default.aspx	−71.755100	42.271000	1995
Cheshire Reservoir, South Basin	MassDEP	Hudson River Basin 1997 Water Quality Assessment Report 11/12/13-AC-1 15.0, January 2000	−73.198397	42.515411	1997
Cheshire Reservoir, South Basin	DCR	DCR Data from ACT, GeoSyntec EM-ACT W-A. Madden FEW	−73.198397	42.515411	1997
Manchaug Pond	MassDEP	Blackstone River Basin 1998 Water Quality Assessment Report 51-AC-1 48.0, May 2011	−71.775873	42.098745	1998
Woolshop Pond	MassDEP	Blackstone River Basin 1998 Water Quality Assessment Report 51-AC-1 48.0, May 2011	−71.748620	42.195842	1998
Newfield Pond	MassDEP	Merrimack River Basin 1999 Water Quality Assessment Report 84-AC-1 52.0, November 2001	−71.389923	42.633784	1999
Long Pond	MassDEP	Merrimack River Basin 1999 Water Quality Assessment Report 84-AC-1 52.0, November 2001	−71.370878	42.692695	1999
Massapoag Pond	MassDEP	Merrimack River Basin 1999 Water Quality Assessment Report 84-AC-1 52.0, November 2001	−71.496539	42.648508	1999
Spectacle Pond	MassDEP	Merrimack River Basin 1999 Water Quality Assessment Report 84-AC-1 52.0, November 2001	−71.516164	42.564009	1999
Spectacle Pond	Individual	From Joy Trahan-Liptak	−71.516164	42.564009	1999
Lake Mascuppic	MassDEP	Merrimack River Basin 1999 Water Quality Assessment Report 84-AC-1 52.0, November 2001	−71.384182	42.677679	1999
Forge Pond	MassDEP	Merrimack River Basin 1999 Water Quality Assessment Report 84-AC-1 52.0, November 2001	−71.489933	42.574972	1999
Gravel Pit Pond	MassDEP	Shawsheen River Watershed 2000 Water Quality Assessment Report 83-AC-2 86.0, July 2003	−71.160460	42.672206	2000

Location	Source	Reference	Longitude	Latitude	Year
Lake Cochituate	USGS	USGS 4/24/2017 http://nas.er.usgs.gov/default.aspx	−71.369000	42.314100	2000
Concord River	MassDEP	SuAsCo Watershed 2001 Water Quality Assessment Report 82-AC-1 CN 92.0, August 2005	−71.308410	42.513430	2001
Clay Pit Pond	USGS	USGS 4/24/2017 http://nas.er.usgs.gov/default.aspx	−71.164500	42.394000	2001
Muddy River	USGS	USGS 4/24/2017 http://nas.er.usgs.gov/default.aspx	−71.092600	42.351500	2001
Charles River	USGS	USGS 4/24/2017 http://nas.er.usgs.gov/default.aspx	−71.120400	42.369400	2001
Charles River	USGS	USGS 4/24/2017 http://nas.er.usgs.gov/default.aspx	−71.089100	42.356900	2001
Unnamed Pond	USGS	USGS 4/24/2017 http://nas.er.usgs.gov/default.aspx	−71.154800	42.389200	2001
Lily Pond	MassDEP	South Shore Coastal Watersheds 2001 Water Quality Assessment Report 94-AC-2 CN 93.0, March 2006	−70.815914	42.224638	2001
Assabet River	USGS	USGS 4/24/2017 http://nas.er.usgs.gov/default.aspx	−71.358500	42.465100	2001
Concord River	USGS	USGS 4/24/2017 http://nas.er.usgs.gov/default.aspx	−71.358300	42.460000	2001
Sudbury River	USGS	USGS 4/24/2017 http://nas.er.usgs.gov/default.aspx	−71.358400	42.465100	2001
Connecticut River	USGS	USGS 4/24/2017 http://nas.er.usgs.gov/default.aspx	−72.556500	42.511100	2001
Prospect Lake	Individual	From Joy Trahan-Liptak	−73.451860	42.195137	2001
Prospect Lake	MassDEP	Housatonic River Watershed 2002 Water Quality Assessment Report 21-AC-4 CN141.5, September 2007	−73.451860	42.195137	2001
Prospect Lake	USGS	USGS 4/24/2017 http://nas.er.usgs.gov/default.aspx	−73.452100	42.195500	2001
Eames Brook	MassDEP	SuAsCo Watershed 2001 Water Quality Assessment Report 82-AC-1 CN 92.0, August 2005	−71.433687	42.287723	2001
Farm Pond	MassDEP	SuAsCo Watershed 2001 Water Quality Assessment Report 82-AC-1 CN 92.0, August 2005	−71.426726	42.281553	2001
Sudbury River	USGS	USGS 4/24/2017 http://nas.er.usgs.gov/default.aspx	−71.419100	42.312300	2001
Forge Pond	MassDEP	South Shore Coastal Watersheds 2001 Water Quality Assessment Report 94-AC-2 CN 93.0, March 2006	−70.880004	42.106224	2001
Connecticut River	USGS	USGS 4/24/2017 http://nas.er.usgs.gov/default.aspx	−72.599500	42.212300	2001
Connecticut River	USGS	USGS 4/24/2017 http://nas.er.usgs.gov/default.aspx	−72.602400	42.212800	2001
Connecticut River	USGS	USGS 4/24/2017 http://nas.er.usgs.gov/default.aspx	−72.599800	42.195700	2001
Connecticut River	USGS	USGS 4/24/2017 http://nas.er.usgs.gov/default.aspx	−72.602400	42.212600	2001

(Continued)

TABLE 7.2
(Continued)

Waterbody	Agency	Report Source	Longitude	Latitude	First Year Reported
Pontoosuc Lake	MassDEP	Housatonic River Watershed 2002 Water Quality Assessment Report 21-AC-4 CN141.5, September 2007	−73.248769	42.496022	2001
Pontoosuc Lake	DCR	DCR Data from Lakes and Ponds, 2008	−73.248769	42.496022	2001
Pontoosuc Lake	USGS	USGS 4/24/2017 http://nas.er.usgs.gov/default.aspx	−73.248800	42.496200	2001
Pontoosuc Lake	USGS	USGS 4/24/2017 http://nas.er.usgs.gov/default.aspx	−73.241100	42.503000	2001
Mill Pond	USGS	USGS 4/24/2017 http://nas.er.usgs.gov/default.aspx	−71.502800	42.535300	2001
Hager Pond	MassDEP	SuAsCo Watershed 2001 Water Quality Assessment Report 82-AC-1 CN 92.0, August 2005	−71.487758	42.349036	2001
Assabet River	MassDEP	SuAsCo Watershed 2001 Water Quality Assessment Report 82-AC-1 CN 92.0, August 2005	−71.442206	42.438442	2001
Lake Cochituate	MassDEP	05-G016–02	−71.367725	42.304477	2001
Lake Cochituate	Individual	From Joy Trahan-Liptak	−71.371426	42.305669	2001
Lake Cochituate	MassDEP	SuAsCo Watershed 2001 Water Quality Assessment Report 82-AC-1 CN 92.0, August 2005	−71.371426	42.305669	2001
Lake Cochituate	DCR	DCR Data from Lakes and Ponds, 2003	−71.371426	42.305669	2001
Lake Cochituate	MassDEP	SuAsCo Watershed 2001 Water Quality Assessment Report 82-AC-1 CN 92.0, August 2005	−71.371867	42.300575	2001
Lake Cochituate	Individual	From Joy Trahan-Liptak	−71.369349	42.288313	2001
Lake Cochituate	DCR	DCR Data from Lakes and Ponds, 2004	−71.367055	42.289488	2001
Lake Cochituate	MassDEP	SuAsCo Watershed 2001 Water Quality Assessment Report 82-AC-1CN 92.0, August 2005	−71.367055	42.289488	2001
Connecticut River	USGS	USGS 4/24/2017 http://nas.er.usgs.gov/default.aspx	−72.611200	42.332000	2001

Assabet River	MassDEP	SuAsCo Watershed 2001 Water Quality Assessment Report 82-AC-1CN 92.0, August 2005	-71.625763	42.333629	2001
Bartlett Pond	MassDEP	SuAsCo Watershed 2001 Water Quality Assessment Report 82-AC-1CN 92.0, August 2005	-71.618375	42.317914	2001
Little Chauncy Pond	College of the Holy Cross	From Robert Bertin	-71.617285	42.306107	2001
Little Chauncy Pond	MassDEP	SuAsCo Watershed 2001 Water Quality Assessment Report 82-AC-1CN 92.0, August 2005	-71.617285	42.306107	2001
Onota Lake	Individual	From Joy Trahan-Liptak	-73.281319	42.470479	2001
Onota Lake	MassDEP	Housatonic River Watershed 2002 Water Quality Assessment Report 21-AC-4 CN141.5, September 2007	-73.281319	42.470479	2001
Onota Lake	DCR	DCR Data from Lakes and Ponds, 2008	-73.281319	42.470479	2001
Onota Lake	USGS	USGS 4/24/2017 http://nas.er.usgs.gov/default.aspx	-73.281300	42.469800	2001
Onota Lake	USGS	USGS 4/24/2017 http://nas.er.usgs.gov/default.aspx	-73.282600	42.471000	2001
Richmond Pond	Individual	From Joy Trahan-Liptak	-73.324741	42.415098	2001
Richmond Pond	Individual	From Joy Trahan-Liptak	-73.324741	42.415098	2001
Richmond Pond	MassDEP	Housatonic River Watershed 2002 Water Quality Assessment Report 21-AC-4CN141.5, September 2007	-73.324741	42.415098	2001
Richmond Pond	USGS	USGS 4/24/2017 http://nas.er.usgs.gov/default.aspx	-73.325500	42.414500	2001
Richmond Pond	USGS	USGS 4/24/2017 http://nas.er.usgs.gov/default.aspx	-73.325000	42.415500	2001
Lily Brook	USGS	USGS 4/24/2017 http://nas.er.usgs.gov/default.aspx	-73.299400	42.335500	2001
Assabet River	MassDEP	06-A027-03	-71.508442	42.411634	2001
Assabet River	MassDEP	SuAsCo Watershed 2001 Water Quality Assessment Report 82-AC-1CN 92.0, August 2005	-71.490454	42.418232	2001
Carding Mill Pond	MassDEP	SuAsCo Watershed 2001 Water Quality Assessment Report 82-AC-1CN 92.0, August 2005	-71.465805	42.362001	2001
Grist Mill Pond	MassDEP	SuAsCo Watershed 2001 Water Quality Assessment Report 82-AC-1CN 92.0, August 2005	-71.479147	42.355473	2001
Sudbury River	MassDEP	06-A006-07	-71.364671	42.396351	2001
Dudley Pond	MassDEP	SuAsCo Watershed 2001 Water Quality Assessment Report 82-AC-1CN 92.0, August 2005	-71.372274	42.330103	2001

(Continued)

TABLE 7.2
(Continued)

Waterbody	Agency	Report Source	Longitude	Latitude	First Year Reported
Card Pond	USGS	USGS 4/24/2017 http://nas.er.usgs.gov/default.aspx	−73.366800	42.326700	2001
Shaker Mill Pond	MassDEP	Housatonic River Watershed 2002 Water Quality Assessment Report 21-AC-4 CN141.5, September 2007	−73.370704	42.339862	2001
Shaker Mill Pond	DCR	DCR Data from National Heritage 2005	−73.370704	42.339862	2001
Shaker Mill Pond	USGS	USGS 4/24/2017 http://nas.er.usgs.gov/default.aspx	−73.369100	42.338100	2001
Pequot Pond	MassDEP	Westfield River Watershed 2001 Water Quality Assessment Report 32-AC-CN 090.0, April 2005	−72.693713	42.184005	2001
Chebacco Lake	MassDEP	North Shore Coastal Watersheds 2002 Water Quality Assessment Report 93-AC-2 CN 138.5, March 2007	−70.811152	42.611984	2002
Mansfield Pond	MassDEP	Housatonic River Watershed 2002 Water Quality Assessment Report 21-AC-4 CN141.5, September 2007	−73.367856	42.203242	2002
Goose Pond Brook	MassDEP	Housatonic River Watershed 2002 Water Quality Assessment Report 21-AC-4 CN141.5, September 2007	−73.226149	42.294173	2002
Laurel Lake	MassDEP	Housatonic River Watershed 2002 Water Quality Assessment Report 21-AC-4 CN141.5, September 2007	−73.269335	42.325463	2002
Laurel Lake	USGS	USGS 4/24/2017 http://nas.er.usgs.gov/default.aspx	−73.269400	42.325600	2002
Flax Pond	MassDEP	North Shore Coastal Watersheds 2002 Water Quality Assessment Report 93-AC-2 CN 138.5, March 2007	−70.951436	42.482703	2002
Lake Garfield	MassDEP	Housatonic River Watershed 2002 Water Quality Assessment Report 21-AC-4 CN141.5, September 2007	−73.197211	42.182427	2002
Lake Garfield	DCR	DCR Data from Lakes and Ponds, 2004	−73.197212	42.182427	2002
Stevens Pond	MassDEP	Housatonic River Watershed 2002 Water Quality Assessment Report 21-AC-4 CN141.5, September 2007	−73.270578	42.180077	2002
Dug Pond	MassDEP	Charles River Watershed 2002–2006 Water Quality Assessment Report 72-AC-4 CN136.5, April 2008	−71.364385	42.276312	2002
Nonesuch Pond	MassDEP	Charles River Watershed 2002–2006 Water Quality Assessment Report 72-AC-4 CN136.5, April 2008	−71.329177	42.324990	2002

Lake Quannapowitt	MassDEP	North Shore Coastal Watersheds 2002 Water Quality Assessment Report 93-AC-2 CN 138.5, March 2007	-71.078556	42.514938	2002
Mirror Lake	MassDEP	Charles river watershed 2002–2006 water quality Assessment report 72-AC-4 CN136.5, April 2008	-71.325293	42.088039	2002
Flannagan Pond	MassDEP	Nashua River Watershed 2003 Water Quality Assessment Report 81-AC-4 CN107.5, August 2008	-71.567523	42.557947	2003
Grove Pond	MassDEP	Nashua River Watershed 2003 Water Quality Assessment Report 81-AC-4 CN107.5, August 2008	-71.584489	42.552566	2003
Center Pond	DCR	DCR Data from Lakes and Ponds, 2003	-73.068746	42.296718	2003
Barton Cove	MassDEP	Connecticut River Basin 2003 Water Quality Assessment Report 34-AC-2 CN 105.5, October 2008	-72.539479	42.605090	2003
Robbins Pond	MassDEP	Nashua River Watershed 2003 Water Quality Assessment Report 81-AC-4 CN107.5, August 2008	-71.604653	42.537585	2003
Lake Bray	MassDEP	Connecticut River Basin 2003 Water Quality Assessment Report 34-AC-2 CN 105.5, October 2008	-72.616671	42.266660	2003
Lake Whalom	College of the Holy Cross	From Robert Bertin	-71.740747	42.574132	2003
Lake Whalom	MassDEP	Nashua River Watershed 2003 Water Quality Assessment Report 81-AC-4 CN107.5, August 2008	-71.740747	42.574132	2003
Porter Lake	MassDEP	Connecticut River Basin 2003 Water Quality Assessment Report 34-AC-2CN 105.5, October 2008	-72.563919	42.073505	2003
Baldpate Pond	MassDEP	Parker River Watershed And Coastal Drainage Area 2004–2008 Water Quality Assessment Report 91-AC-3CN 173.0, March 2010	-71.002371	42.698258	2004
Baldpate Pond	DCR	DCR Data from National Heritage 2004	-71.002371	42.698258	2004
Baldpate Pond	DCR	DCR Data from Lakes and Ponds, 2004	-71.002371	42.698258	2004
Russell Pond	MassDEP	Neponset River Watershed 2004 Water Quality Assessment Report 73-AC-1CN170.4, February 2010	-71.073487	42.236256	2004
Nabnasset Pond	MassDEP	Merrimack River Watershed 2004–2009 Water Quality Assessment Report 84-AC-2 CN179.5, January 2010	-71.426779	42.616987	2004
Upper Mystic Lake	MassDEP	Mystic River Watershed And Coastal Drainage Area 2004–2008 Water Quality Assessment Report 71-AC-2 CN170.2 March 2010	-71.148968	42.437293	2004

(Continued)

TABLE 7.2
(Continued)

Waterbody	Agency	Report Source	Longitude	Latitude	First Year Reported
Horn Pond	MassDEP	Mystic River Watershed And Coastal Drainage Area 2004–2008 Water Quality Assessment Report 71-AC-2CN170.2, March 2010	−71.156593	42.468702	2004
Horn Pond	DCR	DCR Data from Lakes and Ponds, 2004	−71.156593	42.468702	2004
Ware River	MassDEP	05-L002–02	−72.064555	42.391214	2005
Ware River	MassDEP	07-K003–02	−72.064555	42.391214	2005
French River	MassDEP	09-P004–05	−71.885074	42.050945	2005
French River	MassDEP	05-M004–04	−71.885074	42.050945	2005
French River	MassDEP	07-M003–04	−71.885074	42.050945	2005
French River	MassDEP	08-N004–04	−71.885074	42.050945	2005
Sudbury River	MassDEP	06-A034–04	−71.394984	42.338646	2005
Mirror Lake	MassDEP	05-G002–01	−71.609790	42.526149	2005
Fort Pond	MassDEP	05-G001–02	−71.688092	42.523470	2005
Millers River	MassDEP	05-H011–06	−72.341352	42.598155	2005
Millers River	MassDEP	05-H013–06	−72.341352	42.598155	2005
Leonards Pond	MassDEP	05-R003–02	−70.804664	41.750263	2005
Congamond Lakes (Middle Basin)	DCR	DCR Data from Lakes and Ponds, 2005	−72.758734	42.028944	2005
Congamond Lakes (North Basin)	DCR	DCR Data from Lakes and Ponds, 2005	−72.753275	42.045111	2005
Congamond Lakes (South Basin)	DCR	DCR Data from Lakes and Ponds, 2005	−72.764429	42.013980	2005
West River	MassDEP	05-K005–03	−71.601117	42.100361	2005
West River	MassDEP	06-G005–03	−71.601117	42.100361	2005
West River	MassDEP	07-J003–03	−71.601117	42.100361	2005
West River	MassDEP	07-J005–03	−71.601117	42.100361	2005

River	Source	ID	Longitude	Latitude	Year
West River	MassDEP	08-K003–04	−71.601117	42.100361	2005
West River	MassDEP	08-K005–04	−71.601117	42.100361	2005
West River	MassDEP	09-L003–04	−71.601117	42.100361	2005
West River	MassDEP	09-L004–04	−71.601117	42.100361	2005
West River	MassDEP	10-F002–04	−71.601117	42.100361	2005
West River	MassDEP	11-F004–04	−71.601117	42.100361	2005
West River	MassDEP	12-E001–04	−71.601117	42.100361	2005
West River	MassDEP	12-E003–04	−71.601117	42.100361	2005
West River	MassDEP	12-E004–04	−71.601117	42.100361	2005
West River	MassDEP	05-K003–03	−71.601117	42.100361	2005
West River	MassDEP	05-K004–03	−71.601117	42.100361	2005
West River	MassDEP	06-G003–03	−71.601117	42.100361	2005
West River	MassDEP	06-G004–03	−71.601117	42.100361	2005
West River	MassDEP	07-J004–03	−71.601117	42.100361	2005
West River	MassDEP	08-A040–06	−71.601117	42.100361	2005
West River	MassDEP	08-K002–04	−71.601117	42.100361	2005
West River	MassDEP	08-K004–04	−71.601117	42.100361	2005
West River	MassDEP	08-K007–04	−71.601117	42.100361	2005
West River	MassDEP	13-G003–04	−71.601117	42.100361	2005
Ware River	MassDEP	05-L003–04	−72.285755	42.238726	2005
Ware River	MassDEP	09-M004–04	−72.285755	42.238726	2005
Ware River	MassDEP	10-G003–04	−72.285755	42.238726	2005
Ware River	MassDEP	10-G004–04	−72.285755	42.238726	2005
Ware River	MassDEP	12-F004–04	−72.285755	42.238726	2005
Ware River	MassDEP	06-H004–04	−72.285755	42.238726	2005
Ware River	MassDEP	06-H005–04	−72.285755	42.238726	2005
Ware River	MassDEP	07-K004–04	−72.285755	42.238726	2005
Ware River	MassDEP	07-K005–04	−72.285755	42.238726	2005
Ware River	MassDEP	08-L001–04	−72.285755	42.238726	2005

(Continued)

TABLE 7.2
(Continued)

Waterbody	Agency	Report Source	Longitude	Latitude	First Year Reported
Ware River	MassDEP	08-L004–04	−72.285755	42.238726	2005
Ware River	MassDEP	08-L005–04	−72.285755	42.238726	2005
Ware River	MassDEP	08-L006–04	−72.285755	42.238726	2005
Ware River	MassDEP	08-L007–04	−72.285755	42.238726	2005
Ware River	MassDEP	12-F005–04	−72.285755	42.238726	2005
Sudbury River	MassDEP	05-J004–02	−71.397364	42.325435	2005
Sudbury River	MassDEP	06-F004–02	−71.397364	42.325435	2005
Sudbury River	MassDEP	09-N004–02	−71.397364	42.325435	2005
Sudbury River	MassDEP	09-N005–02	−71.397364	42.325435	2005
Webster Lake	MassDEP	05-G014–01	−71.848724	42.054350	2005
Whetstone Brook	MassDEP	05-H011–05	−72.361166	42.594057	2005
Whetstone Brook	MassDEP	05-H013–05	−72.361166	42.594057	2005
Crocker Pond	MassDEP	Taunton River Watershed Water Quality Assessment Report 62-AC-1 CN 94.0, December 2005	−71.294789	42.067691	2005
Nashua River	MassDEP	10-K002–05	−71.593150	42.626477	2006
Nashua River	MassDEP	11-K003–05	−71.593150	42.626477	2006
Nashua River	MassDEP	06-L005–05	−71.593150	42.626477	2006
Nashua River	MassDEP	09-R003–05	−71.593150	42.626477	2006
Nashua River	MassDEP	12-J004–05	−71.593150	42.626477	2006
Nashua River	MassDEP	13-L003–04	−71.593150	42.626477	2006
Assabet River	MassDEP	06-A027–01	−71.586136	42.380309	2006
Assabet River	MassDEP	06-A015–08	−71.586136	42.380309	2006
Jones River	MassDEP	06-C013–05	−70.734114	41.990752	2006

Water body	Source	Station	Latitude	Longitude	Year
Wading River	MassDEP	06-B020-03	42.018856	−71.266702	2006
Assabet River	MassDEP	06-A015-07	42.346501	−71.614603	2006
York Lake	MassDEP	06-E013-01	42.095690	−73.181588	2006
Assabet River	MassDEP	07-L004-01	42.304853	−71.628451	2006
Assabet River	MassDEP	09-N004-01	42.304853	−71.628451	2006
Assabet River	MassDEP	09-N005-01	42.304853	−71.628451	2006
Assabet River	MassDEP	10-H002-01	42.304853	−71.628451	2006
Assabet River	MassDEP	12-G004-01	42.304853	−71.628451	2006
Assabet River	MassDEP	13-I004-01	42.304853	−71.628451	2006
Assabet River	MassDEP	06-F004-01	42.304853	−71.628451	2006
Assabet River	MassDEP	08-M003-01	42.304853	−71.628451	2006
Wading River	MassDEP	06-B020-06	41.951851	−71.223728	2006
Benton Pond	MassDEP	06-E015-02	42.183490	−73.043922	2006
Nashua River	MassDEP	Nashua River Watershed 2003 Water Quality Assessment Report 81-AC-4 CN107.5, August 2008	42.630111	−71.607427	2006
Town Brook	MassDEP	06-C001-04	41.954362	−70.663983	2006
Town Brook	MassDEP	06-C006-04	41.954362	−70.663983	2006
Town Brook	MassDEP	06-C013-04	41.954362	−70.663983	2006
Town Brook	MassDEP	06-C015-04	41.954362	−70.663983	2006
Unnamed Tributary	MassDEP	06-C013-03	41.926027	−70.613532	2006
Sudbury Reservoir	College of the Holy Cross	From Robert Bertin	42.321064	−71.510778	2006
Hop Brook	MassDEP	06-A025-05	42.357086	−71.403137	2006
Hop Brook	MassDEP	06-A034-05	42.357086	−71.403137	2006
Quinsigamond River	MassDEP	12-E003-02	42.230538	−71.708092	2007
Quinsigamond River	MassDEP	07-I003-05	42.230538	−71.708092	2007
Green River	MassDEP	07-D017-09	42.231769	−73.354869	2007
Charles River	MassDEP	07-A010-06	42.210062	−71.351725	2007
Charles River	MassDEP	07-A014-01	42.147718	−71.513910	2007
Charles River	MassDEP	07-A019-01	42.147718	−71.513910	2007

(Continued)

TABLE 7.2
(Continued)

Waterbody	Agency	Report Source	Longitude	Latitude	First Year Reported
Charles River	MassDEP	07-A011-05	-71.325382	42.264400	2007
Ten Mile River	MassDEP	07-R014-05	-71.328462	41.982719	2007
Nashua River	MassDEP	12-I004-04	-71.567765	42.669687	2007
Nashua River	MassDEP	07-P005-04	-71.567765	42.669687	2007
Smallpox Brook	MassDEP	07-G013-10	-70.864659	42.850044	2007
Smallpox Brook	MassDEP	07-G023-10	-70.864659	42.850044	2007
Smallpox Brook	MassDEP	07-G027-10	-70.864659	42.850044	2007
Squannacook River	MassDEP	07-P004-03	-71.643810	42.620942	2007
Charles River	MassDEP	07-A017-01	-71.190333	42.365145	2007
Charles River	MassDEP	10-D005-05	-71.264030	42.326259	2007
Millers River	MassDEP	10-I002-05	-72.093055	42.672322	2007
Millers River	MassDEP	12-I004-05	-72.093055	42.672322	2007
Millers River	MassDEP	07-N004-05	-72.093055	42.672322	2007
Fort River	MassDEP	08-C015-06	-72.520655	42.355650	2008
Bowers Brook	MassDEP	08-G015-08	-71.574278	42.551830	2008
Bowers Brook	MassDEP	08-G028-08	-71.574278	42.551830	2008
Bowers Brook	MassDEP	08-G032-08	-71.574278	42.551830	2008
Still River	MassDEP	08-B005-09	-71.645563	42.455860	2008
Still River	MassDEP	08-G015-04	-71.645563	42.455860	2008
Still River	MassDEP	08-G015-05	-71.632431	42.469172	2008
Still River	MassDEP	08-G003-05	-71.632431	42.469172	2008
Still River	MassDEP	08-G028-05	-71.632431	42.469172	2008
Still River	MassDEP	08-G032-05	-71.632431	42.469172	2008
Dunn Brook	MassDEP	08-B021-08	-72.078248	42.215406	2008
Dunn Brook	MassDEP	08-B002-08	-72.078248	42.215406	2008

Dunn Brook	MassDEP	08-B005-08	-72.078248	42.215406	2008
Dunn Brook	MassDEP	08-B032-08	-72.078248	42.215406	2008
Millers River	MassDEP	09-Q004-01	-72.437840	42.597511	2008
Millers River	MassDEP	12-I005-01	-72.437840	42.597511	2008
Millers River	MassDEP	08-P005-01	-72.437840	42.597511	2008
Catacoonamug Brook	MassDEP	08-G031-03	-71.698654	42.566806	2008
Unnamed Tributary	MassDEP	08-A032-06	-71.787751	42.205410	2008
Unnamed Tributary	MassDEP	08-A026-09	-71.787751	42.205410	2008
Chicopee River	MassDEP	08-B016-05	-72.374590	-72.374590	2008
Ware River	MassDEP	08-B019-07	-72.349834	42.191558	2008
Ware River	MassDEP	08-B033-07	-72.349834	42.191558	2008
East Branch Fever Brook	MassDEP	08-B038-03	-72.245617	42.462583	2008
Beaver Dam Brook	MassDEP	08-F009-02	-70.562879	41.923020	2008
Billington Sea	MassDEP	08-F005-01	-70.681001	41.934830	2008
Unnamed Tributary	MassDEP	08-F005-03	-70.673333	41.926684	2008
Priest Brook	MassDEP	12-I003-04	-72.115037	42.682797	2008
Priest Brook	MassDEP	08-P003-04	-72.115037	42.682797	2008
Priest Brook	MassDEP	08-P004-04	-72.115037	42.682797	2008
Priest Brook	MassDEP	10-J002-04	-72.115037	42.682797	2008
Nashua River	MassDEP	08-G028-11	-71.609769	42.578124	2008
Sevenmile River	MassDEP	08-B002-02	-72.008446	42.249995	2008
Sevenmile River	MassDEP	08-B005-02	-72.008446	42.249995	2008
Quaboag River	MassDEP	08-B002-09	-72.149374	42.228702	2008
Neponset River	MassDEP	09-B010-05	-71.086582	42.269342	2009
Neponset River	MassDEP	09-B012-16	-71.086582	42.269342	2009
Monatiquot River	MassDEP	09-C008-05	-70.980501	42.220580	2009
Massapoag Brook	MassDEP	09-B011-05	-71.145895	42.152208	2009

(Continued)

TABLE 7.2
(Continued)

Waterbody	Agency	Report Source	Longitude	Latitude	First Year Reported
Chase Garden Creek	MassDEP	09-D011–16	−70.200558	41.736337	2009
Quivett Creek	MassDEP	09-D001–12	−70.145314	41.743271	2009
French River	MassDEP	09-P004–04	−71.884182	42.024699	2009
Childs River	MassDEP	09-D007–04	−70.524925	41.593138	2009
Childs River	MassDEP	09-D011–04	−70.524925	41.593138	2009
Childs River	MassDEP	09-D014–04	−70.524925	41.593138	2009
Unnamed Tributary	MassDEP	09-D014–03	−70.563034	41.585417	2009
Stetson Brook	MassDEP	09-G005–03	−70.836605	42.014395	2009
Stetson Brook	MassDEP	10-B009–01	−70.836605	42.014395	2009
Stetson Brook	MassDEP	12-B002–02	−70.836605	42.014395	2009
Stetson Brook	MassDEP	11-B006–03	−70.836605	42.014395	2009
Stetson Brook	MassDEP	12-B006–02	−70.836605	42.014395	2009
Stetson Brook	MassDEP	13-B005–03	−70.836605	42.014395	2009
Stetson Brook	MassDEP	13-B008–02	−70.836605	42.014395	2009
Stetson Brook	MassDEP	14-B002–03	−70.836605	42.014395	2009
Stetson Brook	MassDEP	14-B004–02	−70.836605	42.014395	2009
Stetson Brook	MassDEP	14-B006–02	−70.836605	42.014395	2009
Stetson Brook	MassDEP	15-B004–03	−70.836605	42.014395	2009
Weir River	MassDEP	13-C024–06	−70.859106	42.242234	2009
Weir River	MassDEP	09-C005–03	−70.872251	42.234586	2009
Weir River	MassDEP	09-C014–03	−70.872251	42.234586	2009
Weir River	MassDEP	09-C017–03	−70.872251	42.234586	2009
Weir River	MassDEP	09-C014–02	−70.859055	42.242575	2009
Concord River	MassDEP	09-N003–05	−71.302846	42.641726	2009

Location	Source	Code	Latitude	Longitude	Year
Quashnet River	MassDEP	09-D007-06	41.617753	-70.500578	2009
Pine Tree Brook	MassDEP	09-B011-14	42.269480	-71.072817	2009
Fish Pond	MassDEP	09-G010-01	42.120918	-71.687954	2009
Neponset River	MassDEP	09-B005-02	42.171871	-71.185944	2009
Neponset River	MassDEP	09-B010-02	42.171871	-71.185944	2009
Neponset River	MassDEP	09-B011-02	42.171871	-71.185944	2009
Welsh Pond	MassDEP	09-G009-02	42.150633	-71.789985	2009
Neponset River	MassDEP	09-B006-01	42.147119	-71.255889	2009
Neponset River	MassDEP	09-B005-01	42.147119	-71.255889	2009
Shawsheen River	MassDEP	10-D015-08	42.617047	-71.167895	2010
Middle Pond	MassDEP	10-B006-05	41.668707	-70.415504	2010
Shawsheen River	MassDEP	10-D003-05	42.517079	-71.244712	2010
Swift River	MassDEP	10-G004-03	42.267975	-72.332761	2010
North Brook	MassDEP	10-D007-03	42.412294	-71.650556	2010
Unnamed Tributary	MassDEP	10-D005-06	42.319815	-71.395881	2010
Hinckleys Pond	MassDEP	10-B006-02	41.709592	-70.083929	2010
Cranberry Pond	College of the Holy Cross	From Robert Bertin	42.503270	-71.643570	2010
Quinebaug River	MassDEP	11-T009-08	42.106759	-72.148597	2011
Sassaquin Pond	MassDEP	11-B011-01	41.735321	-70.948919	2011
Quaboag River	MassDEP	11-G004-05	42.181430	-72.263543	2011
White Island Pond (West Basin)	MassDEP	11-B002-02	41.809994	-70.627556	2011
White Island Pond (West Basin)	MassDEP	14-B003-02	41.809994	-70.627556	2011

(Continued)

TABLE 7.2 (Continued)

Waterbody	Agency	Report Source	Longitude	Latitude	First Year Reported
Breakneck Brook	MassDEP	11-D024–04	−72.097147	42.042161	2011
Otter River	MassDEP	11-T005–09	−72.094246	42.633838	2011
Monponsett Pond	MassDEP	12-B002–04	−70.836781	42.001467	2012
Monponsett Pond	MassDEP	14-B004–05	−70.836781	42.001467	2012
Monponsett Pond	MassDEP	15-B007–02	−70.836781	42.001467	2012
Monponsett Pond	MassDEP	13-B008–05	−70.836781	42.001467	2012
Turner Hill Country Club Pond	Individual	From Joy Trahan-Liptak	−70.885150	42.662809	2012
Rumford River	MassDEP	13-C045–03	−71.213447	42.005027	2013
White Island Pond (East Basin)	MassDEP	13-B004–01	−70.617531	41.809183	2013
White Island Pond (East Basin)	MassDEP	14-B003–01	−70.617531	41.809183	2013
Long Pond	Individual	From Joy Trahan-Liptak	−71.470250	42.530021	2014
Windsor Lake	USGS	USGS 4/24/2017 http://nas.er.usgs.gov/default.aspx	−73.092600	42.687100	2014
Rice City Pond	Individual	From Joy Trahan-Liptak	−71.622465	42.100003	2014
Rice City Pond	Individual	From Joy Trahan-Liptak	−71.622465	42.100003	2014
Powwow River	MassDEP	15-C017–05	−70.961591	42.865929	2015
Shawsheen River	MassDEP	15-C022–01	−71.150971	42.652195	2015

Location	Organization	Source	Longitude	Latitude	Year
Blackstone River	Individual	From Joy Trahan-Liptak	−71.553560	42.015699	2015
Beaver Brook	MassDEP	15-C013–07	−71.326335	42.668178	2015
Cow Pond Brook	MassDEP	15-C013–01	−71.506158	42.629730	2015
Monponsett Pond	MassDEP	15-B007–01	−70.845254	42.004596	2015
Reservoir	MassDEP	15-B006–02	−70.855628	42.029884	2015
Mill River	MassDEP	15-C024–02	−71.365440	42.121769	2015
Stetson Pond	MassDEP	15-B003–01	−70.826111	42.028496	2015
Stetson Pond	MassDEP	15-B010–01	−70.826111	42.028496	2015
Stony Brook	MassDEP	15-C013–02	−71.447571	42.597591	2015
Stony Brook	MassDEP	15-C023–02	−71.447571	42.597591	2015
Mossy Pond	DCR	From Jamie Carr	−71.701034	42.417156	2016
Coachlace Pond	DCR	From Jamie Carr	−71.697104	42.413367	2016
South Meadow Pond (East Basin)	DCR	From Jamie Carr	−71.705732	42.415348	2016
South Meadow Pond (West Basin)	DCR	From Jamie Carr	−71.713092	42.414889	2016
Charles River	Individual	From Joy Trahan-Liptak	−71.264994	42.257320	2016
Cheshire Reservoir, Middle Basin	DCR	DCR Data from ACT, GeoSyntec EM-ACT W-A. Madden FEW	−73.191871	42.531451	NA
Cheshire Reservoir, North Basin	DCR	DCR Data from ACT, GeoSyntec EM-ACT W-A. Madden FEW	−73.173964	42.547317	NA
White Pond	College of the Holy Cross	From Robert Bertin	−71.716035	42.513314	NA

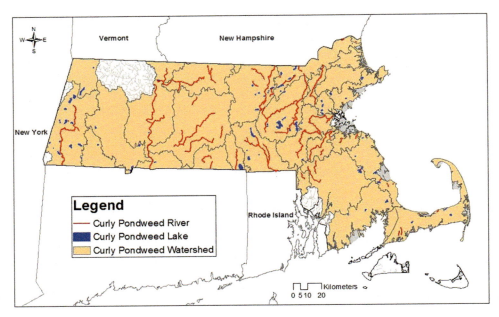

FIGURE 7.3   Distribution map depicting *Potamogeton crispus* in Massachusetts.

released by benthic aquatic organisms can accumulate and dislodge barriers overtime (Barr and Ditomaso 2014). Though benthic barriers need to be routinely maintained to holdfast in their positions, can be expensive, accumulate chemical compounds, and cause the undesired result of chemical pollution to downstream areas, benthic barriers are effective at controlling *P. crispus*. Even so, this method alone is not as effective as combining multiple methods to better manage *P. crispus* outbreaks (Barr and Ditomaso 2014).

CHEMICAL METHODS

Chemical methods may be used to treat curly-leaf pondweed. Endothall, diquat, and fluridone are three effective herbicides (MA DCR2002; Skogerboe and Getsinger 2002; Belgers et al. 2007). Greatest success occurs when herbicides are applied early mid-April–early May before turions are produced (Poovey et al. 2002; Johnson et al. 2012; Bugbee et al. 2015). State permits for herbicides are needed and applications are not suitable to all infestations. Recently, a new treatment method using acetic acid was introduced (Barr and Ditomaso 2014). Acetic acid ($C_2H_4O_2$) is produced by microorganisms naturally in wetland soil. Ultimately, acetic acid, which degrades

to carbon dioxide in aquatic systems, can slow down growth and seed production of invasive species. It is suggested that acetic acid can be combined with benthic barriers to perform as a small-volume high-concentration chemical. More information about chemical control on curly-leaf pondweed and other 18 invasive species can be found in Appendix III It is important to note that the Massachusetts Department of Environmental Protection requires an aquatic nuisance species control license (BRP WM04 Form) before chemical application.

PREVENTION METHODS

Prevention is an efficient and cost-effective approach to stopping the spread of aquatic invasive species. Strategies include boat washing stations, draining water from watercrafts, and disposing of unused bait (Minnesota Department of Natural Resources n.d.). Integrating prevention, education, early detection, control, ecosystem restoration, and environmental law regulation and enforcement will yield the greatest success in protecting waterbodies from invasive species (Wong and Gerstenberger 2015). Public educational programs are important to spread awareness of the impacts and distribution of invasive species. This will facilitate the reporting of new infestations and implementing early season treatment as it is efficient in controlling turion production. Installation of fragment nets and screens is also effective in preventing the spread of *P. crispus* and reducing human-mediated spreading (Mikulyuk and Nault 2009).

So far, no reports on complete eradication of curly-leaf pondweed have been found (Mikulyuk and Nault 2009).

SUCCESS STORY

As stated in the "Managing Actions" section, several herbicides can be an effective tool for the management of *P. crispus*. In this section, two examples of the application of endothall-based herbicides are described which show promising results at reducing the presence of curly-leaf pondweed in targeted and lake wide areas.

The first study occurred in 1999 in a controlled setting consisting of 21 tanks that had curly-leaf pondweed growing in them (Poovey et al. 2002). Prior to this study, herbicide applications mainly occurred in late spring or early summer after turions had already developed. These types of applications were effective at reducing plant biomass but were not effective at long-term management and reducing turion production. It was theorized that applying herbicides in the spring, when water temperature is cooler, could result in more effective long-term control of the invasive plant. Applying herbicides during this time would prove less harmful to native aquatic plants which are generally less active in the spring. Therefore, the treatments often occurred in March, April, and May. Diquat and endothall were applied to the tanks in different rates depending on the ambient temperature. Treatment evaluations were based on recorded plant biomass (the shoots and roots of the plant), turion numbers, and a visual assessment of the plants. Tanks that were treated with diquat showed a shoot biomass decrease of 60% and a root and rhizome decrease of 70% compared to a control tank that received no treatment. Turion numbers decreased by more than 85%. Tanks that were treated with endothall showed a shoot biomass decrease ranging from 60% to 90% and a root

and rhizome decrease ranging from 60% to >90% depending on the temperature. Turion numbers decreased by 55% to >90% depending on the temperature. Poovey et al. suggested the use of short exposure of diquat in spring to reduce plant biomass and turion production, and endothall for larger areas, such as a whole lake treatment. The results showed that utilizing endothall-based herbicides in cooler waters during spring could result in a form of long-term control for curly-leaf pondweed.

A second study occurred between 2006 and 2009 in nine lakes in Minnesota. Treatments occurred in the early spring of each year and consisted of low doses of herbicide (Johnson et al. 2012). Treatments continued for up to five consecutive years. Three additional lakes were identified and left untreated to serve as a control. Of the nine lakes, eight were treated exclusively with endothall, and the remaining lake was treated with fluridone between 2005 and 2007. This remaining lake was then treated with endothall between 2008 and 2009 to allow for the recovery of native plants which had declined due to the fluridone. In this case, a fluridone mixture was dispersed across the entire lake. The endothall treatments were spot treatments, i.e., targeted to specific areas of each lake where curly-leaf pondweed had been identified. Surveys were conducted in each lake to assess the frequency of *P. crispus*, measure plant biomass, and record the number of turions. Results showed that lake wide, early season treatments of both endothall and fluridone had a substantial effect on reducing the frequency of curly-leaf pondweed on a year-to-year basis. Plant biomass also saw a reduction in treated lakes versus the control lakes. Additionally, in lakes that received a treatment, no dense matts of curly-leaf pondweed were observed on the water's surface. Turion production was also greatly reduced as an effect of the herbicide treatments, with >99% reduction in treated lakes when compared to control lakes. There was noticeable variability in the three metrics between certain years, most likely due to differences in climatic conditions.

These results coincide with those of Poovey et al. (2002) implying that treating lakes in early spring in cool water conditions can reduce the biomass of plants and lead to long-term management and control through a reduction in turion production. This approach could also present a lower risk to local aquatic species, which are less active during early spring. Additionally, Johnson et al. (2012) found this treatment approach can work at a lake-wide level, and not just at a localized level within a given waterbody.

Endothall-based herbicide treatments alone will not solve the curly-leaf pondweed issue. Instead, it is necessary to take an integrative and adaptive management approach that utilizes prevention, early season herbicide treatments, and public education. This can help reduce the abundance of curly-leaf pondweed and other invasive aquatic plant species in Massachusetts.

REFERENCES

Acy, C. 2019. AIS spotlight—Curly Leaf Pondweed: The Aquatic LASAGNA plant? *Fox-Wolf Watershed Alliance*. https://fwwa.org/2019/01/23/ais-spotlight-curly-leaf- pondweed/ (Accessed July 3, 2021).

Barr, T.C.III., and J.M. Ditomaso. 2014. Curly leaf pondweed (*Potamogeton crispus*) turion control with acetic acid and benthic barriers. *Journal of Aquatic Plant Management* 52: 31–38.

Belgers, J.D.M., R.J. Van, Lieverloo., L.J.T. Van Der, Pas., and P.J. Van Der, Brink. 2007. Effects of the herbicide 2,4-D on the growth of nine aquatic macrophytes. *Aquatic Botany* 86: 260–268.

Bolduan, B.R., G.C. Van Eeckout, H.W. Quade, and J.E. Gannon. 1994. *Potamogeton crispus*– the other invader. *Lake and Reservoir Management* 10(2): 113–125.

Bugbee, G.J., J.A. Gibbons, and M. June-Wells. 2015. Efficacy of single and consecutive early-season diquat treatments on curlyleaf pondweed and associated aquatic macrophytes: A case study. *Journal of Aquatic Plant Management* 55: 171–177.

Caffrey, J.M., M. Millane, S. Evers, H. Moran, and M. Butler. 2010. A novel approach to aquatic weed control and habitat restoration using biodegradable jute matting. *Aquatic Invasions* 5: 123–129.

Cao, Y., and G. Wang. 2007. Effects of submersed plant *Potamogeton crispus* on suspended mud and sand. *Journal of Ecology and Rural Environment* 23(1): 54–56.

Deng, H., J. Zhang, S. Chen, L. Yang, D. Wang, and S. Yu. 2015. Metal release/accumulation during the decomposition of *Potamogeton crispus* in a shallow macrophytic lake. *Journal of Environmental Sciences* 42: 71–78.

Du, Y., J. Feng, R. Wang, H. Zhang, and J. Liu. 2015. Effects of flavonoids from *Potamogeton crispus* L. on proliferation, migration, and invasion of human ovarian cancer cells. *PLoS One* 10(6). doi: 10.1371/journal.pone.0130685

GISD (Global Invasive Species Database). 2006 Species profile: *Potamogeton crispus*. www.iucngisd.org/gisd/species. php?sc=447 (Accessed July 7, 2017).

Hafez, M.B., N. Hafez, and Y.S. Ramadan. 1992. Uptake of cerium, cobalt, and cesium by *Potamogeton crispus*. *Journal of Chemical Technology and Biotechnology* 54: 337–340.

Hafez, N., S. Abdalla., and Y.S. Ramadan. 1998. Accumulation of phenol by *Potamogeton crispus* from aqueous industrial waste. *Bulletin of Environmental Contamination and Toxicology* 60: 944–948.

IL DNR (Illinois Department of Natural Resources). 2009. Aquatic invasive species: Curlyleaf pondweed. www.in.gov/dnr/files/ CURLYLEAF_PONDWEED.pdf.

ITIS (Integrated Taxonomic Information System). 2021. www.itis. gov/servlet/SingleRpt/SingleRpt?search_topic=TSN&search_value=3 9007#null (Accessed January 29, 2021).

Jensen, D. 2009. Curlyleaf pondweed (Potamogeton crispus). www. seagrant.umn.edu/ais/fieldguide#curlyleaf.

Johnson, J.A., A.R. Jones, and R.M. Newman. 2012. Evaluation of lake wide, early season herbicide treatments for controlling invasive curly leaf pondweed (Potamogeton crispus) in Minnesota lakes. *Lake and Reservoir Management*, 28(4): 346–363.

Kaplan, Z., and J. Fehrer. 2004. Evidence for the hybrid origin of *Potamogeton* x *cooperi* (Potamogetonaceae): Traditional morphology-based taxonomy and molecular techniques in concert. *Folia Geobotanica* 39: 431–453.

MA DCR (Massachusetts Department of Conservation and Recreation). 2002. Curly-leaved Pondweed: An Invasive Aquatic Plant. www.mass.gov/eea/docs/dcr/watersupply/lakepond/factsheet/ curly-leaved- pondweed.pdf (Accessed May 10, 2017).

MA DCR (Massachusetts Department of Conservation and Recreation). 2015. A Guide to Aquatic Plants in Massachusetts, p. 32. www. mass.gov/doc/dcr-guide-to-aquatic-plants-in- massachusetts/ download

Mikulyuk, A., and M.E. Nault. 2009. Curly-leaf Pondweed (*Potamogeton crispus*): A Technical Review of Distribution, Ecology, Impacts, and Management. Wisconsin Department of Natural Resources Bureau of Science Services, PUB-SS-1052 2009. Madison, Wisconsin, USA.

Minnesota Department of Natural Resources. n.d. Curly-leaf pondweed *(Potamogeton crispus)*. www.dnr.state.mn.us/invasives/aquaticplants/curlyleaf_pondweed.html (Accessed December 1, 2020).

Nichols, S.A., and B.H. Shaw. 1986. Ecological life histories of the three aquatic nuisance plants, *Myriophyllum spicatum, Potamogeton crispus* and *Elodea canadensis*. *Hydrobiologia* 131: 3–21.

Poovey, A. 2008. *Curlyleaf pondweed (Potamogeton crispus L.).* Washington, DC: Department of Washington Education.

Poovey, A.G., J.G. Skogerboe, and C.S. Owens. 2002. Spring treatments of diquat and endothall for curly leaf pondweed control. *Journal of Aquatic Plant Management* 40: 63–67.

Ren, D., and S. Zhang. 2008. Separation and identification of the yellow carotenoids in *Potamogeton crispus* L. *Food Chemistry* 106: 410–414.

Skogerboe, J.G., and K.D. Getsinger. 2002. Endothall species selectivity evaluation: northern latitude aquatic plant community. *Journal of Aquatic Plant Management* 40: 1–5.

Stuckey, R.L. 1979. Distributional history of *Potamogeton crispus* (curly pondweed) in North America. *Bartonia* 46: 22–42.

Swistock, B. 2021. Curly-leaf pondweed. https://extension.psu.edu/ curly-leaf-pondweed (Accessed July 2, 2021).

Tomaino, A. 2004. *Potamogeton crispus*: U.S. invasive species impact rank (i-rank). NatureServe, Arlington, VA. www. natureserve.org/explorer (Accessed May 17, 2017).

Washington State Noxious Weed Control Board. 2004. Written findings of the Washington State Noxious Weed Control Board. May 11. www.nwcb.wa.gov/images/weeds/Potamogeton-crispus-2004.pdf (Accessed May 11, 2017).

WI DNR (Wisconsin Department of Natural Resources). 2012. Curly-leaf pondweed (Potamogeton crispus). http://dnr. wi.gov/topic/invasives/fact/curlyleafpondweed.html. (Accessed April, 2013).

Wong, W.H., and S.L. Gerstenberger. 2015. *Biology and Management of Invasive Quagga and Zebra Mussels in the Western United States*. Boca Raton, FL: CRC Press, p. 566.

Wu, J., S. Cheng, W. Liang, F. He, and Z. Wu. 2009. Effects of sediment anoxia and light on turion germination and early growth of Potamogeton crispus. *Hydrobiologia* 628(1): 111–119.

Yeo, R.R. 1966. Yield of propagules of certain aquatic plants. *Weeds* 14: 110–113.

8 Eurasian Milfoil (*Myriophyllum spicatum*) and Variable Milfoil (*Myriophyllum heterophyllum*)

ABSTRACT

Eurasian milfoil (*Myriophyllum spicatum*) and variable milfoil (*Myriophyllum heterophyllum*) are among the most prevalent aquatic invasive species known to occur in Massachusetts waters. These species can become established rapidly once introduced to a new waterbody and are costly to control. As of this writing, there are 115 reported waterbodies infested by Eurasian milfoil and 245 sightings for variable milfoil in Massachusetts. In addition to these two invasive milfoil species, many unidentified (unverified) milfoil species have yet to be categorized as either a native or an invasive milfoil. Therefore, a more comprehensive monitoring of invasive milfoils is needed to increase public awareness, which can be useful in minimizing the spread of invasive milfoils in Massachusetts. Chemicals (fluridone, 2,4-D, florpyrauxifen-benzyl, and triclopyr), physical controls (e.g., lake drawdown and diver-assisted or hand harvest), and biological methods (e.g., American native weevil) are used to manage populations of invasive milfoil species once they become established.

INTRODUCTION

Taxonomy Hierarchy (ITIS 2021)

Kingdom	Plantae
Subkingdom	Viridiplantae
Infrakingdom	Streptophyta
Superdivision	Embryophyta
Division	Tracheophyta
Subdivision	Spermatophytina
Class	Magnoliopsida
Superorder	Saxifraganae
Order	Saxifragales
Family	Haloragaceae
Genus	*Myriophyllum* L.
Species	*Myriophyllum spicatum* L.
Species	*Myriophyllum heterophyllum* Michx. for variable milfoil

BIOLOGY OF EURASIAN MILFOIL AND VARIABLE MILFOIL

There are several milfoils found in Massachusetts: Eurasian milfoil (invasive, aka Eurasian water-milfoil), variable milfoil (invasive, aka variable-leaf milfoil), and low-watermilfoil (*Myriophyllum humile*, a native species) (Figure 8.1). Other native milfoil species, such as *Myriophyllum farwellii*, has also been reported in Massachusetts (Crow and Hellquist 2000). The Eurasian milfoil is a rooted, submerged invasive aquatic plant. This plant has thin stems which can appear either green, brown, or pinkish white and grow 1–3 m in length (Aiken et al. 1979). As the stem moves away from the main growth of the plant, it becomes thinner. Its feathered leaves are blunt-ended and olive green, which are arranged in whorls of three to six (typically four) with gaps of about an inch or more between leaf whorls. Each leaf has 12 or more leaflets per side (Figures 8.2 and 8.3). The flower stalks of Eurasian milfoil grow above the water, and have feather-like bracts that are distinctive. The flowers of Eurasian milfoil, which are small and yellow in color, have four parts and rise 5–10 cm above the surface of the water (Aiken et al. 1979).

Similarly, the variable milfoil is a rooted, submerged invasive plant. The variable milfoil has multiple feather-like leaves that grow 2–4.5 cm long in whorls of four to six around the thick and red stems, giving the plant a resemblance to a "raccoon tail" (Howard 2020) with rounded tips (Figures 8.2 and 8.3). In mid to late summer, the flower stalk of variable milfoil protrudes above the water, with small distinct green bracts on it that are football shaped with toothed edges.

Eurasian and variable milfoil are fairly distinguishable from the native low watermilfoil.

DOI: 10.1201/9781003201106-8

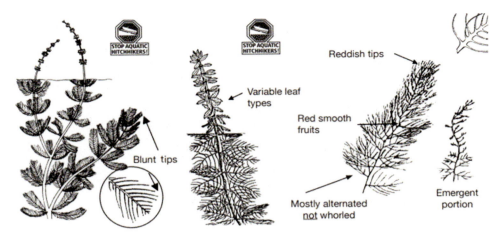

FIGURE 8.1 Eurasian milfoil (*Myriophyllum spicatum*) (*left*), variable milfoil (*Myriophyllum heterophyllum*) (*middle*), and low-watermilfoil (*Myriophyllum humile*) (*right*) (Department of Conservation and Recreation MA 2016).

FIGURE 8.2 Leaves of Eurasian watermilfoil *Myriophyllum spicatum* (*left*) and variable milfoil *Myriophyllum heterophyllum* (*right*). (Photos taken by W. H. Wong.)

Low watermilfoil has stems with limp leaves. In addition, some of these native species have smooth reddish, brownish, or greenish fruits in the axils of their submerged leaves. Low watermilfoil may appear red-tipped with tiny green lobed leaves appearing above the water surface when emergent (Robinson 2002a, 2002b; MA DCR 2016). Because of this, some species of milfoil may be confused with Eurasian milfoil. Even so, invasive Eurasian milfoil can be identified through their leaflet pairs which are usually more than 14, while native species have fewer than 14 pairs (Aiken 1981; Pfingsten et al. 2021). Native milfoils also often have leaves that are staggered along the stem, whereas the invasive milfoils more often have leaves that whorl around the main stem. Additionally, some native species of milfoil produce turion buds (overwintering buds) at the stem tips, while invasive Eurasian milfoil does not. Identification of native milfoils can be tricky, and requires some taxonomic knowledge, and still often genetic verification of the plant if there are no flower stalks or nutlets present.

Like most invasive species, Eurasian milfoil and variable milfoil can grow in various waterbodies as they can tolerate a wide range of temperature, pH, water chemistry, and water nutrient conditions. Both species reside in still or slow-moving waterbodies. However, the

FIGURE 8.3   Stems of Eurasian watermilfoil *Myriophyllum spicatum* (*left*) and variable milfoil *Myriophyllum heterophyllum* (*right*). (Photos taken by W.H. Wong.)

preferences on water chemistry conditions of Eurasian milfoil and variable milfoil are different. Eurasian milfoil is more often found in alkaline waters (pH >7), while variable milfoil is more often found in acidic waters (pH <6.5) (Robinson 2002a; Robinson 2002b). Between a pH of 6 and 7, both species may co-mingle. In terms of their reproductive methods, Eurasian milfoil can reproduce by stem fragmentation and seeds. Over 50% of its plant fragments can resprout, potentially leading to the production of a new plant. Generally, variable milfoil reproduces through fragmentation, rhizome division, budding, and seeds. Stem fragments of the milfoils may attach to boats and other recreational gear, moving plants to other sites. If the watercraft is trailered to another body of water, those hitchhiking fragments can be introduced and become established. Additionally, several hybrids between variable milfoil and native species have been identified. These aggressive hybrid lineages have been found in Connecticut and reproduce vegetatively (Thum and Lennon 2006).

Native Range

Eurasian milfoil and variable milfoil are non-native plants in Massachusetts. Eurasian milfoil is a native plant to Europe, Asia, and North Africa (Robinson 2002b). Variable milfoil is native to Southeast and Midwest of the United States (Howard 2020).

Invasive Range

Both Eurasian milfoil and variable milfoil are on the Massachusetts Prohibited Plants List as they threaten the well-being of ecosystems. Currently, both species are banned from the trade in Massachusetts. The first introduction of the Eurasian milfoil is unknown. Even so, Eurasian milfoil was first observed in North America during the 1800s in New York and Virginia. Since then, this invasive milfoil has been observed in many northern states (Smith and Barko 1990; New York Invasive Species Information 2019).

Variable milfoil is native to many US states but invasive to New England. The first variable milfoil sighting in New England was recorded in 1932 in Bridgeport, Connecticut (Cameron and Wallhead 2004). Since then, variable milfoil can be found in all New England states with recent documentation of variable growth in Vermont (MaineVLMP 2009; VT DEC 2022). It is believed that variable milfoil was introduced to New England by fragments attached to a boat or an aquarium release (Isabel 2011).

ECOLOGICAL IMPACTS

Invasive milfoils have a significant impact on the well-being of native ecosystems. Milfoils can form dense mats in the water column, which can extend to the water surface in as much as 15 ft of water. These mats of invasive milfoil may hinder the water flow, harming other plants and aquatic animals. In addition, the mats can intercept sunlight, which limits native plant growth and alters the chemistry of the water column and sediment (Eiswerth et al. 2000). Invasive milfoils also impact water quality as it increases the nutrient load and temperature, reducing dissolved oxygen. These factors can make the ecosystem less suitable for native species as the removal of native plants threatens both the food source and habitats of native fish and mammals (Eiswerth et al. 2000). This can lead to non-native plants occupying waterbodies, displacing native plants, which can cause fish and other aquatic animals that rely on native plants to relocate or perish, reducing biodiversity.

ECONOMIC IMPACTS

Invasive milfoil infestations can be damaging to local economies, impacting the quality of recreational services, aquaculture, agriculture, and electricity and decreasing property values. Their dense mats, as mentioned earlier, have inhibited boating and fishing as these milfoil mats clog waterways (Invasive Eurasian Watermilfoil 2021). In addition, as a study in Maryland found, invasive milfoil threatens water quality and recreation/industry of fishing, crabbing, oyster production, and clam harvesting (Bayley et al. 1968). Eurasian milfoil can also impact provisioning services such as agriculture and electricity generation by reducing water circulation in irrigation projects and blocking water intakes in power plants (Eiswerth et al. 2000). Furthermore, Olden and Tamayo (2014) found a 19% decline in mean shoreline property sales caused by Eurasian milfoil invasions. Additionally, removing invasive milfoil can be costly as work done by "professional operations range from $100 to $500 per acre" (equivalent to about $400 to $1,200 nowadays) (Rapid Response Plan for Eurasian Watermilfoil n.d.).

MONITORING AND SIGHTINGS

Data about the presence of invasive milfoils in each waterbody were collected by and from a variety of sources, including the Massachusetts Department of Environmental Protection (MassDEP), the Massachusetts Department of Conservation and Recreation (MA DCR), MassDEP Water Quality Assessment Reports, field surveys, personal communications, and EBT Environmental Consults, Inc. Standard operating procedures, including sampling methodology, species identification, and quality control and assurance, were implemented for samples collected by MassDEP (Chase and Wong 2015; MassDEP 2021). Data from other agencies were confirmed by experts. Sites with Aquatic Invasive Species (AIS) were mapped as a data layer with ArcGIS® ArcMap 10.1 (ESRI, Redlands, California).

This chapter mainly addresses the sightings of Eurasian milfoil and variable milfoil. Other unknown milfoil species that were not recognized/hardly distinguishable in Massachusetts waterbodies were documented as "unidentified milfoil." Therefore, characteristics, impacts, and control methods of unidentified milfoil are not discussed in this report because their identification is unknown. The reports and surveys of milfoil locations in Massachusetts have been mapped using Geographic Information Systems (GIS) (Figure 8.4). If regular surveys are conducted to map infestations in the field, GIS can be used to track the spread of invasive milfoils (Figures 8.5–8.8) over time. The map can help to inform future decision-making concerning bodies of water or regulations that could limit the spread of these aquatic invasive species.

Eurasian milfoil and variable milfoil are very aggressive invasive species, and therefore can spread in a short time throughout Massachusetts (Table 8.1). Currently, there are 115, 245, and 32 waterbodies (Tables 8.1 and 8.2) infested by Eurasian milfoil (Figure 8.5), variable milfoil (Figure 8.6), and unidentified milfoil (Figure 8.8), respectively. For these infested waterbodies, 304 are lakes and 32 are rivers. Among these lakes, there are 104 with Eurasian milfoil and 219 with variable milfoil. Meanwhile, 28 lakes were found to support both invasive milfoil species. Among infested rivers, there are 11 and 26 rivers with

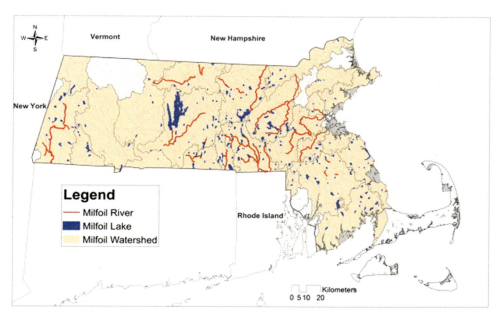

FIGURE 8.4 Distribution map depicting Eurasian milfoil, variable milfoil, and unidentified milfoil in Massachusetts.

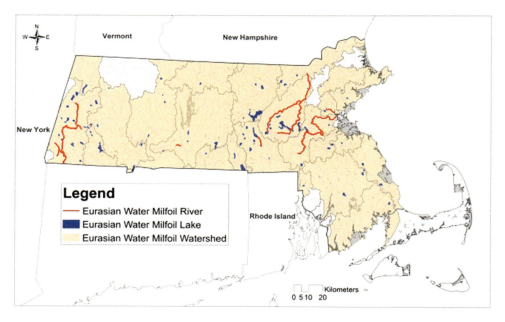

FIGURE 8.5 Distribution maps depicting Eurasian milfoil in Massachusetts.

Eurasian milfoils and variable milfoils. Two rivers were found to have both species. Most watersheds in Massachusetts have been infested by these invasive milfoils, except the following watersheds: Kinderhook River, Bashbish River, Deerfield River, Parker River, Shawsheen River, Narragansett Bay, Cape Cod, and Islands (Figure 8.7 and Table 8.3). While Eurasian milfoil is found statewide (Figure 8.5) and both invasive milfoil species can comingle (Figure 8.7),

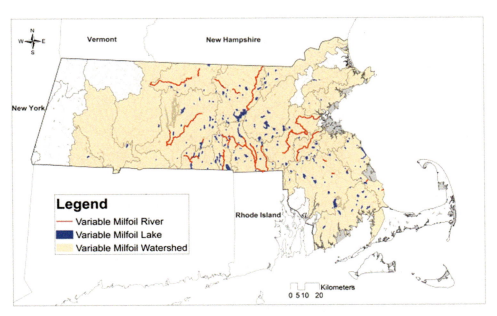

FIGURE 8.6   Distribution maps depicting variable milfoil in Massachusetts.

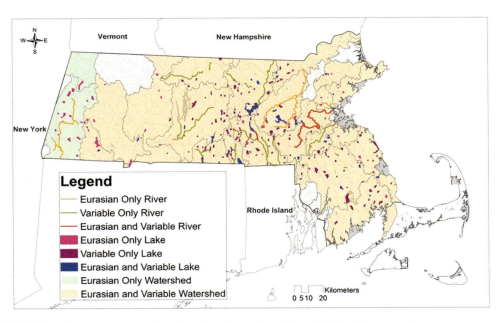

FIGURE 8.7   Distribution maps depicting Eurasian and variable milfoil in Massachusetts.

variable milfoil is mostly detected from eastern to central Massachusetts (Figure 8.6) where waters are relatively lower in pH, alkalinity, or hardness (Smith 1993). This is in agreement with the statement that variable milfoil is often growing in acidic waters (Robinson 2002a; 2002b).

MANAGEMENT

EURASIAN MILFOIL

Many methods have been tested for controlling Eurasian milfoil, including physical, chemical, and biological methods. Physical methods

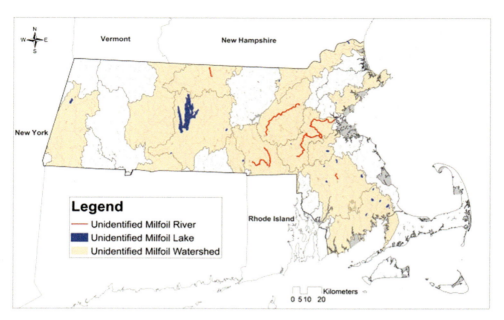

FIGURE 8.8 Distribution maps depicting unidentified milfoil in Massachusetts.

include diver-assisted and hand harvesting, drawdowns, and benthic barriers. Harvesting can effectively reduce small or low-density Eurasian milfoil populations. Because spread of this species occurs by stem pieces, roots, and seeds, harvesting can also cause spread if plant fragments are not carefully managed during harvest events. Fragments may also spread by becoming entangled on boats that are trailered to other bodies of water. Eurasian milfoil fragments are hearty and can survive periods of desiccation while they are transported and unintentionally introduced to a new body of water. Therefore, mechanical harvesting is usually not recommended for plants like milfoil, where fragmentation can be a significant problem by drafting to other sites within the same waterbody or by attaching to boats that can infest other waters. Diver-assisted suction harvesting removes Eurasian milfoil by hand pulling plants and their roots from the bottom sediments. The diver is equipped with a vacuum system on a pontoon boat, and as plants are hand harvested, they are fed up the vacuum tube to a top-side containment method for later disposal. This method can achieve a substantial reduction in milfoil biomass, while having minimal impact on native plants because the diver

may select which plants to harvest. However, it cannot eliminate large-scale or even locally dense milfoil populations in a short time, and requires constant maintenance (Eichler et al. 1994). Similarly, drawdown is a helpful method for removing the rooted plants by lowering the water level. However, it can only be effective when sediments are completely dry or when sediments and plants are removed (Dutartre et al. 2006; Scheers 2016). In addition, to reduce impacts to non-target plants and animals during the growing season, drawdowns are conducted in fall and winter in Massachusetts (Mattson et al. 2004). It is also needed to account for those species that burrow or settle for the winter, and drawdown should be timed and staged to allow for these species to adapt before they become dormant for the season. Following this method, benthic barriers are a control method that limits the plant's growth by creating a cover of clay, silt, sand, or gravel. Benthic barrier covers are usually semipermeable or impermeable "blankets" that are placed over the sediments and held in place with weights or re-bar. Benthic barriers that are fabric can be temporary and removed when plants are controlled. These barriers have been proposed by Massachusetts authorities and are most effective in small areas (such as

TABLE 8.1
Latitude and Longitude for Each Waterbody in Figure 8.1

Table 8.1a Eurasian Milfoil

Waterbody	Agency	Report Source	Longitude	Latitude	First Year Reported
Ashmere Lake	MassDEP	MassDEP/WPP 303d List 2012 extract			
Assabet River	MassDEP	Field Survey 06/26/2017 Wong, Li	−71.449741	42.432064	
Assabet River	MassDEP	13-I005–03	−71.449741	42.432064	
Assabet River	MassDEP	11-H004–03	−71.449741	42.432064	
Assabet River	MassDEP	08-M004–03	−71.449741	42.432064	
Assabet River	MassDEP	06-F005–03	−71.449741	42.432064	
Assabet River	MassDEP	06-F004–03	−71.449741	42.432064	
Assabet River	MassDEP	Field Survey 06/26/2017 Wong, Li	−71.474722	42.423178	
Assabet River	MassDEP	06-A018–04	−71.474722	42.423178	
Assabet River Reservoir	MassDEP	MassDEP/WPP 303d List 2012 extract			
Assabet River Reservoir	College of the Holy Cross	Robert Bertin	−71.646010	42.263618	
Bartlett Pond	MassDEP	MassDEP/WPP 303d List 2012 extract			
Bartons Cove	MassDEP	MassDEP/WPP 303d List 2012 extract			
Bartons Cove	Individual	Joy Trahan-Liptak	−72.537734	42.605267	
Beaver Lake	MassDEP	MassDEP/WPP 303d List 2012 extract			
Benton Pond	MassDEP	MassDEP/WPP 303d List 2012 extract			
Benton Pond	MassDEP	Field Survey 06/20/2017 Wong, Li	−73.043922	42.183490	2006
Benton Pond	MassDEP	06-E015–02	−73.043922	42.183490	2006
Berkshire Pond	MassDEP	MassDEP/WPP 303d List 2012 extract			
Big Bearhole Pond	MassDEP	Taunton River Watershed Water Quality Assessment Report 62-AC-1 CN 94.0, December 2005			
Big Bearhole Pond	DCR	DCR Data from Lakes and Ponds, 2002			
Big Bearhole Pond	DCR	DCR Data from Lakes and Ponds, 2007			
Big Bearhole Pond	DCR	DCR Data from Lakes and Ponds, 2008			
Billington Sea	MassDEP	08-F007–05	−70.681001	41.934830	
Buffumville Lake	College of the Holy Cross	Robert Bertin	−71.914557	42.127838	

Waterbody	Agency	Report Source	Longitude	Latitude	First Year Reported
Buffumville Lake	College of the Holy Cross	Robert Bertin	−71.914557	42.127838	
Center Pond	MassDEP	MassDEP/WPP 303d List 2012 extract			
Center Pond	DCR	DCR Data from Lakes and Ponds, 2003			
Charles River	MassDEP	MassDEP/WPP 303d List 2012 extract			
Charles River	Individual	Joy Trahan-Liptak	−71.264994	42.257320	
Charles River	MassDEP	Field Survey 6/26/2017 Wong, Li	−71.190333	42.365145	2007
Charles River	MassDEP	07-A023–01	−71.190333	42.365145	2007
Charles River	MassDEP	07-A017–01	−71.190333	42.365145	2007
Charles River	MassDEP	Field Survey 6/26/2017 Wong, Li	−71.264030	42.326259	
Charles River	MassDEP	10-D005–05	−71.264030	42.326259	
Charles River	MassDEP	Field Survey 6/26/2017 Wong, Li	−71.254393	42.325843	2007
Charles River	MassDEP	07-A017–06	−71.254393	42.325843	2007
Charles River	MassDEP	MassDEP/WPP 303d List 2012 extract			
Charles River	Individual	Joy Trahan-Liptak	−71.257339	42.353321	
Charles River	Individual	Joy Trahan-Liptak	−71.250210	42.360594	
Charles River	Individual	Joy Trahan-Liptak	−71.259812	42.350683	
Charles River	MassDEP	Field Survey 6/26/2017 Li, Wong	−71.513910	42.147718	2007
Charles River	MassDEP	07-A014–01	−71.513910	42.147718	2007
Chauncy Lake	MassDEP	MassDEP/WPP 303d List 2012 extract			
Cheshire Reservoir, Middle Basin	MassDEP	David Worden			
Cheshire Reservoir, Middle Basin	DCR	DCR Data from ACT,GeoSyntec EM-ACT W-A. Madden FEW			1999
Cheshire Reservoir, North Basin	MassDEP	David Worden			
Cheshire Reservoir, North Basin	DCR	David Worden			1999
Cheshire Reservoir, North Basin	Individual	Joy Trahan-Liptak	−73.173237	42.547261	
Cheshire Reservoir, North Basin	Individual	Joy Trahan-Liptak	−73.173237	42.547261	
Cheshire Reservoir, North Basin	Individual	Joy Trahan-Liptak	−73.173237	42.547261	

(Continued)

TABLE 8.1
Table 8.1a Eurasian Milfoil (Continued)

Waterbody	Agency	Report Source	Longitude	Latitude	First Year Reported
Cheshire Reservoir, South Basin	MassDEP	MassDEP/WPP 303d List 2012 extract			
Cheshire Reservoir, South Basin	DCR	DCR Data from ACT, GeoSyntec EM-ACT W-A. Madden FEW			1999
Chicopee River	MassDEP	Field Survey 06/20/2017 Wong, Li	−72.408085	42.178198	
Chicopee River	MassDEP	08-B012–02	−72.408085	42.178198	
Coes Reservoir	MassDEP	MassDEP/WPP 303d List 2012 extract			
Coes Reservoir	Individual	Joy Trahan-Liptak	−71.841568	42.253969	
Coes Reservoir	College of the Holy Cross	Robert Bertin	−71.841584	42.253967	
Concord River	MassDEP	MassDEP/WPP 303d List 2012 extract			
Concord River	MassDEP	MassDEP/WPP 303d List 2012 extract			
Congamond Lakes, Middle Basin	MassDEP	MassDEP/WPP 303d List 2012 extract			
Congamond Lakes, Middle Basin	DCR	DCR Data from Lakes and Ponds, 2005			2005
Congamond Lakes, North Basin	MassDEP	MassDEP/WPP 303d List 2012 extract			
Congamond Lakes, North Basin	DCR	DCR Data from Lakes and Ponds, 2005			2005
Congamond Lakes, South Basin	MassDEP	MassDEP/WPP 303d List 2012 extract			
Congamond Lakes, South Basin	DCR	DCR Data from Lakes and Ponds, 2005			2005
Cranberry Pond	MassDEP	MassDEP/WPP 303d List 2012 extract			
Dark Brook Reservoir	MassDEP	MassDEP/WPP 303d List 2012 extract			
Dark Brook Reservoir	MassDEP	MassDEP/WPP 303d List 2012 extract			
Dark Brook Reservoir	Individual	Joy Trahan-Liptak	−71.863653	42.193762	
Dark Brook Reservoir	Individual	Joy Trahan-Liptak	−71.863653	42.193762	
Devils Dishfull Pond	MassDEP	MassDEP/WPP 303d List 2012 extract			
Dorothy Pond	MassDEP	MassDEP/WPP 303d List 2012 extract			
Dudley Pond	MassDEP	MassDEP/WPP 303d List 2012 extract			

Waterbody	Agency	Report Source	Longitude	Latitude	First Year Reported
Ellis Pond	MassDEP	MassDEP/WPP 303d List 2012 extract			
Farm Pond	MassDEP	MassDEP/WPP 303d List 2012 extract			
Fisherville Pond	Individual	Joy Trahan-Liptak	−71.692178	42.184408	
Fisherville Pond	Individual	Joy Trahan-Liptak	−71.692178	42.184408	
Fisk Pond	DCR	DCR Data from Lakes and Ponds, 2008			2008
Flint Pond	MassDEP	MassDEP/WPP 303d List 2012 extract			
Flint Pond	MassDEP	MassDEP/WPP 303d List 2012 extract			
Flint Pond	College of the Holy Cross	Robert Bertin	−71.725087	42.241352	
Flint Pond	College of the Holy Cross	Robert Bertin	−71.725087	42.241352	
Flint Pond	College of the Holy Cross	Robert Bertin	−71.725087	42.241352	
Forest Lake	MassDEP	MassDEP/WPP 303d List 2012 extract			
Forge Pond	DCR	DCR Data from ACT,GeoSyntec			
Fort Meadow Reservoir	MassDEP	MassDEP/WPP 303d List 2012 extract			
Fort Meadow Reservoir	DCR	DCR Data from Lakes and Ponds, 2005	−71.547603	42.369941	2005
Framingham Reservoir #1	MassDEP	MassDEP/WPP 303d List 2012 extract			
Framingham Reservoir #3	MassDEP	MassDEP/WPP 303d List 2012 extract			
Goose Pond Brook	MassDEP	MassDEP/WPP 303d List 2012 extract	−73.226149	42.294173	
Greenwater Pond	MassDEP	EBT Environmental Consultants, Inc			
Greenwater Pond	Individual	Joy Trahan-Liptak	−73.149701	42.284945	
Hayden Pond	DCR	DCR Data from Lakes and Ponds, 2002			2002
Hills Pond	MassDEP	MassDEP/WPP 303d List 2012 extract			
Hocomonco Pond	DCR	DCR Data from Lakes and Ponds, 2005			
Horse Pond	MassDEP	MassDEP/WPP 303d List 2012 extract			
Housatonic River	MassDEP	Field Survey 06/16/2017 Wong, Li, Maldonado	−73.240011	42.284636	
Housatonic River	MassDEP	07-D017–03	−73.240011	42.284636	

(Continued)

TABLE 8.1
Table 8.1a Eurasian Milfoil (Continued)

Waterbody	Agency	Report Source	Longitude	Latitude	First Year Reported
Housatonic River	MassDEP	Field Survey 06/16/2017 Wong, Li, Maldonado	−73.354869	42.231769	
Housatonic River	MassDEP	07-D017–09	−73.354869	42.231769	
Hovey Pond	Individual	Joy Trahan-Liptak	−71.714879	42.234214	
Hubbard Brook	MassDEP	MassDEP/WPP 303d List 2012 extract			
Indian Lake	MassDEP	MassDEP/WPP 303d List 2012 extract			
Jamaica Pond	MassDEP	Field Survey 6/26/2017 Li, Wong	−71.119016	42.317541	
Jamaica Pond	MassDEP	09-A010–07	−71.119016	42.317541	
Knops Pond/Lost Lake	MassDEP	MassDEP/WPP 303d List 2012 extract			
Lake Averic	MassDEP	MassDEP/WPP 303d List 2012 extract			
Lake Buel	MassDEP	MassDEP/WPP 303d List 2012 extract			
Lake Cochituate	Individual	Joy Trahan-Liptak	−71.378621	42.319373	
Lake Cochituate	Individual	Joy Trahan-Liptak	−71.378621	42.319373	
Lake Cochituate	MassDEP	MassDEP/WPP 303d List 2012 extract			
Lake Cochituate	DCR	DCR Data from Lakes and Ponds, 2004			2004
Lake Cochituate	Individual	Joy Trahan-Liptak	−71.371426	42.305669	
Lake Cochituate	Individual	Joy Trahan-Liptak	−71.371426	42.305669	
Lake Cochituate	MassDEP	MassDEP/WPP 303d List 2012 extract			
Lake Cochituate	DCR	DCR Data from Lakes and Ponds, 2003			2003
Lake Cochituate	Individual	Joy Trahan-Liptak	−71.371904	42.300685	
Lake Cochituate	MassDEP	MassDEP/WPP 303d List 2012 extract			
Lake Cochituate	Individual	Joy Trahan-Liptak	−71.369349	42.288313	
Lake Cochituate	Individual	Joy Trahan-Liptak	−71.369349	42.288313	
Lake Cochituate	MassDEP	MassDEP/WPP 303d List 2012 extract			
Lake Cochituate	DCR	DCR Data from Lakes and Ponds, 2004			2004
Lake Garfield	MassDEP	MassDEP/WPP 303d List 2012 extract			
Lake Garfield	DCR	DCR Data from Lakes and Ponds, 2004	−73.197212	42.182427	2004

Waterbody	Agency	Report Source	Longitude	Latitude	First Year Reported
Lake Lorraine	MassDEP	Chicopee River Basin 1998 Water Quality Assessment Report 36-AC-2 47.0, April 2001			
Lake Pearl	MassDEP	MassDEP/WPP 303d List 2012 extract			
Lake Quinsigamond	MassDEP	MassDEP/WPP 303d List 2012 extract			
Lake Quinsigamond	College of the Holy Cross	Robert Bertin	−71.749052	42.256062	
Lake Rico	MassDEP	Taunton River Watershed Water Quality Assessment Report 62-AC-1 CN 94.0, December 2005			
Lake Rico	DCR	DCR Data from Lakes and Ponds, 2002			2002
Lake Ripple	College of the Holy Cross	Robert Bertin	−71.697077	42.213605	
Lake Shirley	MassDEP	MassDEP/WPP 303d List 2012 extract			
Lake Shirley	DCR	DCR Data from ACT, GeoSyntec-ACT			
Lake Waban	MassDEP	MassDEP/WPP 303d List 2012 extract			
Lake Whalom	MassDEP	MassDEP/WPP 303d List 2012 extract			
Laurel Lake	MassDEP	MassDEP/WPP 303d List 2012 extract			
Leesville Pond	Individual	Joy Trahan-Liptak	−71.826444	42.229001	
Leverett Pond	MassDEP	MassDEP/WPP 303d List 2012 extract			
Lily Pond, Middle	DCR	Jamie Carr	−71.769540	42.376670	
Little Alum Pond	DCR	DCR Data from Lakes and Ponds			
Long Pond	MassDEP	MassDEP/WPP 303d List 2012 extract			
Mansfield Pond	MassDEP	MassDEP/WPP 303d List 2012 extract			
Middle Pond	MassDEP	Taunton River Watershed Water Quality Assessment Report 62-AC-1 CN 94.0, December 2005			
Middle Pond	DCR	DCR Data from Lakes and Ponds, 2002			2002

(Continued)

TABLE 8.1

Table 8.1a Eurasian Milfoil (Continued)

Waterbody	Agency	Report Source	Longitude	Latitude	First Year Reported
Middle Pond	DCR	DCR Data from Lakes and Ponds, 2007			2007
Monponsett Pond	MassDEP	10-B003–04	−70.836781	42.001467	
Monponsett Pond	MassDEP	14-B002–02	−70.836781	42.001467	
Monponsett Pond	MassDEP	14-B004–05	−70.836781	42.001467	
Monponsett Pond	MassDEP	14-B006–05	−70.836781	42.001467	
Monponsett Pond	MassDEP	14-B008–05	−70.836781	42.001467	
Monponsett Pond	MassDEP	13-B005–06	−70.836781	42.001467	
Monponsett Pond	MassDEP	13-B006–05	−70.836781	42.001467	
Monponsett Pond	MassDEP	13-B008–05	−70.836781	42.001467	
Monponsett Pond	MassDEP	12-B002–04	−70.836781	42.001467	
Monponsett Pond	MassDEP	12-B004–04	−70.836781	42.001467	
Monponsett Pond	MassDEP	12-B006–05	−70.836781	42.001467	
Monponsett Pond	MassDEP	12-B008–05	−70.836781	42.001467	
Monponsett Pond	MassDEP	11-B006–05	−70.836781	42.001467	
Monponsett Pond	MassDEP	11-B008–02	−70.836781	42.001467	
Monponsett Pond	Individual	Joy Trahan-Liptak	−70.837849	42.005701	
Morses Pond	MassDEP	MassDEP/WPP 303d List 2012 extract			
Mystic River	MassDEP	Field Survey 6/26/2017 Li, Wong	−71.100564	42.411965	2009
Mystic River	MassDEP	09-A008–07	−71.100564	42.411965	2009
Newfield Pond	MassDEP	MassDEP/WPP 303d List 2012 extract			
Oldham Pond	MassDEP	South Shore Coastal Watersheds 2001 Water Quality Assessment Report 94-AC-2 CN 93.0, March 2006			
Onota Lake	MassDEP	MassDEP/WPP 303d List 2012 extract			
Onota Lake	DCR	DCR Data from Lakes and Ponds, 2008			2008
Onota Lake	MassDEP	05-G008–01	−73.283941	42.464926	
Onota Lake	Individual	Joy Trahan-Liptak	−73.281319	42.470479	
Orrs Pond	MassDEP	MassDEP/WPP 303d List 2012 extract			
Otis Reservoir	DCR	Data from ACT,GeoSyntec-ACT	−73.040921	42.151148	
Pequot Pond	MassDEP	MassDEP/WPP 303d List 2012 extract			
Plunkett Reservoir	MassDEP	MassDEP/WPP 303d List 2012 extract			

Waterbody	Agency	Report Source	Longitude	Latitude	First Year Reported
Ponkapoag Pond	MassDEP	MassDEP/WPP 303d List 2012 extract			
Pontoosac Lake	MassDEP	MassDEP/WPP 303d List 2012 extract			
Pontoosac Lake	DCR	DCR Data from Lakes and Ponds, 2008			2008
Pontoosac Lake	DCR	DCR Data from ACT, GeoSyntec-ACT			
Purgatory Cove	Individual	Joy Trahan-Liptak	−71.247771	42.358386	
Quaboag Pond	MassDEP	MassDEP/WPP 303d List 2012 extract			
Quaboag Pond	College of the Holy Cross	Robert Bertin	−72.068036	42.197472	
Quacumquasit Pond	MassDEP	MassDEP/WPP 303d List 2012 extract			
Quinsigamond River	Individual	Joy Trahan-Liptak	−71.701242	42.221306	
Quinsigamond River	MassDEP	MassDEP/WPP 303d List 2012 extract			
Reservoir Pond	MassDEP	09-B009–02	−71.126217	42.170467	
Rice City Pond	Individual	Joy Trahan-Liptak	−71.622465	42.100003	
Rice City Pond	Individual	Joy Trahan-Liptak	−71.622465	42.100003	
Richmond Pond	MassDEP	MassDEP/WPP 303d List 2012 extract			
Richmond Pond	Individual	Joy Trahan-Liptak	−73.324741	42.415098	
Richmond Pond	Individual	Joy Trahan-Liptak	−73.324741	42.415098	
Shaker Mill Pond	MassDEP	MassDEP/WPP 303d List 2012 extract			
Shaker Mill Pond	DCR	DCR Data from National Heritage 2005			2005
Shaker Mill Pond	Individual	Joy Trahan-Liptak	−73.369417	42.338953	
Shaw Pond	MassDEP	MassDEP/WPP 303d List 2012 extract			
Silver Lake	DCR	DCR Data from Lakes and Ponds, 2008			2008
Singletary Pond	MassDEP	MassDEP/WPP 303d List 2012 extract			
Sluice Pond	MassDEP	MassDEP/WPP 303d List 2012 extract			
Sluice Pond	DCR	DCR Data from Lakes and Ponds, 2005			2005
Spy Pond	MassDEP	MassDEP/WPP 303d List 2012 extract			
Stetson Pond	MassDEP	Taunton River Watershed Water Quality Assessment Report 62-AC-1 CN 94.0, December 2005			

(Continued)

TABLE 8.1
Table 8.1a Eurasian Milfoil (Continued)

Waterbody	Agency	Report Source	Longitude	Latitude	First Year Reported
Stevens Pond	MassDEP	MassDEP/WPP 303d List 2012 extract			
Stockbridge Bowl	MassDEP	MassDEP/WPP 303d List 2012 extract			
Sudbury Reservoir	College of the Holy Cross	Robert Bertin	−71.510778	42.321064	
Sudbury River	MassDEP	Field Survey 6/26/2017 Li, Wong	−71.397364	42.325435	2008
Sudbury River	MassDEP	08-M003–02	−71.397364	42.325435	2008
Sunset Lake	MassDEP	MassDEP/WPP 303d List 2012 extract			
Thousand Acre Pond	MassDEP	MassDEP/WPP 303d List 2012 extract			
Unionville Pond	DCR	Jamie Carr	−71.842015	42.359704	
Unionville Pond	DCR	Joy Trahan-Liptak	−71.842015	42.359704	
Unionville Pond	MassDEP	MassDEP/WPP 303d List 2012 extract			
Upper Goose Pond	MassDEP	MassDEP/WPP 303d List 2012 extract			
Wachusett Reservoir	DCR	David Worden	−71.787817	42.397107	
Wachusett Reservoir	DCR	David Worden	−71.788251	42.379206	
Wachusett Reservoir	DCR	David Worden	−71.789929	42.389998	
Wachusett Reservoir	DCR	David Worden	−71.787817	42.397107	
Wachusett Reservoir	DCR	David Worden	−71.774821	42.375014	
Wachusett Reservoir	DCR	Jamie Carr	−71.730707	42.380071	
Wachusett Reservoir	DCR	Jamie Carr	−71.789929	42.389998	
Wachusett Reservoir	DCR	Jamie Carr	−71.787817	42.397107	
Wachusett Reservoir	DCR	Jamie Carr	−71.788251	42.379206	
Wachusett Reservoir	DCR	Jamie Carr	−71.795345	42.384187	
Wachusett Reservoir	MassDEP	MassDEP/WPP 303d List 2012 extract			
Wachusett Reservoir	DCR	DCR Data from ACT, GeoSyntec-ACT			
Wachusett Reservoir	College of the Holy Cross	Robert Bertin	−71.787817	42.397107	

Waterbody	Agency	Report Source	Longitude	Latitude	First Year Reported
Wachusett Reservoir	DCR	Joy Trahan-Liptak	−71.787060	42.414454	
Webster Lake	MassDEP	MassDEP/WPP 303d List 2012 extract			
White Island Pond, East Basin	MassDEP	07-T003–02	−70.617531	41.809183	
Whiting Street Reservoir	MassDEP	MassDEP/WPP 303d List 2012 extract			
Willard Brook	MassDEP	MassDEP/WPP 303d List 2012 extract			
Windsor Pond	MassDEP	MassDEP/WPP 303d List 2012 extract			
Windsor Pond	DCR	DCR Data from Lakes and Ponds, 2002			2002
Winning Pond	MassDEP	MassDEP/WPP 303d List 2012 extract			
Woolshop Pond	MassDEP	09-G009–01	−71.749543	42.194774	
Yokum Pond	DCR	DCR Data from Lakes and Ponds, 2003			2003
Yokum Pond	DCR	DCR Data from Lakes and Ponds, 2005			2005

Table 8.1b Variable Milfoil

Waterbody	Agency	Report Source	Longitude	Latitude	First Year Reported
Aaron Reservoir	Individual	Joy Trahan-Liptak	−70.827339	42.206855	
Aaron Reservoir	Individual	Joy Trahan-Liptak	−70.827339	42.206855	
Aaron Reservoir	Individual	Joy Trahan-Liptak	−70.827339	42.206855	
Aldrich Pond	MassDEP	Blackstone River Basin 1998 Water Quality Assessment Report 51-AC-1 48.0, May 2011			
Aldrich Pond	MassDEP	Field Survey 6/26/2017 Wong, Li	−71.740259	42.163314	2009
Aldrich Pond	MassDEP	09-G008–06	−71.740259	42.163314	2009
Ames Long Pond	MassDEP	Taunton River Watershed Water Quality Assessment Report 62-AC-1 CN 94.0, December 2005			
Arcade Pond	MassDEP	Field Survey 6/26/2017 Wong, Li	−71.677520	42.112961	
Arcade Pond	MassDEP	09-G010–02	−71.677520	42.112961	
Arcadia Lake	MassDEP	Connecticut River Basin 2003 Water Quality Assessment Report 34-AC-2 CN 105.5, October 2008			
Ashland Reservoir	MassDEP	SuAsCo Watershed 2001 Water Quality Assessment Report 82-AC-1 CN 92.0, August 2005			

(Continued)

TABLE 8.1

Table 8.1b Variable Milfoil (Continued)

Waterbody	Agency	Report Source	Longitude	Latitude	First Year Reported
Ashland Reservoir	DCR	DCR Data from Lakes and Ponds Pioneer Infestation, 2008	−71.464892	42.240261	2008
Baddacook Pond	MassDEP	05-G001–01	−71.530464	42.618962	
Baddacook Pond	DCR	DCR Data from Lakes and Ponds, 2002	−71.530464	42.618962	
Bailey Road Pond	DCR	Joy Trahan-Liptak	−71.854850	42.343790	
Bare Hill Pond	MassDEP	Nashua River Watershed 2003 Water Quality Assessment Report 81-AC-4 CN107.5, August 2008			
Bare Hill Pond	College of the Holy Cross	Robert Bertin	−71.597616	42.491779	
Bartlett Pond	College of the Holy Cross	Robert Bertin	−71.617660	42.318598	
Beaver Lake	MassDEP	Chicopee River Basin 1998 Water Quality Assessment Report 36-AC-2 47.0, April 2001			
Beaver Pond	MassDEP	Charles River Watershed 2002–2006 Water Quality Assessment Report 72-AC-4 CN136.5, April 2008			
Beaver Pond	DCR	DCR Data from Lakes and Ponds, 2002			
Bennett Pond	Individual	Joy Trahan-Liptak	−72.451889	42.091374	
Billings Street/ East Street Pond	MassDEP	Neponset River Watershed 2004 Water Quality Assessment Report 73-AC-1 CN170.4, February 2010			
Billington Sea	MassDEP	08-F003–05	−70.681001	41.934830	
Black Mountain Pond	MassDEP	South Shore Coastal Watersheds 2001 Water Quality Assessment Report 94-AC-2 CN 93.0, March 2006			
Blackstone River	MassDEP	Field Survey 6/26/2017 Li, Wong	−71.778069	42.202922	
Blackstone River	MassDEP	08-A032–07	−71.778069	42.202922	
Boons Pond	Individual	Joy Trahan-Liptak	−71.502252	42.396447	
Boons Pond	Individual	Joy Trahan-Liptak	−71.502252	42.396447	
Boons Pond	MassDEP	SuAsCo Watershed 2001 Water Quality Assessment Report 82-AC-1 CN 92.0, August 2005			
Bouchard Pond	MassDEP	French and Quinebaug River Watersheds 2004–2008 Water Quality Assessment Report 41/42-AC-2 178.5, November 2009			

Waterbody	Agency	Report Source	Longitude	Latitude	First Year Reported
Brierly Pond	MassDEP	Blackstone River Basin 1998 Water Quality Assessment Report 51-AC-1 48.0, May 2011			
Brigham Pond	DCR	Jamie Carr	−71.998532	42.472474	
Brooks Pond	MassDEP	Chicopee River Basin 1998 Water Quality Assessment Report 36-AC-2 47.0, April 2001			
Brooks Pond	College of the Holy Cross	Robert Bertin	−72.043108	42.302192	
Brooks Pond	College of the Holy Cross	Robert Bertin	−72.043108	42.302192	
Browning Pond	MassDEP	Chicopee River Basin 1998 Water Quality Assessment Report 36-AC-2 47.0, April 2001			
Browning Pond	DCR	DCR Data from Lakes and Ponds, 2007			
Buck Pond	MassDEP	Westfield River Watershed 2001 Water Quality Assessment Report 32-AC-1 CN 090.0, April 2005			
Buffum Pond	MassDEP	French and Quinebaug River Watersheds 2001 Water Quality Assessment Report 41/42-AC-1 51.0, March 2002			
Buffumville Lake	MassDEP	French and Quinebaug River Watersheds 2001 Water Quality Assessment Report 41/42-AC-1 51.0, March 2002			
Buffumville Lake	College of the Holy Cross	Robert Bertin	−71.911424	42.112068	
Burncoat Pond	College of the Holy Cross	Robert Bertin	−71.948645	42.240126	
Carver Pond	MassDEP	Taunton River Watershed Water Quality Assessment Report 62-AC-1 CN 94.0, December 2005			
Cedar Meadow Pond	MassDEP	French and Quinebaug River Watersheds 2001 Water Quality Assessment Report 41/42-AC-1 51.0, March 2002			
Cedar Pond	MassDEP	French and Quinebaug River Watersheds 2001 Water Quality Assessment Report 41/42-AC-1 51.0, March 2002			
Cedar Swamp Pond	MassDEP	Charles River Watershed 2002–2006 Water Quality Assessment Report 72-AC-4 CN136.5, April 2008			

(Continued)

TABLE 8.1
Table 8.1b Variable Milfoil (Continued)

Waterbody	Agency	Report Source	Longitude	Latitude	First Year Reported
Cedar Swamp Pond	Individual	Joy Trahan-Liptak	−71.513609	42.153435	
Cedar Swamp Pond	College of the Holy Cross	Robert Bertin	−71.513609	42.153435	
Chaffin Pond	DCR	Joy Trahan-Liptak	−71.839483	42.331244	
Charles River	MassDEP	Charles River Watershed 2002–2006 Water Quality Assessment Report 72-AC-4 CN136.5, April 2008			
Clark Pond	MassDEP	Neponset River Watershed 2004 Water Quality Assessment Report 73-AC-1 CN170.4, February 2010			
Clear Pond	DCR	DCR Data from Lakes and Ponds, 2005			2005
Cleveland Pond	DCR	DCR Data from Lakes and Ponds, 2008			1999
Coachlace Pond	DCR	Jamie Carr	−71.697104	42.413367	
Cocasset Lake	Individual	Joy Trahan-Liptak	−71.259580	42.058531	
Collins Pond	DCR	DCR Data from Lakes and Ponds, 2008			2008
Comins Pond	College of the Holy Cross	Robert Bertin	−72.195658	42.203643	
Cook Pond	College of the Holy Cross	Robert Bertin	−71.858373	42.285515	
Crane Brook Bog Pond	MassDEP	Buzzards Bay Watershed2000 Water Quality Assessment Report 95-AC-2 085.0, November 2003			
Crocker Pond	DCR	DCR Data from Lakes and Ponds, 2007			2007
Curtis Pond	College of the Holy Cross	Robert Bertin	−71.838502	42.241484	
Dark Brook	MassDEP	Blackstone River Watershed 2003–2007 Water Quality Assessment Report 51-AC-3 CN 240.0, March 2010			
Dark Brook Reservoir	MassDEP	Blackstone River Basin 1998 Water Quality Assessment Report 51-AC-148.0, May 2011			
Dawson Pond	MassDEP	Nashua River Watershed 2003 Water Quality Assessment Report 81-AC-4 CN107.5, August 2008			
Dawson Pond	DCR	Joy Trahan-Liptak	−71.852208	42.335114	

Waterbody	Agency	Report Source	Longitude	Latitude	First Year Reported
Dean Pond	MassDEP	Chicopee River Watershed 2003 Water Quality Assessment Report 36-AC-3 CN 106.5, October 2008			
Dean Pond	DCR	DCR Data from Lakes and Ponds, 2003			2003
Demond Pond	College of the Holy Cross	Robert Bertin	−71.970411	42.352339	
Doane Pond	College of the Holy Cross	Robert Bertin	−72.072370	42.285721	
Eagle Lake	MassDEP	Nashua River Watershed 2003 Water Quality Assessment Report 81-AC-4 CN107.5, August 2008			
Eagle Lake	DCR	Joy Trahan-Liptak	−71.885094	42.356095	
Eagle Lake	College of the Holy Cross	Robert Bertin	−71.884555	42.356373	
East Brimfield Reservoir	MassDEP	French and Quinebaug River Watersheds 2001 Water Quality Assessment Report 41/42-AC-1 51.0, March 2002			
East Brimfield Reservoir	DCR	DCR: Lakes and Ponds, 2007			2007
East Brimfield Reservoir	Individual	Joy Trahan-Liptak	−72.134224	42.109406	
East Brookfield River	MassDEP	Field Survey 06/16/2017 Wong, Li, Maldonado	−72.049996	42.225710	
East Brookfield River	MassDEP	08-B032–05	−72.049996	42.225710	
East Brookfield River	MassDEP	08-B021–05	−72.049996	42.225710	
East Brookfield River	MassDEP	08-B013–05	−72.049996	42.225710	
East Brookfield River	MassDEP	08-B005–05	−72.049996	42.225710	
East Freetown Pond	MassDEP	Taunton River Watershed Water Quality Assessment Report 62-AC-1 CN 94.0, December 2005			
East Head Pond	DCR	DCR Data from Lakes and Ponds, humile 2008			2008
Eddy Pond	MassDEP	Blackstone River Basin 1998 Water Quality Assessment Report 51-AC-1 48.0, May 2011			
Eddy Pond	College of the Holy Cross	Robert Bertin	−71.844001	42.181076	

(Continued)

TABLE 8.1
Table 8.1b Variable Milfoil (Continued)

Waterbody	Agency	Report Source	Longitude	Latitude	First Year Reported
Ellis Pond	MassDEP	Millers River Watershed 2000 Water Quality Assessment Report 35-AC-1 CN089.0, March 2004			
Ellis Pond	College of the Holy Cross	Robert Bertin	−72.204192	42.577244	
Factory Pond	MassDEP	Charles River Watershed 2002–2006 Water Quality Assessment Report 72-AC-4 CN136.5, April 2008			
Falls Pond, South Basin	MassDEP	Ten Mile River Watershed 2002 Water Quality Assessment Report 52-AC-2 CN 137.5, June 2006			
Farrington Pond	MassDEP	Neponset River Watershed 2004 Water Quality Assessment Report 73-AC-1 CN170.4, February 2010			
Federal Pond	MassDEP	Buzzards Bay Watershed2000 Water Quality Assessment Report 95-AC-2 085.0, November 2003			
Field Pond	DCR	DCR Data from Lakes and Ponds, 2007			2007
First Pond	DCR	DCR Data from Lakes and Ponds, 2006			2006
Fish Pond	MassDEP	Blackstone River Basin 1998 Water Quality Assessment Report 51-AC-1 48.0, May 2011			
Fisk Pond	MassDEP	SuAsCo Watershed 2001 Water Quality Assessment Report 82-AC-1 CN 92.0, August 2005			
Fisk Pond	DCR	DCR Data from Lakes and Ponds, 2008			2008
Flannagan Pond	MassDEP	Nashua River Watershed 2003 Water Quality Assessment Report 81-AC-4 CN107.5, August 2008			
Flint Pond	MassDEP	Blackstone River Basin 1998 Water Quality Assessment Report 51-AC-1 48.0, May 2011			
Flint Pond	Individual	Joy Trahan-Liptak	−71.725688	42.240948	
Flint Pond	College of the Holy Cross	Robert Bertin	−71.725087	42.241352	

Waterbody	Agency	Report Source	Longitude	Latitude	First Year Reported
Flint Pond	MassDEP	Merrimack River Basin 1999 Water Quality Assessment Report 84-AC-1 52.0, November 2001			
Flint Pond	DCR	DCR Data from Lakes and Ponds, 2008			2008
Fort Meadow Reservoir	DCR	DCR Data from Lakes and Ponds, 2005	−71.547603	42.369941	2005
Framingham Reservoir #1	MassDEP	SuAsCo Watershed 2001 Water Quality Assessment Report 82-AC-1 CN 92.0, August 2005			
French River	MassDEP	Field Survey 06/16/2017 Wong, Li, Maldonado	−71.885074	42.050945	
French River	MassDEP	10-I002–05	−71.885074	42.050945	
French River	MassDEP	09-P004–05	−71.885074	42.050945	
French River	MassDEP	08-N006–04	−71.885074	42.050945	
French River	MassDEP	08-N005–04	−71.885074	42.050945	
French River	MassDEP	08-N004–04	−71.885074	42.050945	
French River	MassDEP	08-N003–04	−71.885074	42.050945	
French River	MassDEP	07-M005–04	−71.885074	42.050945	
French River	MassDEP	07-M004–04	−71.885074	42.050945	
French River	MassDEP	07-M003–04	−71.885074	42.050945	
French River	MassDEP	07-M001–04	−71.885074	42.050945	
French River	MassDEP	06-J006–04	−71.885074	42.050945	
French River	MassDEP	06-J005–04	−71.885074	42.050945	
French River	MassDEP	06-J004–04	−71.885074	42.050945	
French River	MassDEP	06-J003–04	−71.885074	42.050945	
French River	MassDEP	05-M004–04	−71.885074	42.050945	
French River	MassDEP	05-M003–04	−71.885074	42.050945	
Frye Pond	DCR	DCR Data from Lakes and Ponds, 2008			
Fuller Street Pond	MassDEP	Taunton River Watershed Water Quality Assessment Report 62-AC-1 CN 94.0, December 2005			
Gaston Pond	College of the Holy Cross	Robert Bertin	−72.130135	42.455926	
Gaston Pond	College of the Holy Cross	Robert Bertin	−72.130135	42.455926	
Gavins Pond	MassDEP	Taunton River Watershed Water Quality Assessment Report 62-AC-1 CN 94.0, December 2005			
Gilboa Pond	College of the Holy Cross	Robert Bertin	−71.701442	42.081995	
Glen Echo Pond	MassDEP	Neponset River Watershed 2004 Water Quality Assessment Report 73-AC-1 CN170.4, February 2010			

(Continued)

TABLE 8.1

Table 8.1b Variable Milfoil (Continued)

Waterbody	Agency	Report Source	Longitude	Latitude	First Year Reported
Gore Pond	MassDEP	French and Quinebaug River Watersheds 2001 Water Quality Assessment Report 41/42-AC-1 51.0, March 2002			
Granite Reservoir	MassDEP	French and Quinebaug River Watersheds 2001 Water Quality Assessment Report 41/42-AC-1 51.0, March 2002			
Greenville Pond	College of the Holy Cross	Robert Bertin	−71.922721	42.205981	
Griswold Pond	MassDEP	North Shore Coastal Watersheds 2002 Water Quality Assessment Report 93-AC-2 CN 138.5, March 2007			
Grove Pond	MassDEP	Nashua River Watershed 2003 Water Quality Assessment Report 81-AC-4 CN107.5, August 2008			
Gushee Pond	MassDEP	Taunton River Watershed Water Quality Assessment Report 62-AC-1 CN 94.0, December 2005			
Hamilton Reservoir	MassDEP	French and Quinebaug River Watersheds 2001 Water Quality Assessment Report 41/42-AC-1 51.0, March 2002			
Hardwick Pond	MassDEP	Chicopee River Basin 1998 Water Quality Assessment Report 36-AC-2 47.0, April 2001			
Hardwick Pond	College of the Holy Cross	Robert Bertin	−72.238600	42.313254	
Hobart Pond	MassDEP	Taunton River Watershed Water Quality Assessment Report 62-AC-1 CN 94.0, December 2005			
Hoosicwhisick Pond	Individual	Joy Trahan-Liptak	−71.095558	42.207250	
Hoosicwhisick Pond	DCR	DCR Data from Lakes and Ponds, 2008			2007
Hopkinton Reservoir	MassDEP	SuAsCo Watershed 2001 Water Quality Assessment Report 82-AC-1 CN 92.0, August 2005			
Hopkinton Reservoir	DCR	DCR Data from Lakes and Ponds, 2007			2007
Hopkinton Reservoir	DCR	DCR Data from Lakes and Ponds, 2008			2007
Hopkinton Reservoir	MassDEP	Field Survey 6/26/2017 Li, Wong	−71.513377	42.256563	

Waterbody	Agency	Report Source	Longitude	Latitude	First Year Reported
Hopkinton Reservoir	MassDEP	05-G007–02	−71.513377	42.256563	
Horse Pond	MassDEP	Westfield River Watershed 2001 Water Quality Assessment Report 32-AC-1 CN 090.0, April 2005			
Hovey Pond	Individual	Joy Trahan-Liptak	−71.714879	42.234214	
Howe Pond	DCR	DCR Data from Lakes and Ponds, 2008	−71.999715	42.215269	
Howe Pond	College of the Holy Cross	Robert Bertin	−71.999715	42.215269	
Howe Reservoirs	MassDEP	Blackstone River Basin 1998 Water Quality Assessment Report 51-AC-1 48.0, May 2011			
Indian Brook	MassDEP	Field Survey 7/18/2017 Wong, Maldonado	−70.538045	41.887150	
Indian Brook	MassDEP	08-F009–03	−70.538045	41.887150	
Indian Brook	MassDEP	08-F006–01	−70.538045	41.887150	
Ironstone Reservoir	College of the Holy Cross	Robert Bertin	−71.610900	42.024298	
Jacobs Pond	MassDEP	South Shore Coastal Watersheds 2001 Water Quality Assessment Report 94-AC-2 CN 93.0, March 2006			
Jenks Reservoir	MassDEP	Blackstone River Basin 1998 Water Quality Assessment Report 51-AC-1 48.0, May 2011			
Jewells Pond	MassDEP	Neponset River Watershed 2004 Water Quality Assessment Report 73-AC-1 CN170.4, February 2010			
Kettle Brook	MassDEP	Blackstone River Watershed 2003–2007 Water Quality Assessment Report 51-AC-3 CN 240.0, March 2010			
Knops Pond/Lost Lake	MassDEP	Merrimack River Basin 1999 Water Quality Assessment Report 84-AC-1 52.0, November 2001			
Knops Pond/Lost Lake	DCR	DCR Data from Lakes and Ponds, 2002			2002
Lake Cochituate	Individual	Joy Trahan-Liptak	−71.371426	42.305669	
Lake Cochituate	MassDEP	SuAsCo Watershed 2001 Water Quality Assessment Report 82-AC-1 CN 92.0, August 2005			

(Continued)

TABLE 8.1
Table 8.1b Variable Milfoil (Continued)

Waterbody	Agency	Report Source	Longitude	Latitude	First Year Reported
Lake Cochituate	DCR	DCR Data from Lakes and Ponds, 2003			2003
Lake Cochituate	MassDEP	SuAsCo Watershed 2001 Water Quality Assessment Report 82-AC-1 CN 92.0, August 2005			
Lake Cochituate	Individual	Joy Trahan-Liptak	−71.369349	42.288313	
Lake Cochituate	MassDEP	SuAsCo Watershed 2001 Water Quality Assessment Report 82-AC-1 CN 92.0, August 2005			
Lake Cochituate	DCR	DCR Data from Lakes and Ponds, 2004			2004
Lake Holland	MassDEP	Connecticut River Basin 2003 Water Quality Assessment Report 34-AC-2 CN 105.5, October 2008			
Lake Mattawa	DCR	DCR Data from Lakes and Ponds, 2003			2003
Lake Monomonac	MassDEP	Millers River Watershed 2000 Water Quality Assessment Report 35-AC-1 CN089.0, March 2004			
Lake Pearl	MassDEP	Charles River Watershed 2002–2006 Water Quality Assessment Report 72-AC-4 CN136.5, April 2008			
Lake Quinsigamond	MassDEP	Blackstone River Basin 1998 Water Quality Assessment Report 51-AC-1 48.0, May 2011			
Lake Quinsigamond	College of the Holy Cross	Robert Bertin	−71.749052	42.256062	
Lake Quinsigamond	College of the Holy Cross	Robert Bertin	−71.749052	42.256062	
Lake Rico	DCR	DCR Data from Lakes and Ponds, 2002			2002
Lake Ripple	MassDEP	Blackstone River Basin 1998 Water Quality Assessment Report 51-AC-1 48.0, May 2011			
Lake Rohunta	DCR	DCR Data from Lakes and Ponds, Mattawa			
Lake Sabbatia	MassDEP	Taunton River Watershed Water Quality Assessment Report 62-AC-1 CN 94.0, December 2005			
Lake Samoset	MassDEP	Nashua River Watershed 2003 Water Quality Assessment Report 81-AC-4 CN107.5, August 2008			

Waterbody	Agency	Report Source	Longitude	Latitude	First Year Reported
Lake Shirley	MassDEP	Nashua River Watershed 2003 Water Quality Assessment Report 81-AC-4 CN107.5, August 2008			
Lake Shirley	DCR	DCR Data from ACT, GeoSyntec-ACT			
Lake Whalom	MassDEP	Nashua River Watershed 2003 Water Quality Assessment Report 81-AC-4 CN107.5, August 2008			
Lake Whalom	College of the Holy Cross	Robert Bertin	−71.740747	42.574132	
Lake Winthrop	MassDEP	Charles River Watershed 2002–2006 Water Quality Assessment Report 72-AC-4 CN136.5, April 2008			
Lancaster Millpond	DCR	Jamie Carr	−71.684545	42.407616	
Larner Pond	MassDEP	French and Quinebaug River Watersheds 2001 Water Quality Assessment Report 41/42-AC-1 51.0, March 2002			
Leesville Pond	Individual	Joy Trahan-Liptak	−71.826444	42.229001	
Leesville Pond	College of the Holy Cross	Robert Bertin	−71.827414	42.230332	
Leonards Pond	MassDEP	05-R003–02	−70.804664	41.750263	2005
Lily Pond	MassDEP	South Shore Coastal Watersheds 2001 Water Quality Assessment Report 94-AC-2 CN 93.0, March 2006			
Little Chauncy Pond	MassDEP	SuAsCo Watershed 2001 Water Quality Assessment Report 82-AC-1 CN 92.0, August 2005			
Little Chauncy Pond	College of the Holy Cross	Robert Bertin	−71.617285	42.306107	
Little River	Individual	Joy Trahan-Liptak	−71.883731	42.110570	
Long Island Pond	MassDEP	South Shore Coastal Watersheds 2001 Water Quality Assessment Report 94-AC-2 CN 93.0, March 2006			
Long Pond	MassDEP	Chicopee River Basin 1998 Water Quality Assessment Report 36-AC-2 47.0, April 2001			
Long Pond	DCR	DCR Data from Lakes and Ponds, 2003			2003
Long Pond	DCR	DCR Data from Lakes and Ponds, 2007			2003

(Continued)

TABLE 8.1

Table 8.1b Variable Milfoil (Continued)

Waterbody	Agency	Report Source	Longitude	Latitude	First Year Reported
Long Pond	DCR	DCR Data from Lakes and Ponds, 2008			2003
Long Pond	MassDEP	Taunton River Watershed Water Quality Assessment Report 62-AC-1 CN 94.0, December 2005			
Longwater Pond	MassDEP	Taunton River Watershed Water Quality Assessment Report 62-AC-1 CN 94.0, December 2005			
Lorings Bogs Pond	MassDEP	South Shore Coastal Watersheds 2001 Water Quality Assessment Report 94-AC-2 CN 93.0, March 2006			
Louisa Lake	College of the Holy Cross	Robert Bertin	−71.523061	42.158817	
Low Pond	MassDEP	French and Quinebaug River Watersheds 2001 Water Quality Assessment Report 41/42-AC-1 51.0, March 2002	−70.707816	41.876262	
Manchaug Pond	MassDEP	Blackstone River Basin 1998 Water Quality Assessment Report 51-AC-1 48.0, May 2011			
Mansfield Pond	College of the Holy Cross	Robert Bertin	−71.583462	42.019005	
Maple Spring Pond	DCR	Jamie Carr	−71.882155	42.376553	
Maple Spring Pond	DCR	Joy Trahan-Liptak	−71.882155	42.376553	
Marble Pond	MassDEP	Blackstone River Basin 1998 Water Quality Assessment Report 51-AC-1 48.0, May 2011			
Marble Pond	MassDEP	Field Survey 6/26/2017 Wong, Li	−71.742251	42.165436	
Marble Pond	MassDEP	09-G008−05	−71.742251	42.165436	
Massapoag Lake	MassDEP	Neponset River Watershed 2004 Water Quality Assessment Report 73-AC-1 CN170.4, February 2010			2016
Massapoag Lake	Individual	Joy Trahan-Liptak	−71.177528	42.103486	
Massapoag Lake	Individual	Joy Trahan-Liptak	−71.177528	42.103486	
Massapoag Pond	MassDEP	Merrimack River Basin 1999 Water Quality Assessment Report 84-AC-1 52.0, November 2001			
Merrill Pond No.4	College of the Holy Cross	Robert Bertin	−71.794120	42.137816	

Waterbody	Agency	Report Source	Longitude	Latitude	First Year Reported
Metacomet Lake	MassDEP	Connecticut River Basin 2003 Water Quality Assessment Report 34-AC-2 CN 105.5, October 2008			
Middle River	MassDEP	Field Survey 06/16/2017 Wong, Li, Maldonado	−71.825842	42.239095	
Middle River	MassDEP	11-D006–07	−71.825842	42.239095	
Mill Brook	MassDEP	French and Quinebaug River Watersheds 2001 Water Quality Assessment Report 41/42-AC-1 51.0, March 2002			
Mill Pond	MassDEP	Blackstone River Basin 1998 Water Quality Assessment Report 51-AC-1 48.0, May 2011			
Mill Pond	MassDEP	Buzzards Bay Watershed2000 Water Quality Assessment Report 95-AC-2 085.0, November 2003			
Mill River	MassDEP	Blackstone River Watershed 2003–2007 Water Quality Assessment Report 51-AC-3 CN 240.0, March 2010			
Mill River	MassDEP	Field Survey 6/26/2017 Wong, Li	−71.564216	42.166266	
Mill River	MassDEP	09-G010–06	−71.564216	42.166266	
Millers River	MassDEP	Field Survey 06/20/2017 Wong, Li	−72.341352	42.598155	
Millers River	MassDEP	05-H005–06	−72.341352	42.598155	2005
Miscoe Lake	MassDEP	Blackstone River Basin 1998 Water Quality Assessment Report 51-AC-1 48.0, May 2011			
Monoosnuc Brook	MassDEP	Field Survey 06/20/2017 Wong, Li	−71.737768	42.524956	
Monoosnuc Brook	MassDEP	08-G025–09	−71.737768	42.524956	
Monoosnuc Brook	MassDEP	08-G018–09	−71.737768	42.524956	
Monponsett Pond	MassDEP	14-B002–02	−70.836781	42.001467	
Monponsett Pond	MassDEP	14-B004–05	−70.836781	42.001467	
Monponsett Pond	MassDEP	14-B006–05	−70.836781	42.001467	
Monponsett Pond	MassDEP	14-B008–05	−70.836781	42.001467	
Monponsett Pond	MassDEP	13-B005–06	−70.836781	42.001467	
Monponsett Pond	MassDEP	13-B006–05	−70.836781	42.001467	
Monponsett Pond	MassDEP	13-B008–05	−70.836781	42.001467	
Monponsett Pond	MassDEP	12-B002–04	−70.836781	42.001467	
Monponsett Pond	MassDEP	12-B004–04	−70.836781	42.001467	
Monponsett Pond	MassDEP	11-B006–05	−70.836781	42.001467	

(Continued)

TABLE 8.1

Table 8.1b Variable Milfoil (Continued)

Waterbody	Agency	Report Source	Longitude	Latitude	First Year Reported
Monponsett Pond	MassDEP	11-B008–02	−70.836781	42.001467	
Monponsett Pond	Individual	Joy Trahan-Liptak	−70.837849	42.005701	
Moosehorn Pond	DCR	Jamie Carr	−71.968290	42.469370	
Moosehorn Pond	MassDEP	Chicopee River Basin 1998 Water Quality Assessment Report 36-AC-2 47.0, April 2001			
Moosehorn Pond	Individual	Joy Trahan-Liptak	−71.968290	42.469370	
Moosehorn Pond	College of the Holy Cross	Robert Bertin	−71.968290	42.469370	
Morse Pond	College of the Holy Cross	Robert Bertin	−72.013852	42.031608	
Morses Pond	MassDEP	Charles River Watershed 2002–2006 Water Quality Assessment Report 72-AC-4 CN136.5, April 2008			
Mossy Pond	DCR	Jamie Carr	−71.701034	42.417156	
Nabnasset Pond	MassDEP	Merrimack River Watershed 2004–2009 Water Quality Assessment Report 84-AC-2 CN179.5, January 2010			
Nashua River	MassDEP	Field Survey 6/26/2017 Wong, Li	−71.593150	42.626477	2008
Nashua River	MassDEP	10-K002–05	−71.593150	42.626477	2008
Nashua River	MassDEP	08-Q005–05	−71.593150	42.626477	2008
Nashua River	MassDEP	08-Q004–05	−71.593150	42.626477	2008
Nashua River	MassDEP	Field Survey 6/26/2017 Li, Wong	−71.679300	42.429974	
Nashua River	MassDEP	08-G032–02	−71.679300	42.429974	
Nashua River	MassDEP	08-G028–02	−71.679300	42.429974	
Nashua River	MassDEP	08-G020–02	−71.679300	42.429974	
Nashua River	MassDEP	08-G008–02	−71.679300	42.429974	
Neponset River	MassDEP	Field Survey 7/7/2017 Wong, Maldonado	−71.185944	42.171871	
Neponset River	MassDEP	09-B018–02	−71.185944	42.171871	
Neponset River	MassDEP	09-B016–02	−71.185944	42.171871	
Neponset River	MassDEP	09-B014–02	−71.185944	42.171871	
Neponset River	MassDEP	09-B005–02	−71.185944	42.171871	
New Bedford Reservoir	MassDEP	Buzzards Bay Watershed2000 Water Quality Assessment Report 95-AC-2 085.0, November 2003			
Newton Pond	MassDEP	Blackstone River Basin 1998 Water Quality Assessment Report 51-AC-1 48.0, May 2011			

Waterbody	Agency	Report Source	Longitude	Latitude	First Year Reported
Nipmuck Lake	Individual	Joy Trahan-Liptak	−71.569912	42.096429	
Noannet Pond	MassDEP	Charles River Watershed 2002–2006 Water Quality Assessment Report 72-AC-4 CN136.5, April 2008			
Noquochoke Lake	MassDEP	Buzzards Bay Watershed2000 Water Quality Assessment Report 95-AC-2 085.0, November 2003			
Noquochoke Lake	MassDEP	Buzzards Bay Watershed2000 Water Quality Assessment Report 95-AC-2 085.0, November 2003	−71.043280		
Noquochoke Lake	MassDEP	Buzzards Bay Watershed2000 Water Quality Assessment Report 95-AC-2 085.0, November 2003	−71.523061		
North Pond	MassDEP	Blackstone River Basin 1998 Water Quality Assessment Report 51-AC-1 48.0, May 2011			
North Pond	DCR	DCR Data from Lakes and Ponds, 2008			2008
Norton Reservoir	MassDEP	Taunton River Watershed Water Quality Assessment Report 62-AC-1 CN 94.0, December 2005			
Noyes Pond	MassDEP	Farmington River Watershed 2001 Water Quality Assessment Report 31-AC-2 CN 091.0, January 2005			
Old Oaken Bucket Pond	MassDEP	South Shore Coastal Watersheds 2001 Water Quality Assessment Report 94-AC-2 CN 93.0, March 2006			
Packard Pond	MassDEP	French and Quinebaug River Watersheds 2001 Water Quality Assessment Report 41/42-AC-1 51.0, March 2002			
Paradise Pond	DCR	David Worden	−71.856392	42.504057	
Paradise Pond	MassDEP	Nashua River Watershed 2003 Water Quality Assessment Report 81-AC-4 CN107.5, August 2008			
Paradise Pond	DCR	DCR Data from Lakes and Ponds, 2003	−71.856392	42.504057	
Paradise Pond	DCR	DCR Data from Lakes and Ponds, 2007	−71.856392	42.504057	
Paradise Pond	College of the Holy Cross	Robert Bertin	−71.856392	42.504057	

(Continued)

TABLE 8.1
Table 8.1b Variable Milfoil (Continued)

Waterbody	Agency	Report Source	Longitude	Latitude	First Year Reported
Parker Mills Pond	MassDEP	Buzzards Bay Watershed 2000 Water Quality Assessment Report 95-AC-2 085.0, November 2003			
Partridge Pond	MassDEP	Nashua River Watershed 2003 Water Quality Assessment Report 81-AC-4 CN107.5, August 2008			
Partridgeville Pond	DCR	DCR Data from Lakes and Ponds			2007
Pequot Pond	MassDEP	Westfield River Watershed 2001 Water Quality Assessment Report 32-AC-1 CN 090.0, April 2005			
Pierpoint Meadow Pond	MassDEP	French and Quinebaug River Watersheds 2001 Water Quality Assessment Report 41/42-AC-1 51.0, March 2002			
Pierpoint Meadow Pond	DCR	DCR Data from Lakes and Ponds, 2003			2003
Pierpoint Meadow Pond	College of the Holy Cross	Robert Bertin	−71.915681	42.084317	
Pinewood Pond	MassDEP	Neponset River Watershed 2004 Water Quality Assessment Report 73-AC-1 CN170.4, February 2010			
Ponkapoag Pond	MassDEP	Neponset River Watershed 2004 Water Quality Assessment Report 73-AC-1 CN170.4, February 2010			
Pottapaug Pond	College of the Holy Cross	Robert Bertin	−72.226015	42.406072	
Pratt Pond	MassDEP	Blackstone River Basin 1998 Water Quality Assessment Report 51-AC-1 48.0, May 2011			
Pratt Pond	College of the Holy Cross	Robert Bertin	−71.598913	42.181814	
Quaboag Pond	MassDEP	Chicopee River Basin 1998 Water Quality Assessment Report 36-AC-2 47.0, April 2001			
Quaboag Pond	College of the Holy Cross	Robert Bertin	−72.068036	42.197472	
Quacumquasit Pond	MassDEP	Chicopee River Basin 1998 Water Quality Assessment Report 36-AC-2 47.0, April 2001			

Waterbody	Agency	Report Source	Longitude	Latitude	First Year Reported
Quinebaug River	MassDEP	12-H005–01	−72.118569	42.109562	
Quinebaug River	MassDEP	12-H003–01	−72.118569	42.109562	
Quinebaug River	MassDEP	09-P002–01	−72.118569	42.109562	
Quinebaug River	MassDEP	08-N003–01	−72.118569	42.109562	
Quinebaug River	MassDEP	07-M005–01	−72.118569	42.109562	
Quinebaug River	MassDEP	07-M003–01	−72.118569	42.109562	
Quinebaug River	MassDEP	05-M003–01	−72.118569	42.109562	
Quinsigamond River	MassDEP	Field Survey 6/26/2017 Wong, Li	−71.708092	42.230538	
Quinsigamond River	MassDEP	08-K003–02	−71.708092	42.230538	
Quinsigamond River	MassDEP	Blackstone River Watershed 2003–2007 Water Quality Assessment Report 51-AC-3 CN 240.0, March 2010			
Railroad Pond	MassDEP	French and Quinebaug River Watersheds 2001 Water Quality Assessment Report 41/42-AC-1 51.0, March 2002			
Reservoir	MassDEP	13-B006–06	−70.852628	42.029884	
Reservoir Pond	MassDEP	Neponset River Watershed 2004 Water Quality Assessment Report 73-AC-1 CN170.4, February 2010			
Riverlin Street Pond	MassDEP	Blackstone River Basin 1998 Water Quality Assessment Report 51-AC-1 48.0, May 2011			
Rivulet Pond	MassDEP	Blackstone River Basin 1998 Water Quality Assessment Report 51-AC-1 48.0, May 2011			
Rocky Pond	MassDEP	SuAsCo Watershed 2001 Water Quality Assessment Report 82-AC-1 CN 92.0, August 2005			
Rocky Pond	College of the Holy Cross	Robert Bertin	−71.688349	42.346174	
Salisbury Brook	MassDEP	Field Survey 7/7/2017 Wong, Maldonado	−71.014062	42.079367	
Salisbury Brook	MassDEP	06-B021–04	−71.014062	42.079367	
Salisbury Brook	MassDEP	06-B017–04	−71.014062	42.079367	
Sampson Pond	MassDEP	05-R002–01	−70.752977	41.852065	
Sargent Pond	MassDEP	French and Quinebaug River Watersheds 2001 Water Quality Assessment Report 41/42-AC-1 51.0, March 2002			
Sawmill Pond	MassDEP	Nashua River Watershed 2003 Water Quality Assessment Report 81-AC-4 CN107.5, August 2008			

(Continued)

TABLE 8.1
Table 8.1b Variable Milfoil (Continued)

Waterbody	Agency	Report Source	Longitude	Latitude	First Year Reported
Sawmill Pond	College of the Holy Cross	Robert Bertin	−71.846389	42.544612	
Sewall Pond	DCR	DCR Data from Lakes and Ponds, 2008			2008
Sherman Pond	MassDEP	French and Quinebaug River Watersheds 2001 Water Quality Assessment Report 41/42-AC-1 51.0, March 2002			
Shovelshop Pond	MassDEP	Taunton River Watershed Water Quality Assessment Report 62-AC-1 CN 94.0, December 2005			
Silver Lake	MassDEP	Blackstone River Watershed 2003–2007 Water Quality Assessment Report 51-AC-3 CN 240.0, March 2010			
Silver Lake	Individual	Joy Trahan-Liptak	−71.654971	42.193369	
Silver Lake	College of the Holy Cross	Robert Bertin	−71.654894	42.193409	
Singletary Brook	MassDEP	Blackstone River Watershed 2003–2007 Water Quality Assessment Report 51-AC-3 CN 240.0, March 2010			
Smelt Pond	MassDEP	South Shore Coastal Watersheds 2001 Water Quality Assessment Report 94-AC-2 CN 93.0, March 2006			
South Meadow Pond, East Basin	DCR	Jamie Carr	−71.705732	42.415348	
South Meadow Pond, West Basin	DCR	Jamie Carr	−71.713092	42.414889	
South River	MassDEP	Field Survey 7/7/2017 Wong, Maldonado	−70.717896	42.094494	
South River	MassDEP	06-C014–03	−70.717896	42.094494	
South River	MassDEP	06-C017–03	−70.717896	42.094494	
South River	MassDEP	06-C011–03	−70.717896	42.094494	
South River	MassDEP	06-C008–03	−70.717896	42.094494	
Spectacle Pond	Individual	Joy Trahan-Liptak	−71.516164	42.564009	
Spectacle Pond	MassDEP	Merrimack River Basin 1999 Water Quality Assessment Report 84-AC-1 52.0, November 2001			
Spring Pond	DCR	DCR Data from Lakes and Ponds, 2006			
Stetson Brook	MassDEP	Field Survey 7/7/2017 Wong, Maldonado	−70.836605	42.014395	2011

Waterbody	Agency	Report Source	Longitude	Latitude	First Year Reported
Stetson Brook	MassDEP	15-B007–04	−70.836605	42.014395	2011
Stetson Brook	MassDEP	14-B008–02	−70.836605	42.014395	2011
Stetson Brook	MassDEP	11-B006–03	−70.836605	42.014395	2011
Stoneville Pond	Individual	Joy Trahan-Liptak	−71.846513	42.218616	
Stuart Pond	DCR	David Worden	−71.828492	42.482750	
Stuart Pond	MassDEP	Nashua River Watershed 2003 Water Quality Assessment Report 81-AC-4 CN107.5, August 2008			
Stuart Pond	College of the Holy Cross	Robert Bertin	−71.828492	42.482750	
Stump Pond	DCR	Joy Trahan-Liptak	−71.889471	42.361136	
Stump Pond	MassDEP	Nashua River Watershed 2003 Water Quality Assessment Report 81-AC-4 CN107.5, August 2008			
Sudbury Reservoir	College of the Holy Cross	Robert Bertin	−71.510778	42.321064	
Sudden Pond	DCR	DCR Data from Lakes and Ponds, 2008	−71.063113	42.610012	2008
Swans Pond	MassDEP	Blackstone River Basin 1998 Water Quality Assessment Report 51-AC-148.0, May 2011			
Sweets Pond	MassDEP	Taunton River Watershed Water Quality Assessment Report 62-AC-1 CN 94.0, December 2005			
Swift River	Individual	Joy Trahan-Liptak	−72.333500	42.245300	
Swift River	MassDEP	Field Survey 06/20/2017 Wong, Li	−72.334760	42.243266	
Swift River	MassDEP	08-B015–05	−72.334760	42.243266	
Swift River	MassDEP	Field Survey 06/20/2017 Wong, Li	−72.346422	42.210707	
Swift River	MassDEP	08-B015–06	−72.346422	42.210707	
Sylvestri Pond	MassDEP	French and Quinebaug River Watersheds 2001 Water Quality Assessment Report 41/42-AC-1 51.0, March 2002			
Taft Pond	MassDEP	Blackstone River Watershed 2003–2007 Water Quality Assessment Report 51-AC-3 CN 240.0, March 2010			
The Quag	DCR	David Worden	−71.771138	42.416555	
The Quag	DCR	Jamie Carr	−71.771138	42.416555	
The Quag	DCR	Joy Trahan-Liptak	−71.771138	42.416555	
Thompsons Pond	DCR	DCR Data from Lakes and Ponds, 2008			2008

(Continued)

TABLE 8.1

Table 8.1b Variable Milfoil (Continued)

Waterbody	Agency	Report Source	Longitude	Latitude	First Year Reported
Tinker Hill Pond	MassDEP	Blackstone River Basin 1998 Water Quality Assessment Report 51-AC-1 48.0, May 2011			
Town Brook	MassDEP	Field Survey 7/18/2017 Wong, Maldonado	−70.663983	41.954362	
Town Brook	MassDEP	06-C015–04	−70.663983	41.954362	
Town Brook	MassDEP	06-C013–04	−70.663983	41.954362	
Town Brook	MassDEP	06-C010–04	−70.663983	41.954362	
Town Brook	MassDEP	06-C006–04	−70.663983	41.954362	
Town Brook	MassDEP	06-C001–04	−70.663983	41.954362	
Town River	MassDEP	Field Survey 7/7/2017 Wong, Maldonado	−70.953869	41.997462	
Town River	MassDEP	13-C034–03	−70.953869	41.997462	
Tremont Mill Pond	MassDEP	Buzzards Bay Watershed2000 Water Quality Assessment Report 95-AC-2 085.0, November 2003			
Tuckers Pond	MassDEP	Blackstone River Basin 1998 Water Quality Assessment Report 51-AC-148.0, May 2011			
Turkey Hill Pond	MassDEP	Chicopee River Basin 1998 Water Quality Assessment Report 36-AC-2 47.0, April 2001			
Turkey Hill Pond	College of the Holy Cross	Robert Bertin	−71.947735	42.336725	
Turnpike Lake	MassDEP	Taunton River Watershed Water Quality Assessment Report 62-AC-1 CN 94.0, December 2005			
Turnpike Lake	DCR	DCR Data from Lakes and Ponds, 2004			2004
Uncas Pond	MassDEP	Charles River Watershed 2002–2006 Water Quality Assessment Report 72-AC-4 CN136.5, April 2008			
Uncas Pond	DCR	DCR Data from Lakes and Ponds, 2003			2003
Unionville Pond	DCR	Jamie Carr	−71.842015	42.359704	
Unionville Pond	DCR	Joy Trahan-Liptak	−71.842015	42.359704	
Unnamed Pond	DCR	Jamie Carr	−71.701377	42.411892	
Upper Chandler Pond	MassDEP	South Shore Coastal Watersheds 2001 Water Quality Assessment Report 94-AC-2 CN 93.0, March 2006			

Waterbody	Agency	Report Source	Longitude	Latitude	First Year Reported
Upper Crow Hill Pond	DCR	DCR Data from Lakes and Ponds, 2007			2007
Vandys Pond	MassDEP	Taunton River Watershed Water Quality Assessment Report 62-AC-1 CN 94.0, December 2005			
Wachusett Reservoir	DCR	David Worden	−71.795345	42.384187	
Wachusett Reservoir	DCR	David Worden	−71.787817	42.397107	
Wachusett Reservoir	DCR	Jamie Carr	−71.730707	42.380071	
Wachusett Reservoir	DCR	Jamie Carr	−71.789929	42.389998	
Wachusett Reservoir	DCR	Jamie Carr	−71.787817	42.397107	
Wachusett Reservoir	DCR	Jamie Carr	−71.788251	42.379206	
Wachusett Reservoir	DCR	Jamie Carr	−71.795345	42.384187	
Wachusett Reservoir	DCR	Jamie Carr	−71.795345	42.384187	
Wachusett Reservoir	MassDEP	Nashua River Watershed 2003 Water Quality Assessment Report 81-AC-4 CN107.5, August 2008			
Wachusett Reservoir	DCR	DCR Data from ACT, GeoSyntec-ACT			
Wachusett Reservoir	MassDEP	Richard McVoy	−71.730707	42.380071	
Wachusett Reservoir	MassDEP	Richard McVoy	−71.789929	42.389998	
Wachusett Reservoir	MassDEP	Richard McVoy	−71.795345	42.384187	
Wachusett Reservoir	MassDEP	Richard McVoy	−71.787817	42.397107	
Wachusett Reservoir	MassDEP	Richard McVoy	−71.788251	42.379206	
Wachusett Reservoir	College of the Holy Cross	Robert Bertin	−71.721795	42.385614	
Wachusett Reservoir	DCR	Joy Trahan-Liptak	−71.765499	42.357845	
Waite Pond	College of the Holy Cross	Robert Bertin	−71.891967	42.249054	
Walker Pond	MassDEP	French and Quinebaug River Watersheds 2001 Water Quality Assessment Report 41/42-AC-1 51.0, March 2002			

(Continued)

TABLE 8.1
Table 8.1b Variable Milfoil (Continued)

Waterbody	Agency	Report Source	Longitude	Latitude	First Year Reported
Walker Pond	College of the Holy Cross	Robert Bertin	−72.060053	42.139112	
Ware River	MassDEP	Field Survey 06/16/2017 Wong, Li, Maldonado	−72.159534	42.340072	
Ware River	MassDEP	14-C031–02	−72.159534	42.340072	
Ware River	MassDEP	14-C017–02	−72.159534	42.340072	
Ware River	MassDEP	14-C010–02	−72.159534	42.340072	
Ware River	MassDEP	Field Survey 06/16/2017 Wong, Li, Maldonado	−72.157140	42.343550	
Ware River	MassDEP	08-B031–09	−72.157140	42.343550	
Ware River	MassDEP	Field Survey 06/20/2017 Wong, Li	−72.285755	42.238726	
Ware River	MassDEP	08-L007–04	−72.285755	42.238726	2008
Ware River	MassDEP	08-L006–04	−72.285755	42.238726	2008
Ware River	MassDEP	08-L005–04	−72.285755	42.238726	2008
Ware River	MassDEP	08-L003–04	−72.285755	42.238726	2008
Ware River	MassDEP	08-L001–04	−72.285755	42.238726	2008
Ware River	MassDEP	07-K005–04	−72.285755	42.238726	2008
Ware River	MassDEP	07-K004–04	−72.285755	42.238726	2008
Ware River	MassDEP	06-H005–04	−72.285755	42.238726	2008
Ware River	MassDEP	06-H004–04	−72.285755	42.238726	2008
Ware River	MassDEP	Field Survey 06/20/2017 Wong, Li	−72.278995	42.235461	
Ware River	MassDEP	14-C031–04	−72.278995	42.235461	
Ware River	MassDEP	14-C010–04	−72.278995	42.235461	
Ware River	MassDEP	Field Survey 06/16/2017 Wong, Li, Maldonado	−72.114099	42.380363	
Ware River	MassDEP	08-B031–07	−72.114099	42.380363	
Watson Pond	DCR	DCR Data from Lakes and Ponds, 2003			2003
Watson Pond	DCR	DCR Data from Lakes and Ponds, 2008			
Waushakum Pond	MassDEP	SuAsCo Watershed 2001 Water Quality Assessment Report 82-AC-1 CN 92.0, August 2005			
Webster Lake	MassDEP	French and Quinebaug River Watersheds 2001 Water Quality Assessment Report 41/42-AC-1 51.0, March 2002			
Webster Lake	MassDEP	Field Survey 06/16/2017 Wong, Li, Maldonado	−71.848724	42.054350	
Webster Lake	MassDEP	05-G014–01	−71.848724	42.054350	
Webster Lake	College of the Holy Cross	Robert Bertin	−71.848244	42.038004	

Waterbody	Agency	Report Source	Longitude	Latitude	First Year Reported
Welsh Pond	MassDEP	Blackstone River Basin 1998 Water Quality Assessment Report 51-AC-1 48.0, May 2011			
West Meadow Pond	MassDEP	Taunton River Watershed Water Quality Assessment Report 62-AC-1 CN 94.0, December 2005			
West River	MassDEP	Blackstone River Watershed 2003–2007 Water Quality Assessment Report 51-AC-3 CN 240.0, March 2010			
West River	MassDEP	Field Survey 6/26/2017 Wong, Li	−71.601117	42.100361	2005
West River	MassDEP	05-K005–03	−71.601117	42.100361	2005
West Waushacum Pond	DCR	Joy Trahan-Liptak	−71.764500	42.414394	
West Waushacum Pond	DCR	DCR Data from Lakes and Ponds			
West Waushacum Pond	College of the Holy Cross	Robert Bertin	−71.763731	42.414597	
Westville Lake	College of the Holy Cross	Robert Bertin	−72.060517	42.076893	
White Pond	MassDEP	Nashua River Watershed 2003 Water Quality Assessment Report 81-AC-4 CN107.5, August 2008			
White Pond	DCR	DCR Data from Lakes and Ponds, 2003			2003
White Pond	College of the Holy Cross	Robert Bertin	−71.716035	42.513314	
Whitehall Pond	DCR	DCR Data from Lakes and Ponds, 2007			2007
Whitehall Pond	DCR	DCR Data from Lakes and Ponds, 2008			2008
Whitehall Reservoir	Individual	Joy Trahan-Liptak	−71.573594	42.230720	
Whitehall Reservoir	MassDEP	SuAsCo Watershed 2001 Water Quality Assessment Report 82-AC-1 CN 92.0 August 2005			
Whitehall Reservoir	DCR	DCR Data from Lakes and Ponds, 2008			2008
Whitins Pond	MassDEP	Email communication between David Wong and Therese Beaudoin on July 7, 2017	−71.694011	42.112406	
Whitins Pond	College of the Holy Cross	Robert Bertin	−71.689583	42.113764	
Whitman River	MassDEP	Field Survey 06/20/2017 Wong, Li	−71.866052	42.559799	
Whitman River	MassDEP	11-D022–05	−71.866052	42.559799	

(Continued)

TABLE 8.1
Table 8.1b Variable Milfoil (Continued)

Waterbody	Agency	Report Source	Longitude	Latitude	First Year Reported
Whitman River	MassDEP	11-D007–06	−71.866052	42.559799	
Williamsville Pond	DCR	DCR Data from Lakes and Ponds, 2007			2007
Winnecunnet Pond	DCR	DCR Data from Lakes and Ponds, 2003			2003
Winter Pond	MassDEP	Mystic River Watershed And Coastal Drainage Area 2004–2008 Water Quality Assessment Report 71-AC-2 CN170.2, March 2010			
Woodbury Pond	MassDEP	Blackstone River Basin 1998 Water Quality Assessment Report 51-AC-1 48.0, May 2011			
Woodbury Pond	MassDEP	Field Survey 6/26/2017 Wong, Li	−71.731255	42.164983	
Woodbury Pond	MassDEP	09-G008–07	−71.731255	42.164983	
Woods Pond	MassDEP	Neponset River Watershed 2004 Water Quality Assessment Report 73-AC-1 CN170.4, February 2010			
Wyman Pond	MassDEP	Nashua River Watershed 2003 Water Quality Assessment Report 81-AC-4 CN107.5, August 2008			

Table 8.1c Unidentified Milfoil

Waterbody	Agency	Report Source	Longitude	Latitude	First Reported Year
Assabet River	MassDEP	SuAsCo Watershed 2001 Water Quality Assessment Report 82-AC-1 CN 92.0, August 2005			
Baddacook Pond	MassDEP	05-G001–01	−71.530464	42.618962	
Charles River	MassDEP	Charles River Watershed 2002–2006 Water Quality Assessment Report 72-AC-4 CN136.5, April 2008			
Charles River	MassDEP	Charles River Watershed 2002–2006 Water Quality Assessment Report 72-AC-4 CN136.5, April 2008			

Waterbody	Agency	Report Source	Longitude	Latitude	First Reported Year
East Head Pond	DCR	DCR Data from Lakes and Ponds, humile 2008			2008
First Pond	MassDEP	North Shore Coastal Watersheds 2002 Water Quality Assessment Report 93-AC-2 CN 138.5, March 2007			
Fresh Meadow Pond	MassDEP	Buzzards Bay Watershed2000 Water Quality Assessment Report 95-AC-2 085.0, November 2003			
Hockomock River	MassDEP	13-C022–01	−71.035453	41.988368	
Hockomock River	MassDEP	13-C034–01	−71.035453	41.988368	
Hockomock River	MassDEP	13-C040–01	−71.035453	41.988368	
Houghton Pond	MassDEP	Charles River Watershed 2002–2006 Water Quality Assessment Report 72-AC-4 CN136.5, April 2008			
Lake Gardner	DCR	DCR Data from Lakes and Ponds, 2007			2007
Leonards Pond	MassDEP	05-R003–02	−70.804664	41.750263	2005
Little Rocky Pond	DCR	DCR Data from Lakes and Ponds, 2008			2008
Lower Porter Pond	MassDEP	Taunton River Watershed Water Quality Assessment Report 62-AC-1 CN 94.0, December 2005			
Monponsett Pond	MassDEP	14-B002–02	−70.836781	42.001467	
Monponsett Pond	MassDEP	14-B004–05	−70.836781	42.001467	
Monponsett Pond	MassDEP	14-B006–05	−70.836781	42.001467	
Monponsett Pond	MassDEP	14-B008–05	−70.836781	42.001467	
Monponsett Pond	MassDEP	13-B006–05	−70.836781	42.001467	
Monponsett Pond	MassDEP	13-B008–05	−70.836781	42.001467	
Monponsett Pond	MassDEP	12-B002–04	−70.836781	42.001467	
Monponsett Pond	MassDEP	12-B004–04	−70.836781	42.001467	
Monponsett Pond	MassDEP	12-B006–05	−70.836781	42.001467	
Monponsett Pond	MassDEP	12-B008–05	−70.836781	42.001467	
Monponsett Pond	MassDEP	11-B006–05	−70.836781	42.001467	
Monponsett Pond	MassDEP	11-B008–02	−70.836781	42.001467	
Mumford River	MassDEP	Blackstone River Basin 1998 Water Quality Assessment Report 51-AC-1 48.0, May 2011			

(Continued)

TABLE 8.1

Table 8.1c Unidentified Milfoil (Continued)

Waterbody	Agency	Report Source	Longitude	Latitude	First Reported Year
Mumford River	MassDEP	11-D006–05	−71.737090	42.093107	
Mumford River	MassDEP	11-D015–05	−71.737090	42.093107	
Nine Mile Pond	DCR	DCR Data from Lakes and Ponds, 2003			2003
Onota Lake	MassDEP	05-G008–01	−73.283941	42.464926	
Pottapaug Pond	MassDEP	Chicopee River Basin 1998 Water Quality Assessment Report 36-AC-2 47.0, April 2001			
Priest Brook	MassDEP	Field Survey 06/20/2017 Wong, Li	−72.115037	42.682797	
Priest Brook	MassDEP	08-P004–04	−72.115037	42.682797	
Priest Brook	MassDEP	08-P003–04	−72.115037	42.682797	
Quabbin Reservoir	MassDEP	Chicopee River Basin 1998 Water Quality Assessment Report 36-AC-2 47.0, April 2001			
Reservoir Pond	MassDEP	09-B009–02	−71.126217	42.170467	
Robinsons Pond	DCR	DCR Data from Lakes and Ponds			
Sampson Pond	MassDEP	05-R002–01	−70.752977	41.852065	
Spring Pond	MassDEP	North Shore Coastal Watersheds 2002 Water Quality Assessment Report 93-AC-2 CN 138.5, March 2007			
Thirtyacre Pond	MassDEP	Taunton River Watershed Water Quality Assessment Report 62-AC-1 CN 94.0, December 2005			
Thompsons Pond	DCR	DCR Data from Lakes and Ponds, 2005			2005
Upper Leach Pond	DCR	DCR Data from Lakes and Ponds, 2008			2008
Upper Porter Pond	MassDEP	Taunton River Watershed Water Quality Assessment Report 62-AC-1 CN 94.0, December 2005			
Waldo Lake	MassDEP	Taunton River Watershed Water Quality Assessment Report 62-AC-1 CN 94.0, December 2005			

Waterbody	Agency	Report Source	Longitude	Latitude	First Reported Year
West River	MassDEP	Blackstone River Watershed 2003–2007 Water Quality Assessment Report 51-AC-3 CN 240.0, March 2010			
White Island Pond, East Basin	MassDEP	08-F008–02	−70.617531	41.809183	
White Island Pond, East Basin	MassDEP	14-B005–01	−70.617531	41.809183	
White Island Pond, West Basin	MassDEP	14-B003–02	−70.627556	41.809994	
White Island Pond, West Basin	MassDEP	14-B005–02	−70.627556	41.809994	
Winnecunnet Pond	MassDEP	Taunton River Watershed Water Quality Assessment Report 62-AC-1 CN 94.0, December 2005			

TABLE 8.2
Towns and Infested Waterbodies

Table 8.2a Eurasian Milfoil

Number	Waterbody	Town 1	Town 2
1	Ashmere Lake	Hinsdale	Peru
2	Assabet River	Stow	Hudson/Maynard
3	Assabet River Reservoir	Westborough	–
4	Bartlett Pond	Northborough	–
5	Bartons Cove	Gill	–
6	Beaver Lake	Ware	–
7	Benton Pond	Otis	–
8	Berkshire Pond	Lanesborough	–
9	Big Bearhole Pond	Taunton	–
10	Billington Sea	Plymouth	–
11	Buffumville Lake	Charlton	Oxford
12	Center Pond	Becket	–
13	Charles River	Dover	Needham/Wellesley/Natick
14	Chauncy Lake	Westborough	–
15	Cheshire Reservoir, Middle Basin	Cheshire	Lanesborough
16	Cheshire Reservoir, North Basin	Cheshire	–
17	Cheshire Reservoir, South Basin	Lanesborough	Cheshire
18	Chicopee River	Palmer	Wilbraham/Ludlow

(Continued)

TABLE 8.2
Table 8.2a Eurasian Milfoil (Continued)

Number	Waterbody	Town 1	Town 2
19	Coes Reservoir	Worcester	–
20	Concord River	Concord	Billerica/Carlisle/Bedford
21	Congamond Lakes, Middle Basin	Southwick	–
22	Congamond Lakes, North Basin	Southwick	–
23	Congamond Lakes, South Basin	Southwick	Suffield, CT
24	Cranberry Pond	Sunderland	–
25	Dark Brook Reservoir	Auburn	–
26	Devils Dishfull Pond	Peabody	–
27	Dorothy Pond	Millbury	–
28	Dudley Pond	Wayland	–
29	Ellis Pond	Athol	–
30	Farm Pond	Framingham	–
31	Fisherville Pond	Grafton	–
32	Fisk Pond	Natick	–
33	Flint Pond	Shrewsbury	–
34	Forest Lake	Palmer	–
35	Forge Pond	Westford	Littleton
36	Fort Meadow Reservoir	Marlborough	Hudson
37	Framingham Reservoir #1	Framingham	–
38	Framingham Reservoir #3	Framingham	–
39	Goose Pond Brook	Lee	Tyringham
40	Greenwater Pond	Becket	–
41	Hayden Pond	Otis	–
42	Hills Pond	Arlington	–
43	Hocomonco Pond	Westborough	–
44	Horse Pond	Westfield	–
45	Housatonic River	Stockbridge	Lenox/Lee/Great Barrington
46	Hovey Pond	Grafton	–
47	Hubbard Brook	Sheffield	Egremont/Great Barrington
48	Indian Lake	Worcester	–
49	Jamaica Pond	Boston	–
50	Knops Pond/Lost Lake	Groton	–
51	Lake Averic	Stockbridge	–
52	Lake Buel	Monterey	New Marlborough
53	Lake Cochituate	Wayland	Framingham/Natick
54	Lake Garfield	Monterey	–
55	Lake Lorraine	Springfield	–
56	Lake Pearl	Wrentham	–
57	Lake Quinsigamond	Shrewsbury	Worcester
58	Lake Rico	Taunton	–
59	Lake Ripple	Grafton	–
60	Lake Shirley	Lunenburg	Shirley

Number	Waterbody	Town 1	Town 2
61	Lake Waban	Wellesley	–
62	Lake Whalom	Lunenburg	Leominster
63	Laurel Lake	Lee	Lenox
64	Leesville Pond	Auburn	Worcester
65	Leverett Pond	Leverett	–
66	Lily Pond, Middle	West Boylston	–
67	Little Alum Pond	Brimfield	–
68	Long Pond	Great Barrington	–
69	Mansfield Pond	Great Barrington	–
70	Middle Pond	Taunton	–
71	Monponsett Pond	Halifax	–
72	Morses Pond	Wellesley	Natick
73	Mystic River	Medford	Arlington
74	Newfield Pond	Chelmsford	–
75	Oldham Pond	Pembroke	Hanson
76	Onota Lake	Pittsfield	–
77	Orrs Pond	Attleboro	–
78	Otis Reservoir	Otis	Tolland/Blanford
79	Pequot Pond	Westfield	Southampton
80	Plunkett Reservoir	Hinsdale	–
81	Ponkapoag Pond	Canton	Randolph
82	Pontoosac Lake	Lanesborough	Pittsfield
83	Purgatory Cove	Waltham	Newton
84	Quaboag Pond	Brookfield	East Brookfield
85	Quacumquasit Pond	Brookfield	East Brookfield/Sturbridge
86	Quinsigamond River	Grafton	–
87	Reservoir Pond	Canton	–
88	Rice City Pond	Uxbridge	Northbridge
89	Richmond Pond	Richmond	Pittsfield
90	Shaker Mill Pond	West Stockbridge	–
91	Shaw Pond	Becket	Otis
92	Silver Lake	Wilmington	–
93	Singletary Pond	Sutton	Millbury
94	Sluice Pond	Lynn	–
95	Spy Pond	Arlington	–
96	Stetson Pond	Pembroke	–
97	Stevens Pond	Monterey	–
98	Stockbridge Bowl	Stockbridge	–
99	Sudbury Reservoir	Southborough	Marlborough
100	Sudbury River	Wayland	Sudbury/Framingham
101	Sunset Lake	Braintree	–
102	Thousand Acre Pond	New Marlborough	–
103	Unionville Pond	Holden	–
104	Upper Goose Pond	Lee	Tryingham

(Continued)

TABLE 8.2

Table 8.2a Eurasian Milfoil (Continued)

Number	Waterbody	Town 1	Town 2
105	Wachusett Reservoir	Boylston	West Boylston/Clinton/Sterling
106	Webster Lake	Webster	–
107	White Island Pond, East Basin	Plymouth	Wareham
108	Whiting Street Reservoir	Holyoke	–
109	Willard Brook	Sheffield	–
110	Windsor Pond	Windsor	–
111	Winning Pond	Billerica	–
112	Woolshop Pond	Millbury	–
113	Yokum Pond	Becket	–

Table 8.2b Variable Milfoil

Number	Waterbody	Town 1	Town 2
1	Aaron Reservoir	Cohasset	Hingham
2	Aldrich Pond	Sutton	–
3	Ames Long Pond	Stoughton	Easton
4	Arcade Pond	Northbridge	–
5	Arcadia Lake	Belchertown	–
6	Ashland Reservoir	Ashland	–
7	Baddacook Pond	Groton	–
8	Bailey Road Pond	Holden	–
9	Bare Hill Pond	Harvard	–
10	Bartlett Pond	Northborough	–
11	Beaver Lake	Ware	–
12	Beaver Pond	Franklin	–
13	Bennett Pond	Wilbraham	–
14	Billings Street/East Street Pond	Sharon	–
15	Billington Sea	Plymouth	–
16	Black Mountain Pond	Marshfield	–
17	Blackstone River	Millbury	Worcester/Grafton/Sutton
18	Boons Pond	Stow	Hudson
19	Bouchard Pond	Leicester	–
20	Brierly Pond	Millbury	–
21	Brigham Pond	Hubbardston	–
22	Brooks Pond	North Brookfield	New Braintree/Oakham/Spencer
23	Browning Pond	Oakham	Spencer
24	Buck Pond	Westfield	–
25	Buffum Pond	Charlton	Oxford
26	Buffumville Lake	Charlton	Oxford
27	Burncoat Pond	Leicester	Spencer
28	Carver Pond	Bridgewater	–

Number	Waterbody	Town 1	Town 2
29	Cedar Meadow Pond	Leicester	–
30	Cedar Pond	Sturbridge	–
31	Cedar Swamp Pond	Milford	–
32	Chaffin Pond	Holden	–
33	Charles River	Dover	Needham/Wellesley/Natick
34	Clark Pond	Walpole	–
35	Clear Pond	Lakeville	–
36	Cleveland Pond	Abington	–
37	Coachlace Pond	Clinton	–
38	Cocasset Lake	Foxborough	–
39	Collins Pond	Andover	–
40	Comins Pond	Warren	–
41	Cook Pond	Worcester	–
42	Crane Brook Bog Pond	Carver	–
43	Crocker Pond	Westminster	–
44	Curtis Pond	Worcester	–
45	Dark Brook	Auburn	–
46	Dark Brook Reservoir	Auburn	–
47	Dawson Pond	Holden	–
48	Dean Pond	Brimfield	Monson
49	Demond Pond	Rutland	–
50	Doane Pond	North Brookfield	–
51	Eagle Lake	Holden	–
52	East Brimfield Reservoir	Sturbridge	Brimfield
53	East Brookfield River	East Brookfield	–
54	East Freetown Pond	Freetown	–
55	East Head Pond	Carver	Plymouth
56	Eddy Pond	Auburn	–
57	Ellis Pond	Athol	–
58	Factory Pond	Holliston	–
59	Falls Pond, South Basin	North Attleborough	–
60	Farrington Pond	Stoughton	–
61	Federal Pond	Carver	Plymouth
62	Field Pond	Andover	–
63	First Pond	Saugus	–
64	Fish Pond	Northbridge	–
65	Fisk Pond	Natick	–
66	Flannagan Pond	Ayer	–
67	Flint Pond	Grafton	Worcester/Shrewsbury
68	Fort Meadow Reservoir	Marlborough	Hudson
69	Framingham Reservoir #1	Framingham	–
70	French River	Dudley	Webster
71	Frye Pond	Andover	–

(Continued)

TABLE 8.2

Table 8.2b Variable Milfoil (Continued)

Number	Waterbody	Town 1	Town 2
72	Fuller Street Pond	Middleborough	Carver
73	Gaston Pond	Barre	–
74	Gavins Pond	Sharon	Foxborough
75	Gilboa Pond	Douglas	–
76	Glen Echo Pond	Canton	Stoughton
77	Gore Pond	Dudley	Charlton
78	Granite Reservoir	Charlton	–
79	Greenville Pond	Leicester	–
80	Griswold Pond	Saugus	–
81	Grove Pond	Ayer	–
82	Gushee Pond	Raynham	–
83	Hamilton Reservoir	Holland	–
84	Hardwick Pond	Hardwick	–
85	Hobart Pond	Whitman	–
86	Hoosicwhisick Pond	Milton	–
87	Hopkinton Reservoir	Hopkinton	Ashland
88	Horse Pond	Westfield	–
89	Hovey Pond	Grafton	–
90	Howe Pond	Spencer	–
91	Howe Reservoirs	Millbury	–
92	Indian Brook	Plymouth	–
93	Ironstone Reservoir	Uxbridge	–
94	Jacobs Pond	Norwell	–
95	Jenks Reservoir	Bellingham	–
96	Jewells Pond	Medfield	–
97	Kettle Brook	Leicester	Auburn/Worcester
98	Knops Pond/Lost Lake	Groton	–
99	Lake Cochituate	Natick	Wayland
100	Lake Holland	Belchertown	–
101	Lake Mattawa	Orange	–
102	Lake Monomonac	Winchendon	–
103	Lake Pearl	Wrentham	–
104	Lake Quinsigamond	Shrewsbury	Worcester
105	Lake Rico	Taunton	–
106	Lake Ripple	Grafton	–
107	Lake Rohunta	New Salem	–
108	Lake Sabbatia	Taunton	–
109	Lake Samoset	Leominster	–
110	Lake Shirley	Lunenburg	Shirley
111	Lake Whalom	Lunenburg	Leominster
112	Lake Winthrop	Holliston	–
113	Lancaster Millpond	Clinton	–

Number	Waterbody	Town 1	Town 2
114	Larner Pond	Dudley	–
115	Leesville Pond	Auburn	Worcester
116	Leonards Pond	Rochester	–
117	Lily Pond	Cohasset	–
118	Little Chauncy Pond	Northborough	–
119	Little River	Oxford	–
120	Long Island Pond	Plymouth	–
121	Long Pond	Rutland	–
122	Longwater Pond	Easton	–
123	Lorings Bogs Pond	Duxbury	–
124	Louisa Lake	Milford	–
125	Low Pond	Dudley	–
126	Manchaug Pond	Douglas	Sutton
127	Mansfield Pond	Millville	–
128	Maple Spring Pond	Holden	–
129	Marble Pond	Sutton	–
130	Massapoag Lake	Sharon	–
131	Massapoag Pond	Dunstable	Groton/Tyngsborough
132	Merrill Pond No.4	Sutton	–
133	Metacomet Lake	Belchertown	–
134	Middle River	Worcester	–
135	Mill Brook	Brimfield	–
136	Mill Pond	Upton	–
137	Mill River	Hopedale	Milford/Upton/Mendon
138	Millers River	Orange	Athol/Phillipston/Royalston/Wendell/Erving
139	Miscoe Lake	Wrentham	–
140	Monoosnuc Brook	Leominster	Fitchburg
141	Monponsett Pond	Halifax	–
142	Moosehorn Pond	Hubbardston	–
143	Morse Pond	Southbridge	–
144	Morses Pond	Wellesley	Natick
145	Mossy Pond	Clinton	–
146	Nabnasset Pond	Westford	–
147	Nashua River	Groton	Pepperell/Ayer
148	Neponset River	Walpole	Norwood/Foxborough/Canton/Sharon
149	New Bedford Reservoir	Acushnet	–
150	Newton Pond	Shrewsbury	Boylston
151	Nipmuck Lake	Mendon	–
152	Noannet Pond	Westwood	Dover
153	Noquochoke Lake	Dartmouth	–
154	North Pond	Hopkinton	Milford/Upton
155	Norton Reservoir	Norton	Mansfield
156	Noyes Pond	Tolland	–
157	Old Oaken Bucket Pond	Scituate	–

(Continued)

TABLE 8.2

Table 8.2b Variable Milfoil (Continued)

Number	Waterbody	Town 1	Town 2
158	Packard Pond	Dudley	–
159	Paradise Pond	Princeton	–
160	Parker Mills Pond	Wareham	–
161	Partridge Pond	Westminster	–
162	Partridgeville Pond	Templeton	–
163	Pequot Pond	Westfield	Southampton
164	Pierpoint Meadow Pond	Dudley	Charlton
165	Pinewood Pond	Stoughton	–
166	Ponkapoag Pond	Canton	Randolph
167	Pottapaug Pond	Petersham	Hardwick
168	Pratt Pond	Upton	–
169	Quaboag Pond	Brookfield	East Brookfield
170	Quacumquasit Pond	Brookfield	East Brookfield/Sturbridge
171	Quinebaug River	Sturbridge	Brimfield/Holland
172	Quinsigamond River	Grafton	–
173	Railroad Pond	Charlton	–
174	Reservoir	Hanson	–
175	Reservoir Pond	Canton	–
176	Riverlin Street Pond	Millbury	–
177	Rivulet Pond	Uxbridge	–
178	Rocky Pond	Boylston	–
179	Salisbury Brook	Brockton	–
180	Sampson Pond	Carver	–
181	Sargent Pond	Leicester	–
182	Sawmill Pond	Fitchburg	Westminster
183	Sewall Pond	Boylston	–
184	Sherman Pond	Brimfield	–
185	Shovelshop Pond	Easton	–
186	Silver Lake	Bellingham	–
187	Singletary Brook	Millbury	–
188	Smelt Pond	Kingston	–
189	South Meadow Pond, East Basin	Clinton	–
190	South Meadow Pond, West Basin	Clinton	Lancaster
191	South River	Marshfield	–
192	Spectacle Pond	Littleton	Ayer
193	Spring Pond	Saugus	–
194	Stetson Brook	Halifax	–
195	Stoneville Pond	Auburn	–
196	Stuart Pond	Sterling	–
197	Stump Pond	Holden	–
198	Sudbury Reservoir	Southborough	Marlborough
199	Sudden Pond	North Andover	–

Number	Waterbody	Town 1	Town 2
200	Swans Pond	Sutton	Northbridge
201	Sweets Pond	Mansfield	–
202	Swift River	Belchertown	Ware/Palmer
203	Sylvestri Pond	Dudley	–
204	Taft Pond	Upton	–
205	The Quag	Sterling	–
206	Thompsons Pond	Spencer	–
207	Tinker Hill Pond	Auburn	–
208	Town Brook	Plymouth	–
209	Town River	Bridgewater	–
210	Tremont Mill Pond	Wareham	–
211	Tuckers Pond	Sutton	–
212	Turkey Hill Pond	Rutland	Paxton
213	Turnpike Lake	Plainville	–
214	Uncas Pond	Franklin	–
215	Unionville Pond	Holden	–
216	Unnamed Pond	Clinton	–
217	Upper Chandler Pond	Pembroke	Duxbury
218	Upper Crow Hill Pond	Westminster	–
219	Vandys Pond	Foxborough	–
220	Wachusett Reservoir	Boylston	West Boylston/Clinton/Sterling
221	Waite Pond	Leicester	–
222	Walker Pond	Sturbridge	–
223	Ware River	New Braintree	Hardwick/Ware
224	Watson Pond	Taunton	–
225	Waushakum Pond	Framingham	Ashland
226	Webster Lake	Webster	–
227	Welsh Pond	Sutton	–
228	West Meadow Pond	West Bridgewater	–
229	West River	Upton	Grafton
230	West Waushacum Pond	Sterling	–
231	Westville Lake	Sturbridge	–
232	White Pond	Lancaster	Leominster
233	Whitehall Pond	Rutland	–
234	Whitehall Reservoir	Hopkinton	–
235	Whitins Pond	Northbridge	Sutton
236	Whitman River	Westminster	–
237	Williamsville Pond	Hubbardston	–
238	Winnecunnet Pond	Norton	–
239	Winter Pond	Winchester	–
240	Woodbury Pond	Sutton	–
241	Woods Pond	Stoughton	–
242	Wyman Pond	Westminster	–

(Continued)

Table 8.2c Unidentified Milfoil

Number	Waterbody	Town 1	Town 2
1	Assabet River	Stow	Hudson/Maynard
2	Baddacook Pond	Groton	–
3	Charles River	Medfield	Millis/Norfolk/Medway/Dover/Sherborn/Natick
4	East Head Pond	Carver	Plymouth
5	First Pond	Saugus	–
6	Fresh Meadow Pond	Plymouth	Carver
7	Hockomock River	West Bridgewater	Bridgewater
8	Houghton Pond	Holliston	–
9	Lake Gardner	Amesbury	–
10	Leonards Pond	Rochester	–
11	Little Rocky Pond	Plymouth	–
12	Lower Porter Pond	Brockton	–
13	Monponsett Pond	Halifax	–
14	Mumford River	Uxbridge	Northbridge/Sutton/Douglas
15	Nine Mile Pond	Wilbraham	–
16	Onota Lake	Pittsfield	–
17	Pottapaug Pond	Petersham	Hardwick
18	Priest Brook	Royalston	Winchendon
19	Quabbin Reservoir	Petersham	New Salem/Pelham/Ware/Hardwick/Shutesbury/Belchertown
20	Reservoir Pond	Canton	–
21	Robinsons Pond	Oxford	–
22	Sampson Pond	Carver	–
23	Spring Pond	Saugus	–
24	Thirtyacre Pond	Brockton	–
25	Thompsons Pond	Spencer	–
26	Upper Leach Pond	Sharon	–
27	Upper Porter Pond	Brockton	–
28	Waldo Lake	Avon	Brockton
29	West River	Upton	Northbridge/Uxbridge
30	White Island Pond, East Basin	Plymouth	Wareham
31	White Island Pond, West Basin	Plymouth	Wareham
32	Winnecunnet Pond	Norton	–

beaches) to control invasive species (MA DCR 2016).

Herbicide application is the main chemical method used to control dense beds of Eurasian milfoil. Herbicides contain active ingredients that prove toxic to target plants at recommended concentrations, making this method very effective. However, herbicides can also harm or kill non-target plants and organisms by using the inappropriate type or concentration (Mattson et al. 2004). Several herbicides have been tested or used to control Eurasian milfoil, such as diquat, 2,4-D, fluridone (Hall et al. 1984), bensulfuron-methyl (Getsinger et al. 1994), triclopyr (Madsen and Getsinger 2005), endothall (Netherland et al. 1991), etc. Recovery and

TABLE 8.3

Infested Watersheds

1	Blackstone
2	Boston Harbor: Mystic
3	Boston Harbor: Neponset
4	Boston Harbor: Weymouth and Weir
5	Buzzards Bay
6	Charles
7	Chicopee
8	Concord (SuAsCo)
9	Connecticut
10	Farmington
11	French
12	Housatonic
13	Hudson: Hoosic
14	Ipswich
15	Merrimack
16	Millers
17	Nashua
18	North Coastal
19	Quinebaug
20	South Coastal
21	Taunton
22	Ten Mile
23	Westfield

survival of severely injured Eurasian milfoil in auxin herbicide–treated areas in Gun Lake, Michigan, have been detected and it is therefore suggested that management plans should include field monitoring of efficacy and regrowth whenever possible to best inform adaptive management decision-making (Thum et al. 2017). A recent analysis of aquatic plant communities found significant declines in native plant species in response to lake-wide herbicide treatments in 173 lakes in Wisconsin. This analysis reveals that whole lake chemical treatment can reduce the size of milfoil populations but have non-target effects (Mikulyuk et al. 2020). Furthermore, it is important to try and remove dead biological matter after a chemical treatment, as the matter can cause issues with dissolved oxygen and eutrophication as they decay (Nichols 1975). At the same time, treatment early in the growing season will result in less biomass decomposing.

It is also debatable that if there is no management at all, biomass of invasives would continue to build year by year. So strategically disrupting that cycle with a targeted and timely herbicide treatment can help to limit biomass buildup down the road (Amy Smagula, personal communication).

Biological methods have the objective of achieving control of plants without introducing toxic chemicals or using machinery. For example, the American native weevil (*Euhrychiopsis lecontei*) has been used for Eurasian milfoil control in Massachusetts since 1995, with success being seen in test lake experiments (Mattson et al. 2004). Research has found that weevil larvae significantly reduced plant growth in experiments where Eurasian milfoil exhibited a faster growth rate (Creed and Sheldon 1994). Also, there are instances of Eurasian milfoil populations suffering from declines due to pathogens.

In addition, another example of biological control methods occurred from 1964 to 1967 in Chesapeake Bay as the abundance of Eurasian milfoil declined heavily. This decline was attributed to Northeast and Lake Venice disease (Bayley et al. 1968). While it may be hard to utilize these pathogens for targeted control and management purposes, it is possible populations of Eurasian milfoil will decline naturally when exposed to these diseases. Aquatic macrophyte moth (*Acentria ephemerella*) and the milfoil midge (*Cricotopus myriophylli*) are also known to limit the growth of milfoil growth.

VARIABLE MILFOIL

Like Eurasian milfoil, variable milfoil can be controlled by physical methods, chemical methods, and biological methods. Hand harvesting, drawdowns, and benthic barriers are several common physical ways to manage populations and impede the spread of variable milfoil. Hand harvesting is recommended for use in sparsely infested sites, while benthic barriers are one of the most effective choices in areas with dense populations of variable milfoil (Bailey and Calhoun 2008). Many herbicides have been confirmed to have the potential to control this invasive species and are often used alongside other control methods such as hand pulling. Effective herbicides known for variable milfoil control includes 2,4-D ester, carfentrazone-ethyl (Glomski and Netherland 2007, 2008a), triclopyr, fluridone, penoxsulam, and diquat (Glomski and Netherland 2008b). Variable milfoil shows a high tolerance to diquat from the research of Glomski and Netherland (2008b). In terms of the biological control, weevils are now marketed commercially as a milfoil removal tool, with a recommended stocking rate of 3,000 adults per acre (Mattson et al. 2004). However, it should be noted the effectiveness of milfoil weevils on variable milfoil and Eurasian milfoil may not be the same. Furthermore, weevil introduction may not work for all because every body of water has a different configuration of plants and bugs that make it very difficult to determine if weevils will damage watermilfoil plants. It is suggested that a feasibility pilot study should be conducted before a large-scale implementation to a waterbody.

In recent years, florpyrauxifen-benzyl (active ingredient of ProcellaCOR™) has emerged as a useful tool in target-specific variable milfoil (or Eurasian milfoil) control in the region. More information about chemical control on Eurasian milfoil and variable milfoil, as well as other 17 invasive species can be found in Appendix III. Any chemical used for aquatic invasive species control requires a WM04 Chemical Application License issued by the Massachusetts Department of Environmental Protection.

Additionally, efforts should be made to better educate the public to minimize the risk of spreading invasive aquatic plants between waterbodies. For example, in countries with a history of invasive species control and management, it is common practice to put up educational signage and warnings outside waterbodies with populations of invasive species (Nichols 1975). Boaters should be encouraged to take steps to remove any vegetation and mud from their watercraft, trailer, and recreational equipment to drain all compartments, including the bilge and motor, and allow for a period of drying between operation in different bodies of water. Boaters can also be warned about potential mats of milfoil sitting on the surface of the water that they want to operate in. These mats should be avoided by boats, as the engines can cause great fragmentation of the plant, which could then cause further spreading. It is also suggested to set up restricted use areas or quarantine areas to keep people out of zones of active growth, where fragmentation and spread are possible.

SUCCESS STORY

There are many successful examples of Eurasian milfoil control. A case study of Eurasian Milfoil Removal (Lyman and Eichler 2005) reports that milfoil sites have increased from 3 in 1985 to 127 in 1998 at Lake George, NY. By 2004, around a hundred of these sites have been managed successfully by a combination of physical management techniques, which includes hand pulling, suction/hand harvesting, and benthic barriers. Although there was an increase of two to three sites per year, it was a slow rate of invasive species expansion. This case study proved the physical treatment is an effective management method for Eurasian milfoil control. Another case study that occurred in Washington state, utilizing the herbicide triclopyr, found positive results in the treatment of a river and cove (Getsinger et al. 1997). Eurasian milfoil was reduced by 99% after just four weeks of treatment. Additionally, native populations of aquatic species were able to be restored and rejuvenated because of the reduction in the invasive milfoil, with species richness and plant diversity increasing a year after the treatments had been completed.

For variable milfoil, a field trial was performed by using the herbicide Renovate at Captains Pond, Salem, NH, during the summer of 2008, with promising results (Exotic Aquatic Species Program 2008). Renovate is a systemic aquatic herbicide, which targets both roots and shoots of the plant, effectively controlling the spread of variable milfoil.

CONCLUSION

Though there are many examples of invasive milfoil management, further management plans must be tailored to the regions where invasive outbreaks occur. Therefore, there must be a dissemination of information regarding invasive milfoil, to discuss the appearances and threat of Eurasian and variable milfoil in local waterbodies. With the use of monitoring tools such as ArcGIS, a collaborative effort can be used to track the spread of invasive milfoil and implement strategic and longer-lived management methods and regulations. Successful management of watermilfoil populations requires years of dedicated financial and human resources and frequent population surveys to evaluate the effectiveness of the management approach (Meg Modley, personal communication).

REFERENCES

Aiken, S.G. 1981. A conspectus of Myriophyllum (Haloragaceae) in North America. *Brittonia* 33(1): 57–69.

Aiken, S.G., P.R. Newroth, and I. Wile. 1979. The biology of Canadian weeds. 34. Myriophyllum spicatum L. *Canadian Journal of Plant Science* 59: 201–215.

Bailey, J.E., and A.J.K. Calhoun. 2008. Comparison of three physical management techniques for controlling variable-leaf milfoil in Maine lakes. *Journal of Aquatic Plant Management* 4(6): 163–167.

Bayley, S., H. Rabin, and C.H. Southwick. 1968. Recent decline in the distribution and abundance of Eurasian milfoil in Chesapeake Bay. *Chesapeake Science* 9(3): 173–181.

Cameron, D., and M. Wallhead. 2004. Bulletin #2530, Maine INVASIVE Plants: Variable-Leaf Milfoil, Myriophyllum Heterophyllum (Water MILFOIL family)—cooperative EXTENSION publications—University of Maine cooperative extension. https://extension.umaine.edu/publications/2530e/

Chase, R.F., and W.H. Wong. 2015. Standard Operating Procedure: Field Equipment Decontamination to Prevent the Spread of Invasive Aquatic Organisms. Massachusetts Department of Environmental Protection, Watershed Planning Program. CN 59.6.29 pp.

Creed, R.P. Jr., and S.P. Sheldon. 1994. The effect of two herbivorous insect larvae on Eurasian watermilfoil. *Journal of Aquatic Plant Management* 32: 21–26.

Crow, G.E., and C.B. Hellquist. 2000. Aquatic and Wetland Plants of Northeastern North America (vol. 1) Pteridophytes, Gymnosperms, and Angiosperms: Dicotyledons. The University of Wisconsin Press, Madison WI, USA, p. 480.

Dutartre, A., C. Chauvin, and J. Grange. 2006. Colonisation végétale du canal de Bourgogne à Dijon: bilan 2006, propositions de gestion. irstea. *Rapport Cemagref Bordeaux, France*; pp. 87.

Eichler, L.W., R.T. Bombard, and C.W. Boylen. 1994. *Final Report on the Lake George Eurasian Watermilfoil Survey for 1993.* Troy: Fresh Water Institute Technical Report, pp. 94–91, 53.

Eiswerth, M., G. Susan, B. Donaldson, and W. Johnson. 2000. Potential Environmental Impacts and Economic Damages of Eurasian Watermilfoil (Myriophyllum spicatum) in Western Nevada and Northeastern California. *Weed Technology*, 14(3): 511–518.

Exotic Aquatic Species Program. 2008. Report of the New Hampshire Exotic Aquatic Species Program. New Hampshire Department of Environmental Services, www.des.nh.gov/organization/commissioner/pip/publications/wd/documents/r wd-09-08.pdf (Accessed February 20, 2020).

Getsinger, K.D., G.O. Dick, R.M. Crouch, and L.S. Nelson. 1994. Mesocosm evaluation of bensulfuron methyl activity on Eurasian watermilfoil, vallisneria, and American pondweed. *Journal of Aquatic Plant Management* 32: 1–6.

Getsinger, K.D., E.G. Turner, J.D. Madsen, and M.D. Netherland. 1997. Restoring native vegetation in a Eurasian water-milfoil dominated plant community using the herbicide triclopyr. *Regulated Rivers: Research & Management: An International Journal Devoted to River Research and Management* 13(4): 357–375.

Glomski, L.M., and M.D. Netherland. 2007. Efficacy of diquat and carfentrazone-ethyl on variable-leaf milfoil. *Journal of Aquatic Plant Management* 45: 136–138.

Glomski, L.M., and M.D. Netherland. 2008a. Effect of water temperature on 2,4-D ester and carfentrazone-ethyl applications for control of variable-leaf milfoil. *Journal of Aquatic Plant Management* 46: 119–121.

Glomski, L.M., and M.D. Netherland. 2008b. Efficacy of fluridone, penoxsulam, and bispyribac-sodium on variable-leaf milfoil. *Journal of Aquatic Plant Management* 46(1): 193–196.

Hall, J.F., H.E. Westerdahl, and T.J. Stewart. 1984. Growth response of Myriophyllum spicatum and Hydrilla verticillata when exposed to continuous low concentrations of fluridone. US Army Engineers Waterways Experimental Station, Vicksburg, MS, USA. *Technical Report* A-84–1.

Howard, V. 2020. Myriophyllum heterophyllum Michx.: U.S. Geological Survey, Nonindigenous Aquatic Species Database, Gainesville, FL, https://nas.er.usgs.gov/queries/FactSheet.aspx?speciesID=236, Revision Date: 10/31/2008 (Accessed June 26, 2020).

Invasive Eurasian Watermilfoil: Cutting Back invasive species in Lake George. 2021. from www.lakegeorgeassociation.org/educate/science/lake-george-invasive-species/eurasian-milfoil/ (Accessed June 27, 2021).

Isabel, P. 2011. Variable Watermilfoil, Myriophyllum heterophyllum. https://sites.google.com/a/rsu5.org/invasive/maine-invasive-species/variable-watermilfoil-myriophyllum-heterophyllum (Accessed June 27, 2021).

ITIS (Integrated Taxonomic Information System). 2021. www.itis. gov (Accessed June 26, 2021).

Lyman, L., and L. Eichler. 2005. Successes and Limits of Hand Harvesting, Suction Harvesting, and Benthic Barriers in Lake George, NY. Presentation at the Northeast Aquatic Plant Management Society annual meeting, Saratoga Springs, NY.

MA DCR (Department of Conservation & Recreation, MA). 2016. A guide to aquatic plants in Massachusetts. Department of Conservation & Recreation, MA, http://lakemirimichi.org/ Assets/Aquatic%20life/A%20Guide%20to%20Aquatic%20 Plants_Booklet.PDF (Accessed February 19, 2020).

Madsen, J.D., and K.D. Getsinger. 2005. Selective control of invasive submersed aquatic plants. In: *Invasive Plants: Arming to Defend and Win, Southeast Exotic Pest Plant Council's Seventh Ann. Conf.*, Birmingham, AL, pp. 34–35.

MaineVLMP. 2009. Variable water-milfoil. maine volunteer lake monitoring program. www.mainevlmp.org/mciap/herbarium/ VariableWatermilfoil.php (Accessed January 29, 2020).

MassDEP (Massachusetts Department of Environmental Protection) 2021. Final Massachusetts Integrated List of Waters for the Clean Water Act 2018/2020 Reporting Cycle. Division of Watershed Management, Watershed Planning Program. CN: 505.1. www.mass.gov/ doc/final-massachusetts-integrated-list-of-waters-for-the-clean-water-act-20182020-reporting-cycle/ download.

Mattson, M.D., P.J.G. Barletta, A. Aiello, and K.J. Wagner. 2004. Eutrophication and aquatic plant management in Massachusetts: Final generic environmental impact report. *Commonwealth of Massachusetts, Executive Office of Environmental Affairs Commonwealth of Massachusetts* 1–514. pp. 4–11.

Mikulyuk, A., E. Kujawa, M.E. Nault, S. Van Egeren, K.I. Wagner, M. Barton, J. Hauxwell, and M.J. Vander Zanden. 2020. Is the cure worse than the disease? Comparing the ecological effects of an invasive aquatic plant and the herbicide treatments used to control it. *FACETS* 5: 353–366.

Netherland, M.D., W.R. Green, and K.D. Getsinger. 1991. Endothal concentration and exposure time relationships for the control of Eurasian watermilfoil and hydrilla. *Journal of Aquatic Plant Management* 29: 61–67.

New York Invasive Species (IS) Information. 2019. *Eurasian Watermilfoil*. New York: New York Invasive.

Nichols, S.A. 1975. Identification and management of Eurasian watermilfoil in Wisconsin. *Transactions of the Wisconsin Academy of Science, Arts, and Letters*, Vol. 63 pp. 116–128.

Olden, J.D., and M. Tamayo. 2014. Incentivizing the public to support invasive species. management: Eurasian Milfoil reduces lakefront property values. *PLoS One* 9: e110458. https://doi. org/10.1371/journal.pone.0110458

Pfingsten, I.A., L. Berent, C.C. Jacono, and M.M. Richerson. 2021. Myriophyllum spicatum L.: U.S. Geological Survey, Nonindigenous Aquatic Species Database, Gainesville, FL, https:// nas.er.usgs.gov/queries/factsheet.aspx?SpeciesID=237 (Accessed December 4, 2020).

Rapid Response Plan for Eurasian Watermilfoil. n.d. Westford: ENSR. [RRP.doc]. www.mass.gov/doc/eurasian-milfoil/ download

Robinson, M. 2002a. Variable Milfoil: An Invasive Aquatic Plant Myriophyllum Heterophyllum. MA DCR (Department of Conservation & Recreation, MA) (http://macolap.org/ wp-content/uploads/2015/11/MA-DCR-Variable-Milfoil.pdf (Accessed January 29, 2020).

Robinson, M. 2002b. Eurasian Milfoil: An Invasive Aquatic *Myriophyllum spicatum*. MA DCR (Department of Conservation & Recreation, MA) www.mass.gov/files/documents/2017/09/06/ eurasian-milfoil.pdf (Accessed January 29, 2020).

Scheers, K., L. Denys, J. Packet, and T. Adriaens. 2016. A second population of *Cabomba caroliniana* Gray (Cabombaceae) in Belgium with options for its eradication. *BioInvasions Records* 5(4):227–232.

Smith, C.S., and J.W. Barko. 1990. Ecology of Eurasian Watermilfoil [PDF]. *Journal of Aquatic Plant Management* 28: 55–64.

Smith, D.G. 1993. The Potential for Spread of the Exotic Zebra Mussel (Dreissena polymorpha) in Massachusetts. Report MS-Q-11. Department of Biology, University of Massachusetts Amherst, Massachusetts and Museum of Comparative Zoology, Harvard University, Cambridge, Massachusetts. From www.mass.gov/eea/docs/dcr/watersupply/lakepond/ downloads/zebra-mussel-massdep-ms-q-11.pdf (Accessed June 27, 2018).

Thum, R.A., and J.T. Lennon. 2006. Is hybridization responsible for invasive growth of non-indigenous water-milfoils? *Biological Invasions* 8(5): 1061–1066.

Thum, R.A., S. Parks, J.N. Mcnair, P. Tyning, P. Hausler, L. Chadderton, A. Tucker, and A. Monfils. 2017. Survival and vegetative regrowth of Eurasian and hybrid watermilfoil following operational treatment with auxinic herbicides in Gun Lake, Michigan. *Journal of Aquatic Plant Management* 55: 103–107.

VT DEC (Vermont Department of Environmental Conservation). 2022. Aquatic invasive species control. https://dec.vermont.gov/watershed/lakes-ponds/aquatic-invasives/control (Accessed May 29, 2022).

9 Parrot-feather (*Myriophyllum aquaticum*)

ABSTRACT

Parrot-feather (*Myriophyllum aquaticum*) is an aquatic plant in the watermilfoil family native to South America. It is widely distributed across North America, Europe, Asia, and Africa as an invasive species. Outside of its native range, parrot-feather is regarded as a troublesome weed, due to its ability to grow quickly and clog or stagnate shallow waterways. This chapter describes the basic biology of parrot-feather and presents a listing of known sightings in the state of Massachusetts. Approaches to control this invasive species, such as mechanical removal, chemical treatment (i.e., herbicides), and potential biological agents, are also discussed.

INTRODUCTION

Taxonomic Hierarchy (ITIS 2021)

Kingdom	Plantae
Subkingdom	Viridiplantae
Infrakingdom	Streptophyta
Superdivision	Embryophyta
Division	Tracheophyta
Subdivision	Spermatophytina
Class	Magnoliopsida
Superorder	Saxifraganae
Order	Saxifragales
Family	Haloragaceae
Genus	*Myriophyllum* L.
Species	*Myriophyllum aquaticum* (Vell.) Verdc.

Parrot-feather (*Myriophyllum aquaticum*) is native to the Amazon River Basin in South America, but it has been introduced as an ornamental plant around the world (Wersal et al. 2021). Today, the plant is found on nearly every continent (Lastrucci et al. 2018). Parrot-feather is a type of watermilfoil, classified within the same genus as Eurasian watermilfoil and variable milfoil, two other invasive species. These three species

of watermilfoil, including parrot-feather, are included on the Massachusetts Prohibited Plant List; therefore, it is illegal to import, sell, or trade parrot-feather within the state of Massachusetts (Commonwealth of Massachusetts 2021).

Considered an amphibious and highly adaptive plant species, parrot-feather is heterophyllous in its growth, meaning submerged sections of the plant grow to take on a different form and character to sections of the stem that emerge from the water (Wersal 2010; Wersal et al. 2021). The sections above water are solidly light green and dense with small feather-shaped leaves arranged in whorls of four to six around the stem (Smitha et al. 2007), while the submerged parts of the plant are much sparser, with relatively elongated green leaves around a reddish-brown stem (Figures 9.1 and 9.2). In response to a change in position or in water level, the stem is able to adjust and rapidly change; under experimental conditions, parts of the stem were observed to change from the emergent to the submersed form in a matter of days after a rise in water level (Wersal and Madsen 2011). The majority of the plant's biomass appears to be positioned above the water, but the stems can extend downward up to 2 m, where a rhizome is anchored to the sediment by thin roots (Cook 1985; Murphy 2007; Haberland 2014).

As a dioecious plant, parrot-feather has separate male and female flowers. The plant is capable of achieving pollination through wind and reproducing by seed, but sexual reproduction is thought to be a negligible means of reproduction outside of the plant's native range, since the vast majority of invasive specimens collected appear to be female plants, and flowers are rarely seen (Cook 1985; Murphy 2007). When present, flowers are small, white if female and yellow if male, and they grow along the stem between leaf-whorls (Murphy 2007). Parrot-feather reproduces on a local scale by cloning, with mature plants producing numerous offshoots and allocating resources to younger plants in order to maximize their survival (You

DOI: 10.1201/9781003201106-9

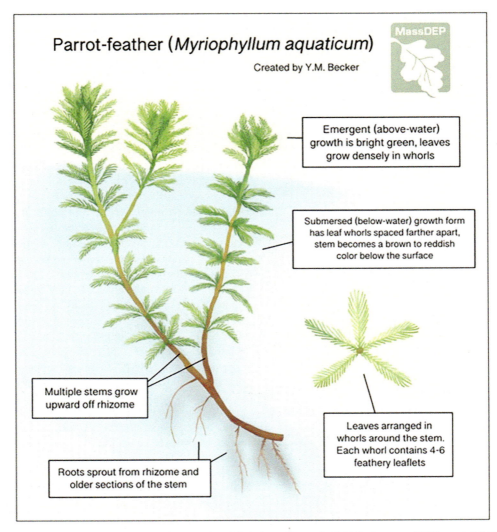

FIGURE 9.1   Illustration of parrot-feather. (Modified from Karwath (2005), Lurvey (2021), and UF IFAS (2021).)

et al. 2013). Since the plant lacks tubers or any other kind of root specialized for nutrient storage and dispersal, these functions are served by the stolons connecting offshoots to their parents (Wersal et al. 2021). Parrot-feather can also reproduce via stem fragmentation, when dislodged or senescent pieces of stem fall into the sediment, where they begin to grow roots and new stems (Figure 9.2).

NATIVE AND INVASIVE RANGE, MECHANISMS OF SPREAD

The earliest recorded specimen of *M. aquaticum* in the United States was collected from New Jersey in 1890, having been introduced to the area as a garden ornamental (Wersal 2010; Wersal et al. 2021). By the 1940s, *M. aquaticum* was well-established in the region of Southern New York, and populations could be found all the way on the west coast, in Washington state (Wersal et al. 2021). Today, this invasive species is established throughout the United States, especially along the coasts and in southern states east of Texas (USGS 2021).

After an initial introduction, parrot-feather is observed to secondarily spread to new habitats mainly by clonal growth and fragmentation (You et al. 2013; Murphy 2007). The species is

FIGURE 9.2 Photos of parrot-feather (King County NWCP 2018). Specimen of a rooted fragment held in hand with roots and young stems growing off fragment (*left*). Examples of dense wild parrot-feather growth (*right*, *top*, and *bottom*).

also capable of flowering and reproducing by seed, but this method of spread is thought to be less common outside of the plant's native range due to the rarity of male introduced specimens (Murphy 2007). Either stem fragments or seeds from an established parrot-feather stand are prone to being carried downstream to colonize new areas, where a population then grows and expands to capacity by producing clonal offshoots. Improperly cleaned boats and traveling waterfowl are also considered vectors for secondary spread, potentially enabling even broader dispersal of plant fragments to new waterways beyond the reach of simple downstream water movement (Murphy 2007; Kuehne et al. 2016).

Systems at the greatest risk of parrot-feather invasion are those which are shallow and prone to disturbance (Lastrucci et al. 2018). Regularly occurring mechanical disturbances facilitate stem fragmentation and create conditions that favor the dominance of *M. aquaticum* in shallow systems (Wersal and Madsen 2011). Since parrot-feather thrives in nutrient-rich environments, pollution and runoff are also risk factors for invasion (Xie et al. 2010; Wersal et al. 2021). Though populations will often fail to establish when faced with severe winters, parrot-feather is overall a hardy plant capable of withstanding moderate frost, short-term changes in water level, and even some degree of saltwater inundation (Wersal and Madsen 2011; Wersal et al. 2021).

ECOLOGICAL AND HUMAN IMPACTS

M. aquaticum is known to be a competitive and fast-growing plant species, often dominating over native plant species as well as influencing the species composition in animal and bacterial communities (Kuehne et al. 2016; Sun et al. 2017; Lastrucci et al. 2018). Parrot-feather has been observed in several instances to compete with and displace native flora, forming dense monospecific stands (Stiers et al. 2011; Lastrucci et al. 2018). Due to the plant's inability to reproduce by seed in its introduced range, these stands are formed through clonal growth. Like other plant species which reproduce in this way, *M. aquaticum* stands are prone to low genetic diversity, which is considered a liability for the plant populations' health and long-term success (Lambertini et al. 2010). Though low genetic diversity does not reflect upon likelihood of invasion success (Roman and Darling 2007; Rollins et al. 2013), the replacement of genetically and taxonomically diverse plant populations with uniform clonal populations threatens the long-term resilience of an ecosystem (Hughes et al. 2008).

Parrot-feather invasion has also been observed to have effects on macroinvertebrate and fish species composition. Studies have found certain nonnative fish and arthropod species living in higher density in areas dominated by *M. aquaticum* as compared to non-invaded areas (Lastrucci et al. 2018; Kuehne et al. 2016). Notably, there is a well-established correlation between the density of *M. aquaticum* and the presence of mosquitoes (Orr and Resh 1989; Lastrucci et al. 2018; Wersal et al. 2021 https://europepmc.org/article/med/2614408). Increased mosquito abundance is a human health risk, given the dangers of mosquito-borne illnesses (Wersal and Madsen 2007; Wersal et al. 2021). In Massachusetts, mosquitoes are known to seasonally carry and spread several diseases, including West Nile virus and eastern equine encephalitis (Cambridge Public Health Department n.d.).

The increased mosquito risk is brought about because of changes parrot-feather makes to the character of the habitats it invades. Parrot-feather grows into thick, dense mats, clogging already shallow waterways so that the flow of the water is reduced or even made stagnant (Hill 2003). The reduction in water movement, along with the protective cover provided by *M. aquaticum*'s emergent growth, creates perfect habitat for mosquitoes to breed and deposit their eggs, and for larvae to mature (Wersal et al. 2021). Blockages in waterways also negatively impact other species, preventing the movement of fish and inhibiting human recreational activities such as swimming and boating (Hill 2003; Lastrucci et al. 2018).

A 2016 study found that dense parrot-feather cover was associated with lower dissolved oxygen (DO) in the water column (Kuehne et al. 2016). In 1999, it was reported that irrigating tobacco with water drawn from sources infested by *M. aquaticum* lead to red discoloration of the harvested tobacco product, resulting in a less profitable harvest (Cilliers 1999), though the exact cause of the discolored crop was unclear. Parrot-feather is also thought to be capable of increasing evapotranspiration, with a higher rate of water loss where the water surface is covered by *M. aquaticum* (Cilliers 1999).

In spite of the risks posed by wild overabundant parrot-feather growth, a study on waste treatment in constructed wetlands found that when *M. aquaticum* is cultivated in water, it is highly efficient in reducing chemical oxygen demand (a measure of organic waste) and enriching communities of denitrifying bacteria (Sun et al. 2017). Though *M. aquaticum* was the only macrophyte included in this study, the authors suggest that impacts on bacterial communities may be specific to certain plant species (Sun et al. 2017). Both *M. aquaticum* and *M. spicatum*, another related milfoil species, have been found to contain allelopathic chemicals capable of inhibiting blue-green algae growth (Nakai et al. 2000; Cheng et al. 2008).

M. aquaticum has also been found to be highly efficient in removing phosphorus, tetracycline, heavy metals, and other pollutants from water, whether for purposes of assessing pollutant concentrations or for ecological restoration (Harguinteguy et al. 2013; Souza et al. 2013; Luo et al. 2017; Colzi et al. 2018; Guo et al. 2019). The plant may not even need to be implanted alive in order to provide these benefits: the dead biomass of *M. aquaticum* has been found useful in removing heavy metal contamination from water, most effectively removing zinc and cadmium specifically (Colzi et al. 2018). These findings point to a potential valuable purpose for the reuse of waste material collected after parrot-feather management efforts, though extreme caution should be taken not to introduce live plant material to high-risk waterways. Complete decomposition of *M. aquaticum* debris will also result in a complex variety of impacts to water quality, involving the re-release of nutrients (Luo et al. 2020).

TABLE 9.1

Sightings and Report Agencies of Infested Waterbodies

Waterbody	Watershed	Agency	Town	Year
Jones River	South Coastal	MassDEP	Kingston	2017
Jones River	South Coastal	MassDEP	Kingston	2006
Jones River	South Coastal	MassDEP	Kingston	2006
Jones River	South Coastal	MassDEP	Kingston	2006
Mill River	Charles	MassDEP	Norfolk	2015
Satucket River	Taunton	MassDEP	East Bridgewater	2017
Satucket River	Taunton	MassDEP	East B ridgewater	2006
Satucket River	Taunton	MassDEP	East Bridgewater	2006
Great Pond	Islands	USGS NAS Database	Tisbury	2020
Burchell's Pond	Islands	USGS NAS Database	Nantucket	2016
Big West Pond	Buzzards Bay	USGS NAS Database	Plymouth	2017

MONITORING AND SIGHTINGS

The data in this report were collected by Massachusetts Department of Environmental Protection (MassDEP) and US Geological Survey (USGS) (Table 9.1). The standard operating procedure was developed for all field surveys by MassDEP, including field sheet development, species identification, and quality control and assurance. MassDEP compiled data from all agencies. Data layers and maps of sites with parrot-feather were created using QGIS open-source software (OSGeo, Beaverton, Oregon) and ArcMap 10.1 (ESRI, Redlands, California).

Parrot-feather (*M. aquaticum*) was detected for the first time in Massachusetts in 2006, in the Jones River at Kingston. Since then, it has been confirmed that a total of six waterbodies in the state, including three rivers and three lakes, have been invaded by this species. These waters are within five watersheds: South Coastal, Charles River, Taunton River, Islands, and Buzzards Bay (Figure 9.3). Most of these waterbodies are in the southeastern region of the state. Monitoring and prevention are needed to better manage this invasive species and prevent its further spread to other waters in Massachusetts. A map depicting all known infested waterbodies and watersheds in the state is included in this report (Figure 9.3). Monitoring for parrot-feather in Massachusetts has so far been carried out by conventional visual assessment and identification. With recent advancements in molecular technologies, monitoring by analyzing environmental DNA (eDNA) is becoming an increasingly accessible and valuable tool for detecting invasive species on a broad scale. Specific identifiers and protocols for the detection of *M. aquaticum* have recently been developed, allowing for this species to be detected and identified through standard PCR-based methods (Shah et al. 2014).

MANAGING ACTIONS

Early detection and response measures need to be developed to monitor the establishment and spread of this invasive species. Like many invasive species, parrot-feather is most able to take hold and become invasive in bodies of water that are already at risk due to other factors—both anthropogenic disturbance and excess nutrients in a system due to pollution are conditions that benefit the proliferation of parrot-feather and eliminating these risk factors is an important step in preventing overabundant growth (Haberland 2014).

Another step to avoid parrot-feather invasion is the prevention of dispersal through human activities. Properly cleaning watercraft, swimming gear, and construction equipment before transporting items between waterbodies will prevent the accidental dispersal of plant fragments from infested systems. Care should be taken not to introduce exotic ornamental garden or aquarium plants to the wild, and the individual selection of native plants for outdoor home or commercial gardens should be encouraged and promoted.

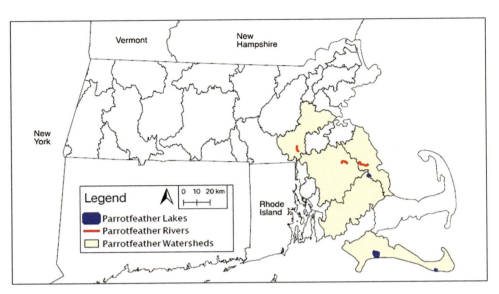

FIGURE 9.3   Infested waterbodies by parrot-feather at Massachusetts.

CHEMICAL CONTROL

Targeted application of chemical herbicides is a popular method of managing nuisance plants and several chemicals are legal to use in Massachusetts. Many different herbicides have been used with success, including glyphosate, endothall, diquat, 2,4-D, triclopyr, and imazapyr (Patten 2003; Hofstra et al. 2006; Souza et al. 2012). In recent years, florpyrauxifen-benzyl (active ingredient of ProcellaCOR™) is a relatively new herbicide for all milfoils, including parrot-feather. Success rates vary across different documented trials, and the effectiveness of a given herbicide application is likely to depend on characteristics of the site, the method of applying the herbicide, and the concentration and composition of the herbicide solution.

Though glyphosate is one of the most commonly used systemic herbicides, opinions vary on its effectiveness against parrot-feather. Glyphosate can be useful, but some advise against it due to its inability to prevent regrowth without multiple applications (Patten 2003; Hofstra et al. 2006; Wersal and Madsen 2007; Kuehne et al. 2018). On the other hand, a study conducted under greenhouse conditions found that certain additives to glyphosate can increase the herbicide's effectiveness on management of *M. aquaticum*. Mixing additives such as

Aterbane® and Veget'oil® to the herbicide solution brought control effectiveness above 90% at certain dosages, and several additives caused reduction in regrowth (Cerveira et al. 2020).

Imazapyr is another chemical herbicide that is considered useful against parrot-feather. As imazapyr tends to break down after a short time when exposed to water and sunlight, it is typically used to control emergent vegetation (Tu et al. 2001). Field experiments comparing the efficacy of various herbicides have found imazapyr-based treatments more effective than those using 2,4-D as a main ingredient (Kuehne et al. 2018). Another experiment using imazapyr resulted in 100% parrot-feather mortality when applying high concentrations of the herbicide (Wersal and Madsen 2007). Imazamox may be useful too because it is the same class of chemicals but made for submergent vegetations, if applied at earlier growths of parrot-feather (Ken Wagner, personal communication).

Another set of experiments were carried out comparing the effects of endothall, triclopyr, dichlobenil, fluridone, clopyralid, and glyphosate on *M. aquaticum* (Hofstra et al. 2006). After initial treatments, clopyralid and fluridone were found to have a minimal impact on parrot-feather biomass and percent cover and were thus rejected. Triclopyr, dichlobenil, and endothall each significantly reduced the biomass of the experimental plants, with triclopyr standing out as the most efficacious, while also being less environmentally persistent than dichlobenil.[1] The second-year trials, conducted in the field, compared only triclopyr and glyphosate, finding triclopyr to provide "substantially" better control than glyphosate (Hofstra et al. 2006).

The relatively new herbicide, ProcellaCOR™, may have higher effectiveness on parrot-feather or other milfoil species without non-target damage. However, there is no data available to support its efficacy in parrot-feather control yet. Methods for application of this product in Massachusetts are restricted to in-water applications and foliar applications from a boat or ground equipment. It should be excluded from use in state-listed aquatic species habitats, unless otherwise authorized in writing on a case-by-case basis by the MA Division of Fisheries and Wildlife pursuant to MA Endangered Species Act (321 CMR 10.14 or 10.18).

Additionally, it is worth noting that, in spite of experimentally demonstrated effectiveness of many herbicides on parrot-feather, some agencies recommend against the use of chemical herbicides on the basis of low efficiency. The Invasive Species Council of Metro Vancouver (2019) lists chemical treatments as "not recommended" due to "poor absorption" and risks to wildlife. King County, WA's Noxious Weed Control Program best practice documentation, mentions that the texture of the stem and leaves demands chemical additives for effective herbicide delivery, and that herbicide applications may need to be repeated over several growing seasons, while altogether the documentation describes chemical management as "the most reasonable option for eradication of large parrotfeather infestations" (King County NWCP 2014). Altogether, there is ample experimental evidence to demonstrate the effectiveness of several chemical herbicides on parrot-feather.

Although herbicides can be very precise and efficient in eliminating target plant species, they may have significant impacts to other species and the environment if not used with care. Proper use is essential to best results in terms of target control and minimized non-target species influences. Therefore, it is critical to apply any herbicide product correctly, read and follow the directions on the product's label, and follow any state or federal laws applicable to herbicide treatments. More information about chemical control on parrot-feather and other 18 invasive species can be found in Appendix III. In the state of Massachusetts, it is required to obtain an aquatic nuisance species control license (BRP WM04 Form) from the Massachusetts Department of Environmental Protection before applying a chemical herbicide.

PHYSICAL CONTROL

Harvesting plant matter, either by hand or by machine, is a simple and intuitive way to reduce the amount of invasive vegetation in a short period of time. However, like many other invasive plant species, parrot-feather is capable of reproducing by fragmentation. This growth strategy poses challenges to physical removal efforts, since mechanically cutting or damaging parrot-feather stems in any way could create loose stem fragments which inadvertently facilitate the spread of the invasive species. A Portugal study comparing mechanical and chemical control found that mechanical removal achieved immediate clearance followed by gradual regrowth back to initial density over the course of the next two years, while herbicide application using glyphosate was slow at first but provided better long-term control (Machado and Rocha 1998).

Another physical technique for removing invasive plants involves making adjustments to water level in an infested area by artificially enacting floods or drawdown. *M. aquaticum* is well-adapted to survive most short-term rises or falls in water level, but experimentation has shown that the species grows poorly under sustained deep flooding conditions (Wersal and Madsen 2011). When applicable, permanently or extensively raising water level could be a good way of mitigating invasive parrot-feather growth. Extended periods of drawdown can also be effective at controlling *M. aquaticum*, though the plants must be kept in dry soil for a long period of time in order to see positive results. A 2013 experiment found an insignificant decrease in parrot-feather survival after a 2-week drawdown, but in contrast survival was down to 18% after a 12-week drawdown (Wersal et al. 2013). The effects of drawdown and flooding on parrot-feather populations are regarded as unpredictable and ill-advised (Metro Vancouver 2019) because this technique is weather-dependent and becoming less reliable with climate change.

Shading and smothering through use of a gas-permeable benthic-barrier is another useful management strategy for aquatic invasive plants, including *M. aquaticum*. Parrot-feather has been found to thrive in partial shade (30–50% shaded) and suffer somewhat in full sunlight. When light is significantly reduced (in conditions of 70% shade), stem length increases while overall biomass decreases (Wersal 2010). Interestingly, when grown in 70% shade, the apical tips of *M. aquaticum* take on the submersed growth form, further indicating stress under insufficient light conditions. Shaded and full-sun conditions also directly impact water temperature, which could contribute to the plant's stress at extremes (Wersal 2010). In Richmond, Vancouver, trials were carried out to determine the effectiveness of shading and smothering on *M. aquaticum* in infested drainage ditches (Metro Vancouver 2019). Barriers were carefully applied to avoid leaving exposed seams that might be exploited by parrot-feather rhizomes.[2] Both the shading and benthic barrier trials were found effective in significantly reducing parrot-feather's biomass within the drainage ditches. However, while inexpensive on the small scale, these methods were acknowledged to be costly and impractical to apply in large open areas (Metro Vancouver 2019).

BIOLOGICAL CONTROL

Biological control of an invasive species entails the intentional release of an organism that is known to feed on the target species, in the hopes that the control agent will either completely eliminate the invader or simply reduce the invasive population so that it will not grow unchecked or dominate over native species. An ideal biological control agent is one that feeds on its target species sufficiently to suppress the target's growth and feeds exclusively or with strong preference on its target species, so that the control agent does not pose a risk to native or non-target species. Several organisms are known to feed on parrot-feather, including some insects and larger herbivores. While no parrot-feather-specific biocontrol agent has been approved for release in the United States, insect biocontrol programs have seen success in other parts of the world (Hill 2003). This section will discuss a range of species known to have potential for parrot-feather biocontrol, as well as the risks and benefits associated with each species.

Insect Control Agents

A bio-control program initiated in South Africa in the early 1990s saw the release of leaf-feeding beetles (*Lysathia* spp.) at 25 infested locations, resulting in a pattern of repeated defoliation and regrowth followed by an ultimate collapse of the parrot-feather population (Cilliers 1999; Hill 2003). While the project saw a degree of success, the control by the *Lysathia* sp. beetles was considered insufficient, and so research has been underway to determine the suitability of a South American weevil (*Listronotus marginicollis*) as another control species (Hill 2003; Oberholzer et al. 2007). *L. marginicollis* was experimentally determined to have host specificity with *M. aquaticum* and deemed safe to release in South Africa (Oberholzer et al. 2007). Additionally, the CABI Invasive Species Compendium recognizes *L. marginicollis* as a "natural enemy" of *M. aquaticum* (CABI 2019).

In 2020, a certain population of *Galerucella grisescens* beetle was found feeding on invasive *M. aquaticum* in Chiba Prefecture, Honshu, Japan (Okamoto et al. 2020). Though the natural host plant of these beetles is a different species, in the family Polygonaceae, reproductive beetles from the Chiba population were found to preferentially lay eggs on *M. aquaticum*, suggesting this species could be useful as biological control if further risk assessment and research to determine suitability are carried out. The authors of this study also note that this beetle population might be viewed not only as a tool for biological control, but perhaps as a hopeful example of a native species broadening its host plant range in order to adapt to the increased presence of an invasive species. If other insect species in Japan and worldwide are commonly able to adapt their feeding preferences in this way, as suggested by other examples of adaptation to non-native food sources or competitors (Carroll 2007; Berthon 2015; Golec et al. 2019), ecosystems around the world could have internal mechanisms for mitigating the threats posed by invasive species. These kinds of emergent biocontrol organisms have clear benefits in the face of ecological invasion.

Fish and Mammal Control Agents

Grass carp (*Ctenopharyngodon idella*) are herbivorous fish native to Asia that are popularly stocked in a controlled manner as a way of reducing invasive vegetation (Reyes 2021). These fish tend to ravenously clear out all plant matter accessible to them, leaving their habitat barren of plants in a relatively short amount of time (Garner et al. 2013). Grass carp can therefore be a useful means of controlling nuisance vegetation or a troublesome force of ecological destruction. Observations on grass carp food preferences have found that the fish prefer to eat most other plant species over parrot-feather, suggesting that they would not be useful as a parrot-feather control agent (Pine and Anderson 1991; Madsen and Wersal n.d.; King County NWCP 2014). However, an in situ study found that grass carp will clear out all available parrot-feather vegetation despite the fact that parrot-feather is not a preferred food source (Garner et al. 2013).

Due to the intense pressure grass carp grazing inflicts on native and non-native vegetation, and the potential for the carp themselves to become established as an invasive species, grass carp should only be considered as a control agent after careful assessment of ecological risk. Upon implementation, a plan should be made for monitoring and regulating grass carp population in the long term, to ensure the success of the invasive species control as well as to prevent damage to local ecosystems. Grass carp are most appropriate as biological control in large, isolated man-made impoundments, with low diversity of native plant species (Reyes 2021). In Massachusetts, the introduction of grass carp is prohibited.

Rather than importing a non-native herbivore to graze on invasive species, one might also achieve management of invasive parrot-feather with the help of a native herbivore: the beaver. Field experiments by Parker et al. (2007) observed the effects of "beaver exclusion" on freshwater plant communities and found that removing beavers from certain plots lead to an extreme disparity in the abundance of the exotic *M. aquaticum*. Where beavers were permitted to graze and construct their nests, *M. aquaticum* biomass was reduced by 90%. This case serves as evidence that reintroducing and protecting native herbivores can act as a preventative measure against noxious plant growth and ecological invasion. In Massachusetts, the practitioners need to check with Division of Fisheries & Wildlife, local Conservation Commissions, or Department of Environmental Protection to get appropriate approvals for an introduction of a vertebrate due to beavers' damming and tree damage.

SUCCESS STORY

Over the course of 14 weeks from August to November 2006, Ryan Wersal and John D. Madsen, at Mississippi State University, performed outdoor mesocosm experiments to test the efficiencies of two systemic herbicides, imazapyr and imazamox, on parrot-feather (Wersal and Madsen 2007). For the first four weeks, parrot-feather was planted and grown in mesocosms (small-scale, simulated ecosystems). Each herbicide was applied to one of three mesocosms in either a low, medium, or high concentration, and the mesocosms were then monitored for ten weeks posttreatment. A non-ionic surfactant called Dyne-Amic® was added to each mixture to help the sprayed herbicide solution stick onto the surface of the plants. Between the two herbicides, imazapyr performed better, achieving 100% mortality of the parrot-feather after eight weeks, at both the medium and high concentrations (Wersal and Madsen 2007).

In 2018, a field study was carried out by Kuehne et al. at four sites along the Chehalis River in Washington State, assessing the efficiency of three chemical herbicide mixtures: imazapyr only, imazapyr mixed with the contact herbicide carfentrazone (known to be ineffective on parrot-feather when used alone), and 2,4-D mixed with carfentrazone (Kuehne et al. 2018). Six weeks after each treatment, the biomass of parrot-feather at the 2,4-D-solution-treated sites had only been reduced 23%, but at each of the sites where an imazapyr solution had been used, parrot-feather cover was reduced by 67–69%, a promising and significant reduction. The concentration of imazapyr used in this field study was at maximum strength, exceeding the highest concentration used in the 2007 Mississippi State study, and a surfactant was also added at twice the volume of that used in the mesocosm study (Kuehne et al. 2018). The incomplete control achieved during the field study is likely the result of variation in outdoor environmental conditions. The wild-grown parrot-feather may also have been tougher on account of being older and better-established than the 4-week-old crop in the mesocosm study. Regardless, a control rate of 67–69% is significant and speaks to the promise of imazapyr solutions as chemical control agents for parrot-feather infestations.

NOTES

1   Dichlobenil is not a permitted chemical in Massachusetts.
2   After benthic barriers were installed to cover parrot-feather in areas of Burchell Pond, Massachusetts, in 2017, the area covered by parrot-feather decreased significantly in 2018. Then, in 2018, benthic barriers reinstalled on areas previously not covered but in 2019, the area covered with parrot-feather increased substantially as the plants grew around the barriers seeking sunlight. After that, the herbicide ProcellaCOR was used to treat the plants and it worked well. After two applications there is hardly any parrot-feather left and the natural vegetation, as well as the animal species inhabiting the pond, does not seem to be impacted (personal communication, Rachel Freeman, dated July 5, 2022).

REFERENCES

Berthon, K. 2015. How do native species respond to invaders? Mechanistic and trait-based perspectives. *Biological Invasions* 17: 2199–2211.

CABI Invasive Species Compendium. 2019. *Listronotus marginicollis*. www.cabi.org/isc/datasheet/28361. (Accessed June 29, 2021).

Cambridge Public Health Department. n.d. Mosquito-borne diseases. www.cambridgepublichealth.org/services/environmental-health/mosquito-borne-diseases/. (Accessed June 29, 2021).

Carroll, S.P. 2007. Brave New World: The epistatic foundations of natives adapting to invaders. *Genetica* 129: 193–204.

Cerveira, W.R., A.F. Silva, J.H.C. Cervoni, C. Cruz, and R.A. Pitelli. 2020. The addition of adjuvants on glyphosate enhances the control of aquatic plant *Myriophyllum aquaticum* (Vell.). *Planta Daninha* 38.

Cheng, W., C. Xuexiu, D. Hongjuan, L. Difu, and L. Junyan. 2008. Allelopathic inhibitory effect of *Myriophyllum aquaticum* (Vell.) Verdc. on *Microcystis aeruginosa* and its physiological mechanism. *Acta Ecologica Sinica* 28(6): 2595–2603.

Cilliers, C.J. 1999. *Lysathia n.* sp. (Coleoptera: Chrysomelidae), a host-specific beetle for the control of the aquatic weed *Myriophyllum aquaticum* (Haloragaceae) in South Africa. *Hydrobiologia* 415: 271–276.

Colzi, I., L. Lastrucci, M. Rangoni, A. Coppi, and C. Gonnelli. 2018. Using *Myriophyllum aquaticum* (Vell.) Verdc. to remove heavy metals from contaminated water: Better dead or alive? *Journal of Environmental Management* 213(1): 320–328.

Commonwealth of Massachusetts. 2021. Massachusetts prohibited plant list. www.mass.gov/massachusetts-prohibited-plant-list. (Accessed June 25, 2021).

Cook, C.D.K. 1985. Worldwide distribution and taxonomy of *Myriophyllum* species. In: *1st International Symposium on Watermilfoil (Myriophyllum spicatum) and Related Haloragaceae Species*. Vancouver, B.C. Canada: Aquatic Plant Management Society.

Garner, A.B., T.J. Kwak, K.L. Manuel, and D.H. Barwick. 2013. High-density grass-carp stocking effects on a reservoir invasive plant and water quality. *Journal of Aquatic Plant Management* 51: 27–33.

Golec, J.R., J.J. Duan, K. Rim, J. Hough-Goldstein, and E.A. Aparicio. 2019. Laboratory adaptation of a native North American parasitoid to an exotic wood-boring beetle: Implications for biological control of invasive pests. *Journal of Pest Science* 92: 1179–1186.

Guo, X., Q. Mu, H. Zhong, P. Li, C. Zhang, D. Wei, and T. Zhao. 2019. Rapid removal of tetracycline by *Myriophyllum aquaticum*: Evaluation of the role and mechanisms of adsorption. *Environmental Pollution* 254: 113101.

Haberland, M. 2014. *Parrot Feather (Myriophyllum aquaticum): A Non-Native Aquatic Plant in New Jersey Waterways*. Hoboken, NJ: New Jersey Agricultural Experiment Station. Rutgers, The State University of New Jersey. https://njaes.rutgers.edu/fs1232/ (Accessed June 25, 2021).

Harguinteguy, C.A., R. Schreiber, and M.L. Pignata. 2013. *Myriophyllum aquaticum* as a biomonitor of water heavy metal input related to agricultural activities in the Xanaes River (Córdoba, Argentina). *Ecological Indicators* 27: 8–16.

Hill, M.P. 2003. The impact and control of alien aquatic vegetation in South African aquatic ecosystems. *African Journal of Aquatic Science* 28(1): 19–24.

Hofstra, D., P.D. Champion, and T.M. Dugdale. 2006. Herbicide trials for the control of parrotsfeather. *Journal of Aquatic Plant Management* 44: 13–18.

Hughes, A.R., B.D. Inouye, M.T.J. Johnson, N. Underwood, and M. Vellend. 2008. Ecological consequences of genetic diversity. *Ecology Letters* 11: 609–623.

ITIS (Integrated Taxonomic Information System). 2021. www.itis.gov/servlet/SingleRpt/SingleRpt?search_topic=TSN&search_value=503904#null (Accessed April 23, 2021).

Karwath, A. 2005. Parrotfeather (Myriophyllum aquaticum). https://en.wikipedia.org/wiki/Myriophyllum_aquaticum#/media/File:Myriophyllum_aquaticum_-_side_(aka).jpg. (Accessed June 30, 2021).

King County Noxious Weed Control Program (NWCP). 2014. Parrotfeather. *Best Management Practices*. https://your.kingcounty.gov/dnrp/library/water-and-land/weeds/BMPs/Parrotfeather-control.pdf. (Accessed June 29, 2021).

King County Noxious Weed Control Program (NWCP). 2018. Parrotfeather photos. https://kingcounty.gov/services/environment/animals-and-plants/noxious-weeds/weed-identification/parrot-feather.aspx. (Accessed June 30, 2021).

Kuehne, L.M., A.K. Adey, T.M. Brownlee, and J.D. Olden. 2018. Field-based comparison of herbicides for control of parrot-feather (*Myriophyllum aquaticum*). *Journal of Aquatic Plant Management* 56: 18–23.

Kuehne, L.M., J.D. Olden, and E.S. Rubenson. 2016. Multi-trophic impacts of an invasive aquatic plant. *Freshwater Biology* 61(11): 1846–1861.

Lambertini, C., T. Riis, B. Olesen, J.S. Clayton, B.K. Sorrell, and H. Brix. 2010. Genetic diversity in three invasive clonal aquatic species in New Zealand. *BMC Genetics* 11: 52.

Lastrucci, L., L. Lazzaro, L. Dell'Olmo, B. Foggi, and F. Cianferoni. 2018. Impacts of *Myriophyllum aquaticum* invasion in a Mediterranean wetland on plant and maco-arthropod communities. *Plant Biosystems* 152(3): 427–435.

Luo, P., F. Liu, X. Liu, X. Wu, R. Yao, L. Chen, X. Li, R. Xiao, and J. Wu. 2017. Phosphorus removal from lagoon-pretreated swine wastewater by pilot-scale surface flow constructed wetlands planted with *Myriophyllum aquaticum*. *Science of the Total Environment* 576: 490–497.

Luo, P., X. Tong, F. Liu, M. Huang, J. Xu, R. Xiao, and J. Wu. 2020. Nutrients release and greenhouse gas emission during decomposition of *Myriophyllum aquaticum* in a sediment water system. *Environmental Pollution* 260: 114015.

Lurvey. 2021. Parrots feather 1 qt. www.lurveys.com/products/lurvey-select-green-goods/pond-plants/AQPLMYRAQU1QT. (Accessed June 30, 2021).

Machado, C., and F. Rocha. 1998. Control of *Myriophyllum aquaticum* in drainage and irrigated channels of Mondego river valley, Portugal. In: *Management and Ecology of Aquatic Plants. Proceedings of the 10th EWRS International Symposium on Aquatic Weeds*. European Weed Research Society, pp. 373–375.

Madsen, J.D., and R.M. Wersal. n.d. Parrotfeather [*Myriophyllum aquaticum* (Vellozo) Verdecourt]. Mississippi State University, Geosystems Research Institute. www.gri.msstate.edu/research/invspec/factsheets/2P/Parrotfeather.pdf. (Accessed June 29, 2021).

Metro Vancouver. 2019. Best Management Practices for Parrot's Feather in the Metro Vancouver Region. www.metrovancouver.org/services/regional-planning/PlanningPublications/ParrotsFeatherBMP.pdf. (Accessed June 29, 2021).

Murphy, K. 2007. *Myriophyllum aquaticum* (parrot's feather). *CABI Invasive Species Compendium*. www.cabi.org/isc/datasheet/34939. (Accessed June 25, 2021).

Nakai, S., Y. Inoue, M. Hosomi, and A. Murakami. 2000. *Myriophyllum spicatum*-released allelopathic polyphenols inhibiting growth of blue-green algae *Microcystis aeruginosa*. *Water Research* 34(11): 3026–3032.

Oberholzer, I.G., D.L. Mafokoane, and M.P. Hill. 2007. The biology and laboratory host range of the Weevil *Listronotus marginicolis* (Hustache) (Coleoptera: Curculionidae), a Natural Enemy of the Invasive Aquatic Weed *Myriophyllum aquaticum* (Velloso) Verde (Haloragaceae) (Parrot's Feather). *Report to the Water Research Commission*, p. 22. www.wrc.org.za/wp-content/uploads/mdocs/KV180.pdf. (Accessed June 29, 2021).

Okamoto, U., S. Shirahama, S. Nasu, H. Miyauchi, and M. Tokuda. 2020. Host range expansion of a Polygonaceae-associated leaf beetle to an invasive aquatic plant *Myriophyllum aquaticum* (Haloragaceae). *Arthropod-Plant Interactions* 14: 491–497.

Orr, B.K., and V.H. Resh. 1989. Experimental test of the influence of aquatic macrophyte cover on the survival of *Anopheles* larvae. *Journal of the American Control Association* 5(4): 579–585.

Parker, J.D., C.C. Caudill, and M.E. Hay. 2007. Beaver herbivory on aquatic plants. *Oecologia* 151: 616–625.

Patten, K. 2003. Evaluating Imazapyr in Aquatic environments. *Agricultural and Environmental News* 205: 77.

Pine, R.T., and L.W.J. Anderson. 1991. Plant preferences of triploid grass carp. *Journal of Aquatic Plant Management* 29: 80–82.

Reyes, A. 2021. Using grass carp to control aquatic vegetation: Practices, uncertainties, and consequences. *Northeast Aquatic Research*, p. 20. https://nysfola.org/wp-content/uploads/Using-Grass-Carp-to-Control-Aquatic-Vegetation-NYSFOLA.pdf. (Accessed June 29, 2021).

Rollins, L.A., A.T. Moles, S. Lam, R. Buitenwerf, J.M. Buswell, C.R. Brandenburger, H. Flores-Moreno, K.B. Nielsen, E. Couchman, G.S. Brown, F.J. Thomson, F. Hemmings, R. Frankham, and W.B. Sherwin. 2013. High genetic diversity is not essential for successful introduction. *Ecology and Evolution* 3(13): 4501–4517.

Roman, J., and J.A. Darling. 2007. Paradox lost: genetic diversity and the success of aquatic invasions. *Trends in Ecology & Evolution* 22(9): 454–464.

Shah, M.A., M.A. Ali, F.M. Al-Hemaid, and Z.A. Reshi. 2014. Delimiting invasive *Myriophyllum aquaticum* in Kashmir Himalaya using a molecular phylogenetic approach. *Genetics and Molecular Research* 13(3): 7564–7570.

Smitha, P., P. Nazeem, T. James, M. Mohan, P.M. Sherif, R. Keshavachandran, P. Nazeem, D. Girija, P. John, and K. Peter. 2007. Micropropagation of the aquarium plant-parrot feather milfoil-*Myriophyllum aquaticum* (Velloso) Verdcourt. In: *Recent Trends in Horticultural Biotechnology*. New India Publishing Agency, pp. 333–336.

Souza, F.A., M. Dziedzic, S.A. Cubas, and L.T. Maranho. 2013. Restoration of polluted waters by phytoremediation using *Myriophyllum aquaticum* (Vell.) Verdc., Haloagaceae. *Journal of Environmental Management* 120: 5–9.

Souza, G.S.F., M.R.R. Pereira, H.S. Vitorino, C.F. Campos, and D. Martins. 2012. Influência da Chuva na Eficácia do Herbicida 2,4-D no Controle de *Myriophyllum Aquaticum*. *Planta Daninha* 30(2): 263–267.

Stiers, I., N. Crohain, G. Josens, and L. Triest. 2011. Impact of three aquatic invasive species on native plants and macroinvertebrates in temperate ponds. *Biological Invasions* 13: 2715–2726.

Sun, H., F. Liu, S. Xu, S. Wu, G. Zhuang, Y. Deng, J. Wu, and X. Zhuang. 2017. *Myriophyllum aquaticum* constructed wetland effectively removes nitrogen in swine wastewater. *Frontiers in Microbiology* 8: 1932.

Tu et al. 2001. Weed control methods handbook. *The Nature Conservancy*. www.invasive.org/gist/products/handbook/17.imazapyr.pdf. (Accessed June 29, 2021).

USGS. 2021. Myriophyllum aquaticum. *NAS—Nonindigenous Aquatic Species*. https://nas.er.usgs.gov/viewer/omap.aspx?SpeciesID=235. (Accessed June 25, 2021).

Wersal, R.M. 2010. *The Conceptual Ecology and Management of Parrotfeather*. Mississippi: Mississippi State University, p. 216.

Wersal, R.M., E. Baker, J. Larson, K. Dettloff, A.J. Fusaro, D.D. Thayer, and I.A. Pfingsten. 2021. *Myriophyllum aquaticum (Vell.) Verdc. U.S. Geological Survey, Nonindigenous Aquatic Species Database*, Gainesville, FL. https://nas.er.usgs.gov/queries/FactSheet.aspx?speciesID=235. (Accessed June 25, 2021).

Wersal, R.M., and J.D. Madsen. 2007. Comparison of imazapyr and imazamox for control of parrotfeather (*Myriophyllum aquaticum* (Vell.) Verdc.). *Journal of Aquatic Plant Management* 45: 132–136.

Wersal, R.M., and J.D. Madsen. 2011. Comparative effects of water level variations on growth characteristics of *Myriophyllum aquaticum*. *Weed Research* 51(4): 386–393.

Wersal, R.M., J.D. Madsen, and P.D. Gerard. 2013. Survival of parrotfeather following simulated drawdown events. *Journal of Aquatic Plant Management* 51: 22–26.

Xie, D., D. Yu, L. Yu, and C. Liu. 2010. Asexual propagations of introduced exotic macrophytes *Elodea nuttallii*, *Myriophyllum aquaticum*, and *M. propinquum* are improved by nutrient-rich sediments in China. *Hydrobiologia* 655: 37–47.

You, W., D. Yu, C. Liu, D. Xie, and W. Xiong. 2013. Clonal integration facilitates invasiveness of the alien aquatic plant *Myriophyllum aquaticum* L. under heterogeneous water availability. *Hydrobiologia* 718: 27–39.

10 Carolina Fanwort (*Cabomba caroliniana*)

ABSTRACT

In Massachusetts, Carolina fanwort (*Cabomba caroliniana*) was first detected in 1930 in Hatfield. The sightings of invasive *C. caroliniana* species were analyzed based on historical data (1983–2017) collected mainly by the Massachusetts Department of Environmental Protection. By now fanwort has become highly invasive in Massachusetts. Currently, there are around 200 waterbodies within 20 watersheds infested by fanwort. This chapter provides a better understanding on the status of its spread in Massachusetts which will help the general public and water users to minimize the further dispersal of *C. caroliniana* in Massachusetts. In addition, this report also provides different methods that have been used or tested for control of *C. caroliniana* such as physical (e.g., drawdown and dredging), chemical (e.g., herbicides), and biological (e.g., weevil or grass carp) control tools.

INTRODUCTION

Taxonomy Hierarchy (ITIS 2021)

Kingdom	Plantae
Subkingdom	Viridiplantae
Infrakingdom	Streptophyta
Superdivision	Embryophyta
Division	Tracheophyta
Subdivision	Spermatophytina
Class	Magnoliopsida
Superorder	Nymphaeanae
Order	Nymphaeales
Family	Cabombaceae
Genus	*Cabomba* Aubl.
Species	*Cabomba caroliniana* A. Gray

BASIC BIOLOGY

Fanwort is a fully submerged aquatic plant that roots in the soft substrates in waterbodies, with stems that branch up to the surface of the water (Figure 10.1). Its fan-like leaves are found in pairs, with each pair of leaves opposite to the other. The underwater leaves are fan-shaped while the floating leaves that support the flowers are diamond-shaped, which are about 5 cm across, and secrete a mucous that coats the submerged portion of the plant (Larson et al. 2020). In late summer, *C. caroliniana* usually blooms with tiny flowers over oval floating leaves (Figure 10.2), while its stems are under the water surface and in thick mats in shallow waters of Massachusetts (MA DCR 2016). The flowers, which are usually less than 2 cm across, can range from a white to pale yellow color, and sometimes can exhibit small amounts of pink or purple coloration, as shown in Figure 10.2 (Larson et al. 2020). The stems will generally have white or reddish-brown hair. Mature plants can reach up to 10 m long and generally fall between 30 and 80 cm, or between 12 and 31 inches (Wilson et al. 2007). The plant produces a leathery, indehiscent fruit with a three-seeded follicle (MA DCR 2005). Fanwort can tolerate a wide range of temperatures, pH levels, water chemistry conditions, and nutrient conditions (MA DCR 2002). However, in waters with high levels of calcium, the growth of the plant can be restrained (MA DCR 2005). While fanwort grows in a wide variety of waters, it does best in acidic water with a pH around 6 (Tobias Bickel, personal communication). *C. caroliniana* usually can be found in slow-moving waterbodies (ponds, lakes, small rivers, etc.) with silty substrates, as well as in rivers and even turbid waterbodies (Larson et al. 2020). There are at least three confirmed phenotypes of fanwort (green, red, and aquarium), of which the invasive green phenotype is the most difficult to control with herbicides (Bultemeier et al. 2009). The phenotype color is related to the leaf coloration, not the flowers, and also that the phenotypes are aquarium cultivars, not the subspecies occurring in its natural environment. More information about the biology

DOI: 10.1201/9781003201106-10

FIGURE 10.1 Fanwort sample from Federal Pond (*left*) and underwater fanwort photo from Arcade Pond (*right*), MA. (Photo by Wai Hing Wong.)

FIGURE 10.2 Fanwort floating leaves and flower buds and flower. (Left photo by Lyn Gettys from https://plants.ifas.ufl.edu; right photo by Tobias O Bickel.)

and taxonomy of fanwort can be obtained from more references: Schneider et al. (1982), Ørgaard (1991), and Mackey and Swarbrick (1997).

NATIVE RANGE

C. caroliniana is a native plant to subtropic–temperate regions of the Americas, specifically to the countries of Brazil, Paraguay, Uruguay, and Argentina, and to the Southern and Eastern United States (Larson et al. 2020), where it lives in slow-flowing and stagnant freshwater (Scheers et al. 2016).

INVASIVE RANGE

C. caroliniana has invaded many regions in the world, such as several European countries,

China, India, Japan, and the United States (Brundu 2015). Fanwort is considered as an invasive species in northern and western United Sates. In Australia, the weed has been declared as a Weed of National Significance due to severe economic and ecological impacts (CHAH 2011).

MECHANISMS/VECTORS OF SPREAD

C. caroliniana is highly adaptive and competitive, and can grow quickly and densely. It can easily spread through seeds and rhizome or stem fragmentation (MA DCR 2002; Scheers et al. 2016). Certain parts of the plant that become fragmented can survive free floating for six to eight weeks, allowing these fragments to easily spread to new waterbodies or repopulate areas in which treatments have been applied (Larson et al. 2020). Unintentional transport on boat trailers is one important way that *C. caroliniana* and other aquatic plants can spread by fragmentation (Rothlisberger et al. 2010; Bruckerhoff et al. 2015). While the plant is sensitive to drying out and generally requires a permanent source of at least shallow water to survive, its fragments can survive for at least 3 hours in a desiccated environment, making it highly resilient and improving its ability to spread (Bickel 2015; Larson et al. 2020).

ECOLOGICAL IMPACTS

As a nuisance species, *C. caroliniana* grows prolifically and forms dense canopy under and over the surface of the water (Figure 10.1). The dense growths of *C. caroliniana* can have a negative influence on the ecosystem. These impacts involve, but are not limited to, reducing biodiversity, displacing native macrophyte species, impeding water flow, and altering nutrient cycling and water quality (Sheldon 1994; Mackey and Swarbrick 1997; Hogsden et al. 2007). Large dense mats of *C. caroliniana* on the water surface can reduce the space that native species have to live because fanwort is highly competitive and creates monocultures that displace other macrophytes. This can affect the fish and other aquatic species that rely on the displaced species, potentially causing them to have to relocate or perish. Further, habitat becomes unsuitable for many native aquatic animals due to the anoxic environment created by the decay of *C. caroliniana* mainly from herbicides or by dense growth of fanwort which is more about the exclusion of light from deeper water, hence no photosynthesis, and that fanwort stands can interfere with water flow (i.e., create stagnant conditions and water doesn't exchange). According to the "Fanwort: An Invasive Aquatic Plant" report by the Department of Conservation and Recreation (2002), *C. caroliniana* mats can intercept the sunlight, excluding other submerged plants. The mats reduce the amount of light to varying degrees in all cases and the amount of light is one of the most important factors that determine aquatic plant distribution in lakes. Additionally, the growth of *C. caroliniana* can also alter the sediment chemistry.

ECONOMIC IMPACTS

Due to the infestation of *C. caroliniana*, different levels of socioeconomic impacts have arisen in many regions. For instance, fishing ability has decreased, tourism and recreational activities have declined, the value of real estate has been influenced, etc.

POSITIVE ASPECTS

Although *C. caroliniana* is an invasive species, there are some positive aspects associated with it. First, it can serve as the habitat for many fish, micro and macro invertebrates, and is a food source for waterfowl and wildlife (Mackey 1996). It should be noted that almost all submersed macrophytes can provide some ecosystem services as what fanwort can do; but it is still preferable to have a diverse macrophyte community than a monoculture. Second, *C. caroliniana* has been widely used in the aquarium industry around the world. However, this is also one of the reasons that fanwort has spread to many regions. The trade of fanwort is banned in Massachusetts.

MONITORING AND SIGHTING DATA

The invasive species data in each waterbody were collected by the Massachusetts Department of Environmental Protection (MassDEP), the Massachusetts Department of Conservation and Recreation (MA DCR), field surveys, Water Quality Assessment Reports, Diagnostic/Feasibility Reports, personal communications, application files, etc. The standard operating procedures, including sampling methodology, species identification, quality control and assurance, were implemented for all samples collected by MassDEP. Data from other agencies have been confirmed by experts. Sites with aquatic invasive species (AIS) were mapped as a data layer with ArcGIS® ArcMap 10.1 (ESRI, Redlands, California). Such a map will help make the general public aware of the presence of fanwort in Massachusetts, hopefully to assist minimizing unintentional transportation between different waterbodies.

Through the reports and surveys of *C. caroliniana*, different locations with fanwort have

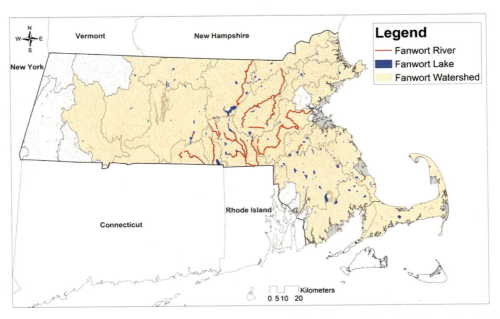

FIGURE 10.3   Distribution map depicting Carolina Fanwort (*Cabomba caroliniana*) in Massachusetts. (See Table 10.1 for site descriptions.)

been mapped using GIS (Figure 10.3). GIS is able to pinpoint the exact area that *C. caroliniana* has spread too, and this can help to inform future decision-making for these waterbodies, or regulations that involve efforts to limit the spread of this aquatic invasive species. *C. caroliniana* is a very aggressive invasive species, and therefore was able to spread in a short period of time throughout Massachusetts (Table 10.1). The first verified sighting by the Department of Environmental Protection in Massachusetts was in 1983 at Whitmans Pond. *C. caroliniana* was detected while completing a Report to Massachusetts Division of Water Pollution Control on Feasibility Study of Lake Restoration for Whitmans Pond. According to MA DCR (2002), Carolina fanwort was detected in Massachusetts as early as 1930 in Hatfield, MA. Currently, there are 196 waterbodies (Table 10.2) within the 23 watersheds in Massachusetts that are infested by *C. caroliniana* (Table 10.3). Among them, 173 are lakes and 23 are rivers. Most of the spread are recorded much later in the 1990s which is mainly due to the fact few monitoring on fanwort were conducted in earlier years.

MANAGEMENT OPTIONS

Currently, this species is banned from the trade in Massachusetts because it is in the prohibited plant list. More actions can be used to prevent the spread of *C. caroliniana*. For example, building a coalition to monitor for *C. caroliniana*, improving the monitoring and enforcement of the distribution and sale of *C. caroliniana*, educating aquarium owners and the industry to be aware of the hazards associated with the spread of invasive species, providing boat washing stations for high-traffic public lake access points, and developing a water recreation vehicle and trailer inspection program are all forms of effective social management (Maki and Galatowitsch 2004; Hussner et al. 2013; Verbrugge 2014; Bickel 2015; State of Michigan 2017).

There are many other different methods that have been tested for a range of fanwort management goals, from early detection and eradication to containment to prevent spread and subsequent maintenance or not doing anything once the infestation is too large or containment is not an option anymore. Effective management

TABLE 10.1

Fanwort Sightings in MA Waters as Shown in Figure 10.3

Waterbody	Agency	Report Source	Longitude	Latitude	First Year Reported
Whitmans Pond	MassDEP	Report to Massachusetts Division of Water Pollution Control on Feasibility Study of Lake Restoration for Whitmans Pond, Weymouth, MA			1983
Forge Pond	BEC	A Diagnostic/Feasibility Study for the Management of Lost Lake/Knopp's Pond Groton, Massachusetts, 1987			1987
Forge Pond	DCR	DCR Data from ACT, GeoSyntec			1987
North Pond	Metcalf & Eddy	Final Report to Town of Hopkinton on Diagnostic Feasibility Study of North Pond Hopkinton MA, 1987			1987
North Pond	DCR	DCR Data from Lakes and Ponds, 2008			1987
Knops Pond/Lost Lake	BEC	A Diagnostic/Feasibility Study for the Management of Lost Lake/Knopp's Pond Groton, Massachusetts, 1992			1992
Knops Pond/Lost Lake	DCR	DCR Data from Lakes and Ponds, 2002			1992
Arcade Pond	MassDEP	09-G010-02	−71.6775195	42.1129609	1994
Arcade Pond	Individual	Joy Trahan-Liptak	−71.67707	42.113303	1994
Arcade Pond	MassDEP	Synoptic lake survey field sheets 1994—Blackstone River Watershed Lakes, 1994			1994
Fish Pond	MassDEP	Synoptic lake survey field sheets 1994—Blackstone River Watershed Lakes, 1994			1994
Fish Pond	MassDEP	09-G010-01	−71.6879542	42.1209182	1994
Girard Pond	MassDEP	Synoptic lake survey field sheets 1994—Blackstone River Watershed Lakes, 1994			1994
Hayes Pond	MassDEP	Synoptic lake survey field sheets 1994—Blackstone River Watershed Lakes, 1994			1994
Kettle Brook	MassDEP	Synoptic lake survey field sheets 1994—Blackstone River Watershed Lakes, 1994			1994
Lake Quinsigamond	MassDEP	Synoptic lake survey field sheets 1994—Blackstone River Watershed Lakes, 1994			1994
Lake Quinsigamond	College of the Holy Cross	Robert Bertin	−71.749052	42.256062	1994
Lake Quinsigamond	College of the Holy Cross	Robert Bertin	−71.749052	42.256062	1994
Lake Quinsigamond	College of the Holy Cross	Robert Bertin	−71.749052	42.256062	1994
Leesville Pond	MassDEP	Synoptic lake survey field sheets 1994—Blackstone River Watershed Lakes, 1994			1994
Leesville Pond	College of the Holy Cross	Robert Bertin	−71.827414	42.230332	1994

(Continued)

TABLE 10.1
(Continued)

Waterbody	Agency	Report Source	Longitude	Latitude	First Year Reported
Leesville Pond	Individual	Joy Trahan-Liptak	−71.826444	42.229001	1994
Mill Pond	MassDEP	Synoptic lake survey field sheets 1994—Blackstone River Watershed Lakes, 1994			1994
Miscoe Lake	MassDEP	Synoptic lake survey field sheets 1994—Blackstone River Watershed Lakes, 1994			1994
Pondville Pond	MassDEP	Synoptic lake survey field sheets 1994—Blackstone River Watershed Lakes, 1994			1994
Pratt Pond	MassDEP	Synoptic lake survey field sheets 1994—Blackstone River Watershed Lakes, 1994			1994
Pratt Pond	College of the Holy Cross	Robert Bertin	−71.598913	42.181814	1994
Quinsigamond River	MassDEP	11-F004–02	−71.7080924	42.2305379	1994
Quinsigamond River	Individual	Joy Trahan-Liptak	−71.701242	42.221306	1994
Quinsigamond River	MassDEP	Synoptic lake survey field sheets 1994—Blackstone River Watershed Lakes, 1994			1994
Stevens Pond	MassDEP	Synoptic lake survey field sheets 1994—Blackstone River Watershed Lakes, 1994			1994
Unnamed Tributary	MassDEP	Synoptic lake survey field sheets 1994—Blackstone River Watershed Lakes, 1994			1994
West River	MassDEP	Synoptic lake survey field sheets 1998—Blackstone River Watershed Lakes, 1998			1994
West River	MassDEP	Synoptic lake survey field sheets 1994—Blackstone River Watershed Lakes, 1994			1994
Whitins Pond	MassDEP	Synoptic lake survey field sheets 1994—Blackstone River Watershed Lakes, 1994			1994
Whitins Pond	College of the Holy Cross	Robert Bertin	−71.689583	42.113764	1994
Whitins Pond	MassDEP	Email communication between David Wong and Therese Beaudoin on July 7, 2017	−71.688119	42.119622	1994
Woodbury Pond	MassDEP	Synoptic lake survey field sheets 1994—Blackstone River Watershed Lakes, 1994			1994
Woodbury Pond	MassDEP	09-G008–07	−71.7312552	42.164983	1994
Cook Pond	College of the Holy Cross	Robert Bertin	−71.858373	42.285515	1995
Cook Pond	College of the Holy Cross	Robert Bertin	−71.858373	42.285515	1995
Curtis Pond	College of the Holy Cross	Robert Bertin	−71.838502	42.241484	1995
Flint Pond	MassDEP	Synoptic lake survey field sheets 1998—Blackstone River Watershed Lakes, 1998			1995
Flint Pond	College of the Holy Cross	Robert Bertin	−71.725087	42.241352	1995
Flint Pond	College of the Holy Cross	Robert Bertin	−71.725087	42.241352	1995

Location	Institution	Collector/Source ID	Source	Longitude	Latitude	Year
Flint Pond	College of the Holy Cross	Robert Bertin		−71.725087	42.241352	1995
Flint Pond	College of the Holy Cross	Robert Bertin		−71.725087	42.241352	1995
Flint Pond	Individual	Joy Trahan-Liptak		−71.725688	42.240948	1995
Flint Pond	MassDEP		Synoptic lake survey field sheets 1998—Blackstone River Watershed Lakes, 1998			1995
Massapoag Brook	MassDEP		The Neponset River Watershed 1994 Resource Assessment Report, 1995			1995
Lake Winthrop	MassDEP		Synoptic lake survey field sheets 1997—Charles River Watersheds lakes			1997
Morses Pond	MassDEP		Synoptic lake survey field sheets 1997—Charles River Watersheds lakes			1997
Arcadia Lake	MassDEP		Synoptic lake survey field sheets 1998—Connecticut River Watershed Lakes			1998
Dark Brook	MassDEP		Synoptic lake survey field sheets 1998—Blackstone River Watershed Lakes, 1998			1998
Hardwick Pond	College of the Holy Cross	Robert Bertin		−72.2386	42.313254	1998
Hardwick Pond	MassDEP		Open lake survey files			1998
Ironstone Reservoir	MassDEP		Synoptic lake survey field sheets 1998—Blackstone River Watershed Lakes, 1998			1998
Ironstone Reservoir	College of the Holy Cross	Robert Bertin		−71.6109	42.024298	1998
Lake Holland	MassDEP		Synoptic lake survey field sheets—Connecticut River Watershed Lakes			1998
Lake Lashaway	Individual	Joy Trahan-Liptak		−72.045061	42.235541	1998
Lake Lashaway	MassDEP		Open lake survey files			1998
Lake Ripple	MassDEP		Synoptic lake survey field sheets 1998—Blackstone River Watershed Lakes, 1998			1998
Lake Ripple	College of the Holy Cross	Robert Bertin		−71.697077	42.213605	1998
Lake Ripple	College of the Holy Cross	Robert Bertin		−71.697077	42.213605	1998
Lake Ripple	MassDEP	09-G008-04		−71.6936628	42.2141715	1998
Lake Ripple	Individual	Joy Trahan-Liptak		−71.696072	42.213862	1998
Manchaug Pond	MassDEP		Synoptic lake survey field sheets 1998—Blackstone River Watershed Lakes, 1998			1998
Manchaug Pond	College of the Holy Cross	Robert Bertin		−71.775563	42.099183	1998
Metacomet Lake	MassDEP		Synoptic lake survey field sheets—Connecticut River Watershed Lakes			1998
Mill River	MassDEP		Synoptic lake survey field sheets 1998—Blackstone River Watershed Lakes, 1998			1998
Mill River	MassDEP		Synoptic lake survey field sheets 1998—Blackstone River Watershed Lakes, 1998			1998
Newton Pond	MassDEP		Synoptic lake survey field sheets 1998—Blackstone River Watershed Lakes, 1998			1998
Newton Pond	College of the Holy Cross	Robert Bertin		−71.747947	42.312773	1998
Quaboag Pond	College of the Holy Cross	Robert Bertin		−72.068036	42.197472	1998

(Continued)

TABLE 10.1 (Continued)

Waterbody	Agency	Report Source	Longitude	Latitude	First Year Reported
Quaboag Pond	MassDEP	Open lake survey files			1998
Quacumquasit Pond	MassDEP	Open lake survey files			1998
Tatnuck Brook	MassDEP	Synoptic lake survey field sheets 1998—Blackstone River Watershed Lakes, 1998			1998
Newfield Pond	MassDEP	DWM 1999 Lake TMDL field sheets			1999
Chebacco Lake	MassDEP	North Coastal Watershed 1997/1998 Water Quality Assessment Report, May 2000			2000
Griswold Pond	MassDEP	North Coastal Watershed 1997/1998 Water Quality Assessment Report, May 2000			2000
Lake Como	MassDEP	Ten Mile River Basin 1997 Water Quality Assessment Report, DWM CN 18.0.2000			2000
Lake Mascuppic	MassDEP	License applications for herbicide treatment: 1996 through 2000.2000			2000
Plain Street Pond	MassDEP	Ten Mile River Basin 1997 Water Quality Assessment Report. DWM CN 18.0.2000			2000
Spring Pond	MassDEP	North Coastal Watershed 1997/1998 Water Quality Assessment Report, May 2000			2000
Spring Pond	DCR	DCR Data from Lakes and Ponds, 2006			2000
Chaffin Pond	MassDEP	Nashua River Basin 1998 Water Quality Assessment Report CN046.0.2001			2001
Chaffin Pond	College of the Holy Cross	Robert Bertin	−71.840022	42.33112	2001
Chaffin Pond	DCR	Joy Trahan-Liptak	−71.839483	42.331244	2001
Dawson Pond	MassDEP	Nashua River Basin 1998 Water Quality Assessment Report CN046.0.2001			2001
Dawson Pond	DCR	Joy Trahan-Liptak	−71.852208	42.335114	2001
Flannagan Pond	MassDEP	Nashua River Basin 1998 Water Quality Assessment Report CN046.0.2001			2001
Grove Pond	MassDEP	Nashua River Basin 1998 Water Quality Assessment Report CN046.0.2001			2001
Lake Shirley	MassDEP	Nashua River Basin 1998 Water Quality Assessment Report CN046.0.2001			2001
Lake Shirley	DCR	DCR Data from ACT, GeoSyntec-ACT			2001
Pentucket Pond	MassDEP	Parker River Watershed Year 3 Assessment Report, June 2001			2001
Plow Shop Pond	MassDEP	Nashua River Basin 1998 Water Quality Assessment Report CN046.0.2001			2001

Waterbody	Source	Reference	Longitude	Latitude	Year
Spectacle Pond	ESS	A Diagnostic/Feasibility Study of Spectacle Pond, Littleton/Ayer, Massachusetts, 2001			2001
Spectacle Pond	DCR	DCR Data from ACT, GeoSyntec-ACT			2001
Wachusett Reservoir	DCR	Water Quality Report: 2001 Wachusett Reservoir and Watershed			2001
Wachusett Reservoir	College of the Holy Cross	Robert Bertin	−71.787817	42.397107	2001
Wachusett Reservoir	DCR	David Worden	−71.787817	42.397107	2001
Wachusett Reservoir	DCR	David Worden	−71.789929	42.389998	2001
Wachusett Reservoir	DCR	David Worden	−71.788251	42.379206	2001
Wachusett Reservoir	DCR	David Worden	−71.774821	42.375014	2001
Wachusett Reservoir	DCR	Jamie Carr	−71.789929	42.389998	2001
Wachusett Reservoir	DCR	Jamie Carr	−71.787817	42.397107	2001
Wachusett Reservoir	DCR	Jamie Carr	−71.788251	42.379206	2001
Wachusett Reservoir	DCR	Jamie Carr	−71.795345	42.384187	2001
Wachusett Reservoir	DCR	DCR Data from ACT, GeoSyntec-ACT			2001
Baddacook Pond	MassDEP	05-G001-01	−71.530464	42.6189623	2002
Baddacook Pond	DCR	DCR Data from Lakes and Ponds, 2002	−71.530464	42.6189623	2002
Bearse Pond	Individual	Joy Trahan-Liptak	−70.33056	41.675875	2002
Bearse Pond	Individual	Joy Trahan-Liptak	−70.33056	41.675875	2002
Bearse Pond	MassDEP	Cape Cod Watershed Water Quality Assessment Report CN 50.0.			2002
Beaver Pond	DCR	Excel spreadsheet of non-native aquatic and wetland plants in Massachusetts lakes and ponds dated January 2005			2002
Beaver Pond	DCR	DCR Data from Lakes and Ponds, 2002			2002
Big Bearhole Pond	MassDEP	Appendix C—Table C1.1996 Taunton River Watershed Lake observations and trophic status estimates. (Taunton River Watershed 2001 Water Quality Assessment Report 62-AC-1 CN 94.0, December 2005)			2002
Big Bearhole Pond	DCR	DCR Data from Lakes and Ponds, 2002			
Big Bearhole Pond	DCR	DCR Data from Lakes and Ponds, 2007			2002
Big Bearhole Pond	DCR	DCR Data from Lakes and Ponds, 2008			2002
Bolivar Pond	MassDEP	Boston Harbor 1999 Water Quality Assessment Report CN049.0.2002			2002
Cedar Meadow Pond	MassDEP	French and Quinebaug River Watersheds 2001 Water Quality Assessment Report 41/42-AC-1 51.0, March 2002			2002

(Continued)

TABLE 10.1
(Continued)

Waterbody	Agency	Report Source	Longitude	Latitude	First Year Reported
Ellis Pond	MassDEP	Boston Harbor 1999 Water Quality Assessment Report CN049.0.2002			2002
Ellis Pond	DCR	DCR Data from Lakes and Ponds, 2005			2002
First Pond	MassDEP	Herbicide license applications for 2002			2002
First Pond	DCR	DCR Data from Lakes and Ponds, 2006			2002
Lake Rico	MassDEP	Appendix C—Table C1.1996 Taunton River Watershed Lake observations and trophic status estimates (Taunton River Watershed 2001 Water Quality Assessment Report 62-AC-1 CN 94.0 December 2005)			2002
Lake Rico	DCR	DCR Data from Lakes and Ponds, 2002			2002
Little Bearhole Pond	DCR	DCR Data from Lakes and Ponds, 2002			2002
Little Bearhole Pond	DCR	DCR Data from Lakes and Ponds, 2007			2002
Little Bearhole Pond	DCR	DCR Data from Lakes and Ponds, 2008			2002
Massapoag Lake	MassDEP	Boston Harbor 1999 Water Quality Assessment Report CN049.0.2002			2002
Massapoag Lake	Individual	Joy Trahan-Liptak	−71.177528	42.103486	2002
Massapoag Lake	Individual	Joy Trahan-Liptak	−71.177528	42.103486	2002
Moulton Pond	College of the Holy Cross	Robert Bertin	−71.954051	42.388392	2002
Neponset Reservoir	MassDEP	Boston Harbor 1999 Water Quality Assessment Report CN049.0.2002			2002
Reservoir Pond	MassDEP	Boston Harbor 1999 Water Quality Assessment Report CN049.0.2002			2002
Swains Pond	MassDEP	Herbicide license applications for 2002			2002
Town Pond	MassDEP	Boston Harbor 1999 Water Quality Assessment Report CN049.0.2002			2002
Turner Pond	MassDEP	Boston Harbor 1999 Water Quality Assessment Report CN049.0.2002			2002
Webster Lake	DCR	DCR Data from Lakes and Ponds, 2002			2002
Webster Lake	MassDEP	05-G014–01	−71.8487243	42.0543495	2002
Webster Lake	College of the Holy Cross	Robert Bertin	−71.848244	42.038004	2002
Webster Lake	College of the Holy Cross	Robert Bertin	−71.848244	42.038004	2002
Wequaquet Lake	Individual	Joy Trahan-Liptak	−70.35087	41.675829	2002

Waterbody	Source	Reference	Longitude	Latitude	Year
Wequaquet Lake	Individual	Joy Trahan-Liptak	−70.35087	41.675829	2002
Wequaquet Lake	Individual	Joy Trahan-Liptak	−70.329647	41.670658	2002
Wequaquet Lake	Individual	Joy Trahan-Liptak	−70.35087	41.675829	2002
Wequaquet Lake	MassDEP	Cape Cod Watershed Water Quality Assessment Report CN 50.0			2002
Ballardvale Impoundment	MassDEP	Appendix D—Table D4.1995 Shawsheen River Watershed summer lake status (Shawsheen River Watershed 2000 Water Quality Assessment Report 83-AC-2 86.0, July 2003)			2003
Federal Pond	MassDEP	Buzzards Bay Watershed 2000 Water Quality Assessment Report 95-AC-2 085.0, November 2003			2003
Fosters Pond	MassDEP	Appendix D—Table D4.1995 Shawsheen River Watershed summer lake status. (Shawsheen River Watershed 2000 Water Quality Assessment Report 83-AC-2 86.0, July 2003)			2003
Fresh Meadow Pond	MassDEP	Buzzards Bay Watershed 2000 Water Quality Assessment Report 95-AC-2 085.0, November 2003			2003
Nashua River	Nashua River Watershed Association	2003 Water Quality Reports for the Nashua River and Tributaries			2003
Pomps Pond	MassDEP	Appendix D—Table D4.1995 Shawsheen River Watershed summer lake status (Shawsheen River Watershed 2000 Water Quality Assessment Report 83-AC-2 86.0, July 2003)			2003
Sampson Pond	MassDEP	05-R002–01	−70.7529767	41.8520645	2003
Sampson Pond	MassDEP	Buzzards Bay Watershed 2000 Water Quality Assessment Report 95-AC-2 085.0, November 2003			2003
Watson Pond	DCR	DCR Data from Lakes and Ponds, 2003			2003
Watson Pond	MassDEP	Appendix C—Table C1.1996 Taunton River Watershed Lake observations and trophic status estimates (Taunton River Watershed 2001 Water Quality Assessment Report 62-AC-1 CN 94.0, December 2005)			2003
Watson Pond	DCR	DCR Data from Lakes and Ponds, 2008			2003
White Island Pond, East Basin	MassDEP	07-T003–02	−70.6175309	41.8091834	2003
White Island Pond, East Basin	MassDEP	08-F008–02	−70.6175309	41.8091834	2003

(Continued)

TABLE 10.1
(Continued)

Waterbody	Agency	Report Source	Longitude	Latitude	First Year Reported
White Island Pond, East Basin	MassDEP	11-B007–01	−70.6175309	41.8091834	2003
White Island Pond, East Basin	MassDEP	14-B005–01	−70.6175309	41.8091834	2003
White Island Pond, East Basin	MassDEP	14-B007–01	−70.6175309	41.8091834	2003
White Island Pond, East Basin	MassDEP	Buzzards Bay Watershed2000 Water Quality Assessment Report 95-AC-2 085.0, November 2003			2003
White Island Pond, West Basin	MassDEP	11-B002–02	−70.6275557	41.8099941	2003
White Island Pond, West Basin	MassDEP	11-B005–02	−70.6275557	41.8099941	2003
White Island Pond, West Basin	MassDEP	11-B007–02	−70.6275557	41.8099941	2003
White Island Pond, West Basin	MassDEP	12-B003–02	−70.6275557	41.8099941	2003
White Island Pond, West Basin	MassDEP	13-B007–02	−70.6275557	41.8099941	2003
White Island Pond, West Basin	MassDEP	13-B009–03	−70.6275557	41.8099941	2003
White Island Pond, West Basin	MassDEP	14-B003–02	−70.6275557	41.8099941	2003
White Island Pond, West Basin	MassDEP	14-B005–02	−70.6275557	41.8099941	2003
White Island Pond, West Basin	MassDEP	14-B007–02	−70.6275557	41.8099941	2003
White Island Pond, West Basin	MassDEP	Buzzards Bay Watershed 2000 Water Quality Assessment Report 95-AC-2 085.0, November 2003			2003

Waterbody	Source	Reference	Longitude	Latitude	Year
Winnecunnet Pond	DCR	DCR Data from Lakes and Ponds, 2003			2003
Baldpate Pond	MassDEP	05-G004-02	−71.0014233	42.6988862	2004
Baldpate Pond	MassDEP	Baseline Lake Survey 2003 Technical Memo. TM-S-16 DWM Control Number CN 205.0.2007			2004
Baldpate Pond	DCR	DCR Data from National Heritage 2004			2004
Baldpate Pond	DCR	DCR Data from Lakes and Ponds, 2004			2004
Bare Hill Pond	College of the Holy Cross	Robert Bertin	−71.597616	42.491779	2004
Edgewater Office Park Pond	MassDEP	Herbicide license applications for 2004			2004
Field Pond	DCR	DCR Data from Lakes and Ponds, 2007			2004
Field Pond	MassDEP	Appendix C—Table C1.1995 Ipswich River Watershed Lake observations and trophic status estimates. (Ipswich River Watershed 2000 Water Quality Assessment Report 92-AC-1 088.0, April 2004)			2004
Lake Ellis	MassDEP	Appendix B—Table B1. Millers River Watershed 1995 Lake Synoptic Survey Data and Trophic Status Estimates (Millers River Watershed 2000 Water Quality Assessment Report 35-AC-1 CN089.0, March 2004)			2004
Lake Ellis	College of the Holy Cross	Robert Bertin	−72.204192	42.577244	2004
Lake Rohunta	MassDEP	Millers River Watershed 2000 Water Quality Assessment Report 35-AC-1 CN089.0, March 2004			2004
Lake Rohunta	MassDEP	Appendix B—Table B1. Millers River Watershed 1995 Lake Synoptic Survey Data and Trophic Status Estimates (Millers River Watershed 2000 Water Quality Assessment Report 35-AC-1 CN089.0, March 2004)			2004
Lake Rohunta	MassDEP	Millers River Watershed 2000 Water Quality Assessment Report 35-AC-1 CN089.0, March 2004			2004
Lake Sabbatia	MassDEP	Appendix C—Table C1.1996 Taunton River Watershed Lake observations and trophic status estimates (Taunton River Watershed 2001 Water Quality Assessment Report 62-AC-1 CN 94.0, December 2005)			2004
Lake Warner	Massachusetts Water Watch Partnership, UMass Extension and the Town of Hadley	Lake Warner 2003 Volunteer Monitoring Program Report			2004

(Continued)

TABLE 10.1 (Continued)

Waterbody	Agency	Report Source	Longitude	Latitude	First Year Reported
Lowe Pond	MassDEP	Appendix C—Table C1.1995 Ipswich River Watershed Lake observations and trophic status estimates (Ipswich River Watershed 2000 Water Quality Assessment Report 92-AC-1 088.0, April 2004)			2004
Martins Pond	MassDEP	Appendix C—Table C1.1995 Ipswich River Watershed Lake observations and trophic status estimates (Ipswich River Watershed 2000 Water Quality Assessment Report 92-AC-1 088.0, April 2004)			2004
Parker Pond	MassDEP	Appendix B—Table B1. Millers River Watershed 1995 Lake Synoptic Survey Data and Trophic Status Estimates (Millers River Watershed 2000 Water Quality Assessment Report 35-AC-1 CN089.0, March 2004)			2004
South Athol Pond	MassDEP	Appendix B—Table B1. Millers River Watershed 1995 Lake Synoptic Survey Data and Trophic Status Estimates (Millers River Watershed 2000 Water Quality Assessment Report 35-AC-1 CN089.0, March 2004)			2004
Tully Lake	College of the Holy Cross	Robert Bertin	−72.215667	42.644638	2004
Tully Lake	College of the Holy Cross	Robert Bertin	−72.215667	42.644638	2004
Tully Lake	College of the Holy Cross	Robert Bertin	−72.215667	42.644638	2004
Turnpike Lake	MassDEP	Appendix C—Table C1.1996 Taunton River Watershed Lake observations and trophic status estimates (Taunton River Watershed 2001 Water Quality Assessment Report 62-AC-1 CN 94.0, December 2005)			2004
Turnpike Lake	DCR	DCR Data from Lakes and Ponds, 2004			2004
White Pond	MassDEP	Appendix B—Table B1. Millers River Watershed 1995 Lake Synoptic Survey Data and Trophic Status Estimates (Millers River Watershed 2000 Water Quality Assessment Report 35-AC-1 CN089.0, March 2004)			2004
Ames Long Pond	MassDEP	Appendix C—Table C1.1996 Taunton River Watershed Lake observations and trophic status estimates (Taunton River Watershed 2001 Water Quality Assessment Report 62-AC-1 CN 94.0, December 2005)			2005
Assabet River	MassDEP	SuAsCo Watershed 2001 Water Quality Assessment Report 82-AC-1 CN 92.0, August 2005			2005

Location	Source	Reference	Longitude	Latitude	Year
Assabet River	Individual	Joy Trahan-Liptak	−71.5393	42.403	2005
Assabet River	MassDEP	SuAsCo Watershed 2001 Water Quality Assessment Report 82-AC-1 CN 92.0, August 2005			2005
Bartlett Pond	College of the Holy Cross	Robert Bertin	−71.61766	42.318598	2005
Bartlett Pond	MassDEP	SuAsCo Watershed 2001 Water Quality Assessment Report 82-AC-1 CN 92.0, August 2005			2005
Barton Cove	DCR	Excel spreadsheet of non-native aquatic and wetland plants in Massachusetts lakes and ponds dated January 2005			2005
Blair Pond	MassDEP	Appendix F—Table F1.1996 Westfield River Watershed Lake observations and trophic status estimates (Westfield River Watershed 2001 Water Quality Assessment Report 32-AC-1 CN 090.0, April 2005)			2005
Brockton Reservoir	MassDEP	Appendix C—Table C1.1996 Taunton River Watershed Lake observations and trophic status estimates (Taunton River Watershed 2001 Water Quality Assessment Report 62-AC-1 CN 94.0, December 2005)			2005
Cleveland Pond	DCR	DCR Data from Lakes and Ponds, 2008			2005
Cleveland Pond	MassDEP	Appendix C—Table C1.1996 Taunton River Watershed Lake observations and trophic status estimates (Taunton River Watershed 2001 Water Quality Assessment Report 62-AC-1 CN 94.0, December 2005)			2005
Cobbs Pond	DCR	Excel spreadsheet of non-native aquatic and wetland plants in Massachusetts lakes and ponds dated January 2005			2005
Concord River	MassDEP	SuAsCo Watershed 2001 Water Quality Assessment Report 82-AC-1 CN 92.0, August 2005			2005
Concord River	MassDEP	SuAsCo Watershed 2001 Water Quality Assessment Report 82-AC-1 CN 92.0, August 2005			2005
Cushing Pond	MassDEP	Appendix C—Table C1.1996 Taunton River Watershed Lake observations and trophic status estimates (Taunton River Watershed 2001 Water Quality Assessment Report 62-AC-1 CN 94.0, December 2005)			2005
Farm Pond	MassDEP	SuAsCo Watershed 2001 Water Quality Assessment Report 82-AC-1 CN 92.0, August 2005			2005
Fort Meadow Reservoir	DCR	DCR Data from Lakes and Ponds, 2005	−71.5476027	42.36994107	2005

(*Continued*)

TABLE 10.1
(Continued)

Waterbody	Agency	Report Source	Longitude	Latitude	First Year Reported
Gushee Pond	MassDEP	Appendix C—Table C1.1996 Taunton River Watershed Lake observations and trophic status estimates (Taunton River Watershed 2001 Water Quality Assessment Report 62-AC-1 CN 94.0, December 2005)			2005
Heard Pond	MassDEP	SuAsCo Watershed 2001 Water Quality Assessment Report 82-AC-1 CN 92.0, August 2005			2005
Island Grove Pond	MassDEP	Appendix C—Table C1.1996 Taunton River Watershed Lake observations and trophic status estimates (Taunton River Watershed 2001 Water Quality Assessment Report 62-AC-1 CN 94.0, December 2005)			2005
Johnson Pond	MassDEP	Appendix C—Table C1.1996 Taunton River Watershed Lake observations and trophic status estimates (Taunton River Watershed 2001 Water Quality Assessment Report 62-AC-1 CN 94.0, December 2005)			2005
Lake Boon	MassDEP	SuAsCo Watershed 2001 Water Quality Assessment Report 82-AC-1 CN 92.0, August 2005			2005
Lake Boon	Individual	Joy Trahan-Liptak	−71.502252	42.396447	2005
Lake Boon	Individual	Joy Trahan-Liptak	−71.502252	42.396447	2005
Lake Mirimichi	MassDEP	Appendix C—Table C1.1996 Taunton River Watershed Lake observations and trophic status estimates. (Taunton River Watershed 2001 Water Quality Assessment Report 62-AC-1 CN 94.0, December 2005)			2005
Lake Nippenicket	MassDEP	Appendix C—Table C1.1996 Taunton River Watershed Lake observations and trophic status estimates (Taunton River Watershed 2001 Water Quality Assessment Report 62-AC-1 CN 94.0, December 2005)			2005
Long Pond	MassDEP	Appendix C—Table C1.1996 Taunton River Watershed Lake observations and trophic status estimates (Taunton River Watershed 2001 Water Quality Assessment Report 62-AC-1 CN 94.0, December 2005)			2005
Long Pond	Individual	Joy Trahan-Liptak	−71.47025	42.530021	2005
Lower Porter Pond	MassDEP	Appendix C—Table C1.1996 Taunton River Watershed Lake observations and trophic status estimates (Taunton River Watershed 2001 Water Quality Assessment Report 62-AC-1 CN 94.0, December 2005)			2005

Waterbody	Source	Reference / Site ID	Longitude	Latitude	Year
Middle Pond	MassDEP	Appendix C—Table C1.1996 Taunton River Watershed Lake observations and trophic status estimates (Taunton River Watershed 2001 Water Quality Assessment Report 62-AC-1 CN 94.0, December 2005)			2005
Middle Pond	DCR	DCR Data from Lakes and Ponds, 2007			2005
Monponsett Pond, East Basin	MassDEP	Appendix C—Table C1.1996 Taunton River Watershed Lake observations and trophic status estimates (Taunton River Watershed 2001 Water Quality Assessment Report 62-AC-1 CN 94.0, December 2005)			2005
Monponsett Pond, East Basin	MassDEP	09-G006–05	−70.836781	42.0014672	2005
Monponsett Pond, East Basin	MassDEP	10-B005–03	−70.836781	42.0014672	2005
Monponsett Pond, East Basin	MassDEP	11-B006–05	−70.836781	42.0014672	2005
Monponsett Pond, East Basin	MassDEP	11-B008–02	−70.836781	42.0014672	2005
Monponsett Pond, East Basin	MassDEP	12-B002–04	−70.836781	42.0014672	2005
Monponsett Pond, East Basin	MassDEP	12-B004–04	−70.836781	42.0014672	2005
Monponsett Pond, East Basin	MassDEP	12-B006–05	−70.836781	42.0014672	2005
Monponsett Pond, East Basin	MassDEP	12-B008–05	−70.836781	42.0014672	2005
Monponsett Pond, East Basin	MassDEP	13-B005–06	−70.836781	42.0014672	2005
Monponsett Pond, East Basin	MassDEP	13-B006–05	−70.836781	42.0014672	2005
Monponsett Pond, East Basin	MassDEP	13-B008–05	−70.836781	42.0014672	2005
Monponsett Pond, East Basin	MassDEP	14-B004–05	−70.836781	42.0014672	2005
Monponsett Pond, East Basin	MassDEP	14-B006–05	−70.836781	42.0014672	2005

(Continued)

TABLE 10.1
(Continued)

Waterbody	Agency	Report Source	Longitude	Latitude	First Year Reported
Monponsett Pond, East Basin	MassDEP	14-B008–05	−70.836781	42.0014672	2005
Monponsett Pond, East Basin	Individual	Joy Trahan-Liptak	−70.837849	42.005701	2005
Monponsett Pond, West Basin	MassDEP	Appendix C—Table C1.1996 Taunton River Watershed Lake observations and trophic status estimates (Taunton River Watershed 2001 Water Quality Assessment Report 62-AC-1 CN 94.0, December 2005)			2005
Monponsett Pond, West Basin	MassDEP	09-G006–01	−70.8452535	42.0045958	2005
Monponsett Pond, West Basin	MassDEP	11-B006–01	−70.8452535	42.0045958	2005
Monponsett Pond, West Basin	MassDEP	12-B008–01	−70.8452535	42.0045958	2005
Monponsett Pond, West Basin	MassDEP	13-B005–02	−70.8452535	42.0045958	2005
Monponsett Pond, West Basin	MassDEP	13-B006–01	−70.8452535	42.0045958	2005
Monponsett Pond, West Basin	MassDEP	13-B008–01	−70.8452535	42.0045958	2005
Monponsett Pond, West Basin	MassDEP	Field Survey 7/7/2017 Wong, Maldonado	−70.845253	42.004596	2005
Mount Hope Mill Pond	MassDEP	Appendix C—Table C1.1996 Taunton River Watershed Lake observations and trophic status estimates (Taunton River Watershed 2001 Water Quality Assessment Report 62-AC-1 CN 94.0, December 2005)			2005
Muddy Pond	MassDEP	Appendix C—Table C1.1996 Taunton River Watershed Lake observations and trophic status estimates (Taunton River Watershed 2001 Water Quality Assessment Report 62-AC-1 CN 94.0, December 2005)			2005

Waterbody	Agency	Reference	Longitude	Latitude	Year
New Pond	MassDEP	Appendix C—Table C1.1996 Taunton River Watershed Lake observations and trophic status estimates (Taunton River Watershed 2001 Water Quality Assessment Report 62-AC-1 CN 94.0, December 2005)			2005
Norton Reservoir	MassDEP	Appendix C—Table C1.1996 Taunton River Watershed Lake observations and trophic status estimates (Taunton River Watershed 2001 Water Quality Assessment Report 62-AC-1 CN 94.0, December 2005)			2005
Richmond Pond	MassDEP	Appendix C—Table C1.1996 Taunton River Watershed Lake observations and trophic status estimates (Taunton River Watershed 2001 Water Quality Assessment Report 62-AC-1 CN 94.0, December 2005)			2005
Russell Pond	MassDEP	05-G010-02	−70.7470003	41.9778691	2005
Savery Pond	MassDEP	Appendix C—Table C1.1996 Taunton River Watershed Lake observations and trophic status estimates (Taunton River Watershed 2001 Water Quality Assessment Report 62-AC-1 CN 94.0, December 2005)			2005
Saxonville Pond	MassDEP	SuAsCo Watershed 2001 Water Quality Assessment Report 82-AC-1 CN 92.0, August 2005			2005
Snipatuit Pond	MassDEP	05-R007-02	−70.8620237	41.7809072	2005
Snipatuit Pond	DCR	DCR Data from Lakes and Ponds, 2007			2005
Thirtyacre Pond	MassDEP	Appendix C—Table C1.1996 Taunton River Watershed Lake observations and trophic status estimates (Taunton River Watershed 2001 Water Quality Assessment Report 62-AC-1 CN 94.0, December 2005)			2005
Upper Porter Pond	MassDEP	Appendix C—Table C1.1996 Taunton River Watershed Lake observations and trophic status estimates (Taunton River Watershed 2001 Water Quality Assessment Report 62-AC-1 CN 94.0, December 2005)			2005
Waldo Lake	MassDEP	Appendix C—Table C1.1996 Taunton River Watershed Lake observations and trophic status estimates (Taunton River Watershed 2001 Water Quality Assessment Report 62-AC-1 CN 94.0, December 2005)			2005
Whitehall Reservoir	MassDEP	SuAsCo Watershed 2001 Water Quality Assessment Report 82-AC-1 CN 92.0, August 2005			2005
Whitehall Reservoir	DCR	DCR Data from Lakes and Ponds, 2008			2005
Whitehall Reservoir	Individual	Joy Trahan-Liptak	−71.573594	42.23072	2005
Whittenton Impoundment	MassDEP	Appendix C—Table C1.1996 Taunton River Watershed Lake observations and trophic status estimates (Taunton River Watershed 2001 Water Quality Assessment Report 62-AC-1 CN 94.0, December 2005)			2005

(Continued)

TABLE 10.1
(Continued)

Waterbody	Agency	Report Source	Longitude	Latitude	First Year Reported
Winnecunnet Pond	MassDEP	Appendix C—Table C1.1996 Taunton River Watershed Lake observations and trophic status estimates (Taunton River Watershed 2001 Water Quality Assessment Report 62-AC-1 CN 94.0, December 2005)			2005
Woods Pond	MassDEP	Appendix C—Table C1.1996 Taunton River Watershed Lake observations and trophic status estimates (Taunton River Watershed 2001 Water Quality Assessment Report 62-AC-1 CN 94.0, December 2005)			2005
Aaron River	MassDEP	South Shore Coastal Watersheds 2001 Water Quality Assessment Report 94-AC-2 CN 93.0, March 2006			2006
Beaver Dam Pond	MassDEP	South Shore Coastal Watersheds 2001 Water Quality Assessment Report 94-AC-2 CN 93.0, March 2006			2006
Briggs Reservoir	MassDEP	South Shore Coastal Watersheds 2001 Water Quality Assessment Report 94-AC-2 CN 93.0, March 2006			2006
Briggs Reservoir	MassDEP	South Shore Coastal Watersheds 2001 Water Quality Assessment Report 94-AC-2 CN 93.0, March 2006			2006
Charles River	Charles River Watershed Association	Draft Upper Charles River Watershed Total Maximum Daily Load Project (Project Nos. 2001–03/104 and 2004–04/319) Volume I: Phase II Final Report and Phase III Data Report			2006
Charles River	Individual	Joy Trahan-Liptak	−71.264994	42.25732	2006
Charles River	Charles River Watershed Association	Draft Upper Charles River Watershed Total Maximum Daily Load Project (Project Nos. 2001–03/104 and 2004–04/319) Volume I: Phase II Final Report and Phase III Data Report			2006
Cooks Pond	MassDEP	South Shore Coastal Watersheds 2001 Water Quality Assessment Report 94-AC-2 CN 93.0, March 2006			2006
Eel River	MassDEP	South Shore Coastal Watersheds 2001 Water Quality Assessment Report 94-AC-2 CN 93.0, March 2006			2006
Forge Pond	MassDEP	South Shore Coastal Watersheds 2001 Water Quality Assessment Report 94-AC-2 CN 93.0, March 2006			2006

Name	Agency	Reference	Longitude	Latitude	Year
Herring Brook	MassDEP	South Shore Coastal Watersheds 2001 Water Quality Assessment Report 94-AC-2 CN 93.0, March 2006			2006
Island Creek Pond	MassDEP	South Shore Coastal Watersheds 2001 Water Quality Assessment Report 94-AC-2 CN 93.0, March 2006			2006
Island Pond	MassDEP	South Shore Coastal Watersheds 2001 Water Quality Assessment Report 94-AC-2 CN 93.0, March 2006			2006
Jacobs Pond	MassDEP	South Shore Coastal Watersheds 2001 Water Quality Assessment Report 94-AC-2 CN 93.0, March 2006			2006
Lily Pond	MassDEP	South Shore Coastal Watersheds 2001 Water Quality Assessment Report 94-AC-2 CN 93.0, March 2006			2006
Long Island Pond	MassDEP	South Shore Coastal Watersheds 2001 Water Quality Assessment Report 94-AC-2 CN 93.0, March 2006			2006
Lower Chandler Pond	MassDEP	South Shore Coastal Watersheds 2001 Water Quality Assessment Report 94-AC-2 CN 93.0, March 2006			2006
Old Oaken Bucket Pond	MassDEP	South Shore Coastal Watersheds 2001 Water Quality Assessment Report 94-AC-2 CN 93.0, March 2006			2006
Pembroke Street South Pond	MassDEP	South Shore Coastal Watersheds 2001 Water Quality Assessment Report 94-AC-2 CN 93.0, March 2006			2006
Reeds Millpond	MassDEP	South Shore Coastal Watersheds 2001 Water Quality Assessment Report 94-AC-2 CN 93.0, March 2006			2006
Silver Lake	College of the Holy Cross	Robert Bertin	−71.654894	42.193409	2006
Silver Lake	Individual	Joy Trahan-Liptak	−71.654971	42.193369	2006
Smelt Pond	MassDEP	South Shore Coastal Watersheds 2001 Water Quality Assessment Report 94-AC-2 CN 93.0, March 2006			2006
Torrey Pond	MassDEP	South Shore Coastal Watersheds 2001 Water Quality Assessment Report 94-AC-2 CN 93.0, March 2006			2006
Wampatuck Pond	MassDEP	South Shore Coastal Watersheds 2001 Water Quality Assessment Report 94-AC-2 CN 93.0, March 2006			2006
Wampatuck Pond	DCR	DCR Data from Lakes and Ponds, 2007	−70.8668332	42.06111681	2006
Wampatuck Pond	DCR	DCR Data from Lakes and Ponds, 2008	−70.8668332	42.06111681	2006
Wampatuck Pond	MassDEP	10-B002-05	−70.8657554	42.0647472	2006
Brooks Pond	College of the Holy Cross	Robert Bertin	−72.043108	42.302192	2007

(Continued)

TABLE 10.1 (Continued)

Waterbody	Agency	Report Source	Longitude	Latitude	First Year Reported
Brooks Pond	College of the Holy Cross	Robert Bertin	−72.043108	42.302192	2007
Harris Pond	College of the Holy Cross	Robert Bertin	−71.509471	42.020057	2007
Houghtons Pond	DCR	DCR Data from Lakes and Ponds, 2007			2007
Houghtons Pond	Individual	Joy Trahan-Liptak	−71.095558	42.20725	2007
Johns Pond	DCR	DCR Data from Lakes and Ponds, 2007			2007
Whitehall Pond	DCR	DCR Data from Lakes and Ponds, 2007			2007
Billington Sea	MassDEP	08-F007–05	−70.681001	41.9348301	2008
Billington Sea	MassDEP	08-F007–05	−70.681001	41.9348301	2008
Billington Sea	MassDEP	08-F005–01	−70.681001	41.9348301	2008
Billington Sea	MassDEP	08-F003–05	−70.681001	41.9348301	2008
Billington Sea	MassDEP	09-G007–06	−70.681001	41.9348301	2008
Billington Sea	MassDEP	09-G004–04	−70.681001	41.9348301	2008
Collins Pond	DCR	DCR Data from Lakes and Ponds, 2008			2008
East Brookfield River	MassDEP	Chicopee River Watershed 2003 Water Quality Assessment Report 36-AC-3 CN 106.5, October 2008			2008
East Head Pond	DCR	DCR Data from Lakes and Ponds, humile 2008			2008
Indian Brook	MassDEP	08-F009–03	−70.538045	41.8871496	2008
Indian Brook	MassDEP	Field Survey 7/18/2017 Wong, Maldonado	−70.538045	41.8871496	2008
Lackey Pond	College of the Holy Cross	Robert Bertin	−71.689012	42.096707	2008
Lackey Pond	Individual	Joy Trahan-Liptak	−71.689618	42.097854	2008
Ponkapoag Pond	DCR	DCR Data from Lakes and Ponds, 2008			2008
Ponkapoag Pond	MassDEP	09-B013–01	−71.0929681	42.1921829	2008
Sewall Pond	DCR	DCR Data from Lakes and Ponds, 2008			2008
Singletary Pond	College of the Holy Cross	Robert Bertin	−71.77917	42.159969	2008

Location	Agency	Source	Longitude	Latitude	Year
State Street Pond	DCR	Excel spreadsheet of non-native aquatic and wetland plants in Massachusetts lakes and ponds dated July 2008			2008
Stetson Pond	DCR	DCR Data from Lakes and Ponds, 2008			2008
Mumford River	MassDEP	09-G010-03	−71.6783156	42.1104152	2009
Mumford River	Individual	Joy Trahan-Liptak	−71.665758	42.109325	2009
Stoneville Pond	College of the Holy Cross	Robert Bertin	−71.846641	42.218518	2009
Stoneville Pond	Individual	Joy Trahan-Liptak	−71.846513	42.218616	2009
Reservoir (White Oak Reservoir)	MassDEP	10-B005-04	−70.8526283	42.0298844	2010
Reservoir (White Oak Reservoir)	MassDEP	12-B004-06	−70.8526283	42.0298844	2010
Reservoir (White Oak Reservoir)	MassDEP	12-B008-06	−70.8526283	42.0298844	2010
Reservoir (White Oak Reservoir)	MassDEP	13-B006-06	−70.8526283	42.0298844	2010
Unnamed Pond	DCR	Jamie Carr	−71.844452	42.445836	2011
Unnamed Pond	DCR	David Worden	−71.844452	42.445836	2011
Auguteback Pond	Individual	Joy Trahan-Liptak	−71.877796	42.117964	2013
French River	Individual	Joy Trahan-Liptak	−71.8832	42.1098	2013
Queen Lake	Individual	Joy Trahan-Liptak	−72.114214	42.53513	2013
Stetson Brook	MassDEP	13-B005-03	−70.8366053	42.0143947	2013
Stetson Brook	MassDEP	13-B006-02	−70.8366053	42.0143947	2013
Blackstone River	Individual	Joy Trahan-Liptak	−71.617101	42.092183	2015
Bryant Pond	DCR	Joy Trahan-Liptak	−71.85485	42.34379	2015
Caprons Pond	Individual	Joy Trahan-Liptak	−71.627869	42.07709	2015
Fisherville Pond	Individual	Joy Trahan-Liptak	−71.692178	42.184408	2015
Hovey Pond	Individual	Joy Trahan-Liptak	−71.714879	42.234214	2015
Sudbury River	Individual	Joy Trahan-Liptak	−71.356919	42.462852	2015
Quinebaug River	MassDEP	Field Survey 06/16/2017 Wong, Li, Maldonado	−72.1185692	42.1095615	2017

TABLE 10.2
Towns and Infested Waterbodies

Number	Waterbody	Town 1	Town 2
1	Aaron River	Cohasset	Hingham
2	Ames Long Pond	Stoughton	Easton
3	Arcade Pond	Northbridge	—
4	Arcadia Lake	Belchertown	
5	Assabet River	Stow	Hudson/Maynard
6	Auguteback Pond	Oxford	—
7	Baddacook Pond	Groton	—
8	Baldpate Pond	Boxford	—
9	Ballardvale Impoundment	Andover	—
10	Bare Hill Pond	Harvard	—
11	Bartlett Pond	Northborough	—
12	Barton Cove	Gill	—
13	Bearse Pond	Barnstable	—
14	Beaver Dam Pond	Plymouth	—
15	Beaver Pond	Franklin	—
16	Big Bearhole Pond	Taunton	—
17	Billington Sea	Plymouth	—
18	Blackstone River	Uxbridge	Millville
19	Blair Pond	Blandford	—
20	Bolivar Pond	Canton	—
21	Briggs Reservoir	Plymouth	—
22	Brockton Reservoir	Avon	—
23	Brooks Pond	North Brookfield	New Braintree/Oakham/Spencer
24	Bryant Pond	Holden	—
25	Caprons Pond	Uxbridge	—
26	Cedar Meadow Pond	Leicester	—
27	Chaffin Pond	Holden	—
28	Charles River	Medfield	Millis/Norfolk/Medway/Dover/Sherborn/Natick
29	Chebacco Lake	Essex	Hamilton
30	Cleveland Pond	Abington	—
31	Cobbs Pond	Walpole	—
32	Collins Pond	Andover	—
33	Concord River	Concord	Billerica/Carlisle/Bedford
34	Cook Pond	Worcester	—
35	Cooks Pond	Plymouth	—
36	Curtis Pond	Worcester	—
37	Cushing Pond	Abington	—
38	Dark Brook	Auburn	—
39	Dawson Pond	Holden	—
40	East Brookfield River	East Brookfield	—
41	East Head Pond	Carver	Plymouth
42	Edgewater Office Park Pond	Wakefield	—

Number	Waterbody	Town 1	Town 2
43	Eel River	Plymouth	–
44	Ellis Pond	Norwood	–
45	Farm Pond	Framingham	–
46	Federal Pond	Carver	Plymouth
47	Field Pond	Andover	–
48	First Pond	Saugus	–
49	Fish Pond	Northbridge	–
50	Fisherville Pond	Grafton	–
51	Flannagan Pond	Ayer	–
52	Flint Pond	Shrewsbury	–
53	Forge Pond	Westford	Littleton
54	Fort Meadow Reservoir	Marlborough	Hudson
55	Fosters Pond	Andover	Wilmington
56	French River	Oxford	–
57	Fresh Meadow Pond	Plymouth	Carver
58	Girard Pond	Sutton	–
59	Griswold Pond	Saugus	–
60	Grove Pond	Ayer	–
61	Gushee Pond	Raynham	–
62	Hardwick Pond	Hardwick	–
63	Harris Pond	Blackstone	–
64	Hayes Pond	Grafton	–
65	Heard Pond	Wayland	–
66	Herring Brook	Cohasset	–
67	Houghtons Pond	Milton	–
68	Hovey Pond	Grafton	–
69	Indian Brook	Plymouth	–
70	Ironstone Reservoir	Uxbridge	–
71	Island Creek Pond	Duxbury	–
72	Island Grove Pond	Abington	–
73	Island Pond	Plymouth	–
74	Jacobs Pond	Norwell	–
75	Johns Pond	Carver	–
76	Johnson Pond	Taunton	–
77	Kettle Brook	Leicester	Auburn/Worcester
78	Knops Pond/Lost Lake	Groton	–
79	Lackey Pond	Uxbridge	Sutton
80	Lake Boon	Stow	Hudson
81	Lake Como	Attleboro	–
82	Lake Ellis	Athol	–
83	Lake Holland	Belchertown	–
84	Lake Lashaway	North Brookfield	East Brookfield
85	Lake Mascuppic	Tyngsborough	Dracut

(Continued)

TABLE 10.2
(Continued)

Number	Waterbody	Town 1	Town 2
86	Lake Mirimichi	Plainville	Foxborough
87	Lake Nippenicket	Bridgewater	Raynham
88	Lake Quinsigamond	Shrewsbury	Worcester
89	Lake Rico	Taunton	–
90	Lake Ripple	Grafton	–
91	Lake Rohunta	New Salem	Athol/Orange
92	Lake Sabbatia	Taunton	–
93	Lake Shirley	Lunenburg	Shirley
94	Lake Warner	Hadley	–
95	Lake Winthrop	Holliston	–
96	Leesville Pond	Auburn	Worcester
97	Lily Pond	Cohasset	–
98	Little Bearhole Pond	Taunton	–
99	Long Island Pond	Plymouth	–
100	Long Pond	Lakeville	Freetown
101	Long Pond	Littleton	–
102	Lowe Pond	Boxford	–
103	Lower Chandler Pond	Duxbury	Pembroke
104	Lower Porter Pond	Brockton	–
105	Manchaug Pond	Douglas	Sutton
106	Martins Pond	North Reading	–
107	Massapoag Brook	Sharon	Canton
108	Massapoag Lake	Sharon	–
109	Metacomet Lake	Belchertown	–
110	Middle Pond	Taunton	–
111	Mill Pond	Upton	–
112	Mill River	Hopedale	Milford/Upton/Mendon
113	Miscoe Lake	Wrentham	–
114	Monponsett Pond, East Basin	Halifax	–
115	Monponsett Pond, West Basin	Halifax	Hanson
116	Morses Pond	Wellesley	Natick
117	Moulton Pond	Rutland	–
118	Mount Hope Mill Pond	Taunton	Dighton
119	Muddy Pond	Carver	–
120	Mumford River	Uxbridge	Northbridge/Sutton/Douglas
121	Nashua River	Groton	Pepperell/Ayer
122	Neponset Reservoir	Foxborough	–
123	New Pond	Easton	–
124	Newfield Pond	Chelmsford	–
125	Newton Pond	Shrewsbury	Boylston

Number	Waterbody	Town 1	Town 2
126	North Pond	Hopkinton	Milford/Upton
127	Norton Reservoir	Norton	Mansfield
128	Old Oaken Bucket Pond	Scituate	–
129	Parker Pond	Gardner	–
130	Pembroke Street South Pond	Kingston	–
131	Pentucket Pond	Georgetown	–
132	Plain Street Pond	Mansfield	–
133	Plow Shop Pond	Ayer	–
134	Pomps Pond	Andover	–
135	Pondville Pond	Auburn	Millbury
136	Ponkapoag Pond	Canton	Randolph
137	Pratt Pond	Upton	–
138	Quaboag Pond	Brookfield	East Brookfield
139	Quacumquasit Pond	Brookfield	East Brookfield/Sturbridge
140	Queen Lake	Phillipston	–
141	Quinebaug River	Sturbridge	Brimfield/Holland
142	Quinsigamond River	Grafton	–
143	Reeds Millpond	Kingston	–
144	Reservoir (White Oak Reservoir)	Hanson	–
145	Reservoir Pond	Canton	–
146	Richmond Pond	Taunton	–
147	Russell Pond	Kingston	–
148	Sampson Pond	Carver	–
149	Savery Pond	Middleborough	–
150	Saxonville Pond	Framingham	–
151	Sewall Pond	Boylston	–
152	Silver Lake	Grafton	–
153	Singletary Pond	Sutton	Millbury
154	Smelt Pond	Kingston	–
155	Snipatuit Pond	Rochester	–
156	South Athol Pond	Athol	–
157	Spectacle Pond	Littleton	Ayer
158	Spring Pond	Saugus	–
159	State Street Pond	Newburyport	–
160	Stetson Brook	Halifax	–
161	Stetson Pond	Pembroke	–
162	Stevens Pond	Sutton	–
163	Stoneville Pond	Auburn	–
164	Sudbury River	Concord	Lincoln/Sudbury/Wayland
165	Swains Pond	Melrose	–
166	Tatnuck Brook	Worcester	Holden
167	Thirtyacre Pond	Brockton	–
168	Torrey Pond	Norwell	–

(Continued)

**TABLE 10.2
(Continued)**

Number	Waterbody	Town 1	Town 2
169	Town Pond	Stoughton	–
170	Tully Lake	Royalston	Athol
171	Turner Pond	Walpole	–
172	Turnpike Lake	Plainville	–
173	Unnamed Pond	Princeton	–
174	Unnamed Tributary	Worcester	–
175	Upper Porter Pond	Brockton	–
176	Wachusett Reservoir	Boylston	West Boylston/Clinton/Sterling
177	Waldo Lake	Avon	Brockton
178	Wampatuck Pond	Hanson	–
179	Watson Pond	Taunton	–
180	Webster Lake	Webster	–
181	Wequaquet Lake	Barnstable	–
182	West River	Upton	Grafton
183	White Island Pond, East Basin	Plymouth	Wareham
184	White Island Pond, West Basin	Plymouth	Wareham
185	White Pond	Athol	–
186	Whitehall Pond	Rutland	–
187	Whitehall Reservoir	Hopkinton	–
188	Whitins Pond	Northbridge	Sutton
189	Whitmans Pond	Weymouth	–
190	Whittenton Impoundment	Taunton	–
191	Winnecunnet Pond	Norton	–
192	Woodbury Pond	Sutton	–
193	Woods Pond	Middleborough	–

measures include drawdown, dredging, benthic barriers, introducing biological control agents such as weevil or grass carp, herbicides, and other control methods.

Drawdown is a lake management tool that tends to reduce the density of susceptible rooted plants, and could be helpful for the control of Carolina fanwort. As this method is a process of lowering the water level, it is more suitable for a waterbody with available structures that allow manipulation of the water level, as well as the acceptability of draining the water to all water users. Additionally, it can only achieve control when sediments are completely dry for an extended period of time or the sediments and plants are completed removed (Dutartre

et al. 2006; Scheers et al. 2016). In order to reduce impacts to non-target plants and animals during the growing season, drawdown in Massachusetts is normally conducted in fall and winter (Mattson et al. 2004).

Dredging and benthic barriers both are effective physical methods to impede the expansion of *C. caroliniana*. Dredging is a sediment removal process that controls the rooted plant growth through light limitation during operations or substrate limitation. Dry dredging appears to result in the removal of more soft sediments and hydraulic dredging usually involves a suction type of dredge to produce slurry which is typically pumped (or barged in rare cases) to a nearby containment area on

TABLE 10.3

Infested Watersheds

Number	Infested Watersheds
1	Blackstone
2	Boston Harbor: Neponset
3	Boston Harbor: Weymouth and Weir
4	Buzzards Bay
5	Cape Cod
6	Charles
7	Chicopee
8	Concord (SuAsCo)
9	Connecticut
10	French
11	Ipswich
12	Merrimack
13	Millers
14	Nashua
15	North Coastal
16	Parker
17	Quinebaug
18	Shawsheen
19	South Coastal
20	Taunton
21	Ten Mile
22	Westfield

shore where the excess water can be separated from the solids by settling (Mattson et al. 2004). The dredging is very efficient if performed well in removing most plants, the regrowth is from missed bits and also if the material is not collected well and starts floating around, and this is an issue with all mechanical harvesting operations. Meanwhile, dredge causes high impacts to the ecosystem as the substrate is disturbed and it is costly and limited to suitable water depths. If the dredge volume is more than 100 cubic yards, a dredging permit is required from MassDEP (see more in the "Regulation and Permitting" section of Chapter 1). Benthic barriers are a control method usually with layers of geofabric or coconut fiber mesh material that is anchored above the weeds. The application of benthic barriers limits the plants growth by a cover of clay, silt, sand, gravel, or different types of mats (Caffrey et al. 2010). But there are some limitations to this method. For instance, it requires constant maintenance and it can harm other benthic organisms (Washington State Department of Ecology n.d.). Other potential methods of management include manual removal or suction harvesting, in which the plants are manually removed either by hand or through a suction apparatus. Harvesting is great for clearing larger areas, like swimming areas as a long-term maintenance option. Hand removal is usually either an emergency tool to control new invasions and used for clearing small areas (e.g., around boat ramps) or as a tool to remove remnants after other control methods. There are many examples of smaller ponds and waterbodies throughout Massachusetts that utilize mechanical harvesting to alleviate the stressors caused by *C. caroliniana* and other invasive species. These waterbodies, such as Wellesley Pond in Wellesley, MA, or Island Creek Pond in Duxbury, MA, generally perform annual mechanical harvests which allow for improved recreation (Mattson et al. 2004). These methods require high levels of manpower and efforts and are generally only successful when treating low to moderate densities of Carolina fanwort (MA DCR 2005). It also depends on the management goal, even if a whole lake is covered by fanwort, it might still be a viable option to use mechanical tools if the goal is simply to maintain public access or swimming areas. These mechanical methods also only result in a temporary solution, and not long-term eradication. Additionally, suction harvesting can result in elevated turbidity in treatments areas, which is often undesired.

Biological control involves the introduction of species such as the aquatic weevil (*Hydrotimetes natans*) or grass carp (*Ctynopharyngodon idella*). In general, biocontrol will always be a tool to reduce invasive plants to a certain level and it is a great tool for management. The weevil is a classical biocontrol tool with a specialist insect, the grass carp is a herbivorous fish that eats all aquatic vegetation. Both tools are used completely different and the limitations and risks are completely different as well. Aquatic weevil is a specialist feeder and it might aid with fanwort control. But aquatic weevils cannot achieve permanent removal and may have an influence on non-target plants

(Louda et al. 2003). Grass carp have to be stocked at the appropriate rate otherwise they don't work. Grass carp will also consume all aquatic plants in a waterbody if stocked incorrectly, so they have to be removed once control is achieved so that macrophytes can reestablish. Grass carp are not approved in Massachusetts as a biological control agent (Mattson et al. 2004). Further, high densities of grass carp can change the sediment composition and limit the habit quality for other organisms (Pípalová 2006).

Many herbicides (e.g., endothall amine salt, fluridone, triclopyr, diquat, 2,4-D, and flumioxazin) have been tested for fanwort control (Nelson et al. 2002). Herbicides contain active and inert ingredients that are toxic to target plants. However, the use of inappropriate herbicides or applying the incorrect dose of herbicide can harm or kill non-target plants and organisms (Mattson et al. 2004). Hunt et al. (2015) found that endothall amine salt could be an effective tool for *C. caroliniana* control, even at low concentrations such as 0.5 mg ae/l. The Department of Conservation and Recreation also recommends the use of fluridone and triclopyr. The recommended doses for the control of fanwort is >10 ppb for fluridone and 0.75–2.5 ppm for triclopyr (MA DCR 2005). Flumioxazin is one

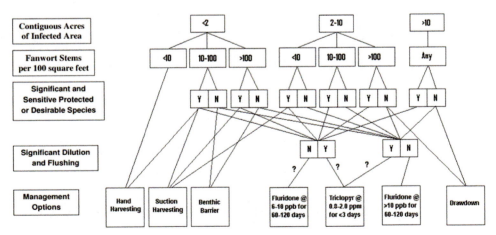

FIGURE 10.4 Decision tree for the control of Carolina fanwort (*Cabomba caroliniana*) (MA DCR, 2005). Hand harvesting and suction harvesting must include root system removal. Benthic barriers should remain in place for 30–60 days. Fluridone is effective at >10 ppb with >60 days exposure; lesser doses and exposure times may yield some control. Triclopyr was approved for use in MA in late 2004; experience is limited in MA. Drawdown use is dependent on many factors, including hydrology and use as a water supply. Moderate to dense growth over an extensive area (>10 acres) may not be appropriate for rapid response consideration.

of the best herbicides to control fanwort with a label rate of 200–400 ppb. More information about chemical control on fanwort and other 18 invasive species can be found in Appendix III. Any chemical used for aquatic invasive species control requires a WM04 Chemical Application License issued by the Massachusetts Department of Environmental Protection.

Other effective methods to suppress the growth of fanwort can also be used, for instance, increasing the shading by trees or shrubs to block the direct sunlight, developing a tall helophyte belt (e.g., Phragmites and Typha) along the shore to reduce the open water area, covering the area with a light-blocking synthetic foil or geo-textile, etc. (Scheers et al. 2016; Schooler 2008).

The MA Department of Conservation and Recreation provided a guideline on how to choose the best technique for *C. caroliniana* control based on different situations (Figure 10.4). This can be helpful when selecting the optimal control method.

SUCCESS STORY

In August 2004, Carolina fanwort was found growing throughout nearly half of Foster's Pond in Andover, MA (ACT 2011). An EPA/DEP registered aquatic herbicide, Sonar AS, which contains 42% fluridone and was chosen for the fanwort control (Mattson et al. 2004). During the control period in 2005, a maintenance herbicide treatment and monitoring program was initiated. This initial whole lake treatment consisted of three low-dose applications of Sonar AS, maintaining 10–20 ppb of the herbicide in the water for more than 60 days. This was followed by a partial lake treatment in 2007. The overall success of the fanwort control program was evaluated in surveys taken in 2008 and 2009. Although fanwort began to reemerge, it covered only 6.1% of Foster Pond, or roughly a ninth of the area it had previously dominated, and there were more native plants found in the pond in 2009 than in 2004. Spot treatments were applied in 2010 to areas where *C. caroliniana* remained and another whole lake treatment with Sonar was proposed for 2011. Therefore, this is a successful example of Carolina fanwort control.

REFERENCES

Aquatic Control Technology, Inc. 2011. Letter titled "Schedule and Program for Proposed 2011 Sonar and Reward Herbicide Treatment of Foster's Pond". *Aquatic Control Technology, INC.* www.fosterspond.org/ACT_2011_Treatment_Plan.pdf (Accessed February 19, 2020).

Bickel, T.O. 2015. A boat hitchhiker's guide to survival: *Cabomba caroliniana* desiccation resistance and survival ability. *Hydrobiologia* 746: 123–134.

Bruckerhoff, L., J. Havel, and S. Knight. 2015. Survival of invasive aquatic plants after air exposure and implications for dispersal by recreational boats. *Hydrobiologia* 746: 113–121.

Brundu, G. 2015. Plant invaders in European and Mediterranean inland waters: Profiles, distribution, and threats. *Hydrobiologia* 746: 61–79.

Bultemeier, B., M. Netherland, J. Ferrell, and W. Haller. 2009. Differential herbicide response among three phenotypes of *Cabomba caroliniana*. *Invasive Plant Science and Management* 2(4): 352–359.

Caffrey J.M., M. Millane, S. Evers, H. Moran, and M. Butler. 2010 A novel approach to aquatic weed control and habitat restoration using biodegradable jute matting. *Aquatic Invasions* 5: 123–129.

Council of Heads of Australasian Herbaria (CHAH). 2011. *Cabomba caroliniana*, https://profiles.ala.org.au/opus/weeds-australia/profile/Cabomba%20caroliniana (Accessed December 10, 2020).

Dutartre, A., C. Chauvin, and J. Grange. 2006. Colonisation végétale du canal de Bourgogne à Dijon: bilan 2006, propositions de gestion. In: *Rapport Cemagref Bordeaux*. Douai, France: Agence de l'Eau Artois-Picardie, p. 87.

Hogsden, K.L., E.P.S. Sager, T.C. Hutchinson. 2007. The impacts of the non-native macrophyte *Cabomba caroliniana* on littoral biota of Kasshabog Lake, Ontario. *Journal of Great Lakes Research* 33(2): 497–504.

Hunt, T.D., T.D. Dugdale, D. Clements, and M. Fridman. 2015. Concentration– exposure time relationships for controlling fanwort (*Cabomba caroliniana*) with endothall amine salt and carfentrazone. *Journal of Aquatic Plant Management* 53: 144–149.

Hussner, A., S. Nehring, and S. Hilt. 2013. From first reports to successful control: A plea for improved management of alien aquatic plant species in Germany. *Hydrobiologia* 737: 321–331.

ITIS (Integrated Taxonomic Information System). 2021. www.itis.gov/servlet/SingleRpt/SingleRpt?search_topic=TSN&search_value=18408#null (Accessed February 28, 2021).

Larson, J.L., L. Berent, and S. Iott. 2020. *Cabomba caroliniana A. Gray: U.S. Geological Survey, Nonindigenous Aquatic Species Database*. Gainesville, FL. https://nas.er.usgs.gov/queries/FactSheet.aspx?SpeciesID=231 (Accessed December 9, 2020).

Louda, S.M., R.W. Pemberton, M.T. Johnson, and P.A. Follett. 2003. Nontarget effects—The Achilles' heel of biological control? Retrospective analyses to reduce risk associated with biocontrol introductions. *Annual Reviews of Entomology* 48: 365–396.

MA DCR (Department of Conservation and Recreation). 2002. Fanwort: An invasive aquatic plant. www.mass.gov/files/documents/2016/08/ne/fanwort.pdf (Accessed February 19, 2020).

MA DCR (Department of Conservation and Recreation). 2005. Rapid Response Plan for Fanwort (*Cabomba caroliniana*) in Massachusetts. www.mass.gov/doc/fanwort-1/download (Accessed December 9, 2020).

MA DCR (Department of Conservation and Recreation). 2016. A Guide to aquatic plants in Massachusetts. http://lakemirimichi.org/Assets/Aquatic%20life/A%20Guide%20to%20Aquatic%20Plants_Booklet.PDF (Accessed February 19, 2020).

Mackey, A.P. 1996. *Cabomba (Cabomba spp.). Pest Status Review Series—Land Protection Branch.* Queensland: Queensland Government Department of Natural Resources and Mines.

Mackey, A.P., and J.T. Swarbrick. 1997. The biology of Australian weeds 32. *Cabomba caroliniana* Gray. *Plant Protection Quarterly* 12: 154–165.

Maki, K., and S. Galatowitsch. 2004. Movement of invasive aquatic plants into Minnesota (USA) through horticultural trade. *Biological Conservation* 118: 389–396.

Mattson, M.D., P.L. Godfrey, R.A. Barletta, and A. Aiello. 2004. *Eutrophication and Aquatic Plant Management in Massachusetts: Final Generic Environmental Impact Report.* Executive Office of Environmental Affairs, Commonwealth of Massachusetts, p. 514.

Nelson, L.S., A.B. Stewart, and K.D. Getsinger. 2002. Fluridone effects on fanwort and water marigold. *Journal of Aquatic Plant Management* 40: 58–63.

Ørgaard, M. 1991. The genus *Cabomba* (Cabombaceae)–A taxonomic study. *Nordic Journal of Botany* 11(2): 179–203.

Pípalová, I. 2006. A review of grass carp use for aquatic weed control and its impact on water bodies. *Journal of Aquatic Plant Management* 44: 1–12.

Rothlisberger, J.R., W.L. Chadderton, J. McNulty, and D.M. Lodge. 2010. Aquatic invasive species transport via trailered boats: what is being moved, who is moving it, and what can be done? *Fisheries* 35: 121–132.

Scheers, K., L. Denys, J. Packet, and T. Adriaens 2016. A second population of *Cabomba caroliniana* Gray (Cabombaceae) in Belgium with options for its eradication. *BioInvasions Records* 5(4): 227–232.

Schneider, E.L., and J.M. Jeter. 1982. Morphological studies of the Nymphaeaceae. XII. The floral biology of *Cabomba caroliniana*. *American Journal of Botany* 69(9): 1410–1419.

Schooler, S.S. 2008. Shade as a management tool for the invasive submerged macrophyte, *Cabomba caroliniana*. *Journal of Aquatic Plant Management* 46: 168–171.

Sheldon, S.P. 1994. Invasions and declines of submersed macrophytes in New England, with particular reference to Vermont lakes and herbivorous invertebrates in New England. *Lake and Reservoir Management* 10(1): 13–17.

State of Michigan. 2017. State of Michigan's Status and Strategy for Carolina Fanwort (*Cabomba caroliniana* A. Gray) Management. www.michigan.gov/documents/deq/wrd-ais-cabomba-caroliniana_499690_7.pdf.

Verbrugge, L.N.H. 2014. Going global: Perceiving, assessing and managing biological invasions. Thesis, Radboud University, Nijmegen, p. 171.

Washington State Department of Ecology. n.d. swollen bladderwort. *Wendy Scholl Invasion Ecology*. http://depts.washington.edu/oldenlab/wordpress/wp-content/uploads/2013/03/Ultricularia-inflata_Scholl_2007R.pdf (Accessed January 22, 2020).

Wilson, C.E., S.J. Darbyshire, and R. Jones. 2007. The biology of invasive alien plants in Canada. 7. *Cabomba caroliniana* A. Gray. *Canadian Journal of Plant Science* 87: 615–638.

11 European Naiad (*Najas minor*)

ABSTRACT

Najas minor is an invasive aquatic plant introduced to the United States from Europe. According to historical data collected mainly by Massachusetts Department of Environmental Protection, there are at least 20 waterbodies in Massachusetts infested by *N. minor*. Understanding the status of the infestation can help raise awareness among the general public and inform those who interact with water sources to help minimize its further spread. This report also describes different methods that have been used for controlling European naiad, including physical (e.g., harvesting and benthic barriers) and chemical treatments (e.g., herbicides).

INTRODUCTION

Taxonomy Hierarchy (ITIS 2021)

Kingdom	Plantae
Subkingdom	Viridiplantae
Infrakingdom	Streptophyta
Superdivision	Embryophyta
Division	Tracheophyta
Subdivision	Spermatophytina
Class	Magnoliopsida
Superorder	Lilianae
Order	Alismatales
Family	Hydrocharitaceae
Genus	*Najas*
Species	*Najas minor* All.

BASIC BIOLOGY

European naiad (*Najas minor*) is also called brittle naiad or brittle water-nymph. *N. minor* is a submersed rooted plant with an annual life cycle. At the height of its growth, *N. minor*'s stem can reach up to 120 cm in length and 1 mm in diameter, before it breaks apart into seeded fragments to reproduce. The leaves are usually stiff, pointed, and recurved, and have a length of 0.5–3.4 cm (Figure 11.1). Leaves are arranged in pairs oppositely at nodes on the stem but appear to grow in bushy whorls at the growing tip of the plant (MN DNR 2021a). There are spines along the leaf margin which are visible to the naked eye. Flowers of *N. minor* present carpellate in light green to purple, growing on the upper axils; purplish seeds are slightly recurved, which are usually 1.5–3.0 mm long and 0.5–0.7 mm wide (Meriläinen 1968; Haynes 1979). The species shares many subtle morphological characteristics with the other seven *Najas* species known to the United States, leading *N. minor* to frequently be misidentified as native naiads such as *N. marina* or *N. gracillima* at different stages of its growth. For this reason, the surest way to confirm the identity is using genetics (Les et al. 2015).

When young, *N. minor* and *N. gracillima* are indistinguishable by appearance, as they both have narrow leaves and truncate to auriculate sheathing leaf bases (Meriläinen 1968). Wentz and Stuckey (1971) also mention that *N. minor* does not have its stiff recurved leaves at the beginning of its growth, so young plants could be easily misidentified as *N. gracillima*. However, Jouko Meriläinen (1968) developed a method for differentiating these two plants based on several morphological characteristics. First, *N. minor* is dark green with a curvy leaf margin, while *N. gracillima* is light green with a straight leaf margin. Second, the length of the fruit is different in each plant: *N. minor*'s fruit is relatively long and *N. gracillima*'s fruit is relatively short. Third, the areolae of seed coat for *N. minor* is transversely elongate, but it is longitudinally elongate for *N. gracillima*. Another often confused species, *N. marina*, can be distinguished in that it has thicker stems and leaves than *N. minor*, and the leaves of *N. marina* are not strongly recurved (MN DNR 2021b). A guide to visually distinguishing *Najas* species of North America (at their adult stages before fragmentation) is included in this report (Figure 11.2).

DOI: 10.1201/9781003201106-11

FIGURE 11.1   Microscopic photos of European Naiad (*left*, 10×) and its stem (*right*, 50×). (Photos by Wai Hing Wong.)

NATIVE AND INVASIVE RANGE

Though relatively rare in its native range in Europe, western Asia, and northern Africa, *N. minor* has become invasive in North America, in some cases aggressively displacing native aquatic plants (Triest 1988; Les et al. 2015; Pfingsten et al. 2021). According to US Geological Survey (Pfingsten et al. 2021), *N. minor* was first found in 1932 at Lake Cardinal, Ashtabula, OH, by Freudenstein and Tadesse.

From the 1930s to the 1960s, the invasive range of *N. minor* expanded from Lake Ontario west to Illinois, and to the south as far as northern Florida (Meriläinen 1968). Currently, it is reported that the number of states populated by *N. minor* has increased to 37 (Pfingsten et al. 2021). The eastern and midwestern regions of the United States are the main areas infected by this species, but specimens of *N. minor* have been uncovered as far as California, which had been previously misidentified as *N. marina* when originally collected in 2003 (Les 2012).

MECHANISMS/VECTORS OF SPREAD

N. minor, as a sexual species, disperses primarily by seeds, and does so very efficiently and prolifically (Les et al. 2015). During the reproductive season, the stem of the plant breaks apart while the seeds remain attached to the leaf axils, so that the fragmented pieces can be easily carried by wind and water to disperse the seeds, which can overwinter in dry conditions (GISD 2021).

The plant itself is also said to tolerate a high-pollutant water environment, allowing *N. minor* to proliferate opportunistically when bodies of water are polluted or degraded. In the literature, *N. minor* and *N. marina* are described as preferring eutrophic, lentic, shallow, and turbid habitats, perhaps with more calcareous water (Baart et al. 2013; Les et al. 2015).

Les et al. (2015) theorize that *N. minor*'s initial introduction to the United States in Lake Cardinal occurred following residential development of the area surrounding the lake, when the plant was used in water gardening by private residents and unintentionally released into the lake. Aside from sporadic aquacultural introduction (i.e., through aquaria), *N. minor* may continue to spread along flyways of waterfowl, with seeds being dispersed by ducks and shore birds who eat them (Meriläinen 1968). As the plant reproduces by splitting into many small, easily transmittable seeded fragments, the seeds could also spread in connection with shipping activities, water and air movement, or by sticking to other live aquatic plants or vegetation that are being transported (Clausen 1936; Martin and Uhler 1939; Chase 1947; McIntyre and Barrett

Visually differentiating *Najas minor* from other North American Najas species (naiads or waternymphs)

Created by Y.M. Becker

Najas minor

"European Naiad" or "Brittle Naiad"

Introduced, invasive.

Dark, spiny, recurved leaves.

Najas marina

"Spiny Naiad" or "Holly-leafed waternymph"

Native, found across most of North America.

Irregularly shaped leaves with very prominent spines.

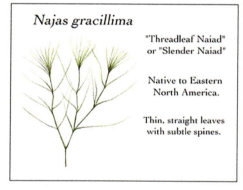

Najas gracillima

"Threadleaf Naiad" or "Slender Naiad"

Native to Eastern North America.

Thin, straight leaves with subtle spines.

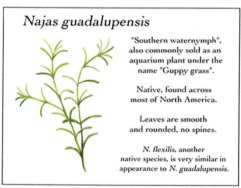

Najas guadalupensis

"Southern waternymph", also commonly sold as an aquarium plant under the name "Guppy grass".

Native, found across most of North America.

Leaves are smooth and rounded, no spines.

N. flexilis, another native species, is very similar in appearance to *N. guadalupensis.*

FIGURE 11.2 Guide to differentiating different North American native and introduced species in the genus *Najas* based on general visual characteristics. The plants are represented here in their adult form by one portion of a growing tip. (Illustrations and descriptions by Y.M. Becker. Photos of *N. minor* (GISD 2021; Native Plant Trust 2021; Lovel 2021; Mehrhoff n.d.; Skawinski n.d.; Na and Choi 2012), *N. marina,* (Topic n.d.; Wunderlin et al. 2021; USDA n.d.b), *N. gracillima,* (USDA n.d.a; Hilty 2019), and *N. guadalupensis* (Dressler 1996; Chayka and Dziuk n.d.) used as reference for illustrations.)

1985). Since *N. minor* can dissipate its seeds in many different ways, the spread of the species is difficult to control once it is established.

ECOLOGICAL IMPACTS

N. minor can thrive in the eutrophic waters, growing in decaying mud on the bottom. While the turbid, anoxic water conditions created by eutrophication harm many plants and animals, *N. minor* has an advantage under these conditions and can outcompete native species (Wentz and Stuckey 1971). According to Robinson (2004), *N. minor*

mats can clog waterways, decrease the sediment level in the water, and intercept the sunlight, which leads to the exclusion of other submerged plants. When large mats of *N. minor* decay at the end of a growing season (because of the plant's annual growth habit), the organisms breaking down the plant matter deplete the water of oxygen, creating a eutrophic, anoxic environment that can cause fish and other aquatic animals to die in large numbers. These eutrophic conditions also allow future populations of *N. minor* to thrive, perpetuating the plant's life cycle and preventing other plants and animals from reestablishing.

N. minor, like other *Najas* species, has other impacts on aquatic systems. Research has found that *N. minor* and *N. marina* contain and exude allelopathic compounds with the ability to suppress algal growth, which might control algal blooms in eutrophic waterbodies (Wang et al. 2010). Additionally, *N. minor* is easily accessible to ducks, waterfowl, and shore birds as a food source since the plant prefers to grow in calm, shallow water (Clausen 1936; Martin and Uhler 1939). Martin and Uhler (1939) mentioned that *Najas* spp. ranked as the eighth most important food sources for wild ducks in the United States and Canada. However, transmission by migrating waterfowl is believed to be one of the ways *N. minor* spreads to new habitats, so consumption by ducks and other birds is also a potential risk factor for invasion (Meriläinen 1968).

ECONOMIC IMPACTS

Dense, bushy *N. minor* in the water may inhibit boating, swimming, and other recreational activities, resulting in economic loss. The spines on *Najas* plants have the potential to irritate skin, leading to an unpleasant swimming experience (Les 2012). With serious infestations, development could be hampered, the value of waterfront real estate could be influenced (Robinson 2004), and the water discharge capacity of channels could be reduced (Pfingsten et al. 2021).

Though rice is generally not a major agricultural product of Massachusetts, *Najas* spp. are often found in association with rice farming and are considered a weed of rice (Pinke et al. 2014). *N. minor* has the potential to cause economic loss if it is introduced as an agricultural pest in rice fields or among other crops that are grown in standing water.

MONITORING AND SIGHTINGS

The invasive species data detected in each waterbody were collected by the Massachusetts Department of Environmental Protection (MassDEP), the Department of Conservation and Recreation (DCR), the College of Holy Cross, as well as one individual, Joy Trahan-Liptak. The standard operating procedures, including sampling methodology, species identification, quality control and assurance, are implemented by

samples collected by MassDEP. Data from other agencies are confirmed by experts. Sites with *N. minor* were mapped as a data layer with ArcGIS® ArcMap 10.1 (ESRI, Redlands, California). This tool can show how *N. minor* has spread, which can help to inform future decisions on use of these waterbodies or regulations to limit the spread of this aquatic invasive species.

Data from different sources are shown in Table 11.1. A map using ArcGIS has been created to show the spatial distribution of *N. minor* in Massachusetts waterbodies (Figure 11.3).

The first sighting of European Naiad (*Najas minor)* in Massachusetts was in the Lily Brook and Stockbridge Bowl in West Stockbridge, Berkshire County, in June 1974 (Hellquist 1977). It was generally thought that *Najas minor* traveled eastward and from there to waters adjacent to the Turnpike. For example, one of the early detections is in the upper basin of Dorothy Pond in Millbury, which functions as a stormwater basin for the Turnpike (Observed by Therese Beaudoin, personal communication). Currently, there are 20 waterbodies in Massachusetts (Table 11.2) with reported sightings of *N. minor* representing eight watersheds (Table 11.3 and Figure 11.3). So far, no invasive European naiad has been detected in Massachusetts rivers, not unexpected as this species prefers to grow in static waters (Clausen 1936; Martin and Uhler 1939). The infested waterbodies are lakes, ponds, and reservoirs mainly located in western, midwestern, and southern Massachusetts. Among the 20 infested bodies of water, 13 of them are reported by MassDEP. According to the data from MassDEP collected between 1994 and 2016, there are 2,050 waterbodies that have been investigated. Therefore, these 13 lakes/ponds represent 0.6% of the investigated waterbodies. Since the sampling is not randomly designed, the sighting ratio likely cannot accurately reflect the true presence of *N. minor* in Massachusetts.

MANAGING ACTIONS

The most efficient way to protect the environment from invasive species is integrated management, including education and prevention, early detection, control, ecosystem restoration, and environmental regulation and enforcement. Among these management options,

TABLE 11.1
Location and Report Source of Each Infested Waterbody in Massachusetts

Waterbody	Agency	Report Source	Longitude	Latitude	First Year Reported
Lake Buel	MassDEP	Housatonic River Watershed 2002 Water Quality Assessment Report 21-AC-4 CN141.5, September 2007			
Laurel Lake	MassDEP	Housatonic River Watershed 2002 Water Quality Assessment Report 21-AC-4 CN141.5, September 2007			2014
Onota Lake	MassDEP	Housatonic River Watershed 2002 Water Quality Assessment Report 21-AC-4 CN141.5, September 2007			
Onota Lake	Individual	Joy Trahan-Liptak	−73.2813	42.4705	
Plunkett Reservoir	MassDEP	Housatonic River Watershed 2002 Water Quality Assessment Report 21-AC-4 CN141.5, September 2007			
Pontoosac Lake	MassDEP	Housatonic River Watershed 2002 Water Quality Assessment Report 21-AC-4 CN141.5, September 2007			
Pontoosac Lake	DCR	DCR Data from Lakes and Ponds, 2008			2008
Pontoosac Lake	DCR	DCR Data from ACT, GeoSyntec-ACT			
Richmond Pond	MassDEP	Housatonic River Watershed 2002 Water Quality Assessment Report 21-AC-4 CN141.5, September 2007			
Leverett Pond	MassDEP	Connecticut River Basin 2003 Water Quality Assessment Report 34-AC-2 CN 105.5, October 2008			
Dark Brook Reservoir	MassDEP	Blackstone River Basin 1998 Water Quality Assessment Report 51-AC-1 48.0, May 2011			
Dorothy Pond	MassDEP	Blackstone River Basin 1998 Water Quality Assessment Report 51-AC-1 48.0, May 2011			
North Pond	DCR	DCR Data from Lakes and Ponds, 2008			2008
Tinker Hill Pond	MassDEP	Blackstone River Basin 1998 Water Quality Assessment Report 51-AC-1 48.0, May 2011			

(Continued)

TABLE 11.1
(Continued)

Waterbody	Agency	Report Source	Longitude	Latitude	First Year Reported
Flint Pond	College of the Holy Cross	Robert Bertin	−71.7251	42.2414	
Lake Shirley	DCR	DCR Data from ACT, GeoSyntec-ACT			
Flint Pond	MassDEP	Merrimack River Basin 1999 Water Quality Assessment Report 84-AC-1 52.0, November 2001			
Martins Pond	MassDEP	Ipswich River Watershed 2000 Water Quality Assessment Report 92-AC-1 088.0, April 2004			
White Island Pond, East Basin	MassDEP	14-B003–01	−70.6175	41.8092	
White Island Pond, East Basin	MassDEP	13-B007–01	−70.6175	41.8092	
White Island Pond, West Basin	MassDEP	12-B003–02	−70.6276	41.8100	
Chicoppee Reservoir	DCR	DCR Data from Lakes and Ponds, 2005	−72.5513	42.1707	2005
Chicoppee Reservoir	DCR	DCR Data from Lakes and Ponds, 2007	−72.5513	42.1707	2007
Lily Pond, Middle	DCR	Jamie Carr	−71.7695	42.3767	
Wachusett Reservoir, Muddy Pond	DCR	Jamie Carr	−71.7871	42.4145	
Lily Pond, West	DCR	Jamie Carr	−71.7720	42.3773	
Lily Pond, East	DCR	Jamie Carr	−71.7671	42.3759	
Lily Brook and Stockbridge Bowl, West Stockbridge	USGS	Hellquist, C.B. (via USGS)	−73.3109	42.3363	1974
The Quag	DCR	Joy Trahan-Liptak	−71.7711	42.4166	
West Waushacum Pond	DCR	Joy Trahan-Liptak	−71.7645	42.4144	

prevention is the most efficient and cost-effective approach in stopping or minimizing the spread of aquatic invasive species. Mapped locations of European naiad provide a spatial understanding of the status of this invasive species in Massachusetts lakes to increase public awareness of this invasive species. For waterbodies that haven't been invaded but may be at risk, proactive prevention should be researched and promoted.

In Massachusetts, waterbodies infested with European naiad will be included in the 303d list of impaired or threatened waters during water quality assessments, fulfilling the requirements of the Clean Water Act. For projects within infested watersheds subject to the jurisdiction of the 401 Water Quality Certification, conditions on invasive species prevention and decontamination are needed in the certificate. In addition, water quality monitoring gear and watercrafts in these infested

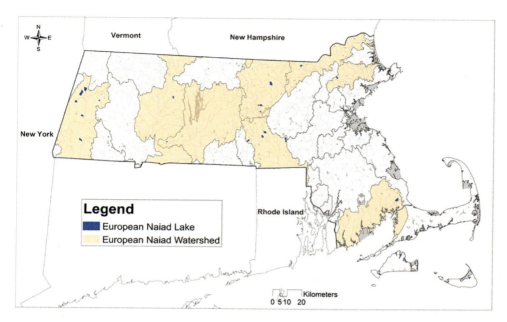

FIGURE 11.3 Distribution map depicting European Naiad (*Najas minor*) in Massachusetts.

TABLE 11.2
Towns and Infested Waterbodies

Number	Waterbody	Town_1	Town_2
1	Chicoppee Reservoir	Chicopee	–
2	Dark Brook Reservoir	Auburn	–
3	Dorothy Pond	Millbury	–
4	Flint Pond	Grafton	Worcester/Shrewsbury
5	Lake Buel	Monterey	New Marlborough
6	Lake Shirley	Lunenburg	Shirley
7	Laurel Lake	Lee	Lenox
8	Leverett Pond	Leverett	–
9	Lily Pond	West Boylston	–
10	Martins Pond	North Reading	–
11	North Pond	Hopkinton	Milford/Upton
12	Onota Lake	Pittsfield	–
13	Plunkett Reservoir	Hinsdale	–
14	Pontoosac Lake	Lanesborough	Pittsfield
15	Richmond Pond	Richmond	Pittsfield
16	The Quag	Sterling	–
17	Tinker Hill Pond	Auburn	–
18	Wachusett Reservoir, Muddy Pond	Sterling	–
19	West Waushacum Pond	Sterling	–
20	White Island Pond	Plymouth	Wareham

TABLE 11.3

Infested Watersheds

Number	Infested Watersheds
1	Blackstone
2	Buzzards Bay
3	Chicopee
4	Connecticut
5	Housatonic
6	Ipswich
7	Merrimack
8	Nashua

watersheds shall be decontaminated before being used in other watersheds to prevent the potential spread of invasive European naiad.

Several control methods for eradicating or reducing the abundance of *N. minor* in places where it is established are provided in this report. Known effective management measures include mechanical/physical control such as harvesting and benthic barriers, and chemical treatments (i.e., herbicides application). However, it is important to use these control methods carefully to avoid unintended consequences and choose a treatment for a specific site that will minimize any potential ecological harm.

When multiple invasive plants occupy the same habitat, a specific plan must be made to combat all invaders simultaneously, to avoid worsening the situation by sparing one or more invasive species from the competitive pressure of the other species (Richardson 2008). For example, if a pond contains three invasive plants in equal numbers, and measures are taken to eliminate only two of those species, the third species might thrive in the absence of its competitors and take over the entire pond, resulting in a less balanced system, and the original problem not being solved by the treatment.

All in all, control is only one small piece of an integrated management plan. It is essential to follow up any measures to eliminate nuisance species from an ecosystem with committed efforts to monitor, restore, and remediate the system. If a body of water is vulnerable to an *N. minor* invasion because ongoing pollution has caused the water to become eutrophic (containing excess nutrients), the site will remain vulnerable even if the invasive species is eliminated. Only by remediating pollution on the site and working to cultivate and reestablish native plant species in place of the ones removed will an infestation of *N. minor* be fully alleviated.

CHEMICAL CONTROL

Chemical control is one of the most popular management approaches for nuisance plants and several chemicals are legal to use in Massachusetts. Endothall, fluridone, and diquat are chemical herbicides that have been recognized as consistently effective against *N. minor* when used with the correct dose, proper formulation, and suitable conditions. Another active ingredient, 2,4-dichlorophenoxyacetic acid (abbreviated 2, 4-D), can also be used but its effectiveness varies as it can be influenced by dose and applying conditions (Mattson et al. 2004). More information about chemical control on *N. minor* and other 18 invasive species can be found in Appendix III.

Herbicides contain active ingredients that have specific biological toxicity to plants with acceptable risks to humans or other animals at the recommended dose. However, to avoid unintended effects, herbicides should be applied per the label and to achieve the concentration to be effective, they should be applied with care to limit exposure to non-target plants which might be indiscriminately damaged by the chemical (Mattson et al. 2004). It is also important to avoid overuse of herbicides because of the emerging possibility of herbicide resistance in aquatic plants (Richardson 2008). Before any chemical application, it is required to obtain an aquatic nuisance species control license (BRP WM04 Form) from Massachusetts Department of Environmental Protection.

PHYSICAL CONTROL

Harvesting of invasive plants can often achieve short-term success. Common harvesting techniques include mechanical harvesting and hand pulling (Mattson et al. 2004). Although this method is easily accessible and, in some cases, cheaper than alternatives, harvesting at the wrong time of the plant's life cycle can be counterproductive and disadvantageous for the removal of *N. minor* since the plant reproduces by breaking into many small seeded fragments (Meriläinen 1968; GISD 2021). Breaking the mature, fruiting plant or moving its pieces around could help the plant reproduce and spread rather than removing it (Robinson 2004). Mechanical removal by cutting or harvesting is therefore not recommended during the reproductive season of the plant (mid to late summer). Before the plant is mature, however, hand pulling is preferable to mechanical harvesting and mainly applicable when small-scale *N. minor* appears in the waterbody, or when *N. minor* is mixed with other aquatic plants.

Benthic barriers, large coverings placed on the bottom of waterbodies to smother benthic organisms, could be an effective physical control against *N. minor* by limiting the invasive population's access to light and oxygen. These barriers could consist of sheets of synthetic gas-impermeable material or layers of sediment such as clay, sand, or gravel deposited over the bottom. As *N. minor* grows submersed and rooted to the ground, if benthic barrier material is thick enough and low in nutrients, it could greatly reduce *N. minor* populations (Mattson et al. 2004). The cost of materials and labor in this technique can be high, and the barrier can often require additional maintenance (Robinson 2004). It is also suggested that, unless covering an entire population, barriers can be moved to cover other infested areas and eventually removed entirely from the waterbody (John McPheDran, personal communication). Benthic barriers also have the potential to

harm non-invasive benthic plants and wildlife (Washington State Department of Ecology n.d.). However, in limited areas that are carefully chosen, such as swimming beaches or densely infested waterbodies, benthic barriers could be a good choice for managing *N. minor*.

BIOLOGICAL CONTROL

As mentioned in a previous section, ducks and other waterfowl are known to feed on *Najas* spp., but their feeding does not lead to reduced abundance of *Najas* spp., rather it helps in its propagation (Martin and Uhler 1939; Meriläinen 1968). Therefore, wild birds are unlikely to be valuable as biological control of *N. minor*.

Introduced populations of grass carp, often genetically modified to be sterile, have often been used to combat invasive aquatic plants, but their introduction can lead to elimination of non-target plant species and other ecological problems (Richardson 2008). The effectiveness of introduced fish or other herbivore species in reducing *N. minor* populations has not been specifically documented.

SUCCESS STORY

In 2016, an aquatic vegetation management survey was developed by SOLitude Lake Management (SLM) for protecting the Borough of Mount Lakes, in Morris, New Jersey. Five nuisance aquatic macrophytes were observed during the survey from June to August in a set of nine waterbodies collectively known as the Mountain Lakes (SLM 2016, 2017, 2018, 2019). *N. minor* was detected among these nuisance plants. As a part of the SLM (2016) plan for dealing with these aquatic invasions, specific herbicides and algaecides were applied at each of the lakes to treat different target plants. Alongside the application of herbicides and algaecides, a water quality monitoring program was conducted under the local regulations.

Naiad species, which includes *N. minor*, were found in Sunset Lake and Wildwood Lake. At these two lakes, a herbicide product called Reward® was used to target naiad species. The active ingredient of Reward® is diquat dibromide and the formulation is intended for use on aquatic plants.

At Wildwood Lake, Reward® was applied over about 8 acres of the lake in mid-July. At Sunset Lake, Reward® was applied later in the season (mid-August) over a smaller area of 4 acres. After the summer of 2016, the effects of the herbicide treatment on *Najas* spp. in Wildwood Lake were limited, and *Najas* spp. were detected growing at nuisance abundance in September and October of 2016. Therefore, SOLitude Lake Management was ready to apply a more aggressive method to control the growth of *Najas* species in their next plan. No specific information was given in this report regarding the abundance of *Najas* species at Sunset Lake within the year following the 2016 treatment, but for the following years through 2019, naiad populations were reported as sparse in Sunset Lake and additional naiad-targeted treatment was not deemed necessary. The 2019 season also saw renewed growth of native bass weed (*Potamogeton amplifolius*) in Sunset Lake, along with reduced algal growth, consistently high dissolved oxygen levels, and low nitrate and phosphate in the water, representing the overall success of the management program. Between 2007 and 2019, Sunset Lake was treated non-aggressively, only with algaecides and herbicides, since water quality in the lake was generally good compared to other lakes falling under the SLM plan throughout this time (SLM 2019).

At Wildwood Lake, *N. minor* continued to grow in years following the treatment, so two new herbicides, Tribune® and Schooner®, were used to control *Najas* species during 2017–2018. Tribune® is another formulation of diquat dibromide, and Schooner® has flumioxazin as the active ingredient. Flumioxazin is a herbicide registered for aquatic use in Massachusetts, but it is also regarded as moderately to acutely toxic to fish and aquatic invertebrates (MDAR/MassDEP 2013). Wildwood Lake was also treated with alum for general nutrient management. However, reduction of *N. minor* in Wildwood Lake was limited, and in 2019, *N. minor* returned in huge densities, taking up an estimated 70% of the lake basin (SLM 2019). Therefore, in 2019, lake management changed tack and a product called Aquastrike® (active ingredients: diquat dibromide and endothall) was used to target *N. minor* instead with two applications, one in July and one in August. According to the 2019 report, Aquastrike®

"proved to be a great management tool for this lake for both naiad, and other submersed aquatic plants for localized management" (SLM 2019).

In this case, though *Najas* levels were eventually reduced from nuisance levels, the plants reappeared each year in spite of treatment, so repeated application of chemical herbicides may be necessary to achieve this reduction. The success found in this case through targeted chemical treatment was also in large part due to persistent and thorough monitoring of macrophyte coverage, phytoplankton, and water quality, which contributed in large part to the improved condition of these lake ecosystems. Chemical treatment alone would likely not have resulted in the same outcome in the Mountain Lakes without the strategic management plan surrounding it.

REFERENCES

Baart, I., S. Hohensinner, Z. Istvan, and T. Hein. 2013. Supporting analysis of floodplain restoration options by historical analysis. *Environmental Science & Policy* 34: 92–102.

Chase, S.S. 1947. Preliminary studies in the genus *Najas* in the United States. Ph.D. dissertation. Cornell University, Ithaca, NY, pp. 51–56.

Chayka, K., and M. Dziuk. n.d. Najas guadalupensis (Southern Waternymph). *Minnesota Wildflowers*. Via www.minnesotawildflowers.info/aquatic/southern-waternymph (Accessed March 23, 2021).

Clausen, R.T. 1936. Studies in the genus *Najas* in the northern United States. *Rhodora* 38: 333–345.

Dressler, K. 1996. University of Florida/IFAS/Center for Aquatic & Invasive Plants. https://plants.ifas.ufl.edu/plant-directory/najas-guadalupensis/. (Accessed March 23, 2021).

Global Invasive Species Database (GISD). 2021. Species profile: *Najas minor*. www.iucngisd.org/gisd/species.php?sc=1560 on 29–01–2021. (Accessed January 29, 2021).

Haynes, R. 1979. Revision of North and Central American *Najas* (Najadaceae). *Sida, Contributions to Botany* 8(1): 34–56.

Hellquist, C.B. 1977. Observations on some uncommon vascular aquatic plants in New England. *Rhodora* 79: 445–452.

Hilty, J. 2019. Thread-Leaf Naiad. Illinois Wildflowers. Via www.illinoiswildflowers.info/wetland/plants/th_naiad.html. (Accessed March 23, 2021).

ITIS (Integrated Taxonomic Information System). 2021. www.itis.gov/servlet/SingleRpt/SingleRpt?search_topic=TSN&search_value=39002#null (Accessed April 22, 2021).

Les, D.H. 2012. Noteworthy collection. *Madroño* 59(4): 230–233.

Les, D.H., E.L. Peredo, N.P. Tippery, L.K. Benoita, H. Razifarda, U.M. Kinga, H.R. Nab, H.K. Choib, L. Chenc, R.K. Shannona, and S.P. Sheldond. 2015. *Najas minor* (Hydrocharitaceae) in North America: A reappraisal. *Aquatic Botany* 126: 60–72.

Lovel, G. 2021. *Brittle Naiad (water nymph)*. Finger Lakes PRISM. http://fingerlakesinvasives.org/invasive_species/brittle-naiad/. (Accessed March 23, 2021).

Martin, A.C., and F.M. Uhler. 1939. *Food of Game Ducks in the United States and Canada*. Washington, DC: United States Department of Agriculture.

Massachusetts Department of Agricultural Resources (MDAR) Pesticide Program and Massachusetts Department of Environmental Protection (MassDEP) Office of Research and Standards. 2013. Flumioxazin. www.mass.gov/doc/flumioxazin/download. (Accessed January 29, 2021).

Mattson, M.D., P.L. Godfrey, R.A. Barletta, and A. Aiello. 2004. Eutrophication and aquatic plant management in Massachusetts: Final generic environmental impact report. *Commonwealth of Massachusetts, Executive Office of Environmental Affairs Commonwealth of Massachusetts* 4: 29–80.

McIntyre, S., and S.C.H. Barrett. 1985. A comparison of weed communities of rice in Australia. *Proc. Ecol. Soc. Aust.* 14: 237–250.

Mehrhoff, L.J. n.d. USGS. Najas minor all. https://nas.er.usgs.gov/queries/FactSheet.aspx?SpeciesID=1118. (Accessed March 23, 2021).

Meriläinen, J. 1968. Najas minor all. In North America. *Rhodora* 70: 161–175.

Minnesota Department of Natural Resources (MN DNR). 2021a. Brittle Naiad (*Najas minor*), also called Brittle Waternymph. www.dnr.state.mn.us/invasives/aquaticplants/brittlenaiad/index.html. (Accessed January 28, 2021).

Minnesota Department of Natural Resources (MN DNR). 2021b. *Najas marina* L. www.dnr.state.mn.us/rsg/profile.html?action=elementDetail&selectedElement=PMNAJ01060. (Accessed January 28, 2021).

Na, H., and H. Choi. 2012. Numerical taxonomic study of Najas L. (Hydrocharitaceae) in Korea. *Korean Journal of Plant Taxonomy* 42(2): 126–140.

Native Plant Trust. 2021. Najas minor. Go Botany (3.4). https://gobotany.nativeplanttrust.org/species/najas/minor/. (Accessed March 23, 2021).

Pfingsten, I.A., L. Cao, and L. Berent. 2021. *Najas Minor All.: U.S. Geological Survey, Nonindigenous Aquatic Species Database.* Ann Arbor, MI and Gainesville, FL: NOAA Great Lakes Aquatic Nonindigenous Species Information System. https://nas.er.usgs.gov/queries/GreatLakes/FactSheet.aspx?SpeciesID=1118 (Accessed January 27, 2021).

Pinke, G., J. Csiky, A. Mesterházy, L. Tari, R.W. Pál, Z. Botta-Dukát, and B. Czúcz. 2014. The impact of management on weeds and aquatic plant communities in Hungarian rice crops. *European Weed Research Society* 54: 388–397.

Richardson, R.J. 2008. Aquatic plant management and the impact of emerging herbicide resistance issues. *Weed Technology* 22(1): 8–15.

Robinson, M. 2004. *European Naiad: An Invasive Aquatic Plant (Najas minor).* Cambridge, MA: Massachusetts Department of Conservation and Recreation; Office of Water Resources; Lakes and Ponds Program.

Skawinski, P. n.d. *Aquatic Invasive Species Quick Guide. Brittle Naiad (Najas minor All.).* Golden Sands Resource Conservation & Development Council, Inc. www.uwsp.edu/cnr-ap/UWEXLakes/Documents/programs/CLMN/AISfactsheets/04BrittleNaiad.pdf. (Accessed March 23, 2021).

SOLitude Lake Management (SLM). 2016. 2016 year-end report lakes management program borough of mountain lakes. https://mtnlakes.org/wp-content/uploads/2014/09/2016-Lakes-Year-End-Report.pdf. (Accessed January 29, 2021).

SOLitude Lake Management (SLM). 2017. 2017 year-end report lakes management program borough of mountain lakes. https://mtnlakes.org/wp-content/uploads/2014/09/MtnLakes2017YEReport_BS.pdf. (Accessed January 29, 2021).

SOLitude Lake Management (SLM). 2018. 2018 year-end report lakes management program borough of mountain lakes. https://mtnlakes.org/wp-content/uploads/2014/09/2018-Lakes-Year-End-Report.pdf. (Accessed January 29, 2021).

SOLitude Lake Management (SLM). 2019. 2019 year-end report lakes management program borough of mountain lakes. https://mtnlakes.org/wp-content/uploads/2014/09/2019-Lakes-Year-End-Report.pdf. (Accessed January 29, 2021).

Topic, J. n.d. USGS. Najas marina L. https://nas.er.usgs.gov/queries/factsheet.aspx?SpeciesID=2674. (Accessed March 23, 2021).

Triest, L. 1988. A revision of the genus *Najas* L. (Najadaceae) in the old world. In: *Academie Royale des Sciences d'Outre-Mer.* Brussels.

U.S. Department of Agriculture (USDA). n.d.a. *Najas gracillima (A. Braun ex Engelm.) Magnus slender waternymph.* Natural Resources Conservation Service. https://plants.usda.gov/core/profile?symbol=NAGR. (Accessed March 23, 2021).

U.S. Department of Agriculture (USDA). n.d.b. *Najas marina L. spiny naiad.* Natural Resources Conservation Service. https://plants.usda.gov/core/profile?symbol=nama. (Accessed March 23, 2020).

Wang, H., H. Zhu, K. Zhang, L. Zhang, and Z. Wu. 2010. Chemical Composition in Aqueous Extracts of *Najas marine* and *Najas minor* and their Algae Inhibition Activity. Conference on Environmental Pollution and Public Health, CEPPH, pp. 806–809.

Washington State Department of Ecology. n.d. swollen bladderwort. *Wendy Scholl Invasion Ecology.* http://depts.washington.edu/oldenlab/wordpress/wp-content/uploads/2013/03/Ultricularia-inflata_Scholl_2007R.pdf. (Accessed March 23, 2020).

Wentz, W.A., and R.L. Stuckey. 1971. The changing distribution of the genus *Najas* (Najadaceae) in Ohio. *Ohio Journal of Science* 71: 292–302.

Wunderlin, R.P., B.F. Hansen, A.R. Franck, and F.B. Essig. 2021. Atlas of Florida Plants (http://florida.plantatlas.usf.edu/). [S. M. Landry and K. N. Campbell (application development), USF Water Institute.] Institute for Systematic Botany, University of South Florida, Tampa. https://florida.plantatlas.usf.edu/Plant.aspx?id=2087&display=photos. (Accessed March 23, 2021).

12 South American Waterweed (*Egeria densa*)

ABSTRACT

According to monitoring data, there are seven waterbodies in Massachusetts that are infested by South American waterweed (*Egeria densa*). This report serves to provide the general public with information about the status of South American waterweed in Massachusetts and help water users to minimize unintentional spread to new waterbodies. This report also provides biological and identification information about the plant, as well as descriptions of methods for control and management.

INTRODUCTION

Taxonomy Hierarchy (ITIS 2021)

Kingdom	Plantae
Subkingdom	Viridiplantae
Infrakingdom	Streptophyta
Superdivision	Embryophyta
Division	Tracheophyta
Subdivision	Spermatophytina
Class	Magnoliopsida
Superorder	Lilianae
Order	Alismatales
Family	Hydrocharitaceae
Genus	*Egeria* Planch.
Species	*Egeria densa* Planch.

South American waterweed or Brazilian waterweed (*Egeria densa*, hereafter referred to by its monotypic genus name, *Egeria*) is a submersed perennial monocot with a tendency for aggressive growth (Robinson 2002). It is a rooted macrophyte usually attached to the mud or sediment at the bottom of water, but it can also be found as a free-floating mat or in the surface water as fragments (Hoshovsky and Anderson 2001). *Egeria*'s stems are covered in small leaves arranged in whorls of four, or eight on a double node, making the whole stem resemble long, fuzzy cylinders, and the stems themselves can reach 1–3 m in length, though they are usually shorter (Cook and Urmi-König 1984; CABI 2019). They often grow until they reach the water surface and, after that, the plants start branching out. Leaves are bright green and longer than they are wide (about 1–3 cm long, Figure 12.1), with a lanceolate shape, smooth surface, and very fine serration on the leaf margins (Robinson 2002). *Egeria* is a dioecious plant, but only male plants have been observed outside of the plant's native range (CABI 2019). Bare floral peduncles, each containing a single flower, emerge from a node pathway down the stem and extend to the surface of the water. The flowers have three white, rounded petals with a yellow center which are the stamens (male reproductive organ) (Figure 12.2).

Egeria densa was previously named *Elodea densa* and classified within the genus *Elodea*. In spite of the species *E. densa* being transferred into the genus *Egeria*, it is still commonly referred to as an "elodea" in some circumstances, which, along with visual similarities, leads to confusion between *Egeria densa* and the currently canonized members of the *Elodea* genus. The singular member of another aquatic plant genus, *Hydrilla verticillata* (L.f.) Royle, is also visually similar to *Egeria* (Robinson 2002). To further complicate things, the genus *Elodea* was previously called *Anacharis*, a now defunct genus name, and all *Elodea*, *Egeria*, and *Hydrilla* species are prone to being marketed within the aquarium trade as "*Anacharis*." Other common names for *Egeria* include Brazilian waterweed, Brazilian Elodea, and dense waterweed (CABI 2019). Frequent reference is made in the aquatic macrophyte literature to a species named *Egeria najas*, but as of 2010, ITIS does not accept *E. najas* as a species (ITIS 2021). Specimens identified as *E. najas* can be keyed for identification as either *Egeria densa* or *Hydrilla verticillata*

DOI: 10.1201/9781003201106-12

245

FIGURE 12.1   South American waterweed (a) and its whorl of leaves (b) and single leaf (c). (Photos by Z. Jakubowska, M. Kiełbasa, and J. Nosek (see Krawczyk and Gabka 2019).)

on account of the formerly accepted waterweed *E. najas* sharing features with each of these accepted species (Catling and Mitrow 2001).

Several visual characteristics can be used to distinguish *Egeria* from *Hydrilla verticillata*, though both are considered invasive in the state of Massachusetts (Figure 12.2; see more about *Hydrilla* in Chapter 6). Generally, *Egeria densa* has four leaves per whorl, whereas *Elodea* spp. have three and *Hydrilla verticillata* five. Hydrilla has slightly shorter leaves, making the stem look thicker and more prominent than the

stem of *Egeria* (Jacono et al. 2021); however, it may not be the reliable difference because, sometimes, leaf length will change depending on environment (density of stand, nutrient availability, light penetration, etc.) of the waterbody (Rosali Smith, personal communication). The flowers of *Hydrilla* are also much smaller than the flowers of *Egeria*, and they are light pink in color with darker pink streaks running through the center of the petals. *Hydrilla* also has tuber-like rhizomes which are absent in *Egeria* (Bossard et al. 2000).

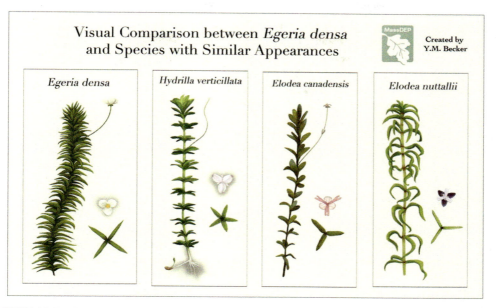

FIGURE 12.2   Guide to native and invasive waterweeds found in North America from the genera *Egeria*, *Hydrilla*, and *Elodea*. Illustrations and written descriptions of vegetated stems, flowers, and leaf whorl cross sections are provided to aid in visual identification. Simplified maps showing native and invasive ranges are also provided with each entry. Photo references used to ensure accuracy of illustrations of *Egeria densa* (Krawczyk and Gabka 2019; King County 2021), *Hydrilla verticillata* (Winterton n.d.; State of Michigan 2021; Houston Advanced Research Center 2021; Finger Lakes PRISM 2021), *Elodea canadensis* (Fischer 2008; Lukavský 2013; Lindman 1917), and *Elodea nuttalli* (Dixon 2019; Fischer 2011).

True *Elodea* spp. can be distinguished from *Egeria* by looking at the number and arrangement of leaves: while *Egeria*'s leaves are arranged in whorls of usually four leaves around each node on the stem, *Elodea* leaf whorls contain only three leaves. *Elodea* flowers are also pink, with thin, stringy petals, being very visually distinct from those of *Egeria*. Three species of *Elodea* are native to the continental United States. More information on differentiating *Egeria* from similar aquatic plants is provided in Figure 12.2 in a visual guide attached at the end of this chapter.

Native and Invasive Range

Egeria is native to Brazil and the surrounding region in South America including parts of Argentina and Peru. Introduced populations can now be found on all continents except Antarctica (Yarrow et al. 2009). The wide-scale naturalization of this weed is believed to be caused by the aquarium trade and possibly exacerbated by boating (Yarrow et al. 2009; Walsh et al. 2012; Rimac et al. 2018).

It is an invasive aquatic plant that has spread to 27 countries and 39 states in the United States (Bugbee et al. 2020). In the United States, *Egeria* was first found in 1893 at Mill Neck in Long Island, New York, according to a journal article by Weatherby (1932). From the early 1900s to present, most states in the United States have been found to contain populations of *Egeria*; the plant has not been found in some northern and midwestern states (Robinson 2002; MaineVLMP 2009).

Spread and Impacts

In its native habitat, *Egeria* commonly spreads by seed; the plants are pollinated mainly by flies traveling between male and female flowers, which grow on a long stalk and sit on the surface of the water. However, in North America,

Egeria is commonly seen to reproduce asexually by fragmentation and regrowth. Similarly, double nodes on stems can either be reproductive, bearing a flower, lateral branch, or root buds or it can be sterile (Yarrow *et al.* 2009).

Stem fragments from mature *Egeria* plants can easily form roots and develop into new dense mats of vegetation (Haynes 1988). *Egeria* is remarkably efficient at reproducing itself in this way and can rapidly establish dominance over an ecosystem, replacing other vegetation (Tanner et al. 1990). Because of the plant's ability to modify its habitat by controlling sediment suspension, nutrients, and algae growth, *Egeria* has been described as an "ecosystem engineer" (Yarrow et al. 2009). This term is commonly used to describe species that strongly modify their surroundings; for example, beavers are considered "ecosystem engineers" because they construct dams to modify the landscape. Corals can also be considered "ecosystem engineers" because of their ability to form coral reefs (Jones et al. 1994).

Rapid asexual reproduction allows *Egeria* to form large monocultural beds in the water, potentially depriving other plants of nutrients, space, and sunlight. *Egeria* also modifies its environment by slowing or preventing the flow of water, sometimes to an extent that is economically detrimental. Dense growth of *Egeria* could clog manmade plumbing, preventing water movement and potentially causing damage (Yarrow et al. 2009). *Egeria* infestation can also prevent larger fish from moving through waterways, especially posing a threat to the well-being of anadromous fish, which rely on passage through rivers and small tributaries to complete their life cycles (Johnson et al. 2006). In this way, the value of waterways for recreational fishing can be reduced. Boating, swimming, and other recreational activities could also be limited due to reduced water flow. Changes made by *Egeria* infestations might also reduce the aesthetic value of waterbodies or reduce the value of surrounding real estate (Robinson 2002).

Though dense, rapid growth of non-native *Egeria* can negatively impact large fish and outcompete native plants, *Egeria* could also have some ecosystem benefits. Small fish and zooplankton could take shelter in *Egeria*'s fronds,

and wild waterfowl have been observed to use *Egeria* as a food source (Yarrow et al. 2009). *Egeria* also draws nutrients from the water column, limiting the growth of phytoplankton and potentially sequestering nutrients sufficiently to reverse the effects of eutrophication when established (Yarrow et al. 2009), though it is not clear that this nutrient sequestering ability is unique to *Egeria*, as related *Elodea* species have also been observed to sequester nitrogen. Mesocosm experiments by Vanderstukken et al. (2011) have directly demonstrated *Egeria*'s ability to suppress phytoplankton growth through nutrient competition, while another common aquatic macrophyte failed to produce an effect.

In spite of costs associated with removing unwanted *Egeria*, the plant has significant economic value, as well. While millions of dollars have been spent toward removing *Egeria* from the wild, a considerable amount of money is also spent by individuals looking to bring the plant into their own home fish tanks. A mere handful of *Egeria* (containing "at least four stems") commands a price of $3.99 through Walmart's website, plus $7.95 shipping (Walmart.com, 2021). Other scientific or aquarium-centered businesses sell *Egeria* stems at even higher prices, though sellers are not always clear about the precise species being sold (often using the names "*Anacharis*," "*Elodea*," "*Egeria*," and numerous other common names interchangeably, with photos of multiple different species attached to a listing). *Egeria* (along with other types of "*Anacharis*") is prized by aquarium keepers for its ecological services: while alive and photosynthesizing, *Egeria* effectively oxygenates water, improving the habitat for captive aquatic animals.

Egeria could also have potential value in industrial agriculture, either in aquaculture (because of its ability to take in nutrients and oxygenate water) or even as livestock feed. Research has found that sun-dried *Egeria* is a good source of nutrients to broiler chicks and human bred waterfowl; this fact combined with the known feasibility of growing *Egeria* quickly, at a low cost, and without taking up a large amount of space, makes *Egeria* appear to be a promising, potentially lucrative animal feed crop (Boyd and McGinty 1981; Maurice et al. 1984; Dillon et al. 1988).

MONITORING AND SIGHTINGS

The invasive species data detected in each waterbody were collected by the Massachusetts Department of Environmental Protection (MassDEP), the Department of Conservation and Recreation (DCR), the College of Holy Cross, as well as by Joy Trahan-Liptak as an individual before she joined DCR. The standard operating procedures, including sampling methodology, species identification, quality control and assurance, are implemented for samples collected by MassDEP. Data from other agencies are confirmed by experts. Sites with *Egeria* were mapped as a data layer with ArcGIS® ArcMap 10.1 (ESRI, Redlands, California). This tool is able to pinpoint the area that *Egeria* has spread, which can help to inform future decisions on processes that can happen in these waterbodies, or regulations and permitting that involve efforts to control or limit the spread of this aquatic invasive species.

Sightings of this species in Massachusetts are shown in (Table 12.1). A map using GIS has been created in order to show the visual distribution of *Egeria* in Massachusetts waterbodies (Figure 12.3). South American waterweed is only found in two lakes out of the 2,050 MassDEP-monitored waterbodies. Five sightings are reported by other agencies. Therefore, there are a total of seven lakes (Table 12.2) within six watersheds in Massachusetts that are infested by *Egeria* (Table 12.3). These infested waterbodies are mainly located in the central and eastern regions of the state. Invasive *Egeria* was first found in Massachusetts in 1939 in the pond at the Memorial Park, Plymouth, and a brook in Furnace Brook Parkway, Norfolk (USGS 2020).

MANAGING ACTIONS

The most efficient way to protect the environment from invasive species is integrated management, including education and prevention,

---

TABLE 12.1

Infested Waterbodies in Massachusetts

Waterbody	Agency	Report Source	Longitude	Latitude	First Year Reported
Clamshell Pond	DCR	Joy Trahan-Liptak	−71.678456	42.399324	
Clamshell Pond	College of the Holy Cross	Robert Bertin	−71.678197	42.399340	
Clamshell Pond	DCR	DCR Data from Lakes and Ponds, 2008			2008
Days Pond	MassDEP	North Shore Coastal Watersheds 2002 Water Quality Assessment Report 93-AC-2 CN 138.5, March 2007			
Hemenway Pond	DCR	DCR Data from ACT, GeoSyntec Barre Hellquist, 1971			1971
Lake Archer	MassDEP	Charles River Watershed 2002–2006 Water Quality Assessment Report 72-AC-4 CN136.5, April 2008			
Lake Archer	DCR	DCR Data from Lakes and Ponds, 2002	−71.338521	42.069854	2002
Lake Archer	DCR	DCR Data from Lakes and Ponds, 2003	−71.338521	42.069854	2003
Lake Cochituate	Individual	Joy Trahan-Liptak	−71.358646	42.284193	
Lake Rico	DCR	DCR Data from Lakes and Ponds, 2007			2007
Lake Rico	DCR	DCR Data from Lakes and Ponds, 2008			2008
North Pond	DCR	DCR Data from ACT, GeoSyntec, 2003			2003

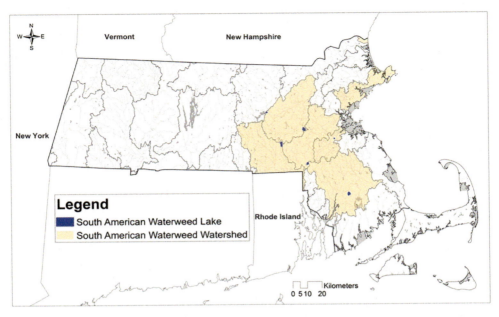

FIGURE 12.3   Distribution map depicting South American Waterweed (*Egeria densa*) in Massachusetts.

TABLE 12.2

Towns and Infested Waterbodies

Number	Waterbody	Town 1	Town 2	Town 3
1	Clamshell Pond	Clinton	–	–
2	Days Pond	Gloucester	–	–
3	Hemenway Pond	Milton	–	–
4	Lake Archer	Wrentham	–	–
5	Lake Cochituate	Framingham	Wayland	Natick
6	Lake Rico	Taunton	–	–
7	North Pond	Florida	–	–

early detection, control, ecosystem restoration, and environmental regulation and enforcement. Among these management options, prevention is the most efficient and cost-effective approach in stopping or minimizing the spread of aquatic invasive species. Mapping sightings of *Egeria* provides a spatial understanding of the status of this invasive species in Massachusetts' lakes to increase public awareness of this invasive species. For waterbodies that haven't been invaded but may be at risk, proactive prevention should be researched and promoted.

In Massachusetts, waterbodies infested with *Egeria* will be included in the 303d list of impaired or threatened waters during water quality assessments to fulfill the requirements of the Clean Water Act. For projects within infested watersheds subject to the jurisdiction of the 401 Water Quality Certification, conditions on invasive species prevention and decontamination must be provided to receive certification. In addition, water quality monitoring gear and watercrafts in these infested watersheds must be decontaminated before being used in other watershed to prevent the potential spread of invasive species.

To combat existing *Egeria* infestations, there are many different methods tested and found

TABLE 12.3

Infested Watersheds

1	Blackstone
2	Boston Harbor
3	Charles
4	North Coastal
5	SuAsCo
6	Taunton

applicable. Effective management measures include chemical, physical, and biological techniques. These techniques are described in the following sections.

CHEMICAL CONTROL

Chemical control is one of the most popular management approaches for nuisance plants and several chemicals are legal to use in Massachusetts. Fluridone, 2,4-dichlorophenoxyacetic acid (abbreviated 2, 4-D), and diquat are chemical herbicides that have been recognized as consistently effective against *Egeria* when used with the correct dose, proper formulation, and suitable conditions (Mattson et al. 2004).

Herbicides usually contain active ingredients that have specific biological toxicity to plants so that they are not dangerous to humans or other animals at the conventional dosage. However, to avoid unintended effects, herbicides should not be used in excess, and they should be applied with care to limit exposure to non-target plants, which might be indiscriminately damaged by the chemical (Mattson et al. 2004). It is also important to avoid overuse of herbicides because of the emerging possibility of herbicide resistance in aquatic plants (Richardson 2008).

Hanson et al. (2006) carried out a microcosm study to observe the effects of oxytetracycline on *Egeria*. The density of the plant was found to decrease in the presence of the chemical, but the decline in growth was attributed to reduced light availability caused by accumulating oxytetracycline by-products, and the research team determined that oxytetracycline does not pose any direct harm to *Egeria*.

More information about chemical control on *Egeria* and other 18 invasive species can be found in Appendix III. Prior to any chemical application to bodies of water, it is required to obtain an aquatic nuisance species control license (BRP WM04 Form) from Massachusetts Department of Environmental Protection.

PHYSICAL CONTROL

Mechanical harvesting, or physical removal, is one of the simplest methods of reducing an invasive plant population in a short period of time. Aquatic plants can be harvested by hand or by machine depending on site conditions (Mattson et al. 2004). Hand pulling is a suitable control method for sparse, emerging, or poorly established *Egeria* stands, especially if non-target plants requiring protection are present, while mechanical harvesting may be practicable where growth is denser, and problems are more severe. Although physical harvesting of *Egeria* is an easily accessible control method, there are some pronounced disadvantages. Since *Egeria* has the capability to regrow by fragments, this being its main reproductive strategy outside of its native range, physically removing *Egeria* stands may inadvertently create plant fragments which facilitate its spread (Curt et al. 2010). If *Egeria* fragments are left behind in the water after harvesting, or if mechanical equipment is not properly decontaminated before beginning work in another waterbody, new infestations could arise following removal efforts.

Benthic barriers, large coverings placed on the bottom of waterbodies to smother benthic organisms, could be an effective physical control against *Egeria*, limiting the whole invasive population's access to light and oxygen. These barriers could consist of sheets of synthetic gas-imperme-able material, or layers of sediment such as clay, sand, or gravel deposited over the bottom. The cost of materials and labor in this technique can be high, and the barrier can often require additional maintenance (Robinson 2002). Benthic barriers also have the potential to harm non-invasive benthic plants and wildlife (Washington State Department of Ecology n.d.). However, in limited areas that are carefully chosen, such as swimming beaches or densely infested waterbodies, benthic barriers could be a good choice for managing aquatic nuisance plants.

Drawdown, or temporarily lowering the water level in a body of water, is potentially a very effective method of eliminating *Egeria* stands, as past experimentation has found that *Egeria* plants cannot survive more than an hour out of the water (Yarrow et al. 2009). Because this method involves removing a large portion of water from the site, it can be difficult to accomplish depending on the volume of water present. Drawdown can also be harmful to non-target aquatic species dependent on water for survival, so care should be taken to mitigate and prevent ecological damage. In order to reduce impacts to non-target plants and animals during the growing season, drawdowns in Massachusetts are normally conducted in fall and winter (Mattson et al. 2004).

BIOLOGICAL CONTROL

The only well-known specialized herbivore of *Egeria* is the leaf-mining fly named *Hydrellia egeriae* Rodrigues-Júnior (Walsh et al. 2012). Risk assessments evaluated under quarantine conditions demonstrated that the fly could not discriminate between *E. densa* and *Elodea canadensis* and as a result, the host range of *H. egeriae* is too broad for use as a biological control agent of *E. densa* in the United States (Pratt et al. 2019). Another study has found that non-target aquatic macrophytes in South Africa cannot sustain multiple generations of a leaf-mining fly population, meaning the fly is not likely to reduce the abundance of any plant other than *Egeria* on a population scale. Based on this risk assessment study, permission has been obtained to release *H. egeriae* for *Egeria* control in South Africa (Smith et al. 2019).

A fungus from the genus *Fusarium* has also been found to have potential as biological control against *Egeria*, causing stem necrosis and foliar chlorosis (Nachtigal and Pitelli 1999; Neto et al. 2004). The use of fungi or pathogens as biological control against nuisance plants is precedented and laboratory experimentation of this method has appeared promising, but field applications have not yet been broadly successful in eliminating target plants (Hussner et al. 2017). Since several other fungi in the genus *Fusarium* are known to be destructive agricultural pathogens, causing a disease known as "Fusaric wilt," additional risk assessment is needed to determine whether the species previously evaluated is safe to release as a mycoherbicide (Singh et al. 2017).

Yarrow et al. (2009) reported observations of waterfowl using *Egeria* as a food source. Allowing ducks or other birds to graze may in fact be an effective means of controlling *Egeria*, demonstrated by the results of a case study in Spain (Curt et al. 2010). This case is discussed in greater detail in the "Success Story" section ahead.

SUCCESS STORIES

CHEMICAL CONTROL

Fence Rock Lake is a private 16.9 acre lake in Guilford, CT (CAES IAPP n.d.; Bugbee et al. 2020). In 2009, the Connecticut Agricultural Experiment Station (CAES) Invasive Aquatic Plant Program (IAPP) developed a survey at this lake, which resulted in documentation of the first appearance of the invasive species *Egeria*. Their vegetation surveys at Fence Rock Lake in the following years (2009–2013) showed that *Egeria* was rapidly expanding to cover large areas of the lake (CAES IAPP n.d.). In 2009, there is only a small amount of *Egeria* growing at the edge of the lake. However, during 2010–2012, *Egeria* spread from the western shoreline of the lake to the lake center, where it spanned to surrounding large areas. By 2013, almost the whole shoreline was covered by *Egeria*, and it was found in great quantities in the deeper water column. Abundances of native species were also decreased in these years (Figure 12.4).

In order to remove *Egeria* completely from Fence Rock Lake, CAES IAPP made a

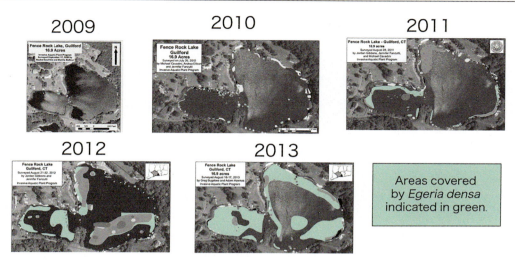

FIGURE 12.4　Maps produced from surveys by CASE IAPP at Fence Rock Lake during 2009–2013. (Modified to highlight areas covered by *Egeria densa* in each consecutive year.)

strategic plan (Bugbee et al. 2020). In 2014, they applied the herbicide diquat (Reward®) across 8.4 acres in the lake using a bottom injection technique. Generally, the effectiveness of herbicides can be seen after several repeated applications. Likewise, after two applications of diquat in Fence Rock Lake, *Egeria* was reportedly successfully controlled, with a survey and aquatic vegetation map from 2016 showing no *Egeria* anywhere in the lake (CAES IAPP n.d.). Not only were nuisance plants eliminated, but the treatment was applied precisely enough so that no native aquatic plants populations were affected. According to the records from the surveys of CAES IAPP, no *Egeria* appeared again at the Fence Rock Lake in the following two years after this chemical treatment (Bugbee et al. 2020). In the most recent survey at the lake, miniscule *Egeria* populations appear at only one point on the lake's shore.

DOMESTIC DUCKS AS BIOLOGICAL CONTROL

Years ago, *Egeria* was accidentally introduced into a small, family-run nursery reservoir in Galicia, northwest Spain (Curt et al. 2010). The introduction was suspected to have occurred

when the nursery began intentionally cultivating water hyacinth *Pontederia crassipes* (formerly *Eichhornia crassipes*, see Chapter 15 of this book); fragments of *Egeria* were believed to have been mixed in with the roots of the *Pontederia* plants. About seven years after the initial introduction, the *Egeria* infestation had become serious, and management efforts began. Unsuccessful physical removal of *Egeria* was carried out for four years, a strategy which was described as "utterly ineffective; indeed, removal seemed to enhance the growth of the waterweed" (Curt et al. 2010). However, the tides changed when two domestic Peking white ducks were incidentally introduced to the property. The ducks were observed to feed on the nuisance plant, preferring the leaves and soft growing tips of the plant. Leafless, fragmented stems left behind by the ducks were unable to propagate and were observed to simply decay. Under specific conditions, even though stems may be leafless, nodes on the stems can still develop into new shoots, roots, etc.; often, damage to shoots stimulates branching of the plant (Mangan and Baars 2016). In the years following the introduction of the waterfowl, the dense infestation of *Egeria* was significantly reduced. Though the invasive plant was not entirely eliminated, its population remained under control as long as the ducks grazed (Curt et al. 2010).

REFERENCES

Bossard, C.C., J.M. Randall, and M.C. Hoshovsky. 2000. *Invasive Plants of California's Wildlands*. Berkeley, CA: University of California Press.

Boyd, C.E., and P.S. McGinty. 1981. Percentage digestible dry matter and crude protein in dried aquatic weeds. *Economic Botany* 35: 296–299.

Bugbee, G.J., C.S. Robb, and S.E. Stebbins. 2020. Efficacy of diquat treatments on Brazilian waterweed, effects on native macrophytes and water quality: A case study. *Journal of Aquatic Plant Management* 58: 83–91.

CABI. 2019. *Egeria densa* (leafy elodea). *Centre for Agriculture and Bioscience International (CABI)*. www.cabi.org/isc/datasheet/20491. (Accessed March 27, 2020).

Catling, P.M., and G. Mitrow. 2001. *Egeria najas* at the Canadian Border and its separation from the related aquatic weeds *Egeria densa* and *Hydrilla verticillata* (Hydrocharitaceae). *Botanical Electronic News 278*. www.ou.edu/cas/botany-micro/ben/ben278.html. (Accessed March 2, 2021).

Connecticut Agricultural Experiment Station, Invasive Aquatic Plant Program (CAES IAPP). n.d. Fence rock lake survey results. https://portal.ct.gov/-/media/CAES/Invasive-Aquatic-Plant-Program/Survey-Results/F/Fence-Rock-Lake/FenceRockLakepdf.pdf. (Accessed March 1, 2021).

Cook, C.D.K., and K. Urmi-König. 1984. A revision of the genus *Egeria* (Hydrocharitaceae). *Aquatic Botany* 19: 73–96.

Curt, M.D., G. Curt, P.L. Aguado, and J. Fernández. 2010. Proposal for the biological control of *Egeria densa* in small reservoirs: A Spanish case study. *Journal of Aquatic Plant Management* 48: 124–127.

Dillon, C.R., D.V. Maurice, and J.E. Jones. 1988. Chemical composition of *Egeria densa*. *Journal of Aquatic and Plant Management* 26: 44–45.

Dixon, C.J. 2019. *Elodea nuttalli*. MAKAQueS. www.makaques.com/gallery.php?sp=2508. (Accessed March 2, 2021).

Finger Lakes PRISM. 2021. Hydrilla. http://fingerlakesinvasives.org/invasive_species/hydrilla/. (Accessed March 2, 2021).

Fischer, C. 2008. Canadian Waterweed. https://en.wikipedia.org/wiki/Elodea_canadensis#/media/File:ElodeaCanadensis.jpg. (Accessed March 2, 2021).

Fischer, C. 2011. Parts of the western waterweed (*Elodea nuttalli*). https://en.wikipedia.org/wiki/Elodea_nuttallii#/media/File:ElodeaNuttallii2.jpg. (Accessed March 2, 2021).

Hanson, M.L., C.W. Knapp, and D.W. Graham. 2006. Field assessment of oxytetracycline exposure to the freshwater macrophytes *Egeria densa* Planch., and *Ceratophyllum demersum* L. *Environmental Pollution* 141: 434–442.

Haynes, R.R. 1988. Reproductive biology of selected aquatic plants. *Annals of the Missouri Botanical Garden* 75: 805–810.

Hoshovsky, M.C., and L. Anderson. 2001. *Egeria densa* Planchon. In C.C. Bossard, J.M. Randall, and M.C. Hoshovsky (eds.), *Invasive Plants of California's Wildlands* (1st edition). Santa Rosa, CA: Pickleweek Press.

Houston Advanced Research Center. 2021. Galveston Bay Field Guide. http://blog.microscopeworld.com/2016/07/hydrilla-verticillatea-leaf-under.html. (Accessed March 2, 2021).

Hussner, A., I. Stiers, M.J.J.M. Verhofstad, E.S. Bakker, B.M.C. Grutters, J. Haury, J.L.C.H. van Valkenburg, G. Brundu, J. Newman, J.S. Clayton, L.W.J. Anderson, and D. Hofstra.

2017. Management and control methods of invasive alien freshwater aquatic plants: A review. *Aquatic Botany* 136: 122–137.

Interagency Taxonomic Information System (ITIS). 2021. Egeria. www.itis.gov/servlet/SingleRpt/SingleRpt?search_topic=TSN&search_value=38972#null. (Accessed March 2, 2021).

Jacono, C.C., M.M. Richerson, H.V. Morgan, and I.A. Pfingsten. 2021. *Hydrilla verticillata (L.f.) Royle: U.S. Geological Survey, Nonindigenous Aquatic Species Database*. Gainesville, FL. https://nas.er.usgs.gov/queries/FactSheet.aspx?SpeciesID=6. (Accessed February 10, 2021).

Johnson, D., M. Carlock, and T. Artz. 2006. *Egeria densa Control Program Second Addendum to 2001 Environmental Impact Report with Five-year Program Review and Future Operations Plan*. California: The State of California Department of Boating and Waterways.

Jones, C.G., J.H. Lawton, and M. Shachak. 1994. Organisms as ecosystem engineers. *Oikos* 69(3): 373–386.

King County, WA. 2021. Egeria (Brazilian elodea) identification and control. www.kingcounty.gov/services/environment/animals-and-plants/noxious-weeds/weed-identification/brazilian-elodea.aspx. (Accessed March 2, 2021).

Krawczyk, R., and M. Gabka. 2019. Egeria densa (Hydrocharitaceae)—Nowy gatunek antropofita we florze Polski. *Fragmenta Floristica et Geobotanica Polonica* 26(1): 41–48.

Lindman, C.A.M. 1917. Elodea Canadensis. https://fr.wikipedia.org/wiki/%C3%89lod%C3%A9e#/media/Fichier:Elodea_canadensis_nf.jpg. (Accessed March 2, 2021).

Lukavský, J. 2013. Canadian Waterweed. www.biolib.cz/en/image/id223880/. (Accessed March 2, 2021).

MaineVLMP. 2009. Brazilian waterweed, Brazilian Elodea, *Anacharis*, *Egeria densa*. *Maine Volunteer Lake Monitoring Program (MaineVLMP)*. www.mainevlmp.org/mciap/herbarium/VariableWatermilfoil.php. (Accessed March 27, 2020).

Mangan R., and J.R. Baars. 2016. Can leaf-mining flies generate damage with significant impact on the submerged weed *Lagarosiphon major*? *BioControl* 61: 803–813.

Mattson, M.D., P.J. Godfrey, R.A. Barletta, and A. Aiello. 2004. Eutrophication and aquatic plant management in Massachusetts: Final generic environmental impact report. *Commonwealth of Massachusetts, Executive Office of Environmental Affairs Commonwealth of Massachusetts* 4: 29–80pp.

Maurice, D.V., J.E. Jones, C.R. Dillon, and J.M. Weber. 1984. Chemical composition and nutritional value of Brazilian elodea (*Egeria densa*) for the chick. *Poultry Science* 63: 317–323.

Nachtigal, G.F., and R.E. Pitelli. 1999. Fusarium sp. as a potential biocontrol agent for Egeria densa and Egeria najas. In: *Program Abstracts, X International Symposium on Biological Control of Weeds*. Bozeman, MT: USDA-ARS and Montana State University, p. 68.

Neto, C.R.B., C.Q. Gorgati, and R.A. Pitelli. 2004. Effects of photoperiod and temperature in the development of disease caused *Fusarium graminearum* on *Egeria densa* and *E. najas*. *Fitopatologia Brasileira* 29(3): 252–258.

Pratt, P.D., J.C. Herr, R.I. Carruthers, and G.C. Walsh. 2019. Complete development on *Elodea canadensis* (Hydrocharitaceae) eliminates *Hydrellia egeriae* (Diptera, Ephydridae) as a candidate biological control agent of *Egeria densa* (Hydrocharitaceae) in the U.S.A. *Biocontrol Science and Technology* 29(4): 405–409.

Richardson, R.J. 2008. Aquatic plant management and the impact of emerging herbicide resistance issues. *Weed Technology* 22(1): 8–15.

Rimac, A., I. Stankovic, A. Alegro, S. Gottstein, N. Koletic, N. Vukovic, V. Segota, and A. Zizic-Nakic. 2018. The Brazilian elodea (*Egeria densa* Planch.) invasion reaches Southeast Europe. *BioInvasions Records* 7(4): 381–389.

Robinson, M. 2002. *South American Waterweed: An Exotic Invasive Plant (Egeria densa)*. Cambridge, MA: Massachusetts Department of Conservation and Recreation; Office of Water Resources; Lakes and Ponds Program.

Scheers, K., L. Denys, J. Packet, and T. Adriaens. 2016. A second population of *Cabomba caroliniana* Gray (Cabombaceae) in Belgium with options for its eradication. *BioInvasions Records* 5(4): 227–232.

Singh, V.K., H.B. Singh, and R.S. Upadhyay. 2017. Role of fusaric acid in the development of '*Fusarium* wilt' symptoms in tomato: Physiological, biochemical and proteomic perspectives. *Plant Physiology and Biochemistry* 118: 320–332.

Smith, R., R. Mangan, and J.A. Coetzee. 2019. Risk assessment to interpret the physiological host range of *Hydrellia egeriae*, a biocontrol agent for *Egeria densa*. *BioControl* 64: 447–456.

State of Michigan. 2021. Hydrilla. www.michigan.gov/invasives/0,5664,7-324-68002_74188-367843--00.html. (Accessed March 2, 2021).

Tanner, C., J. Clayton, and B. Coffey. 1990. Submerged-vegetation changes in lake Rotoria (Hamilton, New Zealand) related to herbicide treatment and invasion by *Egeria densa*. *New Zealand Journal of Marine and Freshwater Research* 24: 45–57.

U.S. Geological Survey (USGS). 2020. *Specimen observation data for Najas minor All., Nonindigenous Aquatic Species Database*. Gainesville, FL, https://nas.er.usgs.gov/queries/CollectionInfo.aspx?SpeciesID=1107. (Accessed March 26, 2020).

Vanderstukken, M., N. Mazzeo, W. Van Colen, S.A.J. Declerck, and K. Muylaert. 2011. Biological control of phytoplankton by the subtropical submerged macrophytes *Egeria densa* and *Potamogeton illinoensis*: A mesocosm study. *Freshwater Biology* 56(9): 1837–1849.

Walmart. 2021. 1 Anacharis Bunch—4+ Stems | *Egeria Densa*—Beginner Tropical Live Aquarium Plant. www.walmart.com/ip/1-Anacharis-Bunch-4-Stems-Egeria-Densa-Beginner-Tropical-Live-Aquarium-Plant/170556602. (Accessed March 2, 2021).

Walsh, G.C., Y.M. Dalto, F.M. Mattioli, R.I. Carruthers, and L.W. Anderson. 2012. Biology and ecology of Brazilian elodea (*Egeria densa*) and its specific herbivore *Hydrellia sp.*, in Argentina. *BioControl* 58: 133–147.

Washington State Department of Ecology. n.d. swollen bladderwort. *Wendy Scholl Invasion Ecology.* http://depts.washington.edu/oldenlab/wordpress/wp-content/uploads/2013/03/Ultricularia-inflata_Scholl_2007R.pdf. (Accessed March 30, 2020).

Weatherby, C.A. 1932. *Anacharis densa* on Long Island. *Rhodora* 34(403): 151–152.

Winterton, S. n.d. Aquarium and pond plants of the world, Edition 3, USDA APHIS PPQ, Bugwood.org. Via www.invasive.org/browse/detail.cfm?imgnum=5563004 (Accessed March 2, 2021).

Yarrow, M., V.H. Marin, M. Finlayson, A. Tironi, L.E. Delgado, and F. Fischer. 2009. The ecology of *Egeria densa* Planchon (Liliopsida: Alismatales): A wetland ecosystem engineer? *Revista Chilena de Historia Natural* 82: 299–313.

13 Swollen Bladderwort (*Utricularia inflata*)

ABSTRACT

In 1990, swollen bladderwort, otherwise known as *Utricularia inflata* (*U. inflata*), was first detected in the Federal Pond in Carver and Plymouth of southeast Massachusetts. Since then, there have been 15 sightings from central to eastern Massachusetts. The documented sightings and findings can provide the public with information regarding *U. inflata* infestations and the effects of infestations in Massachusetts to minimize future spread of this invasive species. This chapter describes the ecological and economic impacts of *U. inflata*. Control and treatment methods such as mechanical removal (e.g., benthic barriers and hands harvesting) and chemical treatment (e.g., herbicides application) are summarized in this chapter.

INTRODUCTION

Taxonomy Hierarchy (ITIS 2021)

Kingdom	Plantae
Subkingdom	Viridiplantae
Infrakingdom	Streptophyta
Superdivision	Embryophyta
Division	Tracheophyta
Subdivision	Spermatophytina
Class	Magnoliopsida
Superorder	Asteranae
Order	Lamiales
Family	Lentibulariaceae
Genus	*Utricularia*
Species	*Utricularia inflata* Walter

Swollen bladderwort is a rootless, carnivorous, free-floating, bushy, submerged plant which can be found just above the sediment (Figure 13.1). Some of its leaves float at the surface of the water. Even so, lateral submerged leaf-like branches fork into two unequal, bushy portions. The leaves are not actually leaves but stems that grow along the main stem (i.e., stolon) (Figure 13.2), which have numerous small bladders (Figure 13.3). In early spring, 3–15 yellow snap-dragon shaped flowers develop on emergent stalks. The stalks are supported by a floating pontoon comprised of four to ten leaves arranged like the spokes of a wheel (Figure 13.2). The pontoon leaves are 1.5 inches long and appear inflated.

Like other *Utricularia* species, swollen bladderwort lures its prey by secreting chemical attractants into its surroundings. They prey on a wide range of organism such as zooplankton (Robinson n.d.), insects (water fleas, worms, insect larvae), and some benthic animals such as arthropods and annelids. In addition, they can eat rotifers and protozoa from water soil (Clark n.d.).

Organisms are lured to the bladders with a sweet scent, and when trigger hairs on the bladder are brushed, a "trap door" opens, and a vacuum force pulls the prey inside to be digested. These traps are laterally inserted on the leaf segments (Figure 13.3). The dynamics of this plant for capturing small aquatic animals is quite similar to neuronal dynamics (Llorens et al. 2012). Glands pump water out of the trap, yielding a negative pressure difference between the plant and its surroundings. The trap door is set into a metastable state and opens quickly as extra pressure is generated by the displacement of a potential prey. As the door opens, the pressure difference sucks the animal into the trap (Figure 13.4). Bladderwort's ability to consume small organisms, in addition to absorbing nutrients directly from the water column, provides a competitive advantage over other species.

U. inflata can thrive due to its ability to reproduce. Swollen bladderwort reproduces by both vegetative methods and seed formation. Vegetatively, *U. inflata* reproduces by stem (stolon) fragmentation which can resprout and grow into new plants. Branches

DOI: 10.1201/9781003201106-13

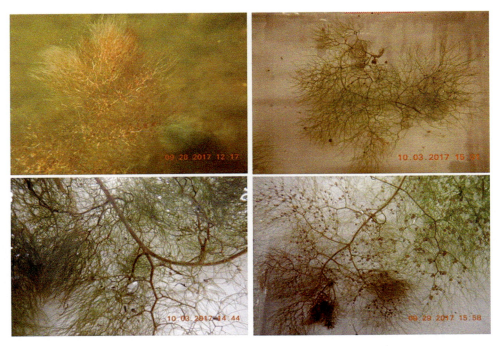

FIGURE 13.1   Swollen bladderwort (*Ultricularia Inflata*) at Indian Brook (*top left*) and samples collected from Monponsett Pond (*top right*), Ponkapog Pond (*bottom left*), and White Island Pond (*bottom right*). (Photos by Wai Hing Wong.)

FIGURE 13.2   Picture of swollen bladderwort stem (*left*), flowers (middle), and sterile fruits (right). (Left microscopic photo by Wai Hing Wong; middle photo by Alan Cressler; right photo modified from Kadono et al. (2019).)

may emerge along the stem and form new individuals when they separate from the original plant (Urban and Dwyer 2016). Seeds can also be produced, resulting in regrowth from seeds remaining in lake or pond sediment. The average flower number is 10–11 (Reinert and Godfrey 1962). However, the morphology of flowers may be different in *U. inflata* depending on their location. For example, the apex of

the spur is obtuse in *U. inflata* in Japan. On the other hand, the apex of the bracts in North American individuals is acute (Taylor 1989; Crow 2014) or subacute to obtuse (Reinert and Godfrey 1962). During the winter season, this species can be under ice and dormant in the frozen lakes of northern climates. Individuals grow under a wide range of water chemistry conditions and can be found in oligotrophic

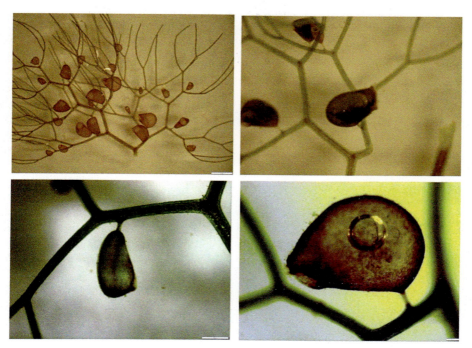

FIGURE 13.3   Microscopic photos of bladderworts of *U. inflata* (*top left:* 10×; *top right:* 25×; *bottom left:* 50×; *bottom right:* 60×. (Photos by Wai Hing Wong.)

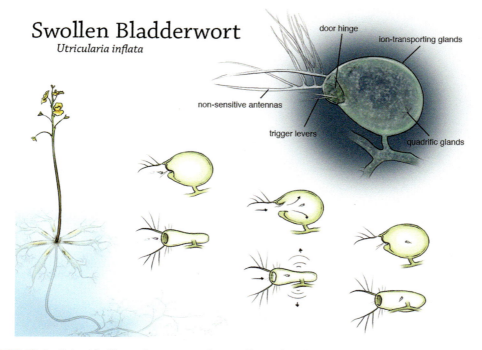

FIGURE 13.4   Scientific illustration on capturing small aquatic animals. (Photo by U-Bin Li.)

(low-nutrient), eutrophic (nutrient-rich), and acidic waters.

Swollen bladderwort prefers slow-moving waters, including lakes, ponds, and slow-moving rivers. In fact, swollen bladderwort is more susceptible to water movement and may spread downstream at a faster rate than common bladderwort (*Utricularia vulgaris*) and attached bladderwort species (Urban and Dwyer 2016). Despite its potential to grow well in shallow water, water movement can prevent the accumulation of *U. inflata* (Urban and Titus 2010).

In Massachusetts, there are other *Utricularia* species such as common bladderwort (*Utricularia vulgaris*) (Figure 13.5), Purple Bladderwort (*Utricularia pupurea*), Little floating bladderwort (*Utricularia radiata*), and Flatleaf Bladderwort (*Utricularia intermedia*). All bladderwort species are rootless and have carnivorous traps (Taylor 1989). Although these bladderwort plants are similar, they have their specific characteristics (MA DCR 2017) and some are as follows:

- Common bladderwort is the biggest bladderwort. Its bladders are large, and the tip of the plant has a tuft of branches with yellow flowers.
- Purple Bladderwort has clusters of branches with the bladders at the very tips of the branches with purple flowers.
- Little floating bladderwort's bladders only appear on underwater leaves. Underwater leaves' forks are more zigzag compared to the common bladderwort. This plan also has yellow flowers.
- Flatleaf Bladderwort has bladders on stems grown into the sediment.

FIGURE 13.5 Photos of common bladderwort (top panel, top right photo was taken under microscope 7×) and Purple Bladderwort (bottom panel). (Top panel photos by Wai Hing Wong; bottom panel photo by Michelle Kirby.)

NATIVE AND INVASIVE RANGE

Swollen bladderwort is native to southern and eastern North America, but it has been confirmed as an invasive species in Pennsylvania, Massachusetts, New York, New Jersey, Rhode Island, and the state of Washington (Urban and Dwyer 2016; Native Plant Trust 2021). This invasive species can dominate the waterbodies in a short time and displace other native species.

VECTORS OF SPREAD

Swollen bladderwort's northern spread is uncertain, but it is likely fragments were carried on boats, boat trailers, or waterfowl (Johnstone et al. 1985; Figuerola and Green 2002; but see Urban and Dwyer 2016). Once introduced, swollen bladderwort can grow by both sexual and asexual reproduction as described earlier. It may also spread by currents, boaters' activities, and waterfowls (Titus and Urban 2013). In addition, it has been documented that water movement and successful asexual reproduction have facilitated swollen bladderwort's downstream spread into Adirondack lakes connected by the Raquette River and the Middle Branch of the Moose River (Titus and Urban 2013).

In a greenhouse experiment comparing the vegetative propagation of swollen bladderwort and common bladderwort (*U. vulgaris*), it was found that swollen bladderwort and common bladderwort both produced potential propagules but exhibited differences in their asexual reproduction (Urban and Dwyer 2016). New common bladderwort branches grew significantly longer but swollen bladderwort fragments exhibited a greater number of new branches. Each new branch has the potential to develop into a new individual as the original stolon decays which may explain how swollen bladderwort is quickly establishing populations in newly colonized lakes (Urban and Dwyer 2016).

ECOLOGICAL IMPACTS

The existence of this harmful invasive species may cause multiple negative environmental impacts. For example, *U. inflata* may deplete nutrients from waterbodies, leading to the decline of native vegetation. Meanwhile, fish and other aquatic animals that rely on these vegetation could not survive. Furthermore, fish and other aquatic animals are also threatened by a low-oxygen environment created by large quantities of decaying swollen bladderwort.

Swollen bladderwort may negatively affect small aquatic animals, as it feeds on a wide range of organism. Therefore, its feeding habits may affect the populations of aquatic species and insects which can result in a decline in native wildlife populations.

Because of swollen bladderwort's strong ability to reproduce, it can cover the water surface creating dense mats. It can also form dense canopies underwater without reaching the water surface (top left panel of Figure 13.1). This phenomenon may increase light attenuation, causing a decline in biomass, even the death of isoetids (Urban et al. 2013). As a result, isoetids that work to oxidize and take up CO_2 from sediments, and therefore balance chemistry of the sediment and water, may be indirectly changed by swollen bladderwort. This may lead to a decline in the sediment's redox potential, more NH_4^+, and the release of mineral nutrients Fe into the water column, and accelerate the growth of algae and *U. inflata* (MA DCR 2007; Urban et al. 2009; Urban et al. 2013). This positive feedback loop, if continued, may further *U. inflata*'s impact on the environment.

ECONOMIC IMPACTS

Economically, massive swollen bladderwort can impede a series of human activities, such as fishing and water recreations, and could reduce the real estate values (MA DCR 2007). In addition, dense mats can limit the ability to navigate waters which interferes with recreational actives. Furthermore, the combination of low-water quality and denes mats can create conditions that may be dangerous for swimmers and recreational boaters (Scholl 2007).

POSITIVE EFFECTS

Although swollen bladderwort is an invasive species, there are a few positive effects of its expansion. A self-stabilizing ecosystem state could be formed in the water column by the proliferation of rootless swollen bladderwort. In addition, many types of wildlife such as turtles,

TABLE 13.1

Locations of Waters Infested by *Utricularia inflata* in Massachusetts

Site Number	Waterbody	First Year Reported	Longitude	Latitude	Report Agency	Report Source
1	Federal Pond	1992	−70.707816	41.876262	Sorrie 1992	Rhodora
2	First Pond	2006	−71.051194	42.475409	DCR	DCR Data from Lakes and Ponds, 2006
3	Long Pond	2008	−71.992134	42.356772	DCR	DCR 2008 (Data from National Heritage, DowCullina)
4	Noquochoke Lake	2008	−71.043280	41.647114	DCR	DCR 2008 (Data from National Heritage, DowCullina)
5	Lake Rico	2008	−70.996974	41.874723	DCR	DCR 2008 (Data from National Heritage, DowCullina)
6	Cooks Pond	2008	−70.665355	41.922381	DCR	DCR 2008 (Data from National Heritage, DowCullina)
7	Whitehall Reservoir	2008	−71.575363	42.229421	DCR	DCR 2008 (Data from National Heritage, DowCullina)
8	North Pond	2008	−71.555872	42.196528	DCR	DCR 2008 (Data from National Heritage, DowCullina)
9	Louisa Lake	2008	−71.523061	42.158817	DCR	DCR 2008 (Data from National Heritage, DowCullina)
10	Ponkapoag Pond	2017	−71.092968	42.192183	MassDEP	MassDEP Field Survey 10/2/17
11	White Island Pond	2017	−70.617531	41.809183	MassDEP	MassDEP Field Survey 9/28/17
12	Sampson Pond	2017	−70.752977	41.852064	MassDEP	MassDEP Field Survey 09/28/17
13	Monponsett Pond	2017	−70.845254	42.004596	MassDEP	MassDEP Field Survey 10/02/2017
14	Turner Pond	2017	−70.970963	41.683796	MassDEP	MassDEP Field Survey 9/28/17
15	Taunton River	2017	−70.942830	41.936230	MassDEP	MassDEP Field Survey 9/28/2017

salamanders, bullfrogs, and even potential prey use Bladderwort as their shelter (USDA n.d.). Further, swollen bladderwort could dominate an area, which may provide greater habitat for phytophilous littoral zone invertebrate, and thus enliven trophic interactions in the waterbodies (Urban et al. 2013).

MONITORING AND SIGHTINGS

The infested waterbody sightings were mainly collected by the Massachusetts Department of Environmental Protection (MassDEP) and Massachusetts Department of Conservation and Recreation (MA DCR) (Table 13.1). The standard

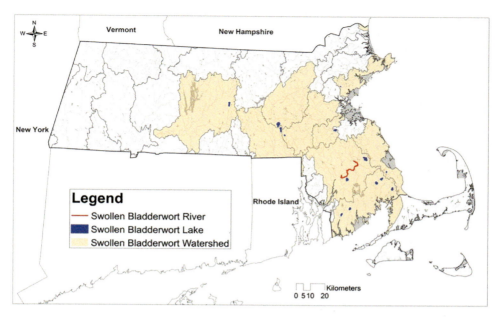

FIGURE 13.6 ArcGIS map depicting distribution of swollen bladderwort (*Utricularia inflata*) in Massachusetts.

operating procedures, including sampling methodology, species identification, quality control and assurance, are implemented by samples collected by MassDEP. Sites with swollen bladderwort were mapped as a data layer with ArcGIS® ArcMap 10.1 (ESRI, Redlands, California). All locations infested by swollen bladderwort are shown in Figure 13.6. The ArcGIS map can pinpoint and confirm swollen bladderwort infestations which can help to inform future decisions on processes and activities that can happen in these bodies of water, or regulations that involve efforts to limit the spread of this aquatic invasive species.

Swollen bladderwort is a very aggressive invasive species. Although it was first detected in Federal Pond in Carver and Plymouth of southeast Massachusetts in 1990 (Sorrie 1992), multiple sightings have been documented since then. In total, it has been detected in 15 waterbodies (Figure 13.6 and Table 13.1). The first sighting documented by MassDEP staff was found in the Charles River in 2001. This can be reviewed in MassDEP's Charles River Watershed 2002–2006 Water Quality Assessment Report. These waterbodies are distributed in the following ten major watersheds: Chicopee, Concord (SuAsCo), Blackstone, Charles River, Taunton, Boston Harbor, South Coastal, Buzzards Bay, North Coastal, and Cape Cod (Table 13.2).

MANAGING ACTIONS

Common management tools for swollen bladderwort include mechanical removal and hand pulling/harvesting, drawdown, benthic barriers, and herbicides application.

Mechanical removal and hand pulling/harvesting may be effective, but these methods may aid in the spread of fragmented swollen bladderwort into new waterbodies if cautions are not taken.

Drawdowns can be used for aquatic plant control by lowering water levels to achieve drying and prompting the freezing of plants. To reduce impacts to non-target plants and animals during the growing season, drawdowns in Massachusetts are normally conducted in fall and winter (Mattson et al. 2004). But mild season or higher level of groundwater seepage may lead to rapid reestablishment of extensive seed beds. Further, lowering the water level may result in a temporary impact to the water quality (e.g., reducing the dissolved oxygen concentration and water clarity level), and a temporary loss of desirable plants, fish, or aquatic animals.

TABLE 13.2

Infested Watersheds and Waterbodies

Watershed	Waterbody
Blackstone	North Pond
Boston Harbor: Neponset	Ponkapoag Pond
Buzzards Bay	Federal Pond
	Noquochoke Lake
	Sampson Pond
	Turner Pond
	White Island Pond
Charles	Louisa Lake
Chicopee	Long Pond
Concord (SuAsCo)	Whitehall Reservoir
North Coastal	First Pond
South Coastal	Cooks Pond
Taunton	Lake Rico
	Monponsett Pond
	Taunton River

Drawdowns are usually useful in dammed waterways, while *U. inflata* is mostly found in quieter ponds. Therefore, it is not the first choice of managing swollen bladderwort.

Benthic barriers are another control method proposed by Massachusetts authorities, which could be applied in small areas such as swimming beaches to control this invasive species (MA DCR 2007). The application of benthic barriers limit plant growth by a cover of clay, silt, sand, or gravel. However, there are some limitations on this method. For instance, it requires constant maintenance, and it causes damage to benthic organisms (Washington State Department of Ecology 2020). Almost all covered plants will be harmed, and the habitat of fish will be influenced. MassDEP endorses the issuance of a Negative Determination of Applicability when these projects are conducted in accordance with the management techniques approved by DEP and described in the DCR Guidance, Standard Operating Procedures for Hand Pulling of Aquatic Vegetation and Benthic Barriers to Control Aquatic Vegetation dated May 15, 2003 (Langley et al. 2004).

Meanwhile, herbicides contain active ingredients that are toxic to target plants, but it may also harm or kill non-target plants and organisms by using inappropriate herbicides or applying high-rate usage (Mattson et al. 2004). In Massachusetts, the following chemicals/herbicides have been used for treating swollen bladderwort and other Bladderwort species: AquaPro, Aquathol K, Captain, Captain XTR, Clearcast, Clipper, Copper Sulfate, ProcellaCOR, Sonar Genesis, and Tribune. More information about chemical control on swollen bladderwort and other 18 invasive species can be found in Appendix III. Before chemicals are applied in Massachusetts waters, it is required to obtain an aquatic nuisance species control license (BRP WM04 Form) from Massachusetts Department of Environmental Protection.

There are no known biological methods for swollen bladderwort control. In La Center Bottoms Park, Washington, some animals such as mallards, wood ducks, muskrats, and turtles eat bladderworts (USDA n.d.). Introducing these animals probably may effectively manage bladderworts, but other native species might also be eaten or impacted and these animals could also spread bladderworts.

SUCCESS STORY

The application of herbicides may be effective in impeding the spread of swollen bladderwort. In 1996, Sonar (fluridone) treatment was tried in

Lake Limerick in Mason County, WA, and it has been shown to be successful on controlling the swollen bladderwort population for around two years (Washington State Department of Ecology 2020). But fluridone can also damage or even kill non-target plants by disrupting carotenoid synthesis (SERA, Inc. 2008). Inappropriate usage of herbicides could cause negative impacts on water quality and human health.

CONCLUSION

As described in this chapter, swollen bladderwort as an invasive species can have a negative impact not only on the ecosystem, but also on economic factors such as the recreational use of Massachusetts waterbodies. Therefore, it is of great importance to continue to raise awareness of the impacts of swollen bladderworts to decrease spread, increase precautionary steps, and implement treatment methods.

REFERENCES

Clark, A. n.d. Bladderwort-Adaptations, uses, habitats, eats (Common Plant). *LotusMagus*. https://lotusmagus.com/bladderwort/

Crow, G.E. 2014. Lentibulariaceae Richard. Flora of North America, Provisional publication. Flora of North America Association. October 27, 2014. http://floranorthamerica.org/files/Lentibulariaceae%20provisional. pdf. (Accessed January 31, 2021).

Figuerola, J., and A.J. Green. 2002. Dispersal of aquatic organisms by waterbirds: A review of past research and priorities for future studies. *Freshwater Biology* 47(3): 483–494.

ITIS (Integrated Taxonomic Information System). 2021. www.itis.gov/servlet/SingleRpt/SingleRpt?search_topic=TSN&search_value=34443#null (Accessed January 29, 2021).

Johnstone, I.M., B.T. Coffey, and C. Howard-Williams. 1985. The role of recreational boat traffic in Interlake dispersal of macrophytes—A New-Zealand case-study. *Journal of Environmental Management* 20(3): 263–279.

Kadono, Y., T. Noda, K. Tsubota, K. Shutoh, and T. Shiga. 2019. The Identity of an Alien Utricularia (Lentibulariaceae) Naturalized in Japan. *Acta Phytotaxonomica et Geobotanica* 70(2): 129–134.

Langley, L., L. Rhodes, and M. Stroman. 2004. *Guidance for Aquatic Plant Management in Lakes and Ponds as It Relates to the Wetlands Protection Act.* Cambridge, MA: Massachusetts Department of Environmental Protection, p. 30.

Llorens, C., M. Argentina, Y. Bouret, P. Marmottant, and O. Vincent. 2012. A dynamical model for the Utricularia trap. *Journal of the Royal Society Interface* 9: 3129–3139.

MA DCR (Massachusetts Department of Conservation & Recreation). 2007. *Swollen bladderwort: An Exotic Aquatic Plant.* Cambridge, MA: MA DCR, p. 3.

MA DCR (Massachusetts Department of Conservation & Recreation). 2017. *A Guide to Aquatic Plants in Massachusetts.* Cambridge, MA: MA DCR, p. 32.

Mattson, M.D., P.J. Godfrey, R.A. Barletta, and A. Aiello. 2004. *Eutrophication and Aquatic Plant Management in Massachusetts: Final Generic Environmental Impact Report.* Cambridge, MA: Commonwealth of Massachusetts, Executive Office of Environmental Affairs Commonwealth of Massachusetts, p. 514.

Native Plant Trust. 2021. Characteristics of Utricularia inflate. *Native Plant Trust,* https://gobotany.nativeplanttrust.org/species/utricularia/inflata/ (Accessed January 29, 2021).

Reinert, G.W., and R.K. Godfrey. 1962. Reappraisal of *Utricularia inflata* and *U. radiata* (Lentibulariaceae). *American Journal of Botany* 49: 213–220.

Robinson, M. n.d. *swollen bladderwort: An Exotic Aquatic Plant Utricularia inflata.* Cambridge, MA: DCR.

Scholl, W. 2007. *Invasion Ecology: swollen bladderwort.* Washington, DC: Washington State Department of Ecology.

SERA, Inc. 2008. *Fluridone Human Health and Ecological Risk Assessment Final Report.* Syracuse Environmental Research Associates, Inc. www.fs.fed.us/foresthealth/pesticide/pdfs/0521002a_Fluridone.pdf

Sorrie, B.A. 1992. *Utricularia inflata Walter* (Lentibulariaceae) in Massachusetts. *Rhodora* 94: 391–392.

Taylor, P. 1989. *The Genus Utricularia: A Taxonomic Monograph. Kew Bulletin Additional Series XIV.* London: HMSO, p. 724.

Titus, J.E., and R.A. Urban. 2013 Invasion in progress: *Utricularia inflata* in Adirondack submersed macrophyte communities. *Journal of the Torrey Botanical Society* 140: 506–516.

United Stated Department of Agriculture (USDA). n.d. Warning! carnivorous plant! Natural resources conservative service, USA, www.nrcs.usda.gov/Internet/FSE_DOCUMENTS/nrcs144p2_033271.pdf (Accessed January 23, 2020).

Urban, R.A., and M.E. Dwyer. 2016. Asexual reproduction and its potential influence on the distribution of an invasive macrophyte. *Northeastern Naturalist* 23: 408–419, https://doi.org/10.1656/045.023.0308

Urban, R.A., and J.E. Titus. 2010. Exposure provides refuge from a rootless invasive macrophyte. *Aquatic Botany* 92: 265–272.

Urban R.A., J.E. Titus, and H.H. Hansen. 2013. Positive feedback favors invasion by a submersed freshwater plant. *Oecologia* 172: 515–523, https://doi.org/10.1007/s00442-012-2496-4

Urban, R.A., J.E. Titus, and W.X. Zhu. 2009. Shading by an invasive macrophyte has cascading effects on sediment chemistry. *Biological Invasions* 11: 265.

Washington State Department of Ecology. 2020. Swollen bladderwort. *Wendy Scholl Invasion Ecology,* http://depts.washington.edu/oldenlab/wordpress/wp-content/uploads/2013/03/Utricularia-inflata_Scholl_2007R.pdf. (Accessed January 22, 2020).

14 Water Chestnut (*Trapa natans*)

ABSTRACT

The earliest record of water chestnut (*Trapa natans*) was documented in Cambridge, MA, in 1877. Since then, sightings of water chestnut have increased to include 18 rivers and 96 lakes located within 17 of Massachusetts' 33 major watersheds. The purpose of this chapter is to present the location and spread of the invasive water chestnut and raise awareness of their ecologic and economic impacts. This process may aid in minimizing the unintentional transportation of water chestnuts between waterbodies. In addition, this chapter provides previous treatment methods that have been useful in controlling the abundance of water chestnut. These methods include physical management (e.g., mechanical harvesting, including pulling by hand, benthic barriers, and dredging), chemical treatments (e.g., fluridone and triclopyr), and biological agents (e.g., leaf beetle).

INTRODUCTION

Taxonomic Hierarchy (ITIS 2021)

Kingdom	Plantae
Subkingdom	Viridiplantae
Infrakingdom	Streptophyta
Superdivision	Embryophyta
Division	Tracheophyta
Subdivision	Spermatophytina
Class	Magnoliopsida
Superorder	Rosanae
Order	Myrtales
Family	Lythraceae
Genus	*Trapa* L.
Species	*Trapa natans* L.

BASIC BIOLOGY

As a floating annual macrophyte, water chestnut has alternate floating leaves with a shiny upper side and fine hairs on the underside (Mass. DCR 2015). Leaf blades are broadly rhomboid, triangular, or deltoid in shape, and can be up to 5 cm wide, with sharp, irregular, serrated margins. The petiole of each floating leaf is inflated like an air bladder to allow the formation of floating rosettes that can reach up to 30 cm in diameter (Crow and Hellquist 2000). Beneath the water surface, feathery, opposite submerged leaves attach to a stem 1–5 m long which connects to the rosettes. Submerged leaves are progressively replaced by fine, pinnately branched, leaf-like adventitious roots (Crow and Hellquist 2000; Mass. DCR 2015). Their slender roots also emerge at the lower stem, usually buried in the substrate, which anchors the plant in the water (Groth et al. 1996; Crow and Hellquist 2000). Flowers have been observed with four white petals emerging in axils of floating leaves from June to September (Hummel and Kiviat 2004; Chorak et al. 2019).

The fruit of water chestnut is a tetrahedral woody nut (or nutlet) with four sharp barbed spines (Crow and Hellquist 2000; Mass. DCR 2015). They are green and malleable when fresh, ultimately turning turn black and hard. Each is 2–4 cm wide and 6 g wet at maturity, and are produced from mid-July to September (Countryman 1978; Gleason and Cronqusit 1991). However, viable plants have been observed to come from nuts only ½ inch (1.27 cm) long (C. Boettner, personal communication). When the nuts mature, they fall off the plant and sink to the bottom of the waterbody (Naylor 2003). Nuts are then anchored by horns underneath the sediment of waterbodies throughout winter and start germination when the water temperature reaches above 8°C (Kurihara and Ikusima 1991), with an estimated germination rate of 86–87% (Kurihara and Ikusima 1991; Cozza et al. 1994). After germination, the remaining spent black casing of the nut may remain in the sediment for several decades or float to the surface. The parent plants decompose in late fall and winter.

DOI: 10.1201/9781003201106-14

After this, the sharp black spent seed casings can often be found along the shoreline of water-bodies, even in winter. Dormant seeds of water chestnut can survive up to 10–12 years, but the viability of seeds is quickly lost if seeds are dried out (Hummel and Kiviat 2004).

Each water chestnut seed can give rise to 10–15 rosettes, with each rosette producing up to 20 seeds (O'Neill 2006). Thus, each seed can, in turn, produce up to 300 seeds in the next generation that can then germinate within one of the following ten years. The density of the rosettes can reach up to 100 rosettes/m² when water chestnut is abundant (Pemberton 2002).

NATIVE RANGE

Water chestnut (*Trapa natans* L.), also known as European Water Chestnut, water caltrop, and bull nut, is native to temperate and tropical rivers, lakes, wetlands, and estuaries in Europe, Asia, and Africa (Hummel and Kiviat 2004; Chorak et al. 2019; Pfingsten et al. 2020). It tends to favor

aquatic environments with abundant nutrients, a pH range between 6.7 and 8.2, moderate alkalinity (12–128 mg/l of calcium carbonate), and soft substrate (Pemberton 2002; Pfingsten et al. 2020). *Trapa natans* L. was introduced to North America in 1874 and cultivated in Asa Gray's botanical garden at Harvard University in Cambridge in 1877 (Countryman 1978; Naylor 2003). Due to a series of detrimental effects on ecosystem services caused by its aggressive growth, water chestnut is now considered a noxious weed in the northeastern United States (Figure 14.1) (Hummel and Kiviat 2004; Chorak et al. 2019).

INVASIVE RANGE

Seeds of water chestnuts can disperse to new sites by using its barbed spines to attach to moving objects, including animals, human clothing, nets, motorboat trailer carpets, life jackets, boats, and other vehicles. Plants can also spread by fragmentation, such as severed rosettes growing new roots and floating downstream. However, humans

FIGURE 14.1   Water chestnut. (Photo by Wai Hing Wong.)

also aid in the spread of water chestnuts through introduction to water gardens and aquarium releases (Mass. DCR 2002; Hummel and Kiviat 2004; Pfingsten et al. 2020). Water chestnut has been reported in states, including Connecticut, Delaware, Maryland, Massachusetts, New Hampshire, New Jersey, New York, Pennsylvania, Vermont, Virginia, and Washington, DC (Pemberton 2002; EDDMapS 2019).

ECOLOGICAL IMPACT

Water chestnut can have negative impacts on ecosystems when they dominate waterbodies. Though sharp fluctuations in water level and swift water flow can prohibit colonization, when conditions are favorable, water chestnuts can cover 100% of the water surface by generating a dense bed, sometimes stacked two to three layers (Pemberton 2002). As a result, these beds can intercept 95% of sunlight and suppress other submerged and floating aquatic plants (Pemberton 2002; Hummel and Kiviat 2004). Compared to native species, water chestnut provides less nutrients as a food source for wildlife once it dominates habitats (Feldman 2001; Wu and Wu 2007). Interestingly, Strayer et al. (2003) found densities of macroinvertebrates were lower in native *Vallisneria* beds than in *Trapa* beds in the Hudson River in New York. However, they speculated that macroinvertebrates in *Trapa* beds may not be available to fish due to low oxygen concentrations. A reduction of dissolved oxygen under water chestnut beds is consistently observed in both laboratories and field settings. Furthermore, these dense beds can reach to hundreds of hectares in size and can induce states of hypoxia or even anoxia (Caraco and Cole 2002). This phenomenon is detrimental to many aquatic invertebrates and fish (Hummel and Findlay 2006; Kornijów et al. 2010). Due to these conditions, water chestnut beds are often sites of low species richness, as few species can tolerate low-oxygen conditions found within them (Feldman 2001). However, a study by Kornijów et al. (2010), which took place in a 120-ha bed of water chestnut in the Hudson River, found high species richness, biodiversity, fish production, and invertebrate production, despite highly hypoxic conditions underneath the bed (Kornijów et al. 2010). A possible explanation that was put forth by Kornijów et al. (2010) stated that the species found inhabiting the bed possessed one or more behavioral or physiological mechanisms that allowed them to survive in hypoxic conditions. Even so, a survey of the fauna under the bed found that the species were not known to possess these traits. Therefore, more research is needed to determine the specific conditions under which water chestnut beds prove either beneficial or harmful to their aquatic environments.

ECONOMIC IMPACT

The infestation of water chestnut has considerable impacts on commercial and recreational water uses (Hummel and Kiviat 2004). In the Chesapeake Bay and other regions, dense water chestnut beds have blocked waterways and impeded activities, including boating and swimming (Pemberton 2002; Wu andWu 2007). Injuries to swimmers and beach visitors caused by the hard spines found on the outside of the nuts have also been reported (Countryman 1978; Giddy 2003). Furthermore, the control and removal of water chestnut can be costly. For example, over 14 million dollars was spent on Lake Champlain, Vermont, from 1982 to 2020 in controlling the spread of water chestnut with some success in the reduction of mechanical harvesting operations to the southern region, where mats continue to present challenges for the long-term ability to eradicate the species. Hand pulling and monitoring operations also continue throughout most of the initial locations where the species was found (K. Jensen, personal communication; Mass. DCR 2002). At the same time, other economic impacts occur due to lost recreational activities such as fishing and boating which cannot be ignored. To summarize, the main economic impacts associated with water chestnut are due to the cost of mechanical and chemical treatments, although these costs are often low if infestations are caught and treated in early stages of water chestnut establishment (Naylor 2003).

POSITIVE ASPECTS

Although recognized as an invasive species, water chestnut exhibits a few positive effects on ecosystems. Water chestnut is capable of phytoremediation by accumulating excess nitrogen and heavy metals in the water, which could present a solution for the treatment of municipal wastewater (Kumar and Chopra 2018). However, this function can only be performed effectively by harvesting the plants before decomposition.

Another potential wastewater treatment solution comes from using the husks of water chestnut plants to create an activated carbon solution that removes chromium compounds from water (Liu et al. 2010). This method entails using dried plant material and could work well with several physical management methods discussed later in this report. In addition, the presence of water chestnut beds also provides shelter and foraging grounds to juvenile fish, insects, and water birds (Hummel and Kiviat 2004). However, these positive effects are outweighed by the overall ecological, economic, and recreational damages caused by water chestnut.

MONITORING AND SIGHTINGS

Data for this project were collected by staff from MassDEP, the Massachusetts Department of Conservation and Recreation (DCR), the US Geological Survey (USGS), the US Fish and Wildlife Service, and other agencies through field visits and individual reports (Table 14.1). Sites with water chestnut were mapped as a data layer with ArcGIS® ArcMap 10.6 (ESRI, Redlands, California).

Since the early observation in Middlesex County, Massachusetts, in 1874, water chestnut has been reported at multiple waterbodies, including both rivers and lakes (Figure 14.2 and Table 14.1). More detailed information about the geographic coordinates of sites, the agencies that reported the sightings, and the specific report sources can be found in Table 14.1. Based on 395 records of water chestnut infestations, water chestnut was detected in 18 rivers and 96 lakes in Massachusetts, totaling 17 out of 33 DWM-WPP watersheds (modified by MassDEP from the MassGIS "Major Basins" layer). Infested watersheds were mainly distributed in the western, central, and northeastern regions of Massachusetts. Watersheds with multiple infested waterbodies or repeatedly reported waterbodies include Blackstone, Boston Harbor, Charles, Connecticut, Concord (SuAsCo),

Housatonic, Mystic, and Nashua. As shown in Figure 14.2 and Tables 14.2 and 14.3, the risk of potential spread from these watersheds/waterbodies to others is quite high. Within the Connecticut River watershed, there are a few recent sightings that may not be included in this chapter.[1]

The distribution pattern of water chestnut in Massachusetts demonstrates the persistence of water chestnut infestations and its ability to spread to new locations. The presence of water chestnut in Sudbury River has been documented from 1879 to 2017 (OARS 2017) and continues to present, at total of 142 years. In terms of water salinity, although Vuorela and Aalto (1982) indicate that seeds fail to germinate when NaCl concentrations exceed 0.1%, water chestnut has been found in both freshwater and brackish water systems like the Mystic River Estuary, demonstrating the plants' highly adaptive nature.

MANAGEMENT ACTIONS

Water chestnut is listed as a prohibited plant and this species is banned from the trade in Massachusetts. The sightings as mapped in this chapter serve as an aid in creating awareness on the presence of water chestnut in Massachusetts, which may be helpful in minimizing unintentional transportation between different waterbodies and alerting local constituencies that management should be undertaken. Waterbodies infested with invasive species such as water chestnuts are reported in the biennial Massachusetts Water quality assessment and 303(d) list water quality. This integrated report is to fulfil the requirement of the federal Clean Water Act. In addition, projects that take place in infested waterbodies/watersheds are subject to the jurisdiction of 401 Water Quality Certifications with conditions to prevent potential spread of invasive species during the project. Furthermore, water quality monitoring gear and watercraft that operate in infested watersheds must be decontaminated before they are used in other watersheds to prevent the potential spread of invasive species such as the water chestnut.

In addition, water chestnut is an annual plant that begins to flower and form seeds in mid-July

TABLE 14.1

Latitude and Longitude for Each Waterbody with Water Chestnut (*Trapa natans*)

Waterbody	Agency	Report Source	N Latitude	W Longitude	First Year Reported
Ashfield Pond	USGS	USGS 6/23/2017 http://nas.er.usgs.gov/default.aspx	42.5313	−72.803	2007
Ashley Pond	USGS	USGS 6/27/2017 http://nas.er.usgs.gov/default.aspx	42.17	−72.6605	2013
Assabet River	MassDEP	SuAsCo Watershed 2001 Water Quality Assessment Report 82-AC-1 CN 92.0, August 2005	42.4	−71.51	2005
Bachelor Brook	USGS	USGS 6/30/2017 http://nas.er.usgs.gov/default.aspx	42.27	−72.595	2013
Bare Hill Pond	USGS	USGS 6/28/2017 http://nas.er.usgs.gov/default.aspx	42.4972	−71.5953	1999
Barrowsville Pond	USGS	USGS 6/30/2017 http://nas.er.usgs.gov/default.aspx	41.95269	−71.2054	2008
Bartlett Pond	College of the Holy Cross	Robert Bertin	42.318598	−71.61766	2009
Barton Cove	MassDEP	USGS 6/27/2017 http://nas.er.usgs.gov/default.aspx	42.6051	−72.5414	2009
Beaver Brook	MassDEP	Charles River Watershed 2002–2006 Water Quality Assessment Report 72-AC-4 CN136.5, April 2008			2008
Blacks Nook	MassDEP	Mystic River Watershed And Coastal Drainage Area 2004–2008 Water Quality Assessment Report 71-AC-2 CN170.2, March 2010			2010
Blackstone River	MassDEP	08-A035–05	42.098719	−71.622314	2008
Carding Mill Pond	MassDEP	SuAsCo Watershed 2001 Water Quality Assessment Report 82-AC-1 CN 92.0, August 2005			2001
Charles River	USGS	USGS 6/30/2017 http://nas.er.usgs.gov/default.aspx	42.360573	−71.181622	1994
Cheshire Reservoir, Middle Basin	DCR	DCR Data from ACT, GeoSyntec EM-ACT W-A. Madden FEW			
Cheshire Reservoir, North Basin	DCR	DCR Data from ACT, GeoSyntec EM-ACT W-A. Madden FEW			
Cheshire Reservoir, South Basin	DCR	DCR Data from ACT, GeoSyntec EM-ACT W-A. Madden FEW			
Chicopee Falls Impoundment	USGS	USGS 6/27/2017 http://nas.er.usgs.gov/default.aspx	42.1604	−72.5678	2007
Clamshell Pond	College of the Holy Cross	Robert Bertin	42.39934	−71.678197	2008
Clark Pond	MassDEP	Neponset River Watershed 2004 Water Quality Assessment Report 73-AC-1 CN170.4, February 2010			2010

(Continued)

**TABLE 14.1
(Continued)**

Waterbody	Agency	Report Source	N Latitude	W Longitude	First Year Reported
Clarks Pond	USGS	USGS 6/30/2017 http://nas.er.usgs.gov/default.aspx	42.86065	−70.92521	2007
Coes Pond	USGS	USGS 6/28/2017 http://nas.er.usgs.gov/default.aspx	42.250027	−71.834364	2001
Coes Reservoir	College of the Holy Cross	Robert Bertin	42.253967	−71.841584	1995
Concord River	USGS	USGS 6/28/2017 http://nas.er.usgs.gov/default.aspx	42.569644	−71.281614	1970
Connecticut River	USGS	USGS 6/27/2017 http://nas.er.usgs.gov/default.aspx	42.212592	−72.602312	2001
Cusky Pond	College of the Holy Cross	Robert Bertin	42.323429	−72.091626	2009
Danks Pond	USGS	USGS 6/30/2017 http://nas.er.usgs.gov/default.aspx	42.2873	−72.6416	2013
Ellis Pond	MassDEP	Neponset River Watershed 2004 Water Quality Assessment Report 73-AC-1 CN170.4, February 2010			2005
Exposition Grounds Pond	USGS	USGS 6/27/2017 http://nas.er.usgs.gov/default.aspx	42.0936	−72.6107	2013
Fisherville Pond	DCR	Joy Trahan-Liptak	42.184408	−71.692178	2015
Fisk Pond	USGS	USGS 6/28/2017 http://nas.er.usgs.gov/default.aspx	42.2819	−71.37107	2004
Flint Pond	USGS	USGS 6/28/2017 http://nas.er.usgs.gov/default.aspx	42.67278	−71.4327	2015
Forge Pond	USGS	USGS 6/28/2017 http://nas.er.usgs.gov/default.aspx	42.273	−72.468	2002
Franklin Reservoir Northeast	USGS	USGS 6/30/2017 http://nas.er.usgs.gov/default.aspx	42.09551	−71.38076	2003
Fresh Pond	USGS	USGS 6/30/2017 http://nas.er.usgs.gov/default.aspx	42.384263	−71.148942	1879
Great Meadows Pond #3	MassDEP	SuAsCo Watershed 2001 Water Quality Assessment Report 82-AC-1 CN 92.0, August 2005			2005
Great Pond	USGS	USGS 6/27/2017 http://nas.er.usgs.gov/default.aspx	42.3938	−72.6063	2013
Greenville Pond	College of the Holy Cross	Robert Bertin	42.205981	−71.922721	2007
Grist Mill Pond	MassDEP	SuAsCo Watershed 2001 Water Quality Assessment Report 82-AC-1 CN 92.0, August 2005			2005
Hardys Pond	MassDEP	Charles River Watershed 2002–2006 Water Quality Assessment Report 72-AC-4 CN136.5, April 2008			2008
Heard Pond	MassDEP	SuAsCo Watershed 2001 Water Quality Assessment Report 82-AC-1 CN 92.0, August 2005			2005

Name	Agency	Reference	Latitude	Longitude	Year
Heart Pond	MassDEP	Email communication between David Wong and Matthew Reardon on 7/12/2017	42.565626	−71.391319	2017
Hockanum Road Pond	USGS	USGS 6/27/2017 http://nas.er.usgs.gov/default.aspx	42.2986	−72.6149	2012
Housatonic River	MassDEP	Housatonic River Watershed 2002 Water Quality Assessment Report 21-AC-4 CN141.5 September 2007	42.394038	−73.240401	2002
Hubbard Brook	MassDEP	Housatonic River Watershed 2002 Water Quality Assessment Report 21-AC-4 CN141.5, September 2007			2007
Hulberts Pond	USGS	USGS 6/27/2017 http://nas.er.usgs.gov/default.aspx	42.2917	−72.644	2003
Ingraham Brook Pond	MassDEP	Connecticut River Basin 2003 Water Quality Assessment Report 34-AC-2 CN 105.5, October 2008			2008
Kingfisher Pond	USGS	USGS 6/28/2017 http://nas.er.usgs.gov/default.aspx	42.1084	−71.3212	2013
Lackey Pond	College of the Holy Cross	Robert Bertin	42.096707	−71.689012	2008
Lake Attitash	DCR	DCR Data from Lakes and Ponds, 2007			2007
Lake Bray	USGS	USGS 6/27/2017 http://nas.er.usgs.gov/default.aspx	42.2666	−72.6164	2013
Lake Cochituate	DCR	DCR Data from Lakes and Ponds, 2005			2005
Lake Quinsigamond	MassDEP	Email communication between David Wong and Barbara Kickham on June 23, 2017	42.296228	−71.754669	2017
Lake Ripple	College of the Holy Cross	Robert Bertin	42.213605	−71.697077	2008
Lake Warner	USGS	USGS 6/27/2017 http://nas.er.usgs.gov/default.aspx	42.3882	−72.5764	2002
Laurel Lake	USGS	USGS 6/23/2017 http://nas.er.usgs.gov/default.aspx	42.325555	−73.269367	2011
Linwood Pond	DCR	Joy Trahan-Liptak	42.101497	−71.651349	2016
Little Pond	USGS	USGS 6/30/2017 http://nas.er.usgs.gov/default.aspx	42.40001	−71.158834	2001
Log Pond Cove	USGS	USGS 6/27/2017 http://nas.er.usgs.gov/default.aspx	42.2148	−72.6117	2005
Long Pond	USGS	USGS 6/30/2017 http://nas.er.usgs.gov/default.aspx	42.598461	−71.251831	2016
Lower Mill Pond	MassDEP	USGS 6/27/2017 http://nas.er.usgs.gov/default.aspx	42.2737	−72.6573	2007
Lower Pond	USGS	USGS 6/27/2017 http://nas.er.usgs.gov/default.aspx	42.2546	−72.5723	2005
Lower Van Horn Park Pond	USGS	USGS 6/27/2017 http://nas.er.usgs.gov/default.aspx	42.1252	−72.5967	2004
Meadow Brook Pond	DCR	Joy Trahan-Liptak	42.078539	−71.592303	2009
Meadow Pond	MassDEP	SuAsCo Watershed 2001 Water Quality Assessment Report 82-AC-1 CN 92.0, August 2005			2005

(Continued)

TABLE 14.1 (Continued)

Waterbody	Agency	Report Source	N Latitude	W Longitude	First Year Reported
Mill River	MassDEP	Parker River Watershed And Coastal Drainage Area 2004–2008 Water Quality Assessment Report 91-AC-3 CN 173.0, March 2010			2010
Mumford River	DCR	Joy Trahan-Liptak	42.109325	−71.665758	2016
Mystic River	MassDEP	09-A012-05	42.417852	−71.118136	2009
Nashawannuck Pond	USGS	USGS 6/27/2017 http://nas.er.usgs.gov/default.aspx	42.2617	−72.6652	2007
Nashua River	MassDEP	08-G027-03	42.66415	−71.577006	2008
North Great Meadows	MassDEP	USGS 6/28/2017 http://nas.er.usgs.gov/default.aspx	42.479261	−71.32867	1983
North Great Meadows	MassDEP	SuAsCo Watershed 2001 Water Quality Assessment Report 82-AC-1 CN 92.0, August 2005			1983
Nutting Lake, East Basin	MassDEP	SuAsCo Watershed 2001 Water Quality Assessment Report 82-AC-1 CN 92.0, August 2005			2005
Onota Lake	USGS	USGS 6/23/2017 http://nas.er.usgs.gov/default.aspx	42.469691	−73.281212	2003
Oxbow	USGS	USGS 6/27/2017 http://nas.er.usgs.gov/default.aspx	42.2877	−72.6375	2004
Oxbow Cutoff	USGS	USGS 6/27/2017 http://nas.er.usgs.gov/default.aspx	42.295	−72.6218	2003
Pepperell Pond	USGS	USGS 6/28/2017 http://nas.er.usgs.gov/default.aspx	42.6547	−71.58197	2000
Pequot Pond	USGS	USGS 6/24/2017 http://nas.er.usgs.gov/default.aspx	42.1888	−72.6957	2007
Pontoosac Lake	USGS	USGS 6/23/2017 http://nas.er.usgs.gov/default.aspx	42.495897	−73.248425	2003
Purgatory Cove	DCR	Joy Trahan-Liptak	42.358386	−71.247771	2012
Rice City Pond	DCR	Joy Trahan-Liptak	42.100003	−71.622465	2011
River Meadow Brook	MassDEP	SuAsCo Watershed 2001 Water Quality Assessment Report 82-AC-1 CN 92.0, August 2005			2005
Riverdale Impoundment	DCR	Joy Trahan-Liptak	42.141777	−71.639295	2015
Rubber Thread Pond	USGS	USGS 6/27/2017 http://nas.er.usgs.gov/default.aspx	42.2647	−72.671	2013
Russell Cove	USGS	USGS 6/27/2017 http://nas.er.usgs.gov/default.aspx	42.2875	−72.604	2002
Russell Millpond	MassDEP	SuAsCo Watershed 2001 Water Quality Assessment Report 82-AC-1 CN 92.0, August 2005			2005

Location	Agency	Source	Latitude	Longitude	Year
Saxonville Pond	USGS	USGS 6/28/2017 http://nas.er.usgs.gov/default.aspx	42.31673	−71.41685	2015
Shaker Mill Pond	DCR	DCR Data from National Heritage, 2005			2005
Shirley Street Pond	MassDEP	Email from David Wong 6/28/2016	42.29933	−71.75352	2016
Silver Lake	DCR	Joy Trahan-Liptak	42.195614	−71.655275	2016
Spectacle Pond	DCR	DCR Data from ACT, GeoSyntec B. Hartzel, 2007			2007
Stearns Mill Pond	MassDEP	SuAsCo Watershed 2001 Water Quality Assessment Report 82-AC-1 CN 92.0, August 2005			2005
Stony Brook	MassDEP	Connecticut River Basin 2003 Water Quality Assessment Report 34-AC-2 CN 105.5, October 2008			2008
Sudbury Reservoir	College of the Holy Cross	Robert Bertin	42.321064	−71.510778	2006
Sudbury River	USGS	USGS 6/28/2017 http://nas.er.usgs.gov/default.aspx	42.46509	−71.35839	1879
Unnamed Pond	USGS	USGS 6/27/2017 http://nas.er.usgs.gov/default.aspx	42.2088	−72.6463	2013
Unnamed Pond	USGS	USGS 6/27/2017 http://nas.er.usgs.gov/default.aspx	42.2238	−72.6424	2006
Unnamed Pond	USGS	USGS 6/27/2017 http://nas.er.usgs.gov/default.aspx	42.2629	−72.6118	2013
Unnamed Pond	USGS	USGS 6/28/2017 http://nas.er.usgs.gov/default.aspx	42.2136	−72.5212	2013
Unnamed Pond	USGS	USGS 6/23/2017 http://nas.er.usgs.gov/default.aspx	42.18486	−72.7474	2016
Unnamed Pond	USGS	USGS 6/27/2017 http://nas.er.usgs.gov/default.aspx	42.2274	−72.6107	2003
Unnamed Pond	USGS	USGS 6/27/2017 http://nas.er.usgs.gov/default.aspx	42.2896	−72.6206	2002
Unnamed Pond	USGS	USGS 6/27/2017 http://nas.er.usgs.gov/default.aspx	42.3523	−72.6665	2013
Unnamed Pond	USGS	USGS 6/27/2017 http://nas.er.usgs.gov/default.aspx	42.2646	−72.6749	2013
Unnamed Pond	USGS	USGS 6/27/2017 http://nas.er.usgs.gov/default.aspx	42.2693	−72.5752	2013
Unnamed Pond	USGS	USGS 6/30/2017 http://nas.er.usgs.gov/default.aspx	42.2095	−72.6017	2003
Unnamed Pond	USGS	USGS 6/30/2017 http://nas.er.usgs.gov/default.aspx	42.2408	−72.5937	2002
Upper Pond	USGS	USGS 6/27/2017 http://nas.er.usgs.gov/default.aspx	42.258095	−72.567401	2001
Warners Pond	MassDEP	SuAsCo Watershed 2001 Water Quality Assessment Report 82-AC-1 CN 92.0, August 2005			2005
West River	MassDEP	09-G010-05	42.092409	−71.602293	2009
Wheeler Pond	DCR	DCR Data from Lakes and Ponds, 2006			2006
Whiting Street Reservoir	USGS	USGS 6/27/2017 http://nas.er.usgs.gov/default.aspx	42.2396	−72.638	2013

(Continued)

TABLE 14.1
(Continued)

Waterbody	Agency	Report Source	N Latitude	W Longitude	First Year Reported
Whitins Pond	MassDEP	Email communication between David Wong and Therese Beaudoin on July 7, 2017	42.119714	−71.688739	2017
Wilton Brook	MassDEP	Connecticut River Basin 2003 Water Quality Assessment Report 34-AC-2 CN 105.5, October 2008			2008
Woods Pond	DCR	DCR Data from MA Department of Fish and Game			

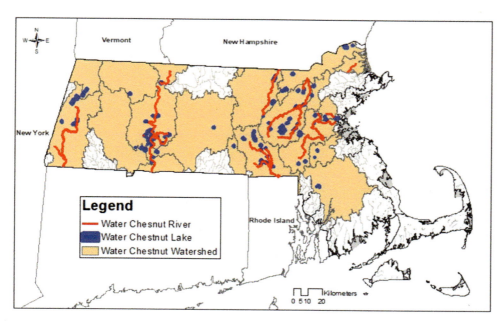

FIGURE 14.2   Distribution map depicting water chestnut (*Trapa natans*) in Massachusetts.

TABLE 14.2

Watersheds Infested with Water Chestnut

Blackstone River

Boston Harbor: Mystic River

Boston Harbor: Neponset River

Charles River

Chicopee River

Concord River (SuAsCo)

Connecticut River

Deerfield River

French River

Housatonic River

Hudson: Hoosic River

Merrimack River

Nashua River

Parker River

Shawsheen River

Taunton River

Westfield River

through fall with each seed producing 10–15 rosettes. Therefore, to manage it more effectively, physical removal or chemical treatments should be performed before seeds develop. This will reduce the number of seeds available for recolonization during the following growing season (Chorak et al. 2019). For best results, most management actions should begin before mid-July, regarding both the four-horned and two-horned variants of water chestnut (Rybicki

TABLE 14.3

Towns and Waterbodies

Town	Waterbody
Amesbury	Clarks Pond
	Lake Attitash
Ashfield	Ashfield Pond
Belmont	Little Pond
Berlin	Wheeler Pond
Billerica	Concord River
	Nutting Lake, East Basin
Boston	Charles River
	Mystic River
Cambridge	Blacks Nook
	Fresh Pond
Carlisle	Meadow Pond
Chelmsford	Heart Pond
	River Meadow Brook
	Russell Millpond
Cheshire	Cheshire Reservoir, Middle Basin
	Cheshire Reservoir, North Basin
Chicopee	Chicopee Falls Impoundment
Clinton	Clamshell Pond
Concord	Concord River
	Great Meadows Pond #3
	North Great Meadows
	Sudbury River
	Warners Pond
Deerfield	Connecticut River
Dover	Charles River
Easthampton	Lower Mill Pond
	Nashawannuck Pond
	Rubber Thread Pond
	Wilton Brook
Framingham	Saxonville Pond
Franklin	Franklin Reservoir Northeast
Gill	Barton Cove
Grafton	Fisherville Pond
	Lake Ripple
	Silver Lake
Granby	Forge Pond
	Ingraham Brook Pond
	Stony Brook
Groton	Nashua River
Hadley	Lake Warner
	Russell Cove

Town	Waterbody
Harvard	Bare Hill Pond
Hatfield	Great Pond
Holyoke	Ashley Pond
	Connecticut River
	Lake Bray
	Log Pond Cove
	Whiting Street Reservoir
Lanesborough	Cheshire Reservoir, South Basin
	Pontoosac Lake
Lee	Laurel Lake
Leicester	Greenville Pond
Lenox	Housatonic River
	Woods Pond
Littleton	Spectacle Pond
Maynard	Assabet River
Medfield	Charles River
Medford	Mystic River
Mendon	Meadow Brook Pond
Montague	Connecticut River
Natick	Fisk Pond
	Lake Cochituate
New Braintree	Cusky Pond
Newton	Charles River
Norfolk	Kingfisher Pond
Northampton	Danks Pond
	Hockanum Road Pond
	Hulberts Pond
	Oxbow
	Oxbow Cutoff
	Bartlett Pond
Northbridge	Blackstone River
	Linwood Pond
	Riverdale Impoundment
	Whitins Pond
Norton	Barrowsville Pond
Norwood	Ellis Pond
Pepperell	Nashua River
	Pepperell Pond
Pittsfield	Onota Lake
Rowley	Mill River
Sheffield	Hubbard Brook
Shirley	Nashua River
Shrewsbury	Lake Quinsigamond
	Shirley Street Pond

(Continued)

TABLE 14.3
(Continued)

Town	Waterbody
South Hadley	Bachelor Brook
	Connecticut River
	Lower Pond
	Upper Pond
Southborough	Sudbury Reservoir
Springfield	Lower Van Horn Park Pond
Stow	Assabet River
Sudbury	Carding Mill Pond
	Grist Mill Pond
	Stearns Mill Pond
Tewksbury	Long Pond
Tyngsborough	Flint Pond
Upton	West River
Uxbridge	Lackey Pond
	Mumford River
	Lackey Pond
	Rice City Pond
Walpole	Clark Pond
Waltham	Beaver Brook
	Hardys Pond
	Purgatory Cove
Wayland	Heard Pond
	Sudbury River
West Springfield	Exposition Grounds Pond
West Stockbridge	Shaker Mill Pond
Westfield	Pequot Pond
Westford	Forge Pond
Worcester	Coes Pond
	Coes Reservoir

2019) and followed up with checks for regrowth and new emergence until late August.

PHYSICAL CONTROL

Hand pulling and mechanical harvesting are commonly applied methods to remove invasive populations of water chestnut (Mass. DCR 2002). Hand pulling is highly selective and well-suited for shallow waterbodies and small-scale infestations or as follow-up once an infestation is reduced using other modes of management (Pearson 2021). Hand removal works especially well when plants are harvested "early and often" starting in early June, and repeated every few weeks to harvest plants when they are smaller and easier to handle (see the description of the Connecticut River watershed hand pulling protocols provided subsequently). A benefit of hand harvesting is that it is feasible to remove the entire plant, including the seed from which it grew. While the high cost of training and labor can be concerning, previous or ongoing control programs in the Potomac River, Lake Champlain, Connecticut River, and Chesapeake Bay utilize(d) less expensive volunteer programs. It is also useful to mention the utility of long handled rakes, as they may be useful in waterbodies where hand pulling may not be feasible and uproot plants (Rybicki 2019). Furthermore, in infested sites with dense and extensive water chestnut beds, mechanical harvesting may serve as an effective approach even though the cost and potential impacts on aquatic fauna may be higher (OARS 2017). Mechanical harvesting utilized by the Connecticut River Watershed Invasive Plant Control Initiative in Log Pond Cove, cost $80,000 in 1999 and $51,000 in 2000 (Massachusetts Aquatic Invasive Species Working Group 2002). Traditional mechanical harvesting entails cutting the plants about 18 inches below the water surface and collecting and disposing of the cut portion. Because plants can regrow and may have time to produce seeds, careful timing of this method, or implementing a second later cut is necessary. Timing can get undermined by low water levels (OARS 2017). However, when successful, mechanical harvesting can reduce a population to a level that can later be managed by hand pulling (OARS 2017).

Regardless of the methods employed, harvesting before the fruits are mature and potentially dropping during the harvesting activity, and harvesting the entire plant, stems, and roots will ensure a successful operation. Care must be taken to ensure the removal of harvested plant parts so that no damaged or fragmented pieces of the plant can continue to grow, fruit, and spread. Once harvested, all plant parts must be disposed of either by composting at an upland site or incinerated (Hummel and Kiviat 2004; Rybicki 2019). Even if a population is thought to be eliminated, due to the unpredictable nature of the long-lived seedbank (seeds that have dropped from previous years), annual monitoring must persist to prevent a resurgence (OARS 2017; Pearson 2021).

Chorak et al. (2019) found that there is a two-horn variety of water chestnut in the Potomac River, Virginia, and suggests that this morphologically similar species may exhibit a different pattern of development. Even though the two-horn water chestnut has not been reported in Massachusetts, it is recommended that an accurate identification of plants should be completed prior to any treatment.

Other physical control methods, including hydro raking, drawdown, dredging, and benthic barriers, can also be used for certain situations but have their own limitations (OARS 2017). For example, dredging can be effective at removing the nuts (or nutlets) from water chestnut that have fallen into the sediment at the bottom of waterbodies, as was the case at Hills Pond in Arlington, MA (Furgo East, Inc. 1996; Mattson et al. 2004). Additionally, since the barbed nuts of water chestnut can stick onto plumage or fur, fencing can be installed around the perimeter of infested waterbodies to keep mammals out and minimize spread (Rybicki 2019); however, waterfowl could still move seeds if they can fly in and out.

BIOLOGICAL CONTROL

Researchers have been looking for biological control agents for water chestnut. An efficient biological control agent should be host-specific to decimate or severely damage its target plant. It is suggested that the leaf beetle *Galerucella birmanica*, originally imported from China, can cause complete defoliation of whole populations of water chestnut and, therefore, is a promising biological control agent of *T. natans* (Ding et al. 2006a, b). *Galerucella birmanica* is only restricted to a quarantine facility (Bernd Blossey, personal communication) and has not been approved for release by the US Department of Agriculture (USDA) or by individual states as a Plant Protection and Quarantine (PPQ) 526 permit from the USDA. Such a permit is required to transport biological control agents across state lines and for release into the environment (OARS 2017). In addition, several other species of insects have potential as possible biological control agents, such as two species of *Nanophyes* weevil and an Italian species of weevil, *Bagous rufimanus* (Naylor 2003).

CHEMICAL CONTROL

Chemical treatment with herbicides is often coupled with physical removal of water chestnut to improve the efficiency of control. 2,4-Dicholorophenoxy acetic acid (2,4-D) and fluridone are currently recognized as being able to control water chestnut, with 2,4-D leading to consistent control and fluridone resulting in partial control of water chestnut according to the Eutrophication and Aquatic Plant Management in Massachusetts Final Generic Environmental Impact Report (Mattson et al. 2004). 2,4-D, a herbicide registered by the EPA for use in aquatic environments since 1959, was successfully used in the Mohawk River, the Hudson River, and Lake Champlain to control water chestnut (Countryman 1978). However, due to toxicity and concerns about migration into groundwater, the use of 2,4-D has been banned in Zone II wellhead protection areas (Aquatic Control Technology 2015). Common alternatives include triclopyr, imazapyr, glyphosate, and the imidazole-based herbicide Clearcast™ (Hummel and Kiviat 2004; OARS 2017). More information about chemical control on water chestnut and other 18 invasive species can be found in Appendix III. Any chemical used for aquatic invasive species control requires a WM04 Chemical Application License issued by the Massachusetts Department of Environmental Protection, as well as an Order of Conditions from the local town's Conservation Commission. Previous control programs that utilized chemical control often involve the monitoring of water quality and nutrient levels to assess the impacts on general water quality conditions from herbicide applications (Harman et al. 2012). Aquatic herbicides must be registered with the EPA (under the Federal Insecticide, Fungicide, and Rodenticide Act) and the Massachusetts Department of Agricultural Resources and approved for legal use in Massachusetts.

SUCCESS STORY

CONNECTICUT RIVER WATERSHED

Water chestnut has been a problem in the Connecticut River watershed since at least the early 1940s when an infestation was documented and management activities began at Lower Pond on Mt. Holyoke College campus in South Hadley (Sigler 2008). An especially dense and persistent infestation exists at Log Pond Cove, a 20-acre cove of the Connecticut River in Holyoke. There, several partnering agencies, including the US Fish and Wildlife Service (specifically, the Silvio O. Conte National Fish and Wildlife Refuge), the City of Holyoke, the Connecticut River Conservancy, and Holyoke Gas and Electric company have tried various modes of mechanical harvesting, chemical control, and integrated management (with augmentation from hand-pullers), with the population fluctuating through time. Control has remained out of reach. The species has spread to scores of other waterbodies, including other coves of the Connecticut River and shallow ponds within the watershed. A volunteer-based initiative

was undertaken about 2000 under the leadership of the Silvio O. Conte National Fish and Wildlife Refuge and later taken over by the Connecticut River Conservancy in 2019. As of early 2022, water chestnut sporadically ranges within the watershed from mid-Vermont and New Hampshire, through Massachusetts and Connecticut, until salinity limits its viability north of Long Island Sound (see Figure 14.1).

The best long-term successes in the Massachusetts portion of the watershed have been through hand pulling efforts. Ideally, the goal is to catch new infestations and physically remove them before they ever have a chance to drop seeds. Unfortunately, most infestations are not detected until they have established a seedbank. Because a single water chestnut plant can produce around 100 seeds, successful management requires greater than 99% plant removal for the population to show any sign of decline.

The US Fish and Wildlife Service, the Connecticut River Conservancy, and other partnering organizations have been seeing very promising results on numerous ponds once they started an "Early and Often" approach to finding and removing as many plants as possible, primarily by hand. This has entailed pulling all visible plants four times during the growing season, starting in early June as soon as rosettes surface. In Massachusetts, this is around the first or second week of June and at that point, rosettes are only about 1–3 inches in diameter. Stewards follow up again two weeks later (late June), and then again two to three weeks after that (in July). A final pull (in mid-August) is conducted to catch any regrowth, late seed emergence, or missed plants and this prevents almost all seed drop. The idea is to be constantly vigilant for small, newly germinating plants and catching any regrowth (from previously snapped stems) long before any seeds can form.

Another benefit of pulling plants while small is the volume of plants handled is reduced. This minimizes the amount of storage area required in the boats, which, in turn, reduces the time required for paddling to offload sites. Depending on the size of the infestation, less people may be required as well (Larger infestations or infestations where water chestnut plants are hidden among other vegetation, can require many more people to find and remove all water chestnut).

Cozza et al. (1994) found that 86% of the water chestnut seedbank germinates the year after seed drop at their study site. Subsequently, 85% of the remaining seeds were able to germinate after a second winter. With this high germination rate the first two years, it should follow that if an intensive effort is put forth to find and remove all plants, it shouldn't take very long to exhaust the bulk of the seedbank and reduce the population significantly. Then it is a matter of catching seeds that will germinate in any one of the next eight to ten years of viability.

This high germination rate observed by Cozza et al. has indeed been observed in data collected on numerous sites in the Connecticut River watershed. One such site in Westfield, MA (Brickyard West), was discovered in September of 2014 after seed drop, extremely dense sections were observed. In 2015, the US Fish and Wildlife Service staff and youth crew, along with many volunteers, did a massive harvest using canoes, kayaks, and a john boat to remove water chestnut by hand in most of the 20 small man-made ponds (14 acres of water). Because of a late start and the large size of plants, they were unable to cover all the ponds that year. In 2016, they increased the effort and completed removal in all the ponds, which required 681 person-hours to hand-pull 18,383 pounds of water chestnut. This exhaustive effort likely prevented seed drop for that year. By the following year (2017), the total plant weight in the ponds had dropped (97.5%) to 454 pounds and then continued to drop by 100 pounds a year until 2021 when only 32 pounds were found within a total of 16 person-hours. The drop in labor since 2016 was 97.7%. Part of the reason for the substantial drop in weight of plants removed is that, starting in 2017, ponds were monitored early in the season and then regularly after that (four total visits per year), so all plants were found and pulled when they were still very small. Even late in the season plants were small because pulled material was typically either newly germinated plants or regrowth from snapped stems (C. Boettner and G. Boettner, personal communication).

Theoretically, an infestation could be extinguished after a total of 12 years, but practically speaking, it may be impossible to find those last hidden plants each year. It is hoped that, for the most part, the Brickyard West water chestnut population may now be kept in check, by just one or two people on the water for 4 hours, four times a year—quite a difference from the labor-intensive effort from just five years previous. Stewards at several other sites using this regimen are also experiencing similar drops in infestation size.

As of this writing in early 2022, there are many other ponds and coves in the Connecticut River watershed that have water chestnut populations. In general, it appears that the sites that are larger and more difficult to cover thoroughly in a growing season are those that have been harder to bring down to a more manageable level. Having a designated site coordinator in charge of planning and promoting harvesting events has proven essential. Challenges to thorough removal at various sites include lack of a site coordinator, garnering adequate human power (or the vessels needed) to effectively remove all plants before they drop seeds, the ability of plants to effectively escape notice in duckweed or other emergent or floating vegetation, and safe access issues due to heavy flow or water quality issues (e.g., cyanobacteria) during the prescribed time for effective plant removal. Each year partners meet to discuss successes and challenges. New reporting mechanisms are being implemented and tested by the Connecticut River Conservancy and the hope is that new infestations will be detected sooner, which will make them easier to bring under control.

CONCLUSION

Water chestnuts often negatively impact the well-being of waterbodies, as well as commercial and recreational water uses. Therefore, more mapping and collecting and publishing of data is needed to aid in implementing appropriate control methods in eradicating invasive water chestnuts. In doing so, it is important to continue to create public awareness of water chestnuts' impacts, as well as implement reporting mechanisms and multifaceted treatments methods such as physical, biological, and chemical controls. Raising awareness about the effectiveness of well-organized and well-timed persistent hand pulling efforts can involve individuals. Even so, more research concerning water chestnut is critical for future management plans.

NOTE

1 According to an email from Summer Stebbins, the Connecticut Agricultural Experiment Station, to David W.H. Wong, Massachusetts Department of Environmental Protection, dated June 21, 2022.

REFERENCES

Aquatic Control Technology. 2015. *2015 Baseline Assessment Sudbury River Survey Findings and Management Recommendations.* Sheridan, CO: Aquatic Control Technology.

Caraco, N.F., and J.J. Cole. 2002. Contrasting impacts of a native and alien macrophyte on dissolved oxygen in a large river. *Ecological Applications* 12: 1496–1509.

Chorak, G.M., L.L. Dodd, N. Rybicki, K. Ingram, M. Buyukyoruk, Y. Kadono, Y.Y. Chen, and R.A. Thum. 2019. Cryptic introduction of water chestnut (Trapa) in the northeastern United States. *Aquatic Botany*, 155: 32–37. doi: 10.1016/j.aquabot.2019.02.006

Countryman, W. 1978. *Nuisance aquatic plants in Lake Champlain: Lake Champlain Basin Study.* Burlington, VT: US Department of Commerce, National Technical Information Service PB-293 439.

Cozza, R., G. Galanti, M.B. Bitonti, and A.M. Innocenti. 1994. Effect of Storage at low Temperature on the Germination of the Waterchestnut (*Trapa natans* L.) Phyton (Horn Austria) 34(2): 315–320.

Crow, G., and C.B. Hellquist. 2000. *Aquatic and Wetland Plants of Northeastern North America.* Madison, WI: University of Wisconsin Press, pp. 206–209.

Ding, J., B. Blossey, Y. Du, and F. Zheng. 2006a. Impact of Galerucella birmanica (Coleoptera: Chrysomelidae) on growth and seed production of Trapa natans. *Biological Control* 37(3): 338–345, doi: 10.1016/j.biocontrol.2005.12.003

Ding, J., B. Blossey, Y. Du, and F. Zheng. 2006b. Galerucella birmanica (Coleoptera: Chrysomelidae), a promising potential biological control agent of water chestnut, Trapa natans. *Biological Control* 36(1): 80–90. doi: 10.1016/j.biocontrol.2005.08.001

EDDMapS (Early Detection & Distribution Mapping System). 2019. The University of Georgia—Center for invasive species and ecosystem health. www.eddmaps.org/ (Accessed December 19, 2019).

Feldman, R.S. 2001. Taxonomic and size structures of phytophilous macroinvertebrate communities in Vallisneria and Trapa beds of the Hudson River, New York. *Hydrobiologia* 452(1): 233–245, doi: 10.1023/A:1011903315998

Furgo East, Inc. 1996. *Summary Report on the Restoration of Hills Pond in Menotomy Rocks Park.* Arlington, MA; Northborough, MA: Furgo East.

Giddy, I.H. 2003. *The Hudson River Water Trail Guide.* Hudson: Hudson River Watertrail Association.

Gleason, H.A., and A. Cronqusit. 1991. *Manual of Vascular Plants of Northeastern United States and Adjacent Canada* (2nd edition). Bronx, NY: New York Botanical Garden.

Groth, A., L. Lovett-Doust, and J. Lovett-Doust. 1996. Population Density and Module Demography in Trapa natans (Trapaceae), an Annual, Clonal Aquatic Macrophyte. *American Journal of Botany* 83, doi: 10.2307/2446095

Harman, W.N., H.A. Waterfield, and M.F. Albright. 2012. *DEC Invasive Species Eradication and Control Grant Final Report 1.* Cooperstown, NY: State University of New York at Oneonta Biological Field Station.

Hummel, M., and S. Findlay. 2006. Effects of water chestnut (Trapa natans) beds on water chemistry in the tidal freshwater Hudson River. *Hydrobiologia* 559(1): 169–181, doi: 10.1007/s10750-005-9201-0

Hummel, M., and E. Kiviat. 2004. Review of world literature on water chestnut with implications for management in North America. *Journal of Aquatic Plant Management* 42: 17–28.

ITIS (Integrated Taxonomic Information System). 2021. www.itis.gov/servlet/SingleRpt/SingleRpt?search_topic=TSN&search_value=27170#null (Accessed January 29, 2021).

Kornijów, R., D.L. Strayer, and N.F. Caraco. 2010. Macroinvertebrate communities of hypoxic habitats created by an invasive plant (*Trapa natans*) in the freshwater tidal Hudson River. *Fundamental and Applied Limnology/Archiv für Hydrobiologie*, 176(3): 199–207.

Kumar, V., and A.K. Chopra. 2018. Phytoremediation potential of water caltrop (*Trapa natans* L.) using municipal wastewater of the activated sludge process-based municipal wastewater treatment plant. *Environmental technology*, 39(1): 12–23.

Kurihara, M., and I. Ikusima. 1991. The ecology of the seed in Trapa natans var. Japonica in a eutrophic lake. *Vegetatio* 97(2): 117–124, doi: 10.1007/BF00035385

Lake Champlain Basin Atlas Water Chestnut. 2019. https://atlas.lcbp.org/issues-in-the-basin/aquatic-invasive-species/water-chestnut/ (Accessed December 15, 2019).

Liu, W., J. Zhang, C. Zhang, Y. Wang, and Y. Li. 2010. Adsorptive removal of Cr (VI) by Fe-modified activated carbon prepared from Trapa natans husk. *Chemical Engineering Journal*, 162(2): 677–684.

Mass. DCR. 2002. *Water Chestnut: An Invasive Aquatic Plant.* Cambridge, MA: Mass. DCR.

Mass. DCR. 2015. *A Guide to Aquatic Plants in Massachusetts.* Cambridge, MA: Mass. DCRL: Massachusetts Department of Conservation and Recreation, p. 19.

Massachusetts Aquatic Invasive Species Working Group. 2002. *Massachusetts Aquatic Invasive Species Management Plan.* Cambridge, MA: Massachusetts Aquatic Invasive Species Working Group.

Mattson, M.D., K.J. Wagner., R.A. Barletta, and A. Aiello. 2004. *Eutrophication and Aquatic Plant Management in Massachusetts: Final Generic Environmental Impact Report.* Cambridge, MA: Commonwealth of Massachusetts, Executive Office of Environmental Affairs Commonwealth of Massachusetts, p. 514.

Naylor, M. 2003. *Water Chestnut (Trapa natans) in the Chesapeake Bay Watershed: A Regional Management Plan.* Lanham, MD: Maryland Department of Natural Resources.

OARS. 2017. Water Chestnut Management Guidance & Five-year Management plan for the Sudbury, Assabet, and Concord River Watersheds. Concord, MA: OARS (Planning and Guidance for Water Chestnut Management in the SuAsCo Watershed).

O'Neill, C.R. 2006. *Water chestnut (Trapa natans) in the Northeast.* New York: New York Sea Grant, Invasive Species Factsheet Series 06–1, Brockport.

Pearson, S. 2021. A Persistent Seed Bank: The Case for Annual Water Chestnut Monitoring Once a Population is Presumed Eradicated. Northeast Aquatic Plant Management Society Annual Meeting. www.neapms.org/2021-virtual-conference-archive

Pemberton, R.W. 2002. *Water Chestnut.* USDA Forest Service Publication FHTET-2002-04. Morgantown, WV: USDA, p. 413.

Pfingsten, I.A., L. Cao, and L. Berent. 2020. *Trapa Natans L.: U.S. Geological Survey, Nonindigenous Aquatic Species Database.* Gainesville, FL. https://nas.er.usgs.gov/queries/factsheet.aspx?SpeciesID=263 (Accessed December 14, 2020).

Rybicki, N. 2019. *Recent Detection and Spread of a new type of Trapa, an Invasive Aquatic Plant, in the Potomac River Watershed.* Washington, DC: U.S. Geological Survey, emerita.

Sigler, M. 2008. Dams and dredging: An overview of Mount Holyoke College's lake history. *Mt. Holyoke College Institutional Digital Archive.* https://ida.mtholyoke.edu/handle/10166/4673

Strayer, D.L.C. Lutz, H.M. Malcom, K. Munger, and W.H. Shaw. 2003. Invertebrate Communities Associated with a Native (Vallisneria americana) and an alien (Trapa natans) macrophyte in a large river. *Freshwater Biology* 48: 1938–1949.

Vuorela, I., and M. Aalto. 1982. Palaeobotanical investigations at a Neolithic dwelling site in southern Finland, with special reference to Trapa natans. *Ann. Bot. Fenn.* 19: 81–92.

Wu, M., and J. Wu. 2007. In-vitro investigations on ultrasonic control of water Chestnut. *Journal Aquatic Plant Manage* 45: 76–83.

15 Water Hyacinth (*Eichhornia crassipes*)

ABSTRACT

Water hyacinth *Eichhornia crassipes* is an invasive free-floating aquatic plant in the United States. Currently, there are only two known infected waterbodies in Massachusetts, Lake Ellis in the Millers Watershed and Mansfield Pond in the Housatonic Watershed. Meanwhile, the spreading capability of *E. crassipes* could not be ignored due to its ecological and economic impacts. Since it is still in an early stage of invasion to Massachusetts waters, rapid response to eradication should be tried as a management strategy. Physical methods such as manual or mechanical harvesting, chemical control such as diquat and glyphosate, and biological control using agent such as weevil species (*Neochetina eichhorniae* and *Neochetina bruchi*) can treat water hyacinth separately or can be implemented complementary according to the status of infestation. The challenge to manage this species continues in many countries worldwide.

INTRODUCTION

Taxonomic Hierarchy (ITIS 2021)

Kingdom	Plantae
Subkingdom	Viridiplantae
Infrakingdom	Streptophyta
Superdivision	Embryophyta
Division	Tracheophyta
Subdivision	Spermatophytina
Class	Magnoliopsida
Superorder	Lilianae
Order	Commelinales
Family	Pontederiaceae
Genus	*Eichhornia* Kunth
Species	*Eichhornia crassipes* (Mart.) Solms

BIOLOGY

Water hyacinth (*Eichhornia crassipes*) is also called floating water hyacinth. *E. crassipes* is mainly a free-floating macrophyte which can be found on the water surface (Figure 15.1). They are also able to survive in wetlands by using their complex root structure to directly absorb nutrients from the soil (Yu et al. 2019). Recently, *Pontederia crassipes* is also used as the scientific name for this species. *E. crassipes* may grow up to 1 m, but it usually appears 0.5 m of the height; its leaves are very thick and waxy with a circular or elliptical size, appearing shiny green; the leaves width are around 10–20 cm, leaf vein are performing numerous and dense, and leaf stalks are inflated and spongy; roots occupy around half of the plant biomass, and the dark violet roots has a feathery appearance; lavender-colored flowers grows in set of 8–15 on each spike, and every flower generally consists of six petals with a yellow blue-lined center (Gopal 1987; Robinson 2003).

When *E. crassipes* is not in the blooming period, it has a similar look with a native plant, frog's-bit (*Limnobium spongia*). They are relatively difficult to distinguish since these two plants both can be found floating on the water surface with thick and rounded green leaves. In addition, frog's-bit can clump up on the surface forming dense mats in the water similar to water hyacinth. When in bloom, the flowers of *L. spongia* are white with three petals and three sepals, and *E. crassipes* flowers are lavender with around six petals (DWFS n.d.). Without the presence of flowers, identification becomes trickier. However, there are key distinctions between the two species found by examining the root stems. *E. crassipes* has dark and feathery roots and inflated spongy leaf stalks (Robinson 2003), while *L. spongia* has whitish roots and slender firm leaf stalks (UF/IFAS n.d.) (Figure 15.2).

DOI: 10.1201/9781003201106-15

FIGURE 15.1 Water hyacinth. (Modified from Patel (2012).)

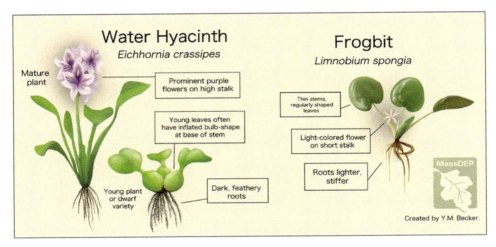

FIGURE 15.2 Water hyacinth comparison to native plant frog's-bit (*Limnobium spongia*). (Photo by Y.M. Becker.)

NATIVE AND INVASIVE RANGE

E. crassipes is native to the northern neotropics of South America. It can easily and rapidly establish itself in the various waterbodies, from small pond to reservoir. In addition, it can also be tolerant to extreme conditions of water level fluctuation, nutrient availability, pH, and toxic substances (Gopal 1987). However, cold weather and freezing temperatures prove challenging for this aquatic plant which may experience winter diebacks (Nesslage et al. 2016).

Currently, it is distributed through North America, South America, eastern Africa, Asia, and Australia, except Europe as a result of the climatic condition (GISD 2006). However, recent report is that this species has invaded in Italy in the first half of the 19th century (Brundu et al. 2013) and other European countries (Uremis et al. 2014; Coetzee et al. 2017). It is reported that *E. crassipes* was first spread from its native habitats to other countries during late 18th century and early 19th century, especially, it was introduced to the United States in the 1880s (Osei-Agyemang 2002).

Due to climate change, *E. crassipes* may be able to spread within and beyond its range as temperatures increase and winters become warmer. Although one study found that even with warming winters, *E. crassipes* populations didn't rebound significantly (Nesslage et al. 2016). Another study showed through two future greenhouse emission scenarios, temperature gradients in the northern hemisphere are too steep for significant geographical range expansion (Kriticos and Brunel 2016). At the same time, it is found in several studies that warming enhanced survival and regrowth of water hyacinth (You et al. 2013; Liu et al. 2016; Yu et al. 2019; Zhang et al. 2021).

Nevertheless, research suggests states plan accordingly and incorporate climate change–related actions and strategies into future management plans as warming creates conditions favorable to various species (Thomas et al. 2008).

MECHANISMS/VECTORS OF SPREAD

In controlling *E. crassipes*, it may be useful to understand reproduction and dispersal mechanisms. *E. crassipes* can reproduce by both sexual and vegetative methods. Vegetative reproduction is the main cause for population increases (Barrett 1980) and the vegetative reproduction of *E. crassipes* occurs as short runner stems (stolons) break apart and float off. The new rosettes (daughter plants) grow from the floating stolon of the original plant (Barrett 1980; Robinson 2003). As plant fragments are easily transported by wind, waves, or boat movement, they can reproduce and infest additional areas, spreading the challenge for managers to continually control this aquatic plant species. Much of the spread may be contained by limiting human activity through infested areas such as avoiding boating activities and movement through infested waters.

Although the most common form of spread is vegetative reproduction, *E. crassipes* is capable of additional vectors of spread through seed and sexual reproduction (Barrett 1980; Pérez et al. 2011a, 2011b). Pollination of the *E. crassipes* is completed by bees between its flowers. In North America, the flowering season can last five to nine months, producing abundant amounts of inflorescences (Barrett 1980). Furthermore, seeds are produced in the mild season, and may be viable for decades (Robinson 2003). However, low light and low temperature can prevent seeds germination, as well as seedling reproduction.

Whether *E. crassipes* reproduces by vegetative method or sexual method, it has the potential to produce large amounts of individuals rapidly making for proper control important to reduce catastrophic impacts.

ECOLOGICAL IMPACTS

E. crassipes is a very noxious aquatic invasive plant because it can thrive in a variety of waters. Once this plant species invades the waterbody and replaces the local vegetation, the natural ecosystems will be affected. According to multiple research (Robinson 2003; Villamagna and Murphy 2010; Koutika and Rainey 2015), the invasion of *E. crassipes* causes various threats and impacts to the ecosystem. Dense mats of *E. crassipes* inhibit the big fish migration, increase the sedimentation and evapotranspiration rates, degrade water quality, and intercept the sunlight which leads to the exclusion of other submerged plants. Massive and thick *E. crassipes* clog canals and waterways, providing a breeding environment for mosquitoes. Moreover, large amount of *E. crassipes* existing in the waterbodies compete with other native species and inhibit the light to the algae. This would cause fish or animals which rely on these native species to relocate or perish. Further, during the decomposition time of *E. crassipes*, fish or other aquatic animals may be harmed or even killed due to the anoxic environment. Therefore, in terms of the spread of *E. crassipes*, it significantly and negatively influences the ecosystem and biodiversity (Waltham et al. 2020). Several reports have revealed impacts or potential impacts from *E. crassipes*, such as the Great Lakes (North America) (Adebayo et al. 2011), Rift-Valley Lake (Ethiopia) (Mengistu et al. 2017), River Nile (Egypt) (Shanab et al. 2010), Lake Dianchi (China) (Wang et al. 2012), and different waters in China (Wang et al. 2012).

ECONOMIC IMPACTS

In addition to the impacts to the environment, the enormous impacts to human and society caused by *E. crassipes* should not be neglected. *E. crassipes* in the water may inhibit the boating and swimming recreational activities, resulting in economic loss. Other water activities such as fishing could also be reduced as plants overgrowth and less fish. Further, tourism development could be hampered, the value of surrounding real estate could be influenced (Robinson 2003), channels and waterways could be blocked, aesthetic value could be decreased, etc. The water use, irrigation, and electric power generation which are associated with agricultural, fishery, and other business would also be indirectly influenced by this nuisance plant (Koutika and Rainey 2015). Because of the thrive of *E. crassipes*, reduced water flow might lead to flooding and damage to canal banks (DWFS n.d.).

The benefits of management greatly outweigh the costs of control. Even after implementation of successful management practices, studies show the importance of long-term maintenance programs (Waltham et al. 2020). Many management decisions are made based on "least-cost control" options and tend to fail the consideration of long-term control effects. As mechanical shredding is a budget-friendly option (Greenfield et al. 2006), there are significant consequences of allowing widespread plant decomposition within an ecosystem. Although alternative treatments such as herbicides may be more expensive initially, over time they may become more cost-effective as the time of regrowth is slower and damaging effects may be minimized (Villamagna and Murphy 2010). Even though treatment may be costly, one study used evidence-based economic analysis which demonstrated that the ecosystem service benefits from water hyacinth management greatly exceed research and control costs. The study performed on a long-term treatment site in Louisiana estimated that peak plant cover would be 76% higher without substantial growth rate suppression (Wainger et al. 2018). The study also found the most cost-effective approach of combined biological and herbicide control programs 34:1 with costs-benefit of ecosystem services from management (Wainger et al. 2018).

UTILIZATION OF WATER HYACINTH

In spite of *E. crassipes* having negative influences to the ecosystem and economy, special structure and biomass may let *E. crassipes* create many benefits. First, it can be used for animals and birds feed as well as organic fertilizers (GISD 2006). Second, prior research (Rommens et al. 2003) found that *E. crassipes* can be applied

to absorb nutrients, such as nitrate, ammonium, and phosphate, for the evaluation of specific water system. This means it can also be applied to trap pollutants for water purification by plant harvesting (Greenfield et al. 2007); thus this benefit might limit the application in soil fertilizing. Third, *E. crassipes* can provide foraging places for Great Egret, Snowy Egret, Tri-colored Heron, Great Blue Heron, etc. large wading birds (Villamagna 2009). Finally, *E. crassipes* as a reusable energy can be used for producing biohydrogen and biomethane (Chuang et al. 2011). A recent review demonstrates that water hyacinth has gained attentions in biochar production, biomethane, biohydrogen, biogas, and utilization in case of wastewater treatment (Gaurav et al. 2020). In regions where permanent eradication proves difficult or impossible, research into finding alternatives for *E. crassipes* may prove more beneficial than focusing all efforts on control techniques (Güereña et al. 2015).

MONITORING AND SIGHTINGS

The invasive species data in this chapter were collected by the Massachusetts Department of Environmental Protection (MassDEP) and the Department of Conservation and Recreation (DCR). The standard operating procedures, including sampling methodology, species identification, quality control and assurance, are implemented by samples collected by MassDEP. Sites with *E. crassipes* were mapped as a data layer with ArcGIS® ArcMap 10.1 (ESRI, Redlands, California). This tool is able to pinpoint the area that *E. crassipes* has spread, which can help to inform future decisions on processes that can happen in these waterbodies, or regulations that involve efforts to limit the spread of this aquatic invasive species.

By using Geographic Information Systems (GIS), the presence of *E. crassipes* has been mapped in order to aid in future decisions that can be made on these waterbodies. From the collected data (Table 15.1), a map using GIS has been created in order to show the distribution of *E. crassipes* in Massachusetts waterbodies (Figure 15.3). Currently, there are two known waterbodies (Table 15.2) infested by *E. crassipes*, and two watersheds in Massachusetts are infested by *E. crassipes* (Table 15.3). These

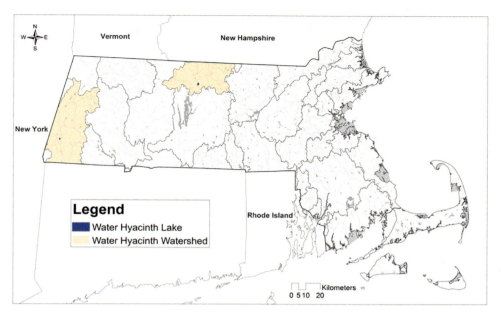

FIGURE 15.3 Distribution map depicting water hyacinth (*Eichhornia crassipes*) in Massachusetts.

TABLE 15.1

Infested Waterbodies in Massachusetts

Waterbody	Agency	Report Source
Lake Ellis	MassDEP	Millers River Watershed 2000 Water Quality Assessment Report 35-AC-1 CN089.0, March 2004
Mansfield Pond	DCR	DCR Data from Lake Association

TABLE 15.2

Towns and Infested Waterbodies

Number	Waterbody	Town 1
1	Lake Ellis	Athol
2	Mansfield Pond	Great Barrington

TABLE 15.3

Infested Watersheds

1	Millers
2	Housatonic

infested waterbodies are located in western and mid-northern Massachusetts.

Through the reports and surveys of *E. crassipes*, it is noted that *E. crassipes* is a very aggressive invasive species, and therefore has the ability to spread in a short period of time throughout Massachusetts. Although there is no massive *E. crassipes* spreading situation present in recent years in Massachusetts, the potential economic loss and ecological impact demonstrate that *E. crassipes* invading could not be neglected. A pointed and effective control method should be applied for infected water-bodies. For waterbodies haven't been invaded or under the risk to be invaded, the prevention methods should be conducted before the expansion of the invasive species.

MANAGEMENT

There are different methods tested for controlling the spread of *E. crassipes*. Effective management measures include physical, chemical, biological, and combined methods. The choice of these methods depends on the invasion range, the water condition, weather, management budget, regulation restrictions, etc.

PHYSICAL

Physical methods include harvesting (manual or mechanical) and in situ cutting (Villamagna and Murphy 2010). Mechanical harvesting is suitable for large-scale *E. crassipes* removal, especially the waterbody is invaded by massive and dense *E. crassipes*. Hand pulling is better than mechanical harvesting when small-scale *E. crassipes* appears in the waterbody, or when *E. crassipes* is mixed with other aquatic plants (Mattson et al. 2004). Although this easily accessible method could solve the problem, some disadvantages need to be noticed. Due to *E. crassipes*' ability to regrow by its fragments, harvesting can cause further spreading. When *E. crassipes* fragments drop in the water, it might be carried by animals or mechanical tools to other waterbodies (Robinson 2003). In situ cutting is another effective mechanical way. By applying this method, *E. crassipes* would die and decompose in the water (Greenfield et al. 2007). But it would release phosphorus, which

might accelerate the eutrophication (Perna and Burrows 2005; Bicudo et al. 2007). Moreover, physical control might be time-consuming and expensive when large-scale *E. crassipes* needs to be removed. Transportation and disposal of large amount of *E. crassipes* may also bring health risk and environment problem owing to its pollution absorption capability (Villamagna and Murphy 2010).

CHEMICAL

Instead of the aforementioned methods, chemical control is also one of the most popular management approaches for undesirable plants control. Diquat, 2, 4-D, and glyphosate have been recognized effectively for controlling *E. crassipes* (Seagrave 1988; Mattson et al. 2004). Diquat and 2, 4-D can be used as consistent control method with correct dose, proper formulation, and suitable conditions (Mattson et al. 2004). Herbicides contain active ingredients that provide the toxicity to target plants but can also harm or kill non-target plants and organisms by using inappropriate herbicides or applying high-rate usage (Mattson et al. 2004). More information about chemical control on water hyacinth and other 18 invasive species can be found in Appendix III. Any chemical used for aquatic invasive species control requires a WM04 Chemical Application License issued by the Massachusetts Department of Environmental Protection.

Water quality conditions, specifically dissolved oxygen, should be minded in arial spray operations. Studies locations show that spraying using a helicopter to control invasive species does not dramatically effect water quality, DO, after mitigation (Tobias et al 2019). Any effects on dissolved oxygen and turbidity are able to recover to regional averages after treatment due to tidal hydrology. However, for wetland cases, this treatment method is appropriate when species cover is minor and restricted to wetland margins. In areas with higher plant cover, with ongoing spraying and low rainfall, hypoxia becomes of concern for coastal wetland function and productivity (Waltham and Fixler 2017). Here, the issue is the rate of consumption of oxygen for decomposition is greater than the amount of oxygen diffused in the water. The suggestion to avoid such cases is to avoid routine spraying and paying close attention to antecedent weather conditions before treatment (Waltham and Fixler 2017).

Additionally, studies show additional concern for plant decay in relation to breeding grounds for insects. Research found within herbicide-treated areas, there was an increase in mosquito larva as plant decay increased. This study suggests the need for integrated approach of invasive weed management with herbicides and larval mosquito control practice (Portilla and Lawler 2020).

BIOLOGICAL

Biological methods mainly involve the introduction of the insects and plant pathogen. Two weevil species (*Neochetina eichhorniae* and *Neochetina bruchi*) have been verified as an effective control method for *E. crassipes* (Robinson 2003; Sosa et al. 2007). These two species can reduce water hyacinth buoyancy so that plants could fall down and decompose (Wilson et al. 2007). However, successful control by these two species can only occur in some waters, other conditions (such as large mats of *E. crassipes* or shallow waters) might influence the control result (Julien 2001; Villamagna and Murphy 2010). Additionally, even with increased weevil densities in waters, not enough damage occurs during cooler seasons for year-round control resulting in the need for other control measures to be implemented in such regions (Hopper et al. 2017).

Another biological control agent, *Megamelus scutellaris* (Hemiptera: Delphacidae), a planthopper, may prove as an effective control method in warmer locations. *M. scutellaris* is native to the subtropical climate of South America and is more effective in warmer waters as peak densities are observed in warmer months (Hopper et al. 2017). Additionally, *M. scutellaris* perform better in shaded or regions of partial cover rather than open areas most likely due to the increased humidity created in such locations (Tipping et al. 2014). Similar to the two weevil species listed earlier, *M. scutellaris* is able to damage leaves to a significant amount (Hopper et al. 2017).

Furthermore, prior research found that some fungus, *Acremonium zonatum*, *Alternaria eichhorniae*, and *Cercospora piaropi*, can control *E. crassipes* with proper conditions (Shabana and Mohamed 2005; Dagno 2011). When pretreated with 3,4-methylenedioxy *trans*-cinnamic acid (MDCA), plants are more susceptible to *A. eichhorniae* as MDCA is a phenylpropanoid pathway inhibitor and weakens the plant's defense system (Shabana and Mohamed 2005). Although caution has been advised as *A. eichhorniae* has been tested to only target *E. crassipes*, effects of both agents on others plants have not been widely studied. But the performance of the application can vary with the change of water activity and temperature, so studies advise the need for understanding how the physical environment will affect the agent's survival (Dagno 2011).

The use of biological control methods coupled with winter diebacks has been found to be extremely effective in some areas (Nesslage et al. 2016). However, biological methods are typically implemented after large quantities of plants are removed by physical or chemical management (Adekoya et al. 1993).

SUCCESS STORY

More successful control locations include infestations found in tropical and subtropical areas as well as infestations from monocultures existing as free-floating mat that are able to sink as a whole when damaged (Driesche et al. 2002). Factors that may accelerate control include wave action, reduced growth due to biological control agents, and high nutrient levels that lead to higher plant quality and thus increased insect population growth (Hill and Olckers 2000). Limiting factors may include the removal of mats by herbicidal or mechanical means or disrupting agent populations, shallow water where damaged plants are unable to sink, ephemeral waterbodies, polluted waters, lower temperatures at high altitude or temperate sites, high nutrients at temperate sites, and smaller frequency releases as opposed to mass or serial releases (Driesche et al. 2002).

In South Africa, biological control methods have been successfully applied to against *E. crassipes* during 1974–1996 (Hill and Olcker

2000). Researchers have tried to release six species, which include five arthropods and one pathogen, for eradicating *E. crassipes*. But only two species were found to contribute the most in this *E. crassipes* control process. These two species are *Neochetina eichhorniae* and *Neochetina bruchi*. Hill and Olcker (2000) highlight that larger waterbodies are more proper for applying this biological method. This is because the greater wave action produced by wind can promote the mortality of plants.

In 1990, 80% area of New Year's Dam in South Africa was found covered by *E. crassipes*; therefore, 200 adult *N. eichhorniae* were released in the dam. In the next few years, the population size of *N. eichhorniae* gradually became larger. By 1994, it was found that in New Year' Dam there was only 10% area that was occupied by *E. crassipes*, which was successfully achieved by the application of biological management with *N. eichhorniae*. The remaining plants were seen in a very small size. But the short-term success did not inhibit the regrowth of *E. crassipes*. This time, other species (*Niphograpta albiguttalis*, *O. terebrantis*, and *E. catarinensis*) were used as a new biological method, but it did not help in *E. crassipes* control. In 1998, the same situation of large area coverage occurred again as increased precipitation accelerated the growth of *E. crassipes*. *N. eichhorniae* was used the second time at the dam, reducing the coverage to 10% in 2000. According to the research analysis, this success was summarized into three important reasons: (1) the dam system is oligotrophic, so less nutrient for *E. crassipes* growth; (2) the life cycle of *N. eichhorniae* was extended by seasonal weather; and (3) biological control was conducted without any pressure of *E. crassipes* removal.

For over a century, water hyacinth has been presented as a problematic invasive species to water systems in Louisiana. Primary treatment strategies up until the 1970s included mechanical removal and herbicide control. Since then, biological control agents have become more prominent. Four insect biological control agents have been introduced in Louisiana: *Neochetina eichhorniae* and *N. bruchi* from 1974 to 1977, *Niphograpta albiguttalis* from 1979 to 1981, and *Megamelus scutellaris* from 2010 to 2016, with the most successful being *N.*

eichhorniae, or the mottled water hyacinth weevil, at establishing and dispersing throughout the state (Wainger et al. 2018). *Neochetina* spp. reduces water hyacinth vigor through larval and adult feeding which reduces production, fertility and spread, and increases susceptibility to herbicides. *Neochetina* spp. remains widespread in Louisiana despite lack of any new release since the late 1980s. Additionally, no other biological control agents have been released since 2017 with the primary management tool currently being herbicide treatments (Wainger et al. 2018).

REMAINING CHALLENGES

Some areas, despite best efforts to eradicate the invasion of *E. crassipes*, have not been successful. New research suggests control efforts move away from eradication and instead focus on utilization of the species. Through an integrated approach of removal and usage, one study poses a solution for Lake Victoria, where even after millions of dollars spent in control programs, water hyacinth remains a serious issue for the East African region (Guerena et al. 2015). While the study points to energy production being a viable usage for water hyacinth, it is also mentioned that current technologies hinder the effectiveness and efficiency as well as produce high operating costs for such a process to be achieved. However, theoretical capabilities of water hyacinth becoming a sustainable fuel supply are very high since an individual plant can produce daughter plants equivalent to 28,000 tons of fresh weight in a single year (Ruan et al. 2016). More research and efforts should be focused on novel uses of biomass in order to produce better management strategies of invaded regions.

REFERENCES

Adebayo, A.A., E. Briski, O. Kalaci, M. Hernandez, S. Ghabooli, B. Beric, F.T. Chan, A. Zhan, E. Fifield, T. Leadley, and H.J. MacIsaac. 2011. Water hyacinth (*Eichhornia crassipes*) and water lettuce (*Pistia stratiotes*) hit the Great Lakes: Playing with fire? *Aquatic Invasions* 6(1): 91–96.

Adekoya, B.B., G.N. Ugwuzor, K.B. Olurin, O.A. Sodeinde, and O.A. Ekpo 1993. A comparative assessment of the methods of control of water hyacinth infestation with regards to fish production. In: 10th Annual Conference of the Fisheries Society of Nigeria (FISON), 16–20 November 1992, Abeokuta, pp. 181–184.

Barrett, S.C.H. 1980. Sexual reproduction in Eichornia crassipes (water hyacinth) II. Seed production in natural populations. *Journal of Applied Ecology* 17: 113–124.

Bicudo, D.D., B.M. Fonseca, L.M. Bini, L.O. Crossetti, C.E.D. Bicudo, and T. Araujo-Jesus. 2007. Undesirable side-effects of water hyacinth control in a shallow tropical reservoir. *Freshwater Biology* 52: 1120–1133.

Brundu, G., M.M. Azzella, C. Blasi, L. Camarda, M. Iberite, and L. Celesti-Grapow. 2013. The silent invasion of *Eichhornia crassipes* (Mart.) Solms. in Italy. *Plant Biosystems-An International Journal Dealing with all Aspects of Plant Biology* 147(4): 1120–1127.

Chuang, Y.S., C.H. Lay, B. Sen, C.C. Chen, K. Gopalakrishnan, J.H. Wu, C.S. Lin, and C.Y. Lin. 2011. Biohydrogen and biomethane from water hyacinth (Eichhornia crassipes) fermentation: effects of substrate concentration and incubation temperature. *International Journal of Hydrogen Energy* 36: 14195–14203.

Coetzee, J.A., M.P. Hill, T. Ruiz-Téllez, U. Starfinger, and S. Brunel. 2017. Monographs on invasive plants in Europe N° 2: Eichhornia crassipes (Mart.) *Solms. Botany Letters* 164(4): 303–326.

Dagno, K., R. Lahlali, M. Diourté, and M.H. Jijakli. 2011. Effect of Temperature and Water Activity on Spore Germination and Mycelial Growth of Three Fungal Biocontrol Agents against Water Hyacinth (Eichhornia Crassipes). *Journal of Applied Microbiology* 110 (2): 521–28.

Department of Wildlife & Fisheries Sciences (DWFS) (n.d.) Frog's-Bit, Limnobium spongia. Department of Wildlife & Fisheries Sciences, Texas A&M AgriLife Extension Service https://agrilife.org/aquaplant/plant-identification/alphabetical-index/frogs-bit/ (Accessed March 30, 2020).

Driesche, F.V., B. Blossey, M. Hoodle, S. Lyon, and R. Reardon. 2002. Biological Control of Invasive Plants in the Eastern United States. United States Department of Agriculture Forest Service, Forest Health Technology Enterprise Team, Morgantown, West Virginia. 413 p.

Gaurav, G.K., T. Mehmood, L. Cheng, J.J. Klemeš, and D.K. Shrivastava. 2020. Water hyacinth as a biomass: A review. *Journal of Cleaner Production* 277: 122214.

Global Invasive Species Database (GISD). 2006. Species profile: Eichhornia crassipes. Downloaded from www.iucngisd.org/gisd/species.php?sc=70 (Accessed March 30, 2020).

Gopal, B. 1987. Water hyacinth. Elsevier Science Publishers, Amsterdam, 477 p.

Greenfield, B.K., M. Blankinship, and T.J. McNabb. 2006. Control costs, operation, and permitting issues for non-chemical plant control: Case studies in the San Francisco Bay-Delta Region, California. *Journal of Aquatic Plant Management* 44: 40–49.

Greenfield, B.K., G.S. Siemering, J.C. Andrews, M. Rajan, S.P. Andrews, and D.F. Spencer. 2007. Mechanical shredding of water hyacinth (Eichhornia crassipes): effects on water quality in the Sacramento-San Joaquin River Delta, California. *Estuaries and Coasts* 30: 627–640.

Güereña, D., H. Neufeldt, J. Berazneva, and S. Duby. 2015. Water hyacinth control in Lake Victoria: Transforming an ecological catastrophe into economic, social, and environmental benefits. *Sustainable Production and Consumption* 3: 59–69.

Hill, M.P., and T. Olckers. 2000. Biological control initiatives against water hyacinth in South Africa: constraining factors, success and new courses of action, M.H. Julien, T.D. Center, M.P. Hill (Eds.), Second IOBC Global Working Group on the Biological and Integrated Control of Water Hyacinth, ACIAR, Beijing, China pp. 33–38.

Hopper, J., P. Pratt, K. Mccue, M. Pitcairn, P. Moran, and J. Madsen. 2017. Spatial and temporal variation of biological control agents associated with Eichhornia crassipes in the Sacramento-San Joaquin River Delta, California. *Biological Control* 111: 13–22.

ITIS (Integrated Taxonomic Information System). 2021. *www.itis.gov/servlet/SingleRpt/SingleRpt#null* (Accessed February 2, 2021).

Julien, M. 2001. Biological control of water hyacinth with arthropods: a review to 2000. In: Julien, M., Hill, M., Center, T., Ding, J. (Eds.), *Proceedings of the Meeting of the Global Working Group for the Biological and Integrated Control of Water Hyacinth, Beijing, China, 9–12* December. Australian Centre for International Agricultural Research, Canberra pp. 8–20.

Koutika, L.S., and H.J. Rainey. 2015. A review of the invasive, biological and beneficial characteristics of aquatic species *Eichhornia Crassipes* and *salvinia molesta. Applied Ecology and Environmental Research*. 13: 263–275.

Kriticos, D.J., and S. Brunel. 2016. Assessing and Managing the Current and Future Pest Risk from Water Hyacinth, (Eichhornia crassipes), an Invasive Aquatic Plant Threatening the Environment and Water Security. *PLoS One* 11(8): e0120054.

Liu, J., X. Chen, Y. Wang, X. Li, D. Yu, and C. Liu. 2016. Response differences of Eichhornia crassipes to shallow submergence and drawdown with an experimental warming in winter. *Aquatic Ecology* 50(2) 307–314.

Mattson, M.D., P.J. Godfrey, R.A. Barletta, A. Aiello, and K.J. Wagner 2004. *Eutrophication and Aquatic Plant Management in Massachusetts: Final Generic Environmental Impact Report.* Washington, DC: Commonwealth of Massachusetts, Executive Office of Environmental Affairs Commonwealth of Massachusetts, vol. 4, pp. 29–80.

Mengistu, B.B., D. Unbushe, and E. Abebe. 2017. Invasion of Water Hyacinth (*Eichhornia crassipes*) Is Associated with Decline in Macrophyte Biodiversity in an Ethiopian Rift-Valley Lake—Abaya. *Open Journal of Ecology* 7(13): 667–681.

Nesslage, G.M., L.A. Wainger, N.E. Harms, and A.F. Cofrancesco. 2016. Quantifying the population response of invasive water hyacinth, Eichhornia crassipes, to biological control and winter weather in Louisiana, USA. *Biological Invasions* 18(7): 2107–2115.

Osei-Agyemang, M. 2002. Introduced Species Summary Project, Water Hyacinth (Eichhorinia crassipies) www.columbia.edu/itc/cerc/danoff-burg/invasion_bio/inv_spp_summ/water%252520hyacinth.html (Accessed March 30, 2020).

Pérez, E.A., J.A. Coetzee, T.R. Tellez, and M.P. Hill. 2011a. A first report of water hyacinth (*Eichhornia crassipes*) soil seed banks in South Africa. *South African Journal of Botany* 77(3): 795–800.

Pérez, E.A., T.R. Téllez, and J.M. Sánchez Guzmán. 2011b. Influence of physico-chemical parameters of the aquatic medium on germination of *Eichhornia crassipes* seeds. *Plant Biology* 13(4): 643–648.

Perna, C., and D. Burrows 2005. Improved dissolved oxygen status following removal of exotic weed mats in important fish habitat lagoons of the tropical Burdekin River floodplain, Australia. *Marine Pollution Bulletin* 51: 138–148.

Portilla, M.A., and S.P. Lawler. 2020. Herbicide treatment alters the effects of water hyacinth on larval mosquito abundance. *Journal of Vector Ecology* 45(1): 69–81.

Robinson, M. 2003. *Potential Invader, Water Hyacinth: An Exotic Aquatic Plant.* Cambridge, MA: Massachusetts Department of Conservation and Recreation; Office of Water Resources; Lakes and Ponds Program, p. 3.

Rommens, W., J. Maes, N. Dekeza, P. Inghelbrecht, T. Nhiwatiwa, E. Holsters, F. Ollevier, B. Marshall, and L. Brendonck 2003. The impact of water hyacinth (*Eichhornia crassipes*) in a eutrophic subtropical impoundment (Lake Chivero, Zimbabwe). I. Water quality. *Archiv Fur Hydrobiologie* 158: 373–388.

Ruan, T., R. Zeng, X. Yin, S. Zhang, and Z. Yang. 2016. Water hyacinth (Eichhornia crassipes) biomass as a biofuel feedstock by enzymatic hydrolysis. *BioResources* 11(1): 2372–2380.

Seagrave, C. 1988. *Aquatic Weed Control*. Surrey: Fishing New Books.

Shabana, Y.M., and Z.A. Mohamed 2005. Integrated control of water hyacinth with a mycoherbicide and phenylpropanoid pathway inhibitor. *Biocontrol Science and Technology* 15: 659–669.

Shanab, S.M.M., E.A. Shalaby, D.A. Lightfoot, and H.A. El-Shemy. 2010. Allelopathic effects of water hyacinth [*Eichhornia crassipes*]. *PLoS One* 5(10): e13200.

Sosa, A.J., H.A. Cordo, and J. Sacco 2007. Preliminary evaluation of *Megamelus seutellaris* Berg (Hemiptera: Delphacidae), a candidate for biological control of water hyacinth. *Biological Control* 42: 129–138.

Thomas, R., A. Kane, and B.G. Bierwagen. 2008. *Effects of Climate Change on Aquatic Invasive Species and Implications for Management and Research*. Lincoln, NE: University of Nebraska—Lincoln.

Tipping, P.W., M.R. Martin, E.N. Pokorny, K.R. Nimmo, D.L. Fitzgerald, F.A. Dray, and T.D. Center. 2014. Current levels of suppression of waterhyacinth in Florida USA by classical biological control agents. *Biological Control* 71: 65–69.

Tobias, V.D., Conrad, J.L., Mahardja, B., and S. Khanna. 2019. Impacts of water hyacinth treatment on water quality in a tidal estuarine environment. *Biological Invasions* 21: 3479–3490.

UF/IFAS, Center for Aquatic and Invasive Plants. n.d. *Limnobium Spongia*. University of Florida, Institute of Food and Agricultural Sciences, Center for Aquatic and Invasive Plants http://plants.ifas.ufl.edu/plant-directory/limnobium-spongia/ (Accessed March 30, 2020).

Uremis, I., A. Uludag, Z.F. Arslan, and O. Abaci. 2014. A new record for the flora of Turkey: *E ichhornia crassipes* (Mart.) Solms (Pontederiaceae). *EPPO Bulletin* 44(1): 83–86.

Villamagna, A.M. 2009. The Ecological Effects of Water Hyacinth (*Eichhornia crassipes*) on Lake Chapala, Mexico. PhD Thesis. Virginia Polytechnic Institute and State University, Blacksburg, VA.

Villamagna, A.M., and B. Murphy. 2010. Ecological and socio-economic impacts of invasive water hyacinth (*Eichhornia crassipes*): A review. *Freshwater Biology* 55: 282–298.

Wainger, L.A., N.E. Harms, C. Magen, D. Liang, G.M. Nesslage, A.M. Mcmurray, and A.F. Cofrancesco. 2018. Evidence-based economic analysis demonstrates that ecosystem service benefits of water hyacinth management greatly exceed research and control costs. *PeerJ* 6: e4824.

Waltham, N.J., L. Coleman, C. Buelow, S. Fry, and D. Burrows. 2020. Restoring fish habitat values on a tropical agricultural floodplain: Learning from two decades of aquatic invasive plant maintenance efforts. *Ocean & Coastal Management* 198: 105355.

Waltham, N.J., and S. Fixler. 2017. Aerial herbicide spray to control invasive water hyacinth (*Eichhornia crassipes*): Water quality concerns fronting fish occupying a tropical floodplain wetland. *Tropical Conservation Science* 10: 1–10.

Wang, H., Q. Wang, P.A. Bowler, and W. Xiong. 2016. Invasive aquatic plants in China. *Aquatic Invasions* 11(1): 1–9.

Wang, Z., Z. Zhang, J. Zhang, Y. Zhang, H. Liu, and S. Yan. 2012. Large-scale utilization of water hyacinth for nutrient removal in Lake Dianchi in China: the effects on the water quality, macrozoobenthos and zooplankton. *Chemosphere* 89(10): 1255–1261.

Wilson, J.R., O. Ajuonu, T.D. Center, M.P. Hill, M.H. Julien, FF. Katagira, P. Neuenschwander, S.W. Njoka, J. Ogwang, R.H. Reeder, and T. Van. 2007. The decline of water hyacinth on Lake Victoria was due to biological control by Neochetina spp. *Aquatic Botany* 87: 90–93.

Wilson, J.R., N. Holst, and M. Rees. 2005. Determinants and patterns of population growth in water hyacinth. *Aquatic Botany* 81(1): 51–67.

You, W., D. Yu, D. Xie, and L. Yu. 2013. Overwintering survival and regrowth of the invasive plant Eichhornia crassipes are enhanced by experimental warming in winter. *Aquatic Biology* 19(1): 45–53.

Yu, H., X. Dong, D. Yu, C. Liu, and S. Fan. 2019. Effects of eutrophication and different water levels on overwintering of *Eichhornia crassipes* at the northern margin of its distribution in China. *Frontiers in Plant Science* 10: 1261.

Yu, H., N. Shen, D. Yu, and C. Liu, C. 2019. Clonal integration increases growth performance and expansion of *Eichhornia crassipes* in littoral zones: A simulation study. *Environmental and Experimental Botany* 159: 13–22.

Zhang, X., H. Yu, T. Lv, L. Yang, C. Liu, S. Fan, and D. Yu. 2021. Effects of different scenarios of temperature rise and biological control agents on interactions between two noxious invasive plants. *Diversity and Distributions* 27(12): 2300–2314.

16 European Waterclover (*Marsilea quadrifolia* L.)

ABSTRACT

European waterclover *Marsilea quadrifolia*, also called water shamrock, is an herbaceous perennial aquatic plant, which grows in slow-moving water environments. *M. quadrifolia* has been established in the northeastern United States for around 100 years. Prior research reported that *M. quadrifolia* spreads at a slow speed and poses a relatively weak threat to the environment (Benson et al. 2004). The distribution of known *M. quadrifolia* species was mapped based on historical data for Massachusetts. More comprehensive monitoring on *M. quadrifolia* has been developed for a better understanding on its distribution, so that further spread of *M. quadrifolia* can be minimized in Massachusetts, and the general public can be aware of the infestations of this species.

INTRODUCTION

Taxonomic Hierarchy (ITIS 2021)

Kingdom	Plantae
Subkingdom	Viridiplantae
Infrakingdom	Streptophyta
Superdivision	Embryophyta
Division	Tracheophyta
Subdivision	Polypodiophytina
Class	Polypodiopsida
Subclass	Polypodiidae
Order	Salviniales
Family	Marsileaceae
Genus	*Marsilea* L.
Species	*Marsilea quadrifolia* L.

BIOLOGY

European waterclover *Marsilea quadrifolia* is an amphibious heterosporous fern that can produce submerged and emergent leaves on the same plant (Kao and Lin 2010) (Figure 16.1). In the aquatic environment, it usually is found in ponds, rivers, streams, muddy shorelines, or roadside ditches (Benson et al. 2004). The species may grow in dense patches or interspersed among other vegetation and be found in shallow waters as well as survive some water-level fluctuation (Kiran et al. 2007). The roots of *M. quadrifolia* are fleshy and creeping; the thin green stems from rhizomes bear only one leaf with four leaflets resembling a four-leaf clover. The mature green clover-like leaves are usually glabrous and 1.5–4 cm in width (Benson et al. 2004; Cao and Berent 2020). Sporocarps form at approximately 1.2 cm above where the leaf branches from the rhizome and are spherical, approximately 4–5 mm long and either hairy or smooth (Holm et al. 1997). They contain both small, male microspores and larger, female megaspores. Additionally, *M. quadrifolia* is able to form dense mats and has the ability to adjust leaflet angle for sunlight optimization, both of which allow it to outcompete native species causing for local concern (Hackett et al. 2018).

M. quadrifolia may be distinguished from other plant species by its four-leaflet arrangement, aquatic habitat, and sporocarps found near the base of the leaves. However, sporocarps typically develop after the leaves and may not always be present, as they do not mature until the leaf has withered for the winter (Johnson 1986).

Additional detection methods may include the use of aerial imagery as well as identifying genetic markers in the aquatic system. Aerial imagery has been used by experts to distinguish emergent and floating aquatic vegetation (Husson et al. 2013). However, due to the size of *M. quadrifolia* in addition to its variable density, detection may be difficult without high spatial resolution. This method may also be limited in its ability to distinguish *M. quadrifolia* from other aquatic vegetation in mixed stands or along shores with tree canopies. In using genetic markers, any genetic material shed by organisms into the environment can be detected early via monitoring and has

DOI: 10.1201/9781003201106-16

FIGURE 16.1   European waterclover. (Photo by A.A. Reznicek (Reznicek et al. 2011).)

been demonstrated in various studies (Hackett et al. 2018). Using such methods has the potential to increase the efficacy of field detection and monitoring of *M. quadrifolia* during the growing season. Although little research has been conducted on identifying genetic material shed from this aquatic plant species, this sampling approach could reduce the need for labor-intensive field surveys until *M. quadrifolia* was already detected in the area (Scriver et al. 2015).

Marsilea quadrifolia might be considered as *Marsilea minuta* or *Marsilea mutica* by mistake because they all have glabrous, fleshy leaves. However, *Marsilea minuta* or *Marsilea mutica* prefer the warmer climate of the southeast, while *Marsilea quadrifolia* has not been identified in south regions (Cao and Berent 2020). In addition, *Marsilea mutica* has obvious characteristics that are different from *Marsilea quadrifolia* (Benson et al. 2004). The *Marsilea mutica* leaves have two colors, lighter in the center and darker toward the margins, and the sporocarps completely lack teeth. However, the *Marsilea quadrifolia* leaves are solid green and the sporocarps bear a tooth.

NATIVE AND INVASIVE RANGE

M. quadrifolia is native to tropical and warm temperate regions from areas of Eurasia to North and South America. The native range includes Caucasia, western Siberia, Afghanistan, southwest India, China, Japan, and Americas (Cao and Berent 2020). However, in some research,

M. quadrifolia was reported as a non-indigenous plant with a century of establishment history in the United States (Benson et al. 2004). While globally classified as a species of least concern, *M. quadrifolia* is labeled as a vulnerable and endangered species in Japan as well as 21 European countries (Strat 2012). In many European countries, such as Italy, Portugal, and Spain, *M. quadrifolia* has been assessed as Endangered or even Critically Endangered (Bruni et al. 2013). Other countries in southeast Asia and the Mediterranean consider *M. quadrifolia* a weed, particularly in rice fields (Holm et al. 1997). In North America, *M. quadrifolia* has been documented in at least 19 states as well as Ontario, Canada. In the United States, it is listed as a priority exotic species on the Michigan "Watch List" as an immediate and significant threat to natural resources (Hackett et al. 2018).

According to the USGS (2020) database, *M. quadrifolia* was first found at Lacey-Keosauqua State Park, Van Buren, Iowa, in 1821. It was first reported in Connecticut at Bantam Lake in 1860 with conflicting ideas of origin and threat; however, it was later found to have spread to parts of Massachusetts and other areas of New England by 1900 (Les and Mehrhoff 1999). Recently, it has been confirmed in 21 states of the United States (USGS 2020). Massachusetts is reported to have the most extensive infestations in the Charles, Concord, Sudbury, and Connecticut Rivers (Benson et al. 2004). The first time *M. quadrifolia* was detected in Massachusetts was at a Botanical Garden in Middlesex county in 1868 (USGS 2020).

Plant species are directly affected by regional rainfall and evaporation patterns as the physiological functions involved in seed dormancy and germination are interrupted through the processes of drying and soaking. Due to climate change, areas around the coasts and wetlands especially experience increased flooding events, changes in rainfall distribution patterns, in addition to more severe downpours, storms, and hurricanes (Kathiresan and Gualbert 2016). In the Cauvery River delta region in India, the average rainfall in recent decades has increased while the rate of evaporation has decreased in comparison to the prior decade (Kathiresan and Gualber 2016). A survey of this region found *M. quadrifolia* invading wetland rice fields, dominating over native weeds such as *Echinochloa* spp. (barnyard grass or cockspur grass) as well as others. This is believed to be due to the species' adaptation and ability to live in alternating flooded and higher soil moisture conditions, which have been prevalent in recent years in the region. These findings suggest, as warming temperatures and rainfall trends continue to increase, *M. quadrifolia*, as well as other amphibious plant species, may have a higher advantage over native species and continue to expand their invasive range (Kathiresan and Gualber 2016).

Furthermore, in the sexual reproduction process of *M. quadrifolia*, it was found that the optimal temperature for male spores was 77–86°F (25–30°C) (Hackett et al. 2018). As waterbodies warm, there may be more optimal conditions for *M. quadrifolia* to spread and reproduce effectively.

MECHANISMS/VECTORS OF SPREAD

M. quadrifolia can reproduce by vegetative and sexual methods (Cao and Berent 2020). The vegetative method relies on its rhizome structure, which is its main reproductive approach. It can also persist throughout the winter seasons because of the underground rhizomes (Benson et al. 2004). During the growing season, *M. quadrifolia* plants are able to adjust the angle of the floating leaflets to optimize access to sunlight and the ability to photosynthesize (Kao and Lin 2010). *M. quadrifolia* can easily spread within connected waterbodies as the rhizome apex sheds and drift around, which allows for reproduction (Hackett et al. 2018). It is unknown what effect asexual reproduction has on long-distance dispersal as research has not concluded how long rhizome fragments can survive desiccation (Johnson 1986).

In addition, *M. quadrifolia* can also reproduce by male and female spore germination and cross-fertilization (Benson et al. 2004). The plant produces sporocarps, which are considered to be a modified leaf or leaflet. As sporocarps are disturbed or lightly abraded while immersed in water, they swell and burst to release spores which are germinated (Soni and Singh 2012). The sporocarp itself doesn't fully mature until the leaves have withered and it is believed light triggers sporocarp germination, which typically occurs after the leaves have matured (Johnson 1986). Furthermore, sporocarps have two external cell layers and internal gelatinous tissue which protect the spores from the environment and premature release in animal digestive tracts (Bloom 1955). The sporocarps can be easily dispersed by waterfowl or remain dormant in the soil for decades (Benson et al. 2004).

ECOLOGICAL IMPACTS

The plant is labeled as a noxious weed in India and becoming an increasing concern while also being considered a vulnerable species in Japan (Luo and Ikeda 2007). There are only few studies that have explored the ecological impacts of *M. quadrifolia*. As previously mentioned, in some European countries, it has been listed as an endangered species. *M. quadrifolia* is reported to have been previously managed in Massachusetts, and is considered as a problem species in at least some instances (Mattson et al. 2004; Wagner 2004). However, there is not enough scientific literature or research found evaluating any negative effects from *M. quadrifolia* on native species, food web, animals, water environment, etc. (SOM 2018).

There are conflicting reports whether *M. quadrifolia* spreads aggressively and how much of a threat it poses ecologically. In one study, *M. quadrifolia* was found to have only migrated 151 ft per year and a total of 1 mile downstream in 35 years (Henry and Myers 1983).

As for possible impacts, *M. quadrifolia* has the ability to adjust the angle of floating leaflets for sunlight optimization, which may allow it to outcompete neighboring species (Kao and Lin 2010). Additionally, *M. quadrifolia* creates dense monoculture of stands that can survive through winter seasons due to underground rhizomes (NOAA n.d.). This strategy proves beneficial in competition against native plants as well.

ECONOMIC IMPACTS

The economic impacts of *M. quadrifolia* have not been quantitatively studied (SOM 2018), but in some instances there might be some expenditure on control or management.

POSITIVE ASPECTS

Plenty of researchers have certified the medical values of *M. quadrifolia* (Longman 1997;

Ripa et al. 2009; Khan and Manzoor 2010; Ashwini et al. 2012; Manjula and Mythili 2012; Mathangi and Prabahakaran 2012; Reddy et al. 2012; Soni and Singh 2012; Uma and Pravin 2013; Gopalakrishnan and Udayakumar 2014; Agarwal et al. 2018). *M. quadrifolia* is high in nutrient content, which can be used to reduce body heat and thirst. It can also be used to treat cough, bronchitis, diabetes, psychiatric diseases, eye diseases, diarrhea, and skin diseases. *M. quadrifolia* has been used in traditional treatments for hypnotic, leprosy, hemorrhoids, fever, insomnia, as a cough expectorant, etc. In addition, this plant was found posing potential antimicrobial activity.

For the environment, *M. quadrifolia* can promote nutrient mitigation from freshwater and contribute to wetland restoration.[1] In studies of sequestering of heavy metal by aquatic plants, *M. quadrifolia* showed the bioaccumulation of cadmium and chromium (SOM 2018).

MONITORING AND SIGHTINGS

The *M. quadrifolia* observation records in Massachusetts waterbodies were collected by the Massachusetts Department of Environmental Protection (MassDEP) and the Department of Conservation & Recreation (DCR). The standard operating procedures, including sampling methodology, species identification, quality control and assurance, were implemented during the sample collections. Sites with *M. quadrifolia* were mapped as a data layer with ArcGIS® ArcMap 10.1 (ESRI, Redlands, California). Mapping the *M. quadrifolia* distribution areas can help to inform future decisions and regulations for controlling the spread of this potentially invasive species.

Data was collected from MassDEP Water Quality Assessment Reports and DCR Lakes and Ponds Program Database (Table 16.1). An ArcGIS map was created to show the distribution of *M. quadrifolia* in Massachusetts (Figure 16.2), which can be used for the future decisions on nuisance plant control and prevention. Currently, there are 9 waterbodies (Table 16.2) within 6 watersheds (Table 16.3) that are infested by *M. quadrifolia* in Massachusetts. Among them, 6 are lakes/ponds and three are rivers (Tables 16.1 and 16.2).

TABLE 16.1

Waterbodies Infested by European Waterclover

Waterbody	Agency	Report Source	First Year Reported
Batemans Pond	MassDEP	SuAsCo Watershed 2001 Water Quality Assessment Report 82-AC-1 CN 92.0, August 2005	2005
Charles River	MassDEP	Charles River Watershed 2002–2006 Water Quality Assessment Report 72-AC-4 CN136.5, April 2008	2008
First Pond	DCR	DCR Data from Lakes and Ponds, 2006	2006
Lake Waban	MassDEP	Charles River Watershed 2002–2006 Water Quality Assessment Report 72-AC-4 CN136.5, April 2008	2008
Lake Waban	DCR	DCR Data from ACT, GeoSyntec-ACT	Not Available
Massapoag Brook	MassDEP	1994 Resource Assessment Report	1994
Prospect Lake	MassDEP	Housatonic River Watershed 2002 Water Quality Assessment Report 21-AC-4 CN141.5, September 2007	2007
Saxonville Pond	MassDEP	SuAsCo Watershed 2001 Water Quality Assessment Report 82-AC-1 CN 92.0, August 2005	2005
Stevens Pond	MassDEP	Ipswich River Watershed 2000 Water Quality Assessment Report 92-AC-1 088.0, April 2004	2004
Sudbury River	Individual	Joy Trahan-Liptak	2013

TABLE 16.2

Towns and Infested Waterbodies

Number	Waterbody	Town
1	Batemans Pond	Concord
2	Charles River	Dover, Newton
3	First Pond	Saugus
4	Lake Waban	Wellesley
5	Massapoag Brook	Sharon, Canton
6	Prospect Lake	Egremont
7	Saxonville Pond	Framingham
8	Stevens Pond	Boxford
9	Sudbury River	Concord, Lincoln, Sudbury, and Wayland

M. quadrifolia is a highly valuable plant for the ecosystem and human health. Currently, this species does not pose aggressive competition and fast spreading. According to the records from MassDEP, there is no significant *M. quadrifolia* distribution that has occurred in recent years in Massachusetts, but the potential economic loss and ecological impact cannot be underestimated. An effective control method should be applied for an infested waterbody. For a waterbody has not been invaded or is at risk to be invaded, prevention methods should be conducted before the expansion of this species.

MANAGING ACTIONS

There are various methods for controlling the spread of *M. quadrifolia*. The management measures range from preventative measures to physical and chemical management methods.

TABLE 16.3

Infested Watersheds

Number	Infested Watersheds
1	Housatonic River
2	Charles River
3	Concord (SuAsCo) River
4	Ipswich River
5	North Coastal Drainage
6	Boston Harbor (Neponset) Drainage

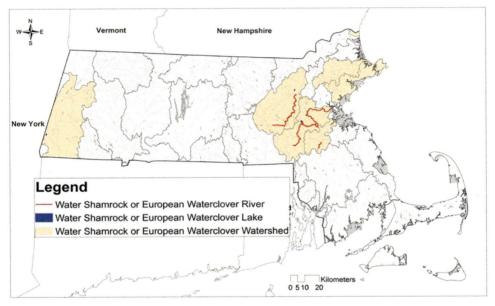

FIGURE 16.2 Distribution map depicting European waterclover (*M. quadrifolia*) in Massachusetts. (See Table 16.1 for site descriptions.)

Biological method of control for *M. quadrifolia* is currently unknown. The choice of management method(s) depends on the invasion range, water quality conditions, weather, management budget, regulatory restrictions, etc.

Preventative measures may assist in minimizing the spread and distribution of a particular invasive species. *M. quadrifolia* is considered a "Watch List species" in certain states. Sometimes species may be subject to restrictions or prohibition on the possession, introduction, importation, or selling of a species (Hackett et al. 2018). These are examples of ways to prevent the spread of an invasive species. In the case of *M. quadrifolia*, further methods include encouraging water

garden and aquarium enthusiasts to use native or non-invasive species and educating user groups of the proper disposal of unwanted plants. The identification and monitoring of waterbodies at higher risk of invasion as well as actively managing known *M. quadrifolia* sites aid in preventative measures as well (Hackett et al. 2018).

Physical management methods include hand pulling, shading, dredging, and benthic barriers (SOM 2018). Hand pulling was found to be as effective as, or even more effective than chemical control methods, in rice field infestations of *M. quadrifolia* (Anwar et al. 2013; Popy et al. 2017). *M. quadrifolia* prefers semi-shade or no shade environment, which means physical shading by completely blocking sunlight can inhibit the plant's growth. Dredging is a sediment removal process that controls the rooted plant growth through light limitation (by adding depth to a waterbody) or substrate limitation (by removing or changing the substrate in which the plant would survive). Benthic barriers are a control method proposed by MassDEP, which could be applied in small areas such as swimming beaches to control this invasive species.

Instead of the above methods, chemical control via herbicide use is also one of the most popular management approaches for nuisance and/or invasive plant control. For the management of *M. quadrifolia*, two types of experiments have been conducted to understand the impacts of chemical treatments on this species. These include the effectiveness of herbicides at controlling *M. quadrifolia* as a rice field weed without damaging rice production and researching the effects of herbicide runoff on growth of *M. quadrifolia* as an endangered species in Eurasia (Bruni et al. 2013). These studies, although different from the intent of this document, may be useful for further investigation of effective methods in *M. quadrifolia* management. Diquat and glyphosate have been recognized as effective for controlling *M. quadrifolia*, but do not achieve complete control and may require higher doses or special environmental conditions (Mattson et al. 2004; Wagner 2004). The State of Michigan reported a summary table of effective herbicide active ingredients for *M. quadrifolia* control (Hackett et al. 2018). The effective active ingredients include Bensulfuron-methyl, Bensulfuron-methyl + pretilachlor, Cyhalofop-butyl, Imazamox + Glyphosate, Mustard crop residue, Pendimethalin, Penoxsulam, Pretilachlor, Profoxydim, and Simetryn. The corresponding herbicide products include Londax™, Londax™ Power, Clincher®, Clearcast® + AquaPro®, Drexel Aquapen®, Viper® India, Aura® (Table 16.4). However, many of the above treatments have not been studied extensively either in field research or their effectiveness on regrowth. In addition, many of the listed chemicals such as Bensulfuron-methyl, Bensulfuron-methyl + pretilachlor, Cyhalofop-butyl, and Pendimethalin are toxic to fish and other invertebrates. Others such as Penoxsulam, Profoxydim, and Imazamox + glyphosate have restrictive usage especially

in aquatic environments. The State of Michigan warns that directions on pesticide labels should always be followed and the appropriate state departments be consulted for up-to-date regulations, restrictions, permitting, licensing, and application information (Hackett et al. 2018). Plant regrowth can usually occur after an initial treatment, therefore repeated applications of herbicides are typically needed. Please note that all the aforementioned management methods have other potential impacts to a waterbody that should be considered when choosing a management approach and may require regulatory approval before proceeding. More information about chemical control on this species and other 18 invasive species can be found in Appendix III. Any chemical used for aquatic invasive species control requires a WM04 Chemical Application License issued by the Massachusetts Department of Environmental Protection.

SUCCESS STORY

Wet direct-sown rice planting is a very popular method of crop establishment among Asian countries in recent years. However, direct-sown rice is more easily affected by *M. quadrifolia* growth when compared to the transplanting system (Saha and Rao 2010). This is because the *M. quadrifolia* has a more similar growth period with wet direct-sown rice, leading to a competitive relationship (Saha 2008). Plant competition may cause a reduction of rice when harvested, therefore an effective management for *M. quadrifolia* control is needed.

In the dry seasons of 2016 and 2017, chemical herbicides were applied in the wet direct-sown rice field for weed control at Cuttack, Odisha, India. Bensulfuron-methyl was applied alone as well as mixed with pretilachlor in the rice field to examine the efficacy. After a month of sowing, the major weed flora in the rice field included *Echinochloa colona* (i.e., jungle rice) (8.7%), *Cyperus difformis* (i.e., variable flatsedge, smallflower umbrella-sedge, or rice sedge) (26.1%), *Fimbristylis miliacea* (i.e., grasslike fimbry) (14.5%), *Sphenochlea zeylanica* (i.e., gooseweed or wedgewort) (27.5%), and *Marsilea quadrifolia* (i.e., European waterclover) (23.2%). Unfortunately, the results showed that there was

TABLE 16.4

Effective Herbicides for European Waterclover Control

Herbicide	Approved for Use in MA	Pros	Cons
Bensulfuron-methyl (Londax™)	No	• Effective in laboratory (90%) when applied at 0.022 pounds ai per acre, seven days after EWC transplant • Less harm to non-target plant species (Selective herbicide)	• Has not been tested in field • Regrowth not examined • Permitted for aquatic use in only rice fields • Moderate toxicity to fish, algae • Resistance can be built • Application in water less than 70°F delays activity
Bensulfuron-methyl + pretilachlor (e.g., Londax™ Power)	No	• Effective in rice field (88.1–95.1%) when applied at 0.054 + 0.54 pounds ai per acre, three days after rice transplant • Slightly more efficient than hand weeding	• Regrowth not examined • Experimented in rice field • Moderate toxicity to fish, aquatic invertebrates, algae
Cyhalofop-butyl (e.g., Clincher®)	No	• Effective in laboratory (80%) when applied at 0.005 ppm. four days after EWC transplant	• Has not been tested in field • Regrowth not examined • Toxic to fish and macroinvertebrates • Permitted for aquatic use in only rice fields • May contaminate groundwater • May harm non-target plant species (broad-spectrum, contact herbicide)
Imazamox + glyphosate (e.g., Clearcast® + AquaPro®)	Yes	• Possibly effective in pond in Michigan (anecdotal)	• Regrowth not yet monitored • Restricted concentration when near potable water intakes • Prohibited for use in waterbodies <½ mile from a potable water intake • May harm non-target species (broad-spectrum, systemic herbicide)
Mustard crop residue	No	• Effective in rice field (83.98–91.73%) at 2.0 ton/ha	• Residue not thoroughly described • Low number of replicates • Non-target effects not measured

Herbicide			
Pendimethalin (e.g., Drexel Aquapen®)	Yes	• Effective in rice field (59.5–73.2%) when applied at 0.071–0.76 pounds ai per acre, two days after rice transplant • Less harm to non-target plant species (selective herbicide)	• Used as pre-emergent with a post-control treatment (e.g., penoxsolum, 2,4-D amine, hand pulling) • Regrowth not examined • Toxic to fish • Permitted for aquatic use in only rice fields • Does not control established weeds
Penoxsulam (e.g., CLEANTRAXX)	Yes	• Effective in rice field (71.3–75.2%) when applied at 0.0179–0.0201 pounds ai per acre, 3–12 days after rice transplant • Effective in laboratory (95%) when applied at 20 ppm • Less harm to non-target plant species (selective herbicide)	• Not effective in pond with glyphosate due to next season regrowth (anecdotal) • Performed at almost same effectiveness as hand pulling • Regrowth not examined • Posttreatment irrigation water restrictions
Pretilachlor	No	• Effective in rice field (88.1–95.1%) when applied at 0.071–0.54 pounds ai per acre, two days after rice transplant • Less harm to non-target plant species (selective herbicide)	• Used as pre-emergent with a post-control treatment (e.g., penoxsolum, 2,4-D amine, hand pulling) • Regrowth not examined • Moderate toxicity to fish, aquatic invertebrates, algae • Permitted for aquatic use in only rice fields
Profoxydim (e.g., Aura®)	No	• Effective in laboratory (82–95%) when applied at 0.02–0.2 ppm, four days after EWC transplant • More effective with Dash™ HC adjuvant • Less harm to non-target plant species (selective herbicide)	• Has not been tested in field • Regrowth not examined • Permitted for aquatic use in only rice fields
Simetryn	No	• Effective in laboratory (significantly different from controls) when applied at 0.0004–40 µmol/l, when inoculated for ten days • Less harm to non-target plant species (selective herbicide)	• Has not been tested in field • Measured fresh weight (not dry) • Regrowth not examined • High toxicity to beneficial algae • Moderate toxicity to fish, aquatic invertebrates

Source: Modified from and specific references in Hackett et al. (2018).

about 50% reduction on the grain yield of rice after herbicide treatment because of the weed competition existing. However, in terms of the efficacy on overall weed control, both Bensulfuron-methyl and Bensulfuron-methyl + pretilachlor performed well. The applications of Bensulfuron-methyl at 60 g/ha and Bensulfuron-methyl + pretilachlor at 50 + 450 g/ha each successfully reduced the total weed density (especially *M. quadrifolia* density) and dry matter production with approximately 90% control.

Furthermore, control measures may be learned through practices and studies performed in areas of conservation. Although the purpose of such studies would be to protect and rebound the *M. quadrifolia* population, it may be discovered what methods have led to its decline adding to information on effective control measurements. One study (Corli et al. 2021), aimed at species recovery in the Po Valley in Northern Italy within paddy fields, discussed the conditions in which *M. quadrifolia* would thrive. While water chemistry was found to have limited effects on the species due to its wide ecological range, it was found shady conditions provided by rice canopies created micro-habitats suitable for *M. quadrifolia* to grow. In this study, *M. quadrifolia* was found to have poor or no capacity to adjust pigment composition to light level, in addition to having lower chlorophyll fluorescence values, suggesting the species undergoes stress in full-light habitats. In management, avoiding opportunity for growth under shady regions and maintaining open spaces within an affected waterbody may contribute to control practices. Additionally, the study found *M. quadrifolia* survived better in organic rice fields (with no herbicide use) as well as at conventional farms with reduced herbicide use. The use of glyphosate during the early development phase of *M. quadrifolia* (mid-May) was found to have possible detrimental effects leading to the die-off of the species. These results show how data from conservation studies may be used in management practice techniques in areas where *M. quadrifolia* is of concern.

NOTE

1 Any invasive species should be inhibited for ecological restoration projects.

REFERENCES

Agarwal, S.K., S. Roy, P. Pramanick, P. Mitra, R. Gobato, and R. Mitra. 2018. Marsilea quadrifolia: A floral species with unique medicinal properties. *Parana Journal of Science and Education* 4(5): 15–20.

Anwar, B.M., A. Hussain, M.A. Ganai, and N.A. Teli. 2013. Efficacy of penoxsulam against weeds in transplanted rice (Oryza sativa L.) under temperate conditions of Kashmir. *Applied Biological Research* 15: 145–148.

Ashwini, G., P. Pranay, G. Thrinath, R.T. Karnaker, and P.V.S. Giri. 2012. Pharmacological evaluation of *Marsilea qudrifolia* plant extracts against Alzheimer's disease. *International Journal of Drug Development and Research* 4(2): 153–158.

Benson, A.J., C.C. Jacono, P.L. Fuller, E.R. McKercher, and M.M. Richerson. 2004. *Summary Report of Nonindigenous Aquatic Species in U.S. Fish and Wildlife Service Region 5*. Arlington, VA: U.S. Fish and Wildlife Service, p. 145.

Bloom, W.W., 1955. Comparative viability of sporocarps of Marsilea quadrifolia L. in relation to age. *Transactions of the Illinois State Academy of Science* 47: 72–76.

Bruni, I., R. Gentili, D.F. Mattia, P. Cortis, G. Rossi, and M. Labra. 2013. A multi-level analysis to evaluate the extinction risk of and conservation strategy for the aquatic fern *Marsilea quadrifolia* L. in Europe. *Aquatic Botany* 111: 35–42.

Cao, L., and L. Berent. 2020. *Marsilea quadrifolia L.: U.S. Geological Survey, Nonindigenous Aquatic Species Database*. Gainesville, FL and Ann Arbor, MI: NOAA Great Lakes Aquatic Nonindigenous Species Information System, https://nas.er.usgs.gov/queries/GreatLakes/FactSheet.aspx?SpeciesID=293 (Accessed May 5, 2020).

Corli, A., S. Orsenigo, R. Gerdol, S. Bocchi, A.P. Smolders, L. Brancaleoni, and G. Rossi. 2021. Coexistence of rice production and threatened plant species: Testing *Marsilea quadrifolia* L. in N-Italy. *Paddy and Water Environment* 19: 395–400.

Gopalakrishnan, K., and R. Udayakumar. 2014. Antimicrobial activity of *Marsilea quadrifolia* (L.) against some selected pathogenic microorganisms. *Microbiology Research Journal International* 4(9): 1046–1056.

Hackett, R.A., B.C. Cahill, and A.K. Monfils. 2018. *2018 Status and Strategy for European Water-clover (Marsilea quadrifolia L.) Management*. Lansing, MI: Michigan Department of Environmental Quality, p. 30.

Henry, R.D., and R.M. Myers. 1983. Spread of *Marsilea quadrifolia* in McDonough County, Illinois. *American Fern Journal* 73(1): 30.

Holm, L., J. Doll, E. Holm, J. Pancho, and J. Herberger. 1997. *World Weeds: Natural Histories and Distribution*. New York, USA: John Wiley and Sons, pp. 455–461.

Husson, E., O. Hagner, and F. Ecke. 2013. Unmanned aircraft systems help to map aquatic vegetation. *Appl Veg Sci* 17: 567–577.

ITIS (Integrated Taxonomic Information System). 2021. *www.itis.gov/servlet/SingleRpt/SingleRpt#null* (Accessed March 27, 2021).

Johnson, D.M. 1986. Systematics of the new world species of Marsilea (Marsileaceae). *Syst Bot Mon* 11: 1–87.

Kathiresan, R., and G. Gualbert. 2016. Impact of climate change on the invasive traits of weeds. *Weed Biology and Management*, 16(2): 59–66.

Kao, W.K., and B.L. Lin. 2010. Phototropic leaf movements and photosynthetic performance in an amphibious fern, *Marsilea quadrifolia*. *Journal of Plant Research* 123(5): 645–653.

Khan, M.A., and A.S. Manzoor. 2010. Studies on biomass changes and nutrient lock-up efficiency in a Kashmir Himalayan wetland ecosystem. *J Ecol Nat Environ* 2(8): 147–153.

Kiran, B.R., E.T. Puttaiah, S. Raghavendra, and M. Ravikumar. 2007. Ecological studies on aquatic macrophytic vegetation in Shivaji Tank, Karnataka, India. *Plant Archives* 7: 637–639.

Les, D.H., and L.J. Mehrhoff, 1999. Introduction of nonindigenous aquatic vascular plants in Southern New England: A historical perspective. *Biological Invasions* 1, 281–300.

Longman, O., 1997. Indian medicinal plants, vol. 4 (Orient Longman Pvt. Ltd., Chennai, India) pp. 5–9.

Luo, X.Y., and H. Ikeda. 2007. Effects of four rice herbicides on the growth of an aquatic fern, *Marsilea quadrifolia* L. *Weed Biology and Management*, 7(4): 237–241.

Manjula, R., and T. Mythili. 2012. Improved phytochemical production using biotic and abiotic elicitors in *Marsilea quadrifolia*. *International Journal of Current Science Research and Review* 98–101.

Mathangi, T., and P. Prabahakaran. 2012. Screening for antimicrobial activity of the leaf extracts of *Marsilea quadrifolia* against various bacterial pathogens. *Herbal Tech Industry* 11–12.

Mattson, M.D., P.J. Godfrey, R.A. Barletta, and A. Aiello. 2004. Eutrophication and Aquatic Plant Management in Massachusetts: Final Generic Environmental Impact Report. Cambridge, MA: Commonwealth of Massachusetts, Executive Office of Environmental Affairs Commonwealth of Massachusetts.

NOAA Great Lakes Environmental Research Laboratory. (n.d.). NOAA National Center for Research on Aquatic Invasive Species (NCRAIS). Retrieved from https://nas.er.usgs.gov/queries/GreatLakes/FactSheet.aspx?SpeciesID=293

Popy, F.S., A.M. Islam, A.K. Hasan, and M.P. Anwar. 2017. Integration of chemical and manual control methods for sustainable weed management in inbred and hybrid rice. *J Bangladesh Agric Univ* 15: 158.

Reddy, K.S., C.S. Reddy, and S. Ganapaty. 2012. Psychopharmacological studies of hydro alcoholic extract of whole plant of *Marsilea quadrifolia*. *Journal of Scientific Research* 4: 279–285.

Reznicek, A.A., E.G. Voss, and B.S. Walters. 2011. MICHIGAN FLORA ONLINE. University of Michigan. Web (Accessed March 28, 2021). http://michiganflora.net/species.aspx?id=1683.

Ripa, F.A., L. Nahar, M. Haque, and M. Islam. 2009. Antibacterial, cytotoxic and antioxidant activity of crude extract of *Marsilea quadrifolia*. *European Journal of Scientific Research* 3: 123–129.

Saha, S. 2008. Sustaining high rice productivity the IWM way. *Indian Farming* 58(3): 18–22.

Saha, S., and K.S. Rao. 2010. Evaluation of bensulfuron-methyl for weed control in wet direct-sown summer rice. *Oryza* 47(1): 38–41.

Scriver, M., A. Marinich, C. Wilson, and J. Freeland. 2015. Development of species-specific environmental DNA (eDNA) markers for invasive aquatic plants. *Aquatic Botany* 122: 27–31.

SOM (State of Michigan). 2018. *Status and Strategy for European Water-Clover (Marsilea quadrifolia L.) Management Scope*. Lansing, MI: State of Michigan, p. 30.

Soni, P., and L. Singh. 2012. *Marsilea quadrifolia* linn—A valuable culinary and remedial fern in Jaduguda, Jharkhand, India. *International Journal of Life Science & Pharma Research* 2(3): 99–104.

Strat, D. 2012. *Marsilea quadrifolia* L. in the protected wetlands from Romania. In: Proceedings of the International Conference on Water Resources and Wetlands, pp. 449–457.

Uma R., and B. Pravin. 2013. Invitro cytotoxic activity of *Marsilea quadrifolia* Linn of MCF-7 cells of human breast cancer. *International Research Journal of Medical Sciences* 1(1): 10–13.

U.S. Geological Survey (USGS). 2020. Specimen observation data for *Marsilea quadrifolia* L., Nonindigenous Aquatic Species Database, Gainesville, FL, https://nas.usgs.gov/queries/CollectionInfo.aspx?SpeciesID=293 (Accessed May 5, 2020).

Wagner K.J. 2004. *The Practical Guide to Lake Management in Massachusetts: A Companion to the Final Generic Environmental Impact Report on Eutrophication and Aquatic Plant Management in Massachusetts*. Boston, MA: Commonwealth of Massachusetts, Executive Office of Environmental Affairs Commonwealth of Massachusetts.

17 Yellow Floating Heart (*Nymphoides peltata*)

ABSTRACT

Yellow floating heart (*Nymphoides peltata*), also called water fringe, is an Eurasian invasive floating-leaved aquatic plant in Massachusetts. Currently, there are six reported sightings in Massachusetts including Scarboro Golf Course Pond in Boston and Lymans Pond in Dover, both within the Charles River Watershed. Since there are not many waterbodies in Massachusetts infested by this species, it is important to prevent the spread to other waterbodies and implement management strategies for the eradication of the *N. peltata*. Physical methods such as hand pulling and dredging and chemical control methods such as the application of glyphosate and diquat are suggested in this chapter.

INTRODUCTION

Taxonomy Hierarchy (ITIS 2021)

Kingdom	Plantae
Subkingdom	Viridiplantae
Infrakingdom	Streptophyta
Superdivision	Embryophyta
Division	Tracheophyta
Subdivision	Spermatophytina
Class	Magnoliopsida
Superorder	Asteranae
Order	Asterales
Family	Menyanthaceae
Genus	*Nymphoides* Hill
Species	*Nymphoides peltata* (S.G. Gmel.) Kuntze

BASIC BIOLOGY

Yellow floating heart (*Nymphoides peltata*), also called Water Fringe, Fringed Water Lily, or floating heart, is an aquatic floating-leaved perennial plant (Figure 17.1). It can grow long branching stolons (up to 2 m) that form new plants with a stem which resembles a rope arising from the roots (Pfingsten et al. 2021). Meanwhile, *N. peltata* has green leaves that appear circular or heart-shaped with a diameter of 3–15 cm (CABI 2019). These leaves grow alternately on the stem but appear oppositely at the flower stalks. *N. peltata* also has ripple margins, with the underside parts of the leaf appearing purple (Nault and Mikulyuk 2009). In addition, margin-fringed yellow flowers (top left panel, Figure 17.1) can be seen on the peduncles arising above the water surface. Each of these flowers has five petals, with a diameter around 3–5 cm (Robinson 2004). The seeds of *N. peltata* are very small (3–5 mm in length and 3 mm in width), with winged margins (Cook 1990; CABI 2019). The seed capsules are shown in the top right panel of Figure 17.1.

N. peltata could be mistakenly identified as the native Little floating heart (*Nymphoides cordata*) and native Yellow Water Lily (also called Yellow Pond Lily, *Nuphar variegata*) as they share some similarities (Figure 17.2). For example, *N. peltata*, *N. cordata*, and *N. variegata* have leaves that float on the water surface, with roots that grow on the water bottom, and flowers that float above their leaves. In addition, they all have dollar-sized, heart-shaped leaves. Even so, there are several distinguishing characteristics that set them apart, as the flowers of *N. cordata* are white without fringed margins, and the roots of *N. cordata* may be found in a bunch on the stem just below the surface (MA DCR 2016). Meanwhile, *N. variegata* has the largest leaves out of the other two species (*N. peltata* and *N. cordata*), which are 6–8 inch (Robinson 2004). Moreover, *N. variegata*'s flowers are cup-shaped, which is different from that of *N. peltata* (Figure 17.2).

NATIVE RANGE

N. peltata is native to Eastern Asia and the Mediterranean (Stuckey 1973). This hardy plant can establish itself in a wide range of

DOI: 10.1201/9781003201106-17

FIGURE 17.1   Yellow floating heart. (Photos by Lyn Gettys.)

waterbodies, but it prefers slow-moving water-bodies such as lakes, ponds, and streams (Robinson 2004). Meusel et al. (1978) reported in their research that *N. peltata* is the only *Nymphoides* species that appears in moderately cold temperature areas. For most circumstances, *N. peltata* is detected growing in neutral and alkaline waterbodies (Darbyshire and Francis 2008). Even so, they can sometimes be found in the acid waterbodies.

INVASIVE RANGE

Given its ability to inhabit a wide range of waterbod-ies, *N. peltata* has spread throughout many regions: Asia, Europe, North America, and Oceania (CABI 2019). In the United States, the earliest record of *N. peltata* invasion was recorded in Winchester, Massachusetts, in 1882 (Stuckey 1973; U.S. Fish and Wildlife Service 2018). There is also a report of *N. peltata* in New York City's Central Park in 1886.

FIGURE 17.2   Yellow water lily (*left*) and little floating heart (*right*). (Left photo by Les Mehrhoff from discoverlife.org, with permission from Dr. John Pickering; right photo by Alexey Zinovjev and Irina Kadis, with permission from salicicola.com)

Other first recordings include Missouri in 1893, Louisiana in 1899, and Pennsylvania in 1905. The earliest record of *N. peltata* in the Hudson River was in 1929 from New York, and its origin was speculated as being an escape from a water garden or pool (Stuckey 1973).

MECHANISMS/VECTORS OF SPREAD

N. peltata could have been introduced by "accident through flooding of ornamental ponds into surrounding natural waterways" (Stuckey 1973; Josefsson and Andersson 2001). Even so, *N. peltata* can reproduce by both vegetative and sexual methods which aids further spread (Robinson 2004). The reproduction by fragments can occur through the stolon, rhizomes, and leaves with part of a stem and part of the stolon. These fragments can then attach on vessels/boat moving from a waterbody to another waterbody, forming new *N. peltata* plant.

Sexual reproduction *N. peltata* also exists due to the production of seeds which are produced by either cross-pollination or self-pollination, but the former method can create more viable seeds than the latter (Ornduff 1966). Like fragments, seeds can also be attached to the boating vessels, causing possible dispersal. In addition, wildlife is another factor that can further the spread of seeds. For example, although seeds can be completely digested by animals, which decreases the possibility of seeds spreading, because of the special structure of the seed, seeds can easily attach to the fur or feather of birds or waterfowls

(Cook 1990). Furthermore, prior research (Van Der Velde and Van Der Heijden 1981) mentioned some insects, such as bees (Apidae), hoverflies (Syrphidae), and shoreflies (Ephydridae) species, play an important role in the pollination among flowers, impacting the seeds production and the sexual reproduction of *N. peltata*.

ECOLOGICAL IMPACTS

N. peltata is a noxious aquatic invasive plant, which can thrive in a variety of waters. Once this plant species invades a waterbody, it may replace the local vegetation and as a result negatively impact the ecosystem. According to multiple research studies (Stuckey 1973; Robinson 2004), the invasion of *N. peltata* might severely impact ecosystems because dense mats of *N. peltata* can lead to an increase in sedimentation, degrade water quality, and intercept the sunlight which leads to the exclusion of other submerged plants. These massive, thick mats could clog drainage systems and waterways, reducing the water flow.

Furthermore, *N. peltata* can threaten food webs. For example, algae which serves as a food source for many species may be shaded out due to the abundance of *N. peltata* (Robinson 2004). Moreover, large amounts of *N. peltata* existing in the waterbodies could compete with other native species. This would cause fish or animals rely on these native species to relocate or perish. Furthermore, decomposing *N. peltata* may lower dissolved oxygen in the water, harming fish or other aquatic animals. Therefore, we can see that *N. peltata* may cause an enormous influence on the ecosystem and biodiversity.

ECONOMIC IMPACTS

N. peltata can have negative economic impacts. *N. peltata* in the water may negatively impact boating and swimming recreational activities, resulting in economic loss. Other water activities such as fishing could also be reduced as ecological habitats are threatened. Furthermore, tourism development could be hampered, the value of surrounding real estate could be influenced (Robinson 2004), channels and waterways could be blocked, and aesthetic value could be decreased. Control methods are expensive (Larson 2007) and time-consuming. For reference, control efforts in Sweden were estimated to cost SEK28,000 (US$4,500)/ha or SEK56,000 (US$9,000) annually (Gren et al. 2007).

POSITIVE ASPECTS

Although *N. peltata* has negative influences on the ecosystem and economy, there are a few benefits. The seeds of *N. peltata* provide food for animals, such as birds and waterfowls, and the plant can be used as a food for *Cataclysta lemnata* (L.) (a moth species) (Van Der Velde 1979). It can also be helpful in algal blooms control and be used to absorb heavy metal or organic pollutants from the wastewater (Wang et al. 2007; Wang et al. 2010). In addition, it is also found that *N. peltata* can be screened for constituents with antitumor activity (Du et al. 2015). However, invasive species are not allowed for any ecological restoration projects in Massachusetts.

MONITORING AND SIGHTINGS

The invasive species data in this chapter were provided in reports by the Massachusetts Department of Environmental Protection (MassDEP) and other agencies (Table 17.1). Sites with *N. peltata* were mapped as a data layer with ArcGIS® ArcMap 10.1 (ESRI, Redlands, California). This tool can pinpoint the area that *N. peltata* has spread to, which can help to inform future decisions on processes that can happen in these waterbodies, or regulations that

involve efforts to limit the spread of this aquatic invasive species.

From the data collected from the Massachusetts Department of Environmental Protection (MassDEP) Water Quality Assessment Reports (Table 17.1), currently, there are two known waterbodies infested by *N. peltata*: Scarboro Golf Course Pond in Boston and Lymans Pond in Dover, MA. Both ponds are within the Charles River Watershed. According to other sources, it is also found in a small pond in the center of the complex of Rosewood Estates on routes 10/202 within the Westfield River Watershed, an ornamental koi pond adjacent to Burchell's Pond along 101 Miacomet Avenue at Nantucket (Table 17.1 and Figure 17.3). According to Stuckey (1973), this invasive species was first detected by C.E. Perkins in September 1882 in Winchester, Massachusetts. This is also the first sighting in the United States. However, the original record/sample was not able to be located. It is possible that the record of this specimen exists, but it has been out on loan for a very long time, or it may have been lost or destroyed.[1] At the same time, Harvard Herbarium have a sample of the little floating heart *N. cordata* (collected by C.E. Perkins, September, 1882) from Winter's Pond in Winchester, MA. So, it is not clear if this is the same specimen and was reidentified later, or if there might be other specimens collected by C.E. Perkins available elsewhere.[2] This invasive species was also reported in Hall's Pond within the Nashua River Watershed; however, no such pond is found in the area; therefore, it is identified as an unconfirmed sighting (Figure 17.3).

MANAGING ACTIONS

There are many different methods tested for controlling the spread of *N. peltata*. The management measures include physical methods, chemical methods, and biological methods. (*Note:* In choosing a particular method, it is important to determine the range or invasion, local water and weather conditions, budget, and restriction)

PHYSICAL METHODS

Physical methods include mechanical harvesting, hand pulling, cutting, dredging, benthic

TABLE 17.1

Sightings of Yellow floating heart in Massachusetts

Waterbody	Watershed	Reporting Agency	Report Source	First Year Reported
Herbarium (not a waterbody)	Mystic River		Stuckey 1973	1882
Hall's Pond	Nashua		Invasive Plant Atlas of New England (IPANE) at the University of Connecticut online database. Reference ID: 2469430. www. eddmaps.org/ipane/	1983
Scarboro Golf Course Pond	Charles River	MassDEP 2008	Charles River Watershed 2002–2006 Water Quality Assessment Report 72-AC-4 CN136.5, April 2008	1997
Rosewood Estates on routes 10/202: small pond in center of the complex	Westfield River		Invasive Plant Atlas of New England (IPANE) at the University of Connecticut online database. Reference ID: 2469431. www. eddmaps.org/ipane/	1998
Lymans Pond	Charles River	MassDEP 2008	Charles River Watershed 2002–2006 Water Quality Assessment Report 72-AC-4 CN136.5, April 2008	2003
Ornamental Koi Pond adjacent to Burchell's Pond, 101 Miacomet Ave.	Islands		Kelly Omand, Nantucket Conservation Foundation. Ref. # 4742657 on www.eddmaps.org/ distribution/	2016

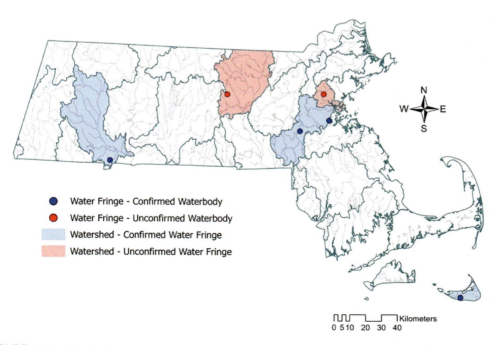

FIGURE 17.3 Distribution map depicting yellow floating heart (*Nymphoides peltata*) in Massachusetts.

barriers, and drawdowns (Robinson 2004; Darbyshire and Francis 2008; DiTomaso et al. 2013). Mechanical harvesting is suitable for large-scale *N. peltata* removal. On a smaller scale, hand pulling can be more efficient than mechanical harvesting, specifically when *N. peltata* is mixed with other aquatic plants (Mattson et al. 2004). However, any physical removal must be done with care to not generate plant fragments and spread the infestation. In addition, repeated cutting might be effective and a viable option (Darbyshire and Francis 2008).

Other physical methods such as dredging and drawdowns may be useful in controlling the spread of *N. peltata*. Dredging is described as a sediment removal process that controls the rooted plant growth through light limitation or substrate limitation. However, dredging sometimes leads to small amounts of regrowth due to fragments formed during the process. Meanwhile, benthic barriers are a control method proposed by Massachusetts authorities, which could be applied in small areas (such as swimming beaches) to control this invasive species. Drawdown is confirmed to successfully control the *N. peltata* when applying benthic barrier methods in the wintertime, but some plants can grow within the barriers margin (Darbyshire and Francis 2008). Although these methods may aid in controlling *N. peltata*, the drawbacks to these methods need to be considered when implemented in waterbodies. For example, physical methods could further spread *N. peltata*, as *N. peltata* can regrow through fragments, roots, and rhizomes (Darbyshire and Francis 2008).

CHEMICAL METHODS

Chemical control is a popular management approach for controlling invasive aquatic plants. Diquat, 2, 4-D, glyphosate, and dichlobenil have been recognized for effectively controlling *N. peltata* (Mattson et al.2004; Darbyshire and Francis 2008). Diquat can be used as consistent control method with correct dosage, proper formulation, and suitable conditions. Meanwhile, 2, 4-D can only achieve partial success (Mattson et al. 2004). Other chemicals such as glyphosate can be used from late spring to mid-late summer with a surfactant or spray (DiTomaso et al. 2013), and dichlobenil can be applied during spring, and

reapplied in the following seasons (Darbyshire and Francis 2008). The relatively new aquatic herbicide ProcellaCOR™ (florpyrauxifen-benzyl) was able to reduce 90% of the total surface coverage of yellow floating heart in a 55-acre drinking water reservoir within 15 days after the treatment, and to less than 3.0 acres within 50 days after the treatment (Lamb et al. 2021) which demonstrates that ProcellaCOR™ has the potential to control yellow floating heart infestations with relatively short-term negative impacts on dissolved oxygen concentrations. More information about chemical control on yellow floating heart and other 18 invasive species can be found in Appendix III. Any chemical used for aquatic invasive species control requires a WM04 Chemical Application License issued by the Massachusetts Department of Environmental Protection.

BIOLOGICAL METHODS

In terms of the biological methods, there has yet to be a significant consensus on biological control methods for *N. peltata* (CEH 2004) as well as research on the results of implemented biological control methods in the United States. Therefore, below is a brief description of species that could be useful in controlling the abundance of *N. peltata*. Waterfowls sometimes feed on *N. peltata* (Steyermark 1963); Eurasian wigeons eat shoots of *N. peltata* (Gaevskaya 1969); Crayfish (*Cambaroides schrenckii*) eats leaves and stalks of *N. peltata* (Gaevskaya 1969); Eurasian coots eat seeds and leaves of *N. peltata*, and mallards also eat *N. peltata* seeds (Van Der Velde et al. 1982); Oligochaeta (earthworms) and Mollusca (clams) contribute in decomposing the *N. peltata* plants (Brock 1984; Xie et al. 2005). Although waterfowls or wigeons may provide some natural control in some areas, it is not suggested as a good idea as a tool of biological control because it might not control the yellow floating heart effectively and also might lead to the spread of *N. peltata* if its seeds are moved in feathers of waterfowls.

Harms and Nachtrieb (2019) as well as Stutz (2019) studied whether *N. peltata* is a suitable target for classical biological control in the United States. Harms concluded that *Nymphoides peltata* and *Nymphoides cristata* are appropriate targets for biological control in the United States. On the other hand, the Stutz study concluded that further surveys are necessary to find natural enemies of *N. peltata*. Before foreign insects are permitted by USDA APHIS for introduction as biological controls in the United States, they must undergo strict host-specificity testing and safety assessment. Therefore, it may take some time to begin implementing biological controls and recording significant results in the United States.

SUCCESS STORY

Lake Carl Blackwell is a mid-sized lake of 3,350 acres in Oklahoma in the United States, which is the only reservoir in Oklahoma known to be infested with *N. peltata* in recent years (Rigsby 2019). To eradicate this invasive species, OSU (Oklahoma State University) biologists created an aquatic plant management plan targeted to *N. peltata*. As a result, in July 2019, a new active and selective herbicide ProcellaCOR SC was applied at Lake Carl Blackwell. The special active ingredient of ProcellaCOR SC effectively achieved fast and long-term treatments. This herbicide is also an approved product by the EPA (US Environmental Protection Agency), designed as "partially non-toxic" to human and animals (SePRO n.d.).

During the application of ProcellaCOR SC, water samples were taken every day and water quality was tested weekly in 2019. The result showed that the concentration of the herbicide was lower than the detection level. At the end of the chemical treatment at Lake Carl Blackwell, there was only around 3 acres of the *N. peltata* found through satellite and drone images, and almost 96% of *N. peltata* was reduced by this treatment.

According to feedback from later observations, some regrowth occurred in the second growing season (Rigsby 2019). Some areas of regrowth were concentrated in narrow spaces where the boats could not easily access. For further control, OSU biologists are researching some alternative methods to test in the coming summer (July 2020).

CONCLUSION

N. peltata is currently in limited waters in Massachusetts. Given its ability to damage water-bodies and ecosystems, it is imperative to continue to gather information on the spatial range of *N. peltata* to limit the spread of the species to other neighboring waterbodies. For documented sightings, if infested area is manageable, it is recommended to eradicate this invasive species using one of the tools recommended in this document.

NOTES

1 According to an email from Matthew Pace, Assistant Curator of the Herbarium at New York Botanical Garden, to Davis Miller, MassDEP Intern, on September 23, 2021.

2 According to an email from Karen Merguerian, Northeastern University Library, to Davis Miller, MassDEP Intern, on September 17, 2021.

REFERENCES

Brock, T.C.M. 1984. Aspects of the decomposition of *Nymphoides peltata* (Gmel.) O. Kuntze (Menyanthaceae). *Aquat. Bot.* 19: 131–156.

CABI. 2019. Nymphoides peltata (yellow floating-heart). Centre for Agriculture and Bioscience International (CABI), www.cabi.org/isc/datasheet/107746#tosummaryOfInvasiveness (Accessed April 21, 2020).

CABI. 2017. *Nymphoides peltata (yellow floating-heart) [original text by A. Mikulyuk]. In Invasive Species Compendium.* Wallingford: CAB International. www.cabi.org/isc/data-sheet/107746. (Accessed April, 2018).

Centre for Ecology and Hydrology (CEH). 2004. Information Sheet 6: Fringed Waterlily. Natural Environment Research Council, Centre for Aquatic Plant Management, p. 2.

Cook, C.D.K. 1990. Seed dispersal of Nymphoides peltata (S.G. Gmelin) O. Kuntze (Menyanthaceae). *Aquatic Botany* 37(4): 325–340.

Darbyshire, S.J., and A. Francis 2008. The Biology of Invasive Alien Plants in Canada. 10. Nymphoides peltata (S. G. Gmel.) Kuntze. *Canadian Journal of Plant Science* 88: 811–829.

DiTomaso, J.M., G.B. Kyser, S.R. Oneto, R.G. Wilson, S.B. Orloff, L.W. Anderson, S.D. Wright, J.A. Roncoroni, T.L. Miller, T.S. Prather, C. Ransom, K.G. Beck, C. Duncan, K.A. Wilson, and J.J. Mann. 2013. Weed Control in Natural Areas in the Western United States. Weed Research and Information Center, University of California, p. 544.

Du, Y., R. Wang, H. Zhang, and J. Liu. 2015. Antitumor constituents of the wetland plant *Nymphoides peltata*: A case study for the potential utilization of constructed wetland plant resources. *Natural Product Communications* https://doi.org/10.1177/1934578X1501000203.

Gaevskaya, N.S. 1969. The role of higher aquatic plants in the nutrition of the animals of fresh-water basins. National Lending Library for Science and Technology, Boston Spa, UK, p. 629.

Gren, I., L. Isacs, and M. Carlsson. 2007. Calculation of costs of alien invasive species in Sweden. Technical report. Uppsala, Sweden: Swedish University of Agricultural Sciences.

Harms, N.E., and J.G. Nachtrieb. 2019. Suitability of Introduced Nymphoides spp. (Nymphoides cristata, N. peltata) as Targets for Biological Control in the United States. https://apps.dtic.mil/sti/pdfs/AD1068632.pdf.

ITIS (Integrated Taxonomic Information System). 2021. www.itis.gov/servlet/SingleRpt/SingleRpt?search_topic=TSN&search_value=29998#null (Accessed April 22, 2021).

Josefsson, M., and B. Andersson. 2001. The environmental consequences of alien species in the Swedish Lakes Mälaren, Hjälmaren, Vänern and Vättern. A Journal of the human Environment 30(8): 514–521.

Lamb, B.T., A.A. McCrea, S.H. Stoodley, and A.R. Dzialowski. 2021. Monitoring and water quality impacts of an herbicide treatment on an aquatic invasive plant in a drinking water reservoir. *Journal of Environmental Management* 288: 112444.

Larson, D. 2007. Reproduction strategies in introduced *Nymphoides peltata* populations revealed by genetic markers. *Aquat. Bot* 86: 402–406.

MA DCR (Department of Conservation & Recreation). 2016. A Guide to Aquatic Plants in Massachusetts. 32 P.

MassDEP. 2008. Charles River Watershed 2002–2006 Water Quality Assessment Report. Report Number 72-AC-4/DWM Control Number CN136.5. Massachusetts Department of Environmental Protection, Division of Watershed Management, Worcester, Massachusetts. 210 P.

Mattson, M.D., P.L. Godfrey, R.A. Barletta, and A. Aiello. 2004. Eutrophication and Aquatic Plant Management in Massachusetts: Final Generic Environmental Impact Report. Commonwealth of Massachusetts, Executive Office of Environmental Affairs Commonwealth of Massachusetts 4: 29–80.

Meusel, H., E. Ja?ger, S. Rauschert, and E. Weinert. 1978. Vergleichende Chorologie der zentraleuropa?ischen. *Flora. Band.* 2. Verl. G. Fischer, Jena.

Nault, M.E., and A. Mikulyuk. 2009. Yellow floating heart (Nymphoides peltata): A Technical Review of Distribution, Ecology, Impacts, and Management. Wisconsin Department of Natural Resources Bureau of Science Services, PUB-SS-1051 2009. Madison, Wisconsin, USA.

Ornduff, R. 1996. The origin of dioecism from heterostyly in Nymphoides (Menyanthaceae). *Evolution* 20(3): 309–314.

Pfingsten, I.A., D.D. Thayer, L. Berent, and V. Howard. 2021. Nymphoides peltata (S.G. Gmel.) Kuntze: U.S. Geological Survey, Nonindigenous Aquatic Species Database, Gainesville, FL, https://nas.er.usgs.gov/queries/FactSheet.aspx?speciesID=243, Revision Date: 3/13/2018, Peer Review Date: 3/23/2016 (Accessed April 21, 2020).

Rigsby, S. 2019. Treatment of Yellow floating heart at Lake Carl Blackwell exceeds expectations. Oklahoma State University (OSU)https://news.okstate.edu/articles/communications/2019/treatment_of_yfh_exceeds_expectations_at_lcb.html (Accessed April 21, 2020).

Robinson, M. 2004. *Yellow floating heart: An Exotic Aquatic Plant, Nymphoides Peltata.* Cambridge, MA: Massachusetts Department of Conservation and Recreation; Office of Water Resources; Lakes and Ponds Program.

SePRO Corporation. n.d. ProcellaCOR®, Long-term control of milfoil, hydrilla, and more. www.sepro.com/aquatics/procellacor (Accessed April 21, 2020).

Steyermark, J.A. 1963. *Flora of Missouri.* Iowa: Iowa State University Press, p. 1725.

Stuckey, R.L. 1973. The introduction and distribution of Nymphoides peltata (Menyanthaceae) in North America. *Bartonia* 42: 14–23.

Stutz, S. 2019. *Identifying the origin of yellow floating heart, nymphoides peltata.* CABI.org. www.cabi.org/projects/identifying-the-origin-of-yellow-floating-heart-nymphoides-peltata/.

U.S. Fish and Wildlife Service. 2018. Yellow Floatingheart (*Nymphoides peltata*) Ecological Risk Screening Summary. www.fws.gov/fisheries/ans/erss/highrisk/ERSS-Nymphoides-peltata-FINAL.pdf.

Van Der Velde, G. 1979. Nymphoides peltata (Gmel.) O. Kuntze (Menyanthaceae) as a food plant for Cataclysta lemnata (L.) (Lepidoptera, pyralidae). *Aquatic Botany* 7: 301–304.

Van Der Velde, G., and L.A. Van Der Heijden. 1981. The floral biology and seed production of Nymphoides peltata (Gmel) O. Kuntze (Menyanthaceae). *Aquat. Bot* 10: 261–293.

Van Der Velde, G., L.A. Van Der Heijden, P.A.J. Van Grunsven, and P.M.M. Bexkens. 1982. Initial decomposition of Nymphoides peltata (Gmel.) O. Kuntze (Menyanthaceae), as studied by the leaf-marking method. *Hydrobiol. Bull.* 16: 51–60.

Wang, X., G. Shi, Q. Xu, and J. Hu. 2007. Exogenous polyamines enhance copper tolerance of Nymphoides peltatum. *Journal of Plant Physiology* 164: 1062–1070.

Wang, L., G. Wang, X.Y. Tang, and Z.Y. Chen. 2010. Inhibitory effect of Nymphoides peltatum on *Microcystis aerugnosa* and its mechanism. *Journal of Ecology and Rural Environment* 26: 257–263.

Xie, Z., T. Tang, K. Ma, R. Liu, X. Qu, J. Chen, and Q. Cai. 2005. The influence of environmental variables on macro- invertebrates in a macrophyte-dominated Chinese lake, with emphasis on the relationships between macrophyte heterogeneity and macroinvertebrate patterns. *J. Freshwater. Ecol.* 20: 503–512.

18 Common Reed Grass (*Phragmites australis*)

ABSTRACT

Common reed grass (*Phragmites australis*) locations recorded from 1994 to 2008 in Massachusetts lakes and rivers are summarized for a better overview of the status of this species in the state. As of 2008, there are 189 reported sites (representing 110 waterbodies) that contain common reed grass in Massachusetts. Brief description on the biology of common reed grass and its management are provided in this report. The mapped sightings will raise the general public's awareness of the presence of reed grass in Massachusetts, which may be helpful in minimizing unintentional spreading of invasive reed grass to other waterbodies. This report also describes different methods that have been used to control this invasive species, including physical (e.g., hand pulling, flooding, and mowing), chemical (e.g., glyphosate and imazapyr), and biological control (e.g., insects and livestock grazing). Depending on the site to be treated, optimal control of invasive common reed grass usually involves a combination of control methods.

INTRODUCTION

Taxonomic Hierarchy (ITIS 2021)

Kingdom	Plantae
Subkingdom	Tracheobionta
Superdivision	Spermatophyta
Division	Magnoliophyta
Class	Liliopsida
Subclass	Commelinidae
Order	Cyperales
Family	Poaceae-Gramineae
Genus	*Phragmites*
Species	*Phragmites australis* (Cav.) Trin. ex Steud.

The common reed, or common reed grass, *Phragmites australis*, is a perennial plant with a cosmopolitan distribution (Figure 18.1) (Packer et al. 2017). Typically seen by roadsides or in disturbed wetlands, *Phragmites australis* is a monocot in the Poaceae (Gramineae) family (Crow and Hellquist 2000). It is a wetland species that is able to tolerate its base and rhizome being submerged in water and conditions with high salinity. It is also capable of growing 2–4 m tall, prominently affecting the appearance of a landscape (Figure 18.2). The mature stalks of the plant have a distinctive brushy inflorescence, or panicle, on top containing seeds. The panicle, typically measuring between 15–38 cm long, is a deep purple when it first emerges, transitioning to a tan color as seeds mature (Del Tredici 2020). *Phragmites* plants have long, pointed dark green leaves that can grow to a length of 60 cm and a width of 5 cm. The leaves form sheaths around the stalks. *Phragmites* grows in large colonies, spreading rapidly via rhizomes as well as by seed (Kelly 1999). *Phragmites* also can produce up to 2,000 seeds annually (Figure 18.3), which can be carried by the wind or transported by wildlife to a new location (Avers et al. 2007). Seed germination is affected by flooding, salinity, and temperature. A depth of two inches or more of water and salinities above 20 parts per thousand (ppt) prevent germination, whereas salinities below 10 ppt do not affect germination (Tiner 1998). Germination usually increases with temperature (e.g., from 16–25°C), while sprouting time decreases with rising temperatures (e.g., from 25 to 10 days, respectively, for that temperature range). Optimal pH for common reed is circumneutral, from 5.5 to 7.5, with the most robust stands occurring in this range. However, it can tolerate acidic conditions as low as pH 3.6 and alkaline conditions up to pH 8.6 (Tiner 1998).

DOI: 10.1201/9781003201106-18

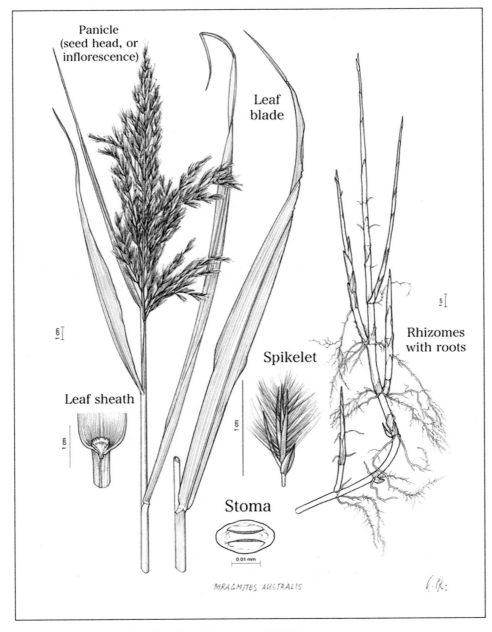

Panicle (seed head, or inflorescence)

Leaf blade

Spikelet

Leaf sheath

Rhizomes with roots

Stoma

0.01 mm

PHRAGMITES AUSTRALIS

FIGURE 18.1 *Phragmites.* (Modified from Packer et al. (2017).)

The common reed grows under a wide range of environmental conditions, from salt to fresh marshes, to dry upland sites or shallow open water. This broad ecological amplitude has probably facilitated its nearly circumglobal occurrence. It grows in pristine wetlands as well as in altered wetlands, dredged material disposal sites (former wetlands), impoundment dikes, disturbed uplands, powerline rights-of-way, roadside ditches, railroad embankments, sandy soils, mine waste areas (refuse piles, acid spoil areas, and tipple areas), and in and around mine slurry impoundments. It can grow as a mat across shallow water and may be found in

FIGURE 18.2 *Phragmites australis* on the roadside of Massachusetts Turnpike (Interstate 90) (*left*) and along the shoreline of a pond nearby Interstate 90 (right). (Photos by W.H. Wong.)

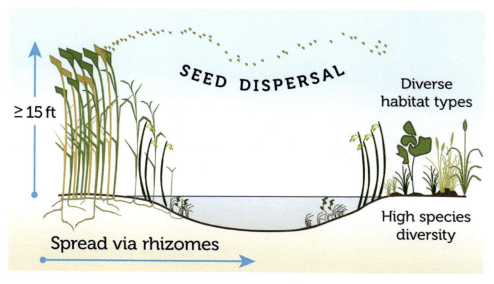

FIGURE 18.3 *Phragmites* spreads to new areas through seed and rhizome dispersal. (From Greatlakesphragmites.net.)

the shallows of lakes and ponds in water about 3 ft deep (Tiner 1998). Found throughout temperate regions of the world, common reed is quite likely the most widely distributed plant (Saltonstall and Meyerson 2016). Common reed quickly becomes the most dominate plant in areas where it occurs, forming a characteristic tall-grass monoculture throughout its global range.

Though it has been unclear in the past whether *P. australis* is a native or an invasive species in the United States, it is now understood that both native and invasive varieties can be found here. Both fossil and archaeological evidence have confirmed that native varieties of *Phragmites* have been present on the continent for thousands of years. The plant has long been important to Native American culture. Indigenous people have used reed grass for food, weaving, and medicine, and as components of tools, including pipes and musical instruments (Kiviat and Hamilton 2001; Tewksbury et al. 2002; Del Tredici 2020). In the early 19th century, the non-native variety, most likely European in origin, appeared in coastal ports in the eastern United States (US FWS n.d.). Prior to the 1900s, *Phragmites* was a relatively uncommon plant in North America. Since the 1960s, however, the plant has become extremely abundant, spreading aggressively and overwhelming wetland ecosystems. It is hypothesized that this rapid spread was a result of the introduction of a non-native variety (Saltonstall 2002). Historic and molecular evidence supports this theory. In 1877, a Eurasian strain of *Phragmites* was documented to have arrived in Philadelphia in ship's ballast, when a young common reed plant was found growing in a ballast dumping area. Burk (1877) wrote of the plant, "If this is not a distinct species, it is a very singular variety." This plant and others carried over in the same way are thought to have proliferated on the east coast and spread inland over time, competing with North American *Phragmites* varieties and other native wetland plants. The invasive behavior seen in the Eurasian *Phragmites* strain may have been exacerbated by increased development in and human disturbances to wetland habitats, including highway construction and pollution of various kinds (Jodoin et al. 2008).

Modern genetic analytical techniques have allowed for a better understanding of the introduction and spread of Eurasian *Phragmites*. A study by Saltonstall (2002) significantly supported the theory of an introduced aggressive *Phragmites* strain to explain the apparent sudden invasiveness of the plant in North America. By sequencing regions of the *Phragmites* genome with intraspecific variation from modern and herbarium-preserved plant samples, Saltonstall (2002) identified several native *Phragmites* haplotypes,[1] including 11 haplotypes A-H, S, Z, and AA, and haplotype I, found predominantly in the Gulf Coast, which have together been identified as a separate subspecies, *Phragmites australis*

subsp. *americanus* (Saltonstall et al. 2004). The most prominent Eurasian haplotype was designated haplotype M. Preserved reed grass samples from earlier than 1910 show widespread abundance of native haplotypes and a limited range of haplotype M in North America. While modern-day analysis still shows the presence of native *P. australis americanus* in the United States, since 1960 haplotype M has expanded its range dramatically, to the point where it can be found across the country (Saltonstall 2002).

The Eurasian haplotype M has spread to an alarming extent, outcompeting native wetland species, including native *Phragmites australis* subsp. *Americanus*, and decreasing overall biodiversity (Tewksbury et al. 2002). It was originally reported that the three haplotypes of *Phragmites australis* subsp. *americanus* native to southern and central New England had been driven to extinction, including haplotype AA, which was unique to New England (Saltonstall 2002). While it is possible that some native strains, such as haplotype AA, have gone extinct due to competition from invasive common reed grass, native varieties can still be found in some areas of New England, lending hope to the survival of original North American varieties. For instance, populations of native *Phragmites australis* were discovered in Falmouth, Massachusetts, in 2004 and 2005 (Payne and Blossey 2007). In some locations, it is believed that there are native *Phgramites* such as those with red tint stems along the Carlisle section of the Concord River (Figure 18.4) but molecular analysis is still required for confirmation.

Native *Phragmites australis* subsp. *americanus* can be visually differentiated from invasive *P. australis* in several ways (Figure 18.5). The American subspecies is considered by some researchers to be sufficiently different from its Eurasian counterpart to justify its elevation to a fully separate species, under the name *Phragmites americanus*, a change which may be made official in the future (Haines 2010; Blossey et al. 2019). The stems of native *P. australis* are reddish, smooth, shiny, and flexible, compared to the tan, rough, dull, and rigid stems of invasive *P. australis*. Leaves of native *P. australis* are lighter yellow or green (Avers et al. 2007). The culms of native *P. australis* remain exposed during the winter, while invasive culms are not exposed. Finally, while invasive *P. australis* often

FIGURE 18.4 *Phragmites australis* along the Carlisle section of the Concord River that are about 21 ft tall. (Photo by Edward Reiner.)

occurs as a monoculture (stands with low genetic diversity resulting from a mainly clonal mode of growth), native *P. australis* rarely occurs as such (Saltonstall et al. 2004). Molecular analysis can be useful in distinguishing native and invasive strains as well as in tracking the spread of *Phragmites* between habitats. The invasive strain was unfortunately not differentiated from the native strain in the sightings included in this report.

Common reed has been reported in every county of Massachusetts. Drainage ditches found along the side of highways and major roads provide a suitable habitat for *Phragmites*. The plant is often found along roadsides. Increasing levels of road construction over the past hundred years may have created corridors facilitating the spread of *Phragmites*. Though roadside stands of *Phragmites* could be beneficial, potentially protecting roads from structural damage caused by ice and snow, abundant roadside *Phragmites* threatens to encroach on wetlands found near roads (Jodoin et al. 2008). The application of de-icing salts on roads goes hand in hand with *Phragmites*' intrusion as a threat to freshwater bog and fen habitats, as the increased salinity caused by road salt runoff harms many noncoastal wetland plant species while also creating conditions that are more favorable to salt-tolerant *Phragmites* (Richburg et al. 2001; Jodoin et al.

2008). In Jodoin et al.'s (2008) study, it was also found that *Phragmites* plants sampled on roadsides from Quebec were majorly composed of haplotype M, the invasive subspecies.

Phragmites seems to grow more abundantly in correlation with both warmer temperatures and higher levels of nutrients in agricultural runoff zones, especially if the density of surrounding competitive plant species is reduced by these factors (Minchinton and Bertness 2003; Jodoin et al. 2008; Legault et al. 2018). *Phragmites* can spread to new areas either in the form of seeds carried by wind or wildlife, with one individual capable of producing 2,000 seeds annually (Figures 18.1, 18.3, and 18.5), or vegetatively from rhizome fragments, which can be carried by water currents, animals, or on construction equipment (Tewksbury et al. 2002; Avers et al. 2007). Once established, reed grass is very difficult to remove or control. According to Martin and Blossey (2013), more than 4 billion dollars is spent annually by different municipal, state, federal, land trust, and private organizations to control common reed grass. Despite the high cost associated with eradication efforts, significant control has not been achieved with current conventional methods, which typically include chemical herbicides, mowing, burning, hand removal, and flooding.

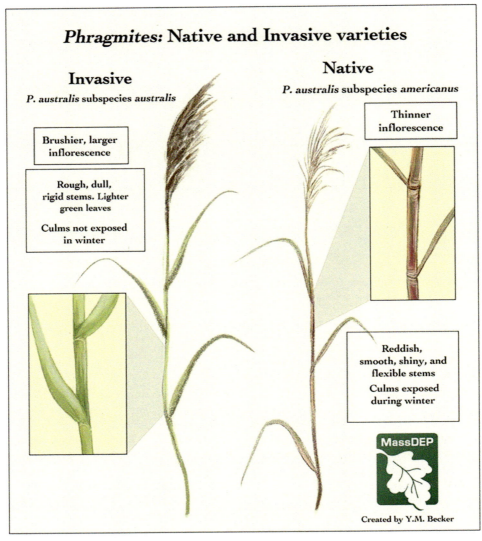

FIGURE 18.5   Differentiating native and invasive *Phragmites australis*. (Photos and written information from within this report and one external source (Tip of the Mitt Watershed Council n.d.) referenced.)

While introduced *Phragmites* is detrimental to many waterbodies, it has also been documented that these plants may be valuable for storm damage prevention and pollution attenuation (i.e., carbon storage and nutrient binding) (Bonham 1980; Gersberg et al. 1986; Gray and Biddlestone 1995; Weis et al. 2021). In the United States, *Phragmites australis* provides sustenance for certain native and introduced species of herbivores, but specialized herbivores of the plant are significantly less abundant in North America than they are in Europe. Effective long-term control of *Phragmites* invasion is required after *Phragmites* stands are removed, stable plant communities must be established in their place. Native wetlands species such as *Spartina alterniflora* (Medeiros et al. 2013) or freshwater bog species can maintain a competitive foothold against *Phragmites* while providing some of the same landscape and ecological benefits. Following efforts to remove *Phragmites*, it is also essential that salt and agricultural pollution are remediated to prevent its reemergence. In addition, it is documented that differences in native plant species abundance were attributed to invasive plant species-specific characteristics, and differences in species richness and composition were attributed to physical location (zonation) in the tidal Hudson River freshwater marshes (McGlynn 2008). "Invasive" status of a dominant plant species was less important in invasive plant–native plant interactions than species-specific characteristics and zonation.

MONITORING AND SIGHTINGS

The data in this report were collected by the Massachusetts Department of Environmental Protection (MassDEP) and the Massachusetts Department of Conservation and Recreation (MA DCR) during staff field visits. MassDEP compiled the data, which represent a portion of the 2,050 Massachusetts waterbodies surveyed by MassDEP from 1994 to 2008. Data layers and maps of sites with common reed grass were created using ArcGIS® ArcMap 10.1 (ESRI, Redlands, California).

Between 1994 and 2008, common reed grass has been reported in multiple waterbodies in Massachusetts, including both rivers and lakes (Figure 18.6 and Table 18.1). More detailed information about the latitude and longitude of specific sites and first reporting years can be found in Appendix A. Based on the MassDEP's records, common reed grass was detected in 82 of the 2,050 waterbodies assessed in Massachusetts. Within DCR's data, 26 additional waterbodies in Massachusetts are reported to contain common reed grass, for a total of 110 waterbodies in this report. For these 110 waterbodies, there are 189 reported sightings; 34 out of 110 waterbodies have been surveyed multiple times while the remaining waterbodies have been surveyed only once. Among these 110 waterbodies, 98 are lakes, 10 are rivers, and 2 are estuaries.

The map displays 110 waterbodies that contain common reed grass (Figure 18.6). It should be noted that this report excludes reed grass in wetlands and reed grass growing in other suitable conditions such as roadside habitats (Figure 18.2) created by road salt implementation (Jodoin et al. 2008). Thus, the sightings in this report underestimate the true presence of reed grass in Massachusetts. Future mapping needs to include information from wetlands, other areas, and other reports such as those reported in Falmouth, Massachusetts (Payne and Blossey 2007). Molecular data and information about the separate distribution of native and invasive haplotypes are also not included in this report, so some sightings may represent native reed grass.

MANAGING ACTIONS

The best results for controlling invasive *Phragmites* involve the use of multiple control methods together and depend on the site to be treated. Before applying treatments, the site must be evaluated to determine the level of the infestation and the density of the invasive *Phragmites* stands.

CHEMICAL CONTROL

Chemical control is one of the most frequently used methods to manage *Phragmites* (Marks 1994; Avers et al. 2007). Although herbicides can be very precise and efficient in eliminating target plant species, they can have detrimental effects on other species and on the environment if not used with care. Therefore, it is critical to apply any herbicide product correctly, read and follow the directions on the product's label, and follow any state or federal laws applicable to herbicide treatments.

Glyphosate is a general-use herbicide that works by targeting the shikimic acid pathway in plants. It does not typically travel through groundwater as it binds tightly to soil where it is broken down within a year of application. Pure glyphosate is not toxic to animals, but depending on the commercial mixture used, it may be combined with other ingredients that are toxic to humans or wildlife (Henderson et al. 2010). An aqueous form of glyphosate, made specifically to be used on plants growing in wetland resource areas water, is sold under the brand name Rodeo™ and has been found to be effective against *P. australis* (Marks 1994).

Another key chemical known to be useful in controlling *Phragmites* is imazapyr (Marks 1994; Avers et al. 2007). Recent studies have shown that imazapyr, either applied alone or in

TABLE 18.1

List of Town Names and Waterbodies that Contain *Phragmites australis*

Town	Waterbody	Town	Waterbody	Town	Waterbody
Ashland	Ashland Reservoir	Holland	Hamilton Reservoir	Pembroke	Furnace Pond
Barnstable	Hamblin Pond				Great Sandy Bottom Pond
Becket	Robin Hood Lake	Holliston	Bogastow Brook	Petersham	Quabbin Reservoir
	Long Bow Lake	Holyoke	Clear Pond		Town Barn Beaver Pond
	Silver Shield Pond	Hopedale	Mill River	Plainville	Wetherells Pond
Bellingham	Box Pond	Kingston	Foundry Pond	Plymouth	Billington Sea
	Jenks Reservoir		Tussock Brook		Ship Pond
	Silver Lake	Lakeville	Little Quittacas Pond		White Island Pond (East Basin)
Billerica	Richardson Pond North	Leicester	Southwick Pond		Five Mile Pond
Boston	Muddy River	Littleton	Fort Pond		Island Pond
Bourne	Flax Pond		Long Pond	Quincy	Blue Hills Reservoir
Brewster	Cliff Pond	Ludlow	Haviland Pond	Rochester	Turner Pond
	Little Cliff Pond	Lynn	Floating Bridge Pond	Rockport	Cape Pond
Brockton	Thirty Acre Pond	Marlborough	Fort Meadow Reservoir	Rutland	Long Pond
Canton	Forge Pond	Mashpee	Duxbury Bay		Moulton Pond
Charlton	Buffum Pond		Mashpee Pond		Muschopauge Pond
	Sibley Pond	Mattapoise	Tinkham Pond		Whitehall Pond
Chelmsford	Newfield Pond	Middleboro	Fuller Street Pond	Salem	Forest River
Cheshire	Cheshire Reservoir (Middle Basin)	Millbury	Hathaway Pond	Sandwich	Peters Pond
Chicopee	Chicoppee Reservoir	Milton	Gulliver Creek	Scituate	Musquashcut Pond
Clinton	Coachlace Pond	Monterey	Lake Garfield	Sheffield	Mansfield Pond
Cohasset	Aaron River Reservoir	Nantucket	Long Pond	Somerset	Somerset Reservoir
Danvers	Porters Pond	New Marlboro	York Lake	S.Hampton	Tighe Carmody Reservoir
	Crane River	New Salem	Lake Rohunta	Springfield	Bass Pond
	Seaplane Basin	N. Andover	Sudden Pond		Five Mile Pond
	Waters River	N. Attleboro	Sevenmile River	Sturbridge	Pistol Pond

Dighton	Broad Cove		Whiting Pond		Kings Pond
Douglas	Dudley Pond	N. Hampton	Roberts Meadow Reservoir		Segreganset River
	Wallum Lake	Otis	Benton Pond		Ponds
Dudley	Shepherd Pond		Creek Pond	Tewksbury	Round Pond
Fall River	Cook Pond		Otis Reservoir	Uxbridge	Lackey Pond
Gloucester	Mill Pond	Oxford	Hudson Pond		Rice City Pond
	Niles Pond	Paxton	Asnebumskit Pond	Wareham	Dicks Pond
Halifax	Monponsett Pond	Peabody	Cedar Pond		Parker Mills Pond
Hanover	Forge Pond		Crystal Pond	Warren	Snipatuit Pond
Hanson	Wampatuck Pond		Pierces Pond	Wrentham	Crocker Pond
Holden	Unionville Pond				Lake Archer

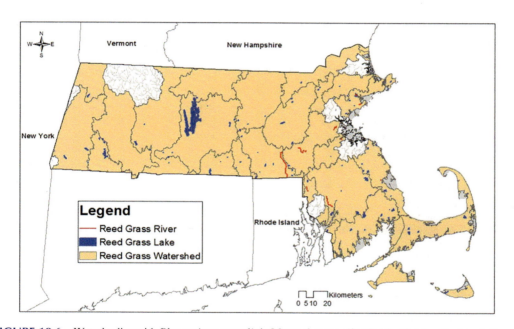

FIGURE 18.6 Waterbodies with *Phragmites australis* in Massachusetts. (See Table 18.1 for site descriptions.)

Note: This map underrepresents the full extent of *Phragmites* throughout of the states (see text for more information).

combination with glyphosate, achieves in more efficient control of *Phragmites* than glyphosate alone (Getsinger et al. 2007). However, while very effective at eliminating *Phragmites*, imazapyr persists in the environment for a longer period than glyphosate. When exposed to light in surface water, the chemical breaks down within days, but in soil it can persist for up to five months (Wisconsin DNR 2012). Stie monitoring after herbicide treatment has also indicated that heavier applications of imazapyr early in the growing season could be somewhat detrimental to recolonization by native plants (Mozdzer et al. 2008). Imazapyr can also move through soil and affect woody tree species in areas adjacent to the application. Therefore, applying an excessive amount of herbicide should be avoided.

The seasonal timing of herbicide application is an important factor in the treatment's success. Imazapyr is most effective on plants that are in stages of active growth, and should be applied to *Phragmites* after full leaf elongation in late spring and before plants have fully matured (Avers et al. 2007; Wisconsin DNR 2012). Glyphosate is more effective on mature plants once they have stopped growing and are in full bloom, i.e., in late summer (Avers et al. 2007). Therefore, for most effective chemical control, it is recommended to treat in early summer with imazapyr, in late summer with glyphosate, or in both seasons with each respective herbicide if necessary. Since imazapyr is more effective than glyphosate and requires fewer applications for complete removal, imazapyr application is generally recommended over glyphosate (Marks et al. 1994; Mozdzer et al. 2008).

Herbicides can be applied via a number of different methods. These include manually injecting stems, directly to a cut stem, spraying from a backpack, or aerial application in extreme cases. A practitioner on the Cape Cod developed a method of cutting *Phragmites* at about waist height, bundling the cut stems together, and wiping them with herbicide. Care should always be taken to avoid exposing nontarget plants to general-use herbicides to prevent environmental degradation. Common reed grass stems growing among or near desirable plants can be hand wiped (not something to do on acres of *Phragmites*). In this situation, applicator can wear an absorbent cotton glove over a rubber glove, then moistens the outer glove with herbicide, and each stem is then wiped from the bottom-up (Joan Deely, personal communication). Environmental monitoring is also needed for posttreatment assessment. It should also be noted that single-year applications of herbicide are never effective in fully eliminating *Phragmites*. A typical treatment cycle would be deployed over at least three growing seasons. Additionally, when herbicide regimens are not directly followed by replanting and cultivation of native plants, *Phragmites* often reemerges at the same site from remnant rhizomes or seeds (Hazelton et al. 2014). Therefore, chemical treatment plans should involve repetition of treatment for multiple years along with specific plans to simultaneously restore native flora. It is also found that native plant regeneration proceeds well in wetlands once the *Phragmites* has been removed and this is especially the case in salt marshes, where seed bank supplies are abundant, and ground moisture is continuous (Joan Deely, personal communication).

More information about chemical control on *Phragmites* and other 18 invasive species can be found in Appendix III. In the state of Massachusetts, an aquatic nuisance species control license (BRP WM04 Form) must be obtained from the Massachusetts Department of Environmental Protection before applying a chemical herbicide to *Phragmites* stands that are growing in or adjacent to open water.

PHYSICAL CONTROL

Controlled burns and mechanical removal are often suggested as methods of control to be used in connection with herbicide treatment. It is not recommended to use prescribed fire treatments or mechanical removal (such as mowing or cutting vegetation either by hand or with machinery) without first applying herbicides, as physical destruction of aboveground plant matter promotes the growth of invasive *Phragmites* (Avers et al. 2007; Hazelton et al. 2014). Physical removal is recommended for eliminating dead plant material after herbicide treatment. When excess plant matter is removed from a site, it is also easier to apply follow-up herbicide spot treatments to remaining live *Phragmites*.

Fire treatments, if applicable and implemented safely, can be a cost-effective way to remove dead material and are of low risk to surrounding native plant life, as fires can promote general plant growth. However, *Phragmites* fires can be dangerous to humans and property, as dead or dry reed stalks are very flammable and can cause spot fires at unexpected locations (Marks et al. 1994), so burns should be planned carefully. The best timing for a controlled burn is during the driest part of the year. However, fires can spread as underground peat burns, which are difficult to control.

When physical removal is implemented without chemical treatment, mostly by cutting stems, higher rates of successful *Phragmites* eradication following physical removal have been reported when plastic is placed on top of the cut stems for the rest of the growing season (Hazelton et al. 2014). Flooding has also been found to be effective in preventing regrowth after manual removal. A study by Smith (2005) demonstrated that in flooded areas, manually cutting *Phragmites* stems below the water line prevents the plant from being able to transport oxygen efficiently. This method successfully prevented regrowth: 1.5 years after breaking the stems, reed grass was gone from all but one of the ponds in the trial (Smith 2005). Though techniques involving cutting and smothering stems are effective, they can be time-consuming and require many people to complete. Smothering can also have detrimental effects on soil biota, essentially sterilizing it and necessitating soil remediation efforts.

Deliberately increasing tidal flooding can also be a powerful method of *Phragmites* control, as it can raise salinity to a level that *Phragmites* cannot tolerate. This is especially useful in restoring salt marshes overtaken by *Phragmites*, since the native salt marsh grass *Spartina alterniflora* tolerates higher salinity than *Phragmites* (Medeiros et al. 2013). In the Buzzards Bay estuary, a culvert replacement intended to raise salinity by increasing tidal flow resulted in an ecological shift: where *Phragmites* initially made up 44% of the marsh vegetation, it made up 38% of the marsh five years after the culvert construction, and the portion of marsh covered by *Spartina* rose from 20% to 32% (US EPA 2007). A similar effort at the Trustees of Reservation Argilla road culvert replacement in Ipswich, Massachusetts, increased tidal flooding and saw the proportion of cover made up of *Phragmites* fall from 40% to 22% in a year, alongside the increased abundance of *Spartina alterniflora* (Buchsbaum et al. 2006). Flooding may additionally be effective if applied following herbicide and prescribed burning treatment, or together with other mechanical control methods.

One problem that is not easily solved by physical control is the extensive root system left behind after *Phragmites* stalks are removed (Figure 18.7). The roots can extend up to 2 m underground (Marks et al. 1994), making it very difficult to remove by pulling. Reed grass root systems or remnant aerial shoots can prevent the regrowth of other plant species because of the space they occupy in the soil. Furthermore, since *Phragmites* can propagate through its

FIGURE 18.7 The root and rhizome system of the common reed grass. (Photo by Robert E. Meadows.)

rhizome, the plant can produce new stalks even after all aboveground components of the plant have been removed. To effectively eradicate the plant, the roots must be completely starved or destroyed via flooding, smothering, or by repeated removal of stalks. The site must be well-monitored to watch for and quickly respond to new shoots emerging from the rhizome. In situations where complete removal on a very large scale is deemed necessary, and if sufficient time and funding are available, machine excavation might be considered (Hazelton et al. 2014). However, any such operation involving dredging on or adjacent to wetlands in the state of Massachusetts would require the landowners submit a 401 Water Quality Cert. application and receive clearance through the MassDEP.

BIOLOGICAL CONTROL

Many new avenues for biological control of introduced *Phragmites* have emerged in recent years. In 2002, Tewksbury et al. identified many organisms native to North America or to Europe (and introduced to North America) that feed either exclusively or in part on *Phragmites*,. Over the past two decades, through rigorous study and testing, the list was narrowed from nine to four, and then to two candidates for biological control. Finally, in 2018, Blossey et al. recommended the introduction of two stem-boring moths, *Archanara geminipuncta* and *Archanara neurica*, for the biological control of *Phragmites*, given the two species' extreme dietary specificity. Blossey et al. petitioned the USDA and equivalent regulatory agencies in Canada to approve the release of *A. geminipuncta* and *A. neurica* for this purpose. In April, 2019, the two species were officially "recommended for release" by the USDA (Blossey et al. 2019; USDA 2020).

These developments were met with some backlash that sparked debate in the scientific community (Kiviat et al. 2019; Blossey et al. 2019). There has been hesitation and concern surrounding the introduction of new non-native species as biological control, as unintended consequences may occur that are difficult to manage once introduction has been carried out (Tewksbury et al. 2002; Hazelton et al. 2014; Kiviat et al. 2019). Kiviat et al. (2019) expressed specific concern that *A. geminipuncta* and

A. neurica will feed indiscriminately on the native *Phragmites australis americanus* along with the introduced strain, but Blossey et al. (2019) believe that this is not likely to occur, as accidentally introduced monophagous *Phragmites australis* herbivores in North America have not been observed to feed on native *P. australis americanus*. The effectiveness of these two promising moths has not yet been tested in the field, and neither *Archanara* species is yet commercially available, but the USDA has officially recommended *A. geminipuncta* and *A. neurica* for the biological control of introduced *Phragmites*. Though it has its risks, biological control is potentially cheaper and less labor-intensive than conventional methods of invasive plant control involving herbicides (Hazelton et al. 2014; Blossey et al. 2019).

Mammals have, at times, been employed as a form of biological control, thinning and removing *Phragmites* stands by grazing. Livestock have reportedly been grazed on *Phragmites* in its native Europe, but results have been mixed in the United States. Some field tests involving goats found that goats preferentially grazed on all plant matter except for *Phragmites* in an area (Hazelton et al. 2014). However, when goats are restricted to areas without alternative food sources, the goats can reduce *Phragmites* cover significantly (Hazelton et al. 2014; Silliman et al. 2014). Cows and horses may also readily consume this plant (Silliman et al. 2014). However, sites where grazing is implemented might suffer from expected problems such as trampling and compaction of soil (Hazelton et al. 2014). Phragmites may also be nutritionally insufficient to livestock as their only food source, and it is suggested other nutrients should also be made available to grazing animals such as goats (Sun et al. 2008).

Experience has shown that the most important and best proven ecological control of *Phragmites* is restoration of native plant assemblages. When *Phragmites* is removed from an area by any means, it is essential that the area is repopulated with native species to prevent another invasion and repair ecological damage caused by its aggressive growth. In estuarine habitats, *Phragmites* competes with the much shorter native salt-marsh grass *Spartina alterniflora* by shading it from the sunlight (Getsinger et al. 2007; Medeiros et al. 2013), but

S. alterniflora has the advantage of being able to withstand higher salinity than *Phragmites*. In salt marsh and tidal habitats, *Phragmites* may be effectively combated by increasing tidal flow to raise salinity while simultaneously supporting the regrowth of *S. alterniflora* in its place (Medeiros et al. 2013).

Restoration of native mammal habitat may also be a valuable method of ecological control. Small native mammals, though not typically considered for control of nuisance plants, may play an important role in limiting their expansion. Specifically in low-salinity marshes, small mammals were found to preferentially graze on *Phragmites*, not eliminating the reed grass but preventing its spread (Gedan et al. 2009). Muskrats, iconic inhabitants of New England's wetlands, have been observed to feed on *Phragmites* (Hazelton et al. 2014). Reintroduction of native mammals to degraded habitats, or restoration of habitats suitable to muskrats by installing muskrat platforms, could both prevent reed grass invasion and also be beneficial to wetland ecosystems. This method may be especially useful in freshwater wetlands, in areas of low salinity unfavorable to the reintroduction of *Spartina alterniflora*, or where intentional tidal flow increase is neither appropriate nor feasible.

GENERAL MANAGEMENT PLANS

The following general chemical and mechanical management plans were published by Avers et al. (2007) to be selected depending on characteristics of the site and the density of the *Phragmites* stand.

For large, dense stands at either a dry or wet site, the advised treatment is as follows:

1. Apply herbicide in early or late summer depending on what herbicides are being used (see the earlier section pertaining to chemical treatments for guidance) and wait at least three weeks to allow for plants to be exposed to the chemicals.
2. In the following year, conduct a prescribed fire in either late summer (mid-July through August) or winter (January to prior spring green-up) if a fire cannot be applied in the summer.

3. Assess the site the following growing season and spot-treat with herbicide as deemed necessary.

For large, dense stands in impoundments, the advised treatment is as follows:

1. Apply herbicide in late summer (late August and September) followed immediately by flooding to a minimum depth of 6 inches.
2. Dewater the site in late July of the following year.
3. Keep the site as dry as possible until mid-August and then treat with a prescribed fire.
4. Immediately following the burn, reflood the site to a minimum depth of 6 inches, maintaining this water level for at least a year.
5. Assess the site the following growing season and spot-treat with herbicide as deemed necessary.

For low-density stands at either a dry or wet site, the advised treatment is as follows:

1. Apply herbicide in early or late summer depending on what herbicides are being used (see the earlier section pertaining to chemical treatments for guidance) and wait at least two weeks to allow for plants to be exposed to the chemicals.
2. Mechanically treat the site either in late summer or fall prior to spring green-up. If the site is a wet site, wait until the ground is frozen. Use hand tools, weed whips, or small mowers on denser stands.
3. Assess the site the following growing season and spot-treat with herbicide as deemed necessary.

Using the methods already outlined, in combination with long-term monitoring and annual maintenance, invasive *Phragmites* can be controlled, and in some cases even eradicated from certain sites. An example of three different management plans over a three-year period can be found in Diagram 18.1. Due to the hardiness of the

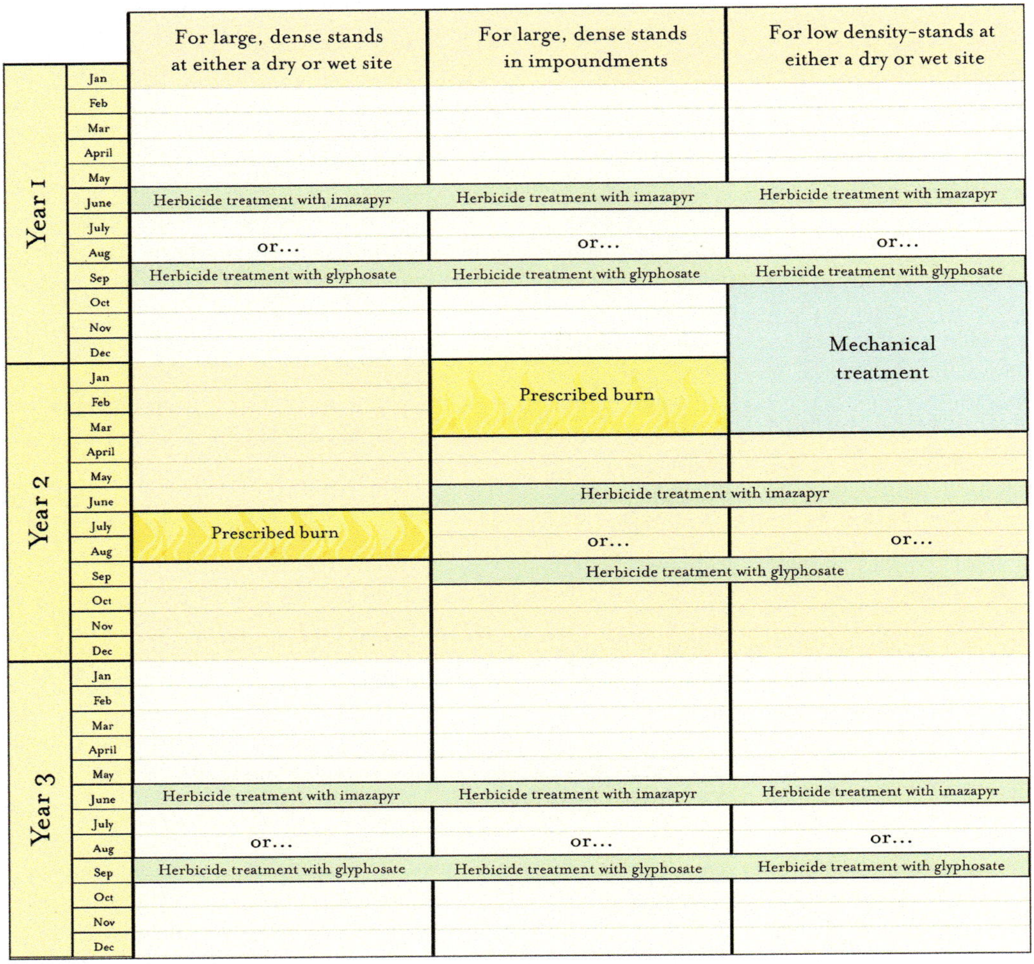

		For large, dense stands at either a dry or wet site	For large, dense stands in impoundments	For low density-stands at either a dry or wet site
Year 1	Jan			
	Feb			
	Mar			
	April			
	May			
	June	Herbicide treatment with imazapyr	Herbicide treatment with imazapyr	Herbicide treatment with imazapyr
	July	or...	or...	or...
	Aug			
	Sep	Herbicide treatment with glyphosate	Herbicide treatment with glyphosate	Herbicide treatment with glyphosate
	Oct			Mechanical treatment
	Nov			
	Dec			
Year 2	Jan		Prescribed burn	
	Feb			
	Mar			
	April			
	May			
	June		Herbicide treatment with imazapyr	Herbicide treatment with imazapyr
	July	Prescribed burn	or...	or...
	Aug			
	Sep		Herbicide treatment with glyphosate	Herbicide treatment with glyphosate
	Oct			
	Nov			
	Dec			
Year 3	Jan			
	Feb			
	Mar			
	April			
	May			
	June	Herbicide treatment with imazapyr	Herbicide treatment with imazapyr	Herbicide treatment with imazapyr
	July	or...	or...	or...
	Aug			
	Sep	Herbicide treatment with glyphosate	Herbicide treatment with glyphosate	Herbicide treatment with glyphosate
	Oct			
	Nov			
	Dec			

DIAGRAM 18.1   Three suggested management approaches over a three-year period. (Modified from Avers et al. (2007).)

species and its ability to spread, it is critical to commit to long-term management, repeated over multiple years, so that new patches and stands of *Phragmites* do not develop. This requires the cooperation of different agencies and organizations, as well as a specific restoration plan. An example of a successful chemical and mechanical management plan carried out with positive results is discussed in the "Success Story" section.

SUCCESS STORY

A study conducted in the Adirondack Park in upstate New York sought to manage invasive *Phragmites* using a glyphosate-based herbicide, with 346 populations receiving treatment over a seven-year period (Quirion et al. 2018). Prior to the beginning of treatment, 114 sites were monitored for three to six years to assess the annual increase in patch area and patch coverage in the absence of any control methods. Initial patches ranged from 0.36 to 4,314 m^2 with an initial site coverage ranging from 38% to 75%. For smaller patches of *Phragmites*, herbicide was applied by foliar spray, while large patches used a combination of foliar spray and stem injection (or cut and drip). A combination approach was also used in areas where the invasive species were found near native plants to reduce negative spill-over effects that may occur when using a foliar spray application. Over the course of the study, a site-specific adaptive management approach was used; most sites required an additional treatment of herbicide following the first treatment. If resprouts were abundant in a patch, a foliar spray was applied; if resprouts were infrequent, isolated, or near native plants, stem injection was used. If no new growth was found, treatment at that site did not occur in that year. Any singular *Phragmites* plant found was removed manually.

A total of 243 sites were treated with foliar spray, with an average of 4.5 treatments per site. These smaller sites were generally treated intermittently, as sprouts would be present one year but absent the next. A total of 91 sites were treated with the combined application of foliar spray and stem injections. Eradication was determined to be achieved when a minimum of three consecutive years of *Phragmites* absence had been achieved. Following the seven-year study period, eradication had been achieved in 104 sites, with 88 of these sites receiving exclusively intermittent foliar spray. The probability of eradication decreased as the pretreatment patch size increased: the probability of eradicating smaller sites was 0.83, 0.26 for medium sites, and 0.02 for large sites.

Quirion et al. (2018) concluded that eradication of smaller sites was achievable, but with larger, well-established sites, eradication would be more difficult and require longer project periods and the extensive use of resources. They recommended responding rapidly and administering treatment to small and new patches of *Phragmites* in order to reduce its spread. They also noted a surprising lack of posttreatment evaluations, making it hard to determine what types of treatments are effective in different scenarios. They urged therefore that future projects aimed at the management of invasive species evaluate and publish their results, promoting accountability and aiding in decision-making for future projects. These takeaways, in conjunction with the goals of this project to map existing *Phragmites* locations in Massachusetts, can be used in the future to strategically pick which sites to administer treatment where it will be most effective. This method will also ensure that resources are not wasted by trying to treat patches where the probability of eradication are very slim. At existing sites, the methods laid out in the "Managing Actions"

section can be used to treat existing sites, coupled with long-term monitoring.

NOTE

1 A haplotype is a set of DNA variations that are inherited together from a single parent.

REFERENCES

Avers, B., R. Fahlsing, E. Kafcas, J. Schafer, T. Collin, L. Esman, E. Finnell, A. Lounds, R. Terry, J. Hazelman, J. Hudgins, K. Getsinger, and D. Schuen. 2007. A Guide to the Control and Management of Invasive Phragmites. *Michigan Department of Environmental Quality, Lansing.*

Blossey, B., S.B. Endriss, R. Casagrande, P. Häfliger, H. Hinz, A. Dávalos, C. Brown-Lima, L. Tewksbury, and R.S. Bourchier. 2019. When misconceptions impede best practices: evidence supports biological control of invasive *Phragmites. Biological Invasions* 22: 873–883.

Bonham, A.J. 1980. Bank protection using emergent plants against boat wash in rivers and canals. Hydraulics Research Station #IT 206, UK, 28 pp.

Buchsbaum, R.N., J Catena, E. Hutchins, and M. James-Pirri. 2006. Changes in salt marsh vegetation, *Phragmites australis*, and nekton in response to increased tidal flushing in a New England salt marsh. *Wetlands* 26: 544–557.

Burk, I. 1877. List of Plants Recently Collected on Ships' Ballast in the Neighborhood of Philedalphia. *Proceedings of the Academy of Natural Sciences of Philedalphia* 29: 105–109.

Crow, G.E., and C.B. Hellquist. 2000. Aquatic and Wetland Plants of Northeastern North America: A Revised and Enlarged Edition of Norman C. Fassett's A Manual of Aquatic Plants (Volume Two) Angiosperms: Monocotyledons. University of Wisconsin Press, Madison WI, USA, p. 456.

Del Tredici, P. 2020. *Wild Urban Plants of the Northeast: A Field Guide.* Cornell University Press.

Gedan, K.M., C.M. Crain, and M.D. Bertness. 2009. Small-mammal herbivore control of secondary succession in New England tidal marshes. *Ecology* 90(2): 430–440.

Gersberg, R.M., B.V. Elkins, S.R. Lyon, and C.R. Goldman. 1986. Role of aquatic plants in wastewater treatment by artificial wetlands. *Water Research* 20(3): 363–368.

Getsinger, K.D., L.S. Nelson, L.M. Glomski, E. Kafcas, J. Schafer, S. Kogge, and M. Nurse. 2007. *Control of Phragmites in a Michigan Great Lakes Marsh.* U.S. Army Engineer Research and Development Center, Vicksburg, MS, p. 120.

Gray, K.R., and A.J. Biddlestone. 1995. Engineered reed-bed systems for wastewater treatment. *Trends in Biotechnology* 13(7): 248–252.

Haines, A. 2010. Botanical Notes 13: 5. *Stantec.* www.scribd.com/document/37804531/Botanical-Notes-13. (Accessed January 22, 2021).

Hazelton, E.L.G., T.J. Mozdzer, D. Burdick, K.M. Kettenring, and D. Whigham D. 2014. *Phragmites australis* management in the United States: 40 years of methods and outcomes. *Annals of Botany PLANTS* 6.

Henderson, A.M., J.A. Gervais, B. Luukinen, K. Buhl, D. Stone, A. Cross, and J. Jenkins. 2010. Glyphosate General Fact Sheet. *National Pesticide Information Center, Oregon State University Extension Services.* http://npic.orst.edu/factsheets/glyphogen.html. (Accessed January 20, 2021).

ITIS (Integrated Taxonomic Information System). 2021. www.itis.gov/servlet/SingleRpt/SingleRpt?search_topic=TSN&search_value=41072#null (Accessed April 9, 2021).

Jodoin, Y., C. Lavoie, P. Villeneuve, M. Theriault, J. Beaulieu, and F. Belzile. 2008. Highways as corridors and habitats for the invasive common reed *Phragmites australis* in Quebec, Canada. *Journal of Applied Ecology* 45 (2): 459–466.

Kelly, W. 1999. A Guide to Aquatic Plants in Massachusetts. New England Aquarium, Massachusetts Department of Environmental Management, Lakes and Ponds Program. Boston MA, USA, p. 32.

Kiviat, E., and E. Hamilton. 2001. *Phragmites* use by Native North Americans. *Aquatic Botany* 69(2–4): 341–357.

Kiviat, E., L.A. Meyerson, T.J. Mozdzer, W.J. Allen, A.H. Baldwin, G.P. Bhattarai, H. Brix, J.S. Caplan, K.M. Kettenring, C. Lambertini, J. Weis, D.F. Whigham, and J.T. Cronin. 2019. Evidence does not support the targeting of cryptic invaders at the subspecies level using classical biological control: the example of *Phragmites*. *Biological Invasions* 21: 2529–2541.

Legault, R., G.P. Zogg, and S.E. Travis. 2018. Competitive interactions between native *Spartina alterniflora* and non-native *Phragmites australis* depend on nutrient loading and temperature. *PLoS One* 13(2): e0192234.

Marks, M., B. Lapin, and J. Randall. 1994. *Phragmites australis* (*P. Communis*): Threats, management, and monitoring. *Natural Areas Journal* 14: 285–294.

Martin, L.J., and B. Blossey. 2013. The Runaway Weed: Costs and Failures of *Phragmites australis* Management in the USA. *Estuaries and Coasts* 36 (3): 626–32.

McGlynn, C.A. 2008. Native and invasive plant interactions in wetlands and the minimal role of invasiveness. *Biological Invasions* 11: 1929–1939.

Medeiros, D.L., D.S. White, and B.L. Howes. 2013. Replacement of *Phragmites australis* by *Spartina alterniflora:* The role of competition and salinity. *Wetlands* 33: 421–430.

Minchinton, T.E., and M.D. Bertness. 2003. Disturbance-mediated competition and the spread of *Phragmites australis* in a coastal marsh. *Ecological Applications* 13: 1400–1416.

Mozdzer, T.J., C.J. Hutto, P.A. Clarke, and D.P. Field. 2008. Efficacy of Imazapyr and Glyphosate in the Control of Non-Native *Phragmites australis*. *Restoration Ecology* 16(2): 221–224.

Packer, J.G., L.A. Meyerson, H. Skálová, P. Pyšek, and C. Kueffer. 2017. Biological Flora of the British Isles: *Phragmites australis*. *Journal of Ecology* 105(4): 1123–1162.

Payne, R.E., and B. Blossey. 2007. Presence and Abundance of Native and Introduced *Phragmites Australis* (Poaceae) in Falmouth, Massachusetts. *Rhodora* 109: 96–100.

Quirion, B., Z. Simek, A. Dávalos, and B. Blossey. 2018. Management of invasive *Phragmites australis* in the Adirondacks: a cautionary tale about prospects of eradication. *Biological Invasions* 20: 49–73.

Richburg, J.A., W.A. Patterson, and F. Lowenstein. 2001. Effects of Road Salt and *Phragmites australis* invasion on the vegetation of a western Massachusetts calcareous lake-basin fen. *Wetlands* 21(2): 247–255.

Saltonstall, K. 2002. Cryptic Invasion by a Non-native Genotype of the Common Reed, *Phragmites australis*, into North America. *Proceedings of the National Academy of Sciences of the United States of America* 99 (4): 2445–449.

Saltonstall, K., and L.A. Meyerson. 2016. *Phragmites australis*: from genes to ecosystems. *Biological Invasions* 18: 2415–2420.

Saltonstall, K., P.M. Peterson, and R.J. Soreng. 2004. Recognition of *Phragmites australis subsp.* americanus (Poaceae: Arundinoideae) in North America: Evidence from Morphological and Genetic Analysis. *SIDA, Contributions to Botany*, 683–692.

Silliman, B.R., T. Mozdzer, C. Angelini, J.E. Brundage, P. Esselink, J.P. Bakker, K.B. Gedan, J. van de Koppel, and A.H. Baldwin. 2014. Livestock as a potential biological control agent for an invasive wetland plant. *Peer J* 2: e567 10.7717/peerj.567.

Smith, S.M. 2005. Manual Control of *Phragmites australis* in Freshwater Ponds of Cape Cod National Seashore, Massachusetts, USA. *Journal of Aquatic Plant Management* 43: 50–53.

Sun, Z.W., D.W. Zhou, L.M.M. Ferreira, Q.Z. Zhong, and Y.J. Lou. 2008. Diet composition, herbage intake and digestibility in Inner Mongolian Cashmere goats grazing on native *Leymus chinensis* plant communities. *Livestock Science* 116 (1–3): 155–146.

Tewksbury, L., R. Casagrande, B. Blossey, P. Hälfliger, and M. Schwarzländer. 2002. Potential for biological control of *Phragmites australis* in North America. *Biological Control* 23(2): 191–212.

Tiner, R. 1998. Managing Common Reed (*Phragmites australis*) in Massachusetts. *U.S. Fish and Wildlife Service*. Boston, MA.

Tip of the Mitt Watershed Council. n.d. *Phragmites*. Photos via www.watershedcouncil.org/phragmites.html. (Accessed March 23, 2021).

US EPA (US Environmental Protection Agency). 2007. As Salt Water Returns to Winsegansett Marsh, So Does Natural Vegetation. www.epa.gov/sites/production/files/2015-11/documents/ma_winsegansett.pdf (Accessed March 15, 2017).

US FWS (US Fish and Wildlife Services). n.d. PHRAGMITES: Questions and Answers. www.fws.gov/gomcp/pdfs/phragmitesqa_factsheet.pdf (Accessed September 22, 2020).

USDA (US Department of Agriculture). 2020. Technical Advisory Group for Biological Control Agents of Weeds TAG Petitions—APHIS Actions. www.aphis.usda.gov/plant_health/permits/tag/downloads/TAGPetitionAction.pdf. (Accessed January 22, 2021).

Weis, J.S., E.B. Watson, B. Ravit, C. Harman, and M. Yepsen. 2021. The status and future of tidal marshes in New Jersey faced with sea level rise. *Anthropocene Coasts* 4: 168–192.

Wisconsin Department of Natural Resources (DNR). 2012. Imazapyr chemical fact sheet. https://dnr.wi.gov/lakes/plants/factsheets/ImazapyrFactsheet.pdf. (Accessed January 20, 2021).

APPENDIX A
Latitude and Longitude for Each Water body with *Phragmites australis*

Waterbody	Agency	Report Source	Longitude	Latitude	First Year Reported
Gulliver Creek	MassDEP	The Neponset River Watershed 1994 Resource Assessment Report	−71.05115043	42.27154821	1994
Long Pond	DCR	Complied Data from DEP Herbicide Files 98–05 (1995)	−70.17939013	41.27627396	1995
Broad Cove	MassDEP	Appendix C—DWM 1996 and 2001 Lake Survey Data in the Taunton River Watershed	−71.12887741	41.79206496	1996
Crocker Pond	MassDEP	Appendix C—DWM 1996 and 2001 Lake Survey Data in the Taunton River Watershed	−71.29478902	42.06769128	1996
Kings Pond	MassDEP	Appendix C—DWM 1996 and 2001 Lake Survey Data in the Taunton River Watershed	−71.05515697	41.93537918	1996
Little Quittacas Pond	MassDEP	Appendix C—DWM 1996 and 2001 Lake Survey Data in the Taunton River Watershed	−70.9161824	41.79260103	1996
Segreganset River	MassDEP	Appendix C—DWM 1996 and 2001 Lake Survey Data in the Taunton River Watershed	−71.16488734	41.85364572	1996
Somerset Reservoir	MassDEP	Appendix C—DWM 1996 and 2001 Lake Survey Data in the Taunton River Watershed	−71.13922277	41.78275579	1996
Thirty Acre Pond	MassDEP	Appendix C—DWM 1996 and 2001 Lake Survey Data in the Taunton River Watershed	−71.04550893	42.09554728	1996
Cape Pond	MassDEP	North Coastal Watershed 1997/1998 Water Quality Assessment Report	−70.630667	42.63992974	1997
Crane River	MassDEP	North Coastal Watershed 1997/1998 Water Quality Assessment Report	−70.93004499	42.55576425	1997
Floating Bridge Pond	MassDEP	North Coastal Watershed 1997/1998 Water Quality Assessment Report	−70.93984302	42.4853812	1997
Fuller Street Pond	MassDEP	Ten Mile River Basin 1997 Water Quality Assessment Report	−70.81318104	41.90716261	1997
Mill Pond	MassDEP	North Coastal Watershed 1997/1998 Water Quality Assessment Report	−70.67647849	42.63015993	1997
Mill River	MassDEP	Charles River Watershed 1997/1998 Water Quality Assessment Report			1997
Muddy River	MassDEP	Charles River Watershed 1997/1998 Water Quality Assessment Report 72-AC-3 16.0 February 2000	−71.10608544	42.34361819	1997
Niles Pond	MassDEP	North Coastal Watershed 1997/1998 Water Quality Assessment Report			1997
Porters Pond	MassDEP	North Coastal Watershed 1997/1998 Water Quality Assessment Report	−70.92704635	42.56318893	1997
Seaplane Basin	MassDEP	North Coastal Watershed 1997/1998 Water Quality Assessment Report	−71.01247661	42.43276228	1997
Sevenmile River	MassDEP	Ten Mile River Basin 1997 Water Quality Assessment Report	−71.34364142	41.94411787	1997
Waters River	MassDEP	North Coastal Watershed 1997/1998 Water Quality Assessment Report	−70.93863051	42.54647	1997
Wetherells Pond	MassDEP	Ten Mile River Basin 1997 Water Quality Assessment Report	−71.33702134	42.00018511	1997
Whiting Pond	MassDEP	Ten Mile River Basin 1997 Water Quality Assessment Report	−71.33522823	41.99363083	1997
Asnebumskit Pond	MassDEP	Appendix A—1998 DEP DWM Nashua River Basin QA/QC Report	−71.90908184	42.32300705	1998

(Continued)

APPENDIX A
(Continued)

Waterbody	Agency	Report Source	Longitude	Latitude	First Year Reported
Box Pond	MassDEP	DEP WQAR (1998) NW	−71.49254045	42.09303602	1998
Coachlace Pond	MassDEP	Appendix A—1998 DEP DWM Nashua River Basin QA/QC Report	−71.69718294	42.413375	1998
Five Mile Pond	MassDEP	Chicopee River Basin 1998 Water Quality Assessment Report	−72.51066309	42.14168286	1998
Hathaway Pond	MassDEP	DEP WQAR (1998) NW	−71.75522937	42.17778252	1998
Haviland Pond	MassDEP	Chicopee River Basin 1998 Water Quality Assessment Report	−72.47399733	42.17244881	1998
Lackey Pond	MassDEP	Blackstone River Basin 1998 Water Quality Assessment Report	−71.6903591	42.09496427	1998
Moulton Pond	MassDEP	Chicopee River Basin 1998 Water Quality Assessment Report	−71.95411692	42.38905736	1998
Muschopauge Pond	MassDEP	Appendix A—1998 DEP DWM Nashua River Basin QA/QC Report	−71.92177215	42.38361307	1998
Quabbin Reservoir	MassDEP	Chicopee River Basin 1998 Water Quality Assessment Report 36-AC-2 47.0, April 2001	−72.30953364	42.39167411	1998
Roberts Meadow Reservoir	MassDEP	Connecticut River Basin 1998 Water Quality Assessment Report	−72.71269532	42.3511603	1998
Southwick Pond	MassDEP	DEP WQAR (1998) NW	−71.89129683	42.28423712	1998
Tighe Carmody Reservoir	MassDEP	Connecticut River Basin 1998 Water Quality Assessment Report	−72.7824297	42.22766881	1998
Town Barn Beaver Pond	MassDEP	Chicopee River Basin 1998 Water Quality Assessment Report	−72.1818303	42.47819577	1998
Unionville Pond	MassDEP	Appendix A—1998 DEP DWM Nashua River Basin QA/QC Report	−71.84205587	42.35973521	1998
Blue Hills Reservoir	MassDEP	Boston Harbor 1999 Water Quality Assessment Report	−71.05058786	42.22840255	1999
Dudley Pond	DCR	Compiled Data from DEP Herbicide Files 98-05 1999	−71.73810589	42.05104264	1999
Forge Pond	MassDEP	Boston Harbor 1999 Water Quality Assessment Report	−71.14075992	42.15619228	1999
Newfield Pond	MassDEP	Merrimack River Basin 1999 Water Quality Assessment Report	−71.3899233	42.63378436	1999
Bogastow Brook	MassDEP	Shawsheen River Watershed 2000 Water Quality Assessment Report	−71.39378189	42.1974852	2000
Cheshire Reservoir (Middle Basin)	MassDEP	Hudson River Basin 1997 Water Quality Assessment Report 11/12/13-AC-1 15.0, January 2000	−73.19187111	42.53145142	2000
Crystal Pond	MassDEP	Ipswich River Watershed 2000 Water Quality Assessment Report	−70.99999865	42.54906234	2000
Dicks Pond	MassDEP	Buzzards Bay Watershed 2000 Water Quality Assessment Report	−70.65766372	41.76031051	2000
Lake Rohunta	MassDEP	Millers River Watershed 2000 Water Quality Assessment Report 35-AC-1 CN089.0, March 2004	−72.27094837	42.53048841	2000
Pierces Pond	MassDEP	Ipswich River Watershed 2000 Water Quality Assessment Report	−70.996911	42.52940234	2000

Richardson Pond North	MassDEP	Shawsheen River Watershed 2000 Water Quality Assessment Report	-71.24628015	42.58771504	2000
Round Pond	MassDEP	Shawsheen River Watershed 2000 Water Quality Assessment Report	-71.23213478	42.602819	2000
Sudden Pond	MassDEP	Ipswich River Watershed 2000 Water Quality Assessment Report	-71.06311289	42.61001223	2000
Tinkham Pond	MassDEP	Buzzards Bay Watershed 2000 Water Quality Assessment Report	-70.85926672	41.68372684	2000
Benton Pond	MassDEP	Farmington River Watershed 2001 Biological Assessment	-73.04739091	42.1841285	2001
Buffum Pond	MassDEP	French and Quinebaug River Watershed 2001 Water Quality Assessment Report	-71.90231756	42.11611701	2001
Clear Pond	MassDEP	Westfield River Watershed 2001 Water Quality Assessment Report	-72.65945886	42.18564647	2001
Creek Pond	MassDEP	Farmington River Watershed 2001 Biological Assessment	-73.04844921	42.22894348	2001
Duxbury Bay	MassDEP	11-E008-01	-70.69884371	41.98662005	2001
Forge Pond	MassDEP	South Shore Coastal Watersheds 2001 Water Quality Assessment Report, South Coastal Watersheds DWM Year 2001 Water Quality Monitoring Data (PG. C2)	-70.88000386	42.10622436	2001
Fort Meadow Reservoir	MassDEP	Concord Watershed 2001 DWM Water Quality Monitoring Data	-71.54760267	42.36994107	2001
Foundry Pond	MassDEP	South Shore Coastal Watersheds 2001 Water Quality Assessment Report	-70.70910461	41.98428943	2001
Great Sandy Bottom Pond	MassDEP	South Shore Coastal Watersheds 2001 Water Quality Assessment Report, South Coastal Watersheds DWM Year 2001 Water Quality Monitoring Data (PG. C2)	-70.83153067	42.05287367	2001
Hudson Pond	MassDEP	French and Quinebaug River Watershed 2001 Water Quality Assessment Report	-71.83403839	42.14317709	2001
Long Bow Lake	MassDEP	Westfield River Watershed 2001 Water Quality Assessment Report	-73.09099528	42.25328255	2001
Mashpee Pond	MassDEP	05-G011-01	-70.4869729	41.66106655	2001
Monponsett Pond	MassDEP	09-G006-01	-70.8452535	42.0045958	2001
Musquashcut Pond	MassDEP	South Coastal Watersheds DWM Year 2001 Water Quality Monitoring Data (PG. C2)	-70.75814195	42.22765855	2001
Pistol Pond	MassDEP	French and Quinebaug River Watershed 2001 Water Quality Assessment Report	-72.06952873	42.11567299	2001
Robin Hood Lake	MassDEP	Westfield River Watershed DWM 2001 Water Quality Monitoring Data	-73.06232945	42.24891028	2001
Shepherd Pond	MassDEP	French and Quinebaug River Watershed 2001 Water Quality Assessment Report			2001
Ship Pond	MassDEP	South Shore Coastal Watersheds 2001 Water Quality Assessment Report, South Coastal Watersheds DWM Year 2001 Water Quality Monitoring Data (PG. C2)	-70.53373331	41.86995472	2001
Sibley Pond	MassDEP	French and Quinebaug River Watershed 2001 Water Quality Assessment Report	-72.00809455	42.14567651	2001
Silver Shield Pond	MassDEP	Farmington River Watershed 2001 Biological Assessment	-73.090534	42.24786439	2001
Turner Pond	MassDEP	10-B004-01	-70.9709629	41.6837958	2001
Forest River	MassDEP	North Shore Coastal Watersheds 2002 Water Quality Assessment Report	-70.88759081	42.49586698	2002
Lake Archer	DCR	DCR Data from Lakes and Ponds, 2002	-71.33852068	42.06985374	2002

(Continued)

APPENDIX A
(Continued)

Waterbody	Agency	Report Source	Longitude	Latitude	First Year Reported
Snipatuit Pond	MassDEP	05-R010–02	−70.8620237	41.7809072	2002
Bass Pond	DCR	Complied Data from DEP Herbicide Files 98–05 2003	−72.5011394	42.10680081	2003
Fort Pond	DCR	Complied Data from DEP Herbicide Files 98–05 (2003)	−71.4659799	42.50759896	2003
Furnace Pond	MassDEP	Baseline Lake Survey 2003 Technical Memo	−70.82578292	42.05600568	2003
Island Pond	DCR	Complied Data from DEP Herbicide Files 98–05 (2003)	−70.64061553	41.90022056	2003
Jenks Reservoir	DCR	Complied Data from DEP Herbicide Files 98–05 (2003)	−71.4652706	42.03098393	2003
Lake Garfield	MassDEP	Baseline Lake Survey 2003 Technical Memo	−73.19721196	42.18242683	2003
Long Pond	DCR	DCR Data from Lakes and Ponds, 2003	−71.9921338	42.35677269	2003
Long Pond	DCR	Complied Data from DEP Herbicide Files 98–05 (2003)	−71.47077877	42.52992969	2003
Mill Pond	DCR	Complied Data from DEP Herbicide Files 98–05 (2003)	−73.36785604	42.20324203	2003
Parker Mills Pond	MassDEP	Buzzards Bay Watershed 2000 Water Quality Assessment Report 95-AC-2 085.0, November 2003	−70.71941257	41.77585128	2003
Wallum Lake	DCR	DCR Data from Lakes and Ponds, 2003	−71.76891342	42.01626694	2003
White Island Pond (East Basin)	MassDEP	Buzzards Bay Watershed2000 Water Quality Assessment Report 95-AC-2 085.0 November 2003			2003
Hamblin Pond	DCR	DCR Data from Lakes and Ponds, 2004	−70.40942032	41.66826289	2004
Hamilton Reservoir	DCR	Complied Data from DEP Herbicide Files 98–05 (2004)	−72.15477622	42.05095943	2004
Monponsett Pond	MassDEP	09-G005–04	−70.836781	42.0014672	2004
Peters Pond	DCR	DCR Data from Lakes and Ponds, 2004	−70.49118083	41.6888922	2004
Silver Lake	DCR	Complied Data from DEP Herbicide Files 98–05 VM (2004)	−71.46498695	42.06298288	2004
Billington Sea	MassDEP	08-F007–05	−70.68653428	41.93449681	2005
Chicoppee Reservoir	DCR	DCR Data from Lakes and Ponds, 2005	−72.55132737	42.17068898	2005
Cliff Pond	DCR	DCR Data from Lakes and Ponds, 2005	−70.02404774	41.75805904	2005
Little Cliff Pond	DCR	DCR Data from Lakes and Ponds, 2005	−70.01460736	41.75708527	2005
Tussock Brook	MassDEP	11-E014–05	−70.7221562	42.003955	2005

Name	Source	Description	Longitude	Latitude	Year
York Lake	DCR	DCR Data from Lakes and Ponds, 2005	−73.18290621	42.09800289	2005
Cedar Pond	MassDEP	North Shore Coastal Watersheds 2002 Water Quality Assessment Report 93-AC-2 CN 138.5, March 2007	−70.98205992	42.5183998	2007
Cook Pond	DCR	DCR Data from Lakes and Ponds, 2007	−71.17385703	41.67628843	2007
Flax Pond	DCR	DCR Data from Lakes and Ponds, 2007	−70.02643866	41.76683915	2007
Wampatuck Pond	DCR	DCR Data from Lakes and Ponds, 2007	−70.86683315	42.06111681	2007
Whitehall Pond	DCR	DCR Data from Lakes and Ponds, 2007	−71.99553739	42.37075916	2007
Aaron River Reservoir	DCR	DCR Data from Lakes and Ponds Pioneer Infestation, 2008	−70.82739822	42.20687841	2008
Ashland Reservoir	DCR	DCR Data from Lakes and Ponds, 2008	−71.46489203	42.2402613	2008
Five Mile Pond	DCR	DCR Data from Lakes and Ponds, 2008			2008
Otis Reservoir	DCR	Data from ACT, GeoSyntec EM-ACT	−73.04092109	42.15114776	NA
Rice City Pond	DCR	DCR Data from BWRA	−71.62321852	42.10152849	NA

19 Purple Loosestrife (*Lythrum salicaria*)

ABSTRACT

This study compiles information on 390 water-bodies in Massachusetts where purple loosestrife (*Lythrum salicaria*) has been documented in the surrounding riparian areas. These sightings are mapped to have a spatial understanding of the status of this species in the adjacent wetlands and/or riparian zones of Massachusetts' lakes and rivers. This enhances public awareness of the presence of purple loosestrife in Massachusetts which may be helpful in minimizing unintentional spreading of this species to other waterbodies. This chapter also provides different methods that have been used to control this invasive species which include physical (e.g., digging, hand pulling, and cutting), chemical treatments (e.g., triclopyr and rodeo), and biological control (e.g., beetles).

INTRODUCTION

Taxonomic Hierarchy

Kingdom	Plantae—plantes, Planta, Vegetal, plants
Subkingdom	Viridiplantae—green plants
Infrakingdom	Streptophyta—land plants
Superdivision	Embryophyta
Division	Tracheophyta—vascular plants, tracheophytes
Subdivision	Spermatophytina—spermatophytes, seed plants, phanérogames
Class	Magnoliopsida
Superorder	Rosanae
Order	Myrtales
Family	Lythraceae—loosestrife
Genus	*Lythrum* L.—loosestrife
Species	*Lythrum salicaria* L.—purple lythrum, rainbow weed, spiked loosestrife, purple loosestrife

BASIC BIOLOGY

Purple loosestrife *L. salicaria*, also called loosestrife, spiked loosestrife, salicaire, bouquet violet, and rainbow weed, is an invasive species in North America but a native species to Europe and Asia (Cao et al. 2020). Generally, it grows to 1–3 m tall (Figures 19.1 and 19.2). The root mass is fibrous with a canopy usually comprised of four to six stems. Root systems usually grow together as a bulk-sized block. If the thicker portion of the root becomes cut, it is capable of vegetatively reproducing another purple loosestrife plant (Cygan 2003). Each plant has on average 1–15 flowering stems, although a single plant can produce up to 50 stems (Figure 19.1) (Warne 2016). Matured stems are usually hard and woody, while young stems are soft, smooth, hairless, and green in color. Stems will have four to six sides and form a square shape. Purple loosestrife has surface aerenchyma cells that enable the species to live in standing water where there is little oxygen, which is typical of a wetland environment. Purple loosestrife tolerates slightly acidic soil (as low as pH 4) as well as soil that is low in nutrients (Shamsi and Whitehead 1974; Henderson 1987). Leaves are narrowly lanceolate with a cordate leaf base (Lesica 2012). The cordate, lanceolate leaf of purple loosestrife, is generally 13 cm long and 2 cm wide, and changes throughout the day in order to maximize light availability. The leaves grow on the stem in opposite directions or in a whorled arrangement from the bottom (Figure 19.1). Leaves of purple loosestrife have no stalk and instead attach directly to the stem of the plant (Warne 2016). Flowers of purple loosestrife are usually deep pink or purple, although there are on occasion lighter pink or white varieties (Figure 19.1) (Warne 2016). Each flower has five to seven petals and is arranged in a dense terminal spike cluster. The terminal

DOI: 10.1201/9781003201106-19

FIGURE 19.1   Purple loosestrife flower (*left*), stem (*top right*), and seed (*bottom right*). (Left photo by Owen Williams (Warne 2016); top right photo by Rob Routledge, from invadingspecies.com; bottom right, Photo by Donna MacKenzie (Warne 2016).)

spikes are usually 15–30 cm high. Three different floral morphs have been identified, known as tristyly, each with a different length and height. Purple loosestrife reproduces both sexually and asexually. The plant can generate 2.5–2.7 million seeds annually starting at the end of July and continuing until autumn, with a single stem being able to produce up to 1,000 seed pods (Figure 19.1). Senescence begins in winter and its stems will holdfast throughout the rest of year, sometimes standing for one or two years

(Jacobs 2008). The very small, reddish to brown seeds average 1 mm in length and are found in rounded pods which range 3–6 mm in length. Seeds of purple loosestrife are buoyant, allowing them to be carried by water currents, and can attach to clothing (Regional Lythrum salicaria Working Group 2004). Seeds can also be carried by wetland animals. It takes eight to ten weeks with a temperature ranging from 15.5°C to 18.3°C (60–65°F) for the seeds to germinate (Cygan 2003). The preferred locations for purple

FIGURE 19.2   A plant community infested by purple loosestrife in the riparian zone of Otsego Lake, New York. (Photo by Wai Hing Wong.)

loosestrife to germinate are generally in sunny spots during late spring into early summer. In Minnesota, late-summer emerging purple loosestrife seedlings have survived overwinter (Katovich et al. 2003). After the 19th century, increased highway system speeds created strong wind currents which facilitated the transportation of dry *L. salicaria* seeds.

Considered a beautiful and vigorous plant, *L. salicaria* was used in the horticultural business in the 17th century (Thompson et al. 1987). In his 1814 book *Flora Americae Spetentrionalis*, Frederick T. Pursh recorded purple loosestrife in New England (Balogh 1986). *L. salicaria* survives across the northern hemisphere between the 39th parallel (Italy) and the 65th parallel (Norway) (Jacobs 2008). It is a very popular plant and can be found almost everywhere in the UK, the main islands of Japan, New Zealand, and eastern Africa. Besides its ornamental value, tonics made from the flowering branches, leaves, and roots are used medicinally to cure wounds, diarrhea, dysentery, sores, and

ulcers (Thompson et al. 1987; Malecki et al. 1993). Beekeepers also profit from the abundant amount of nectar this plant produces which helped bee colonies survive through the winter (Pellet 1977; Malecki et al. 1991). In Europe and Asia, native insect species, like the beetle *Galerucella calmariensis*, feed on *L. salicaria* (Chandler et al. n.d.).

During the 19th century, *L. salicaria* seeds were unintentionally mixed with moist soil used as ballast to maintain the equilibrium of merchandise on cargo ships traveling from Europe to North America (Jacobs 2008). When the cargo arrived at North American shorelines, the seeds were given a chance to germinate. The plant was well-established in the New England area by the 1830s. The first herbarium specimen of purple loosestrife was collected in 1831 from Massachusetts (Lavoie 2009). Although purple loosestrife was accidentally introduced from the ballasts of ships, it was also introduced to North America intentionally because of its medicinal value (Lindgren and Clay 1993). It is possible that eradication of purple loosestrife could result in an economic loss for the honey industry (Thompson et al. 1987). However, if native plants are allowed to regain their habitat in local wetland systems, they could provide sustenance for bees, and the pollination provided by the bees in turn can help increase native plant populations. Purple loosestrife is also believed to have been introduced through imported livestock, fleece, for use in gardens, and as a beekeeping resource (Warne 2016).

ECOLOGICAL AND ECONOMIC IMPACTS

Regardless of the perception of horticultural, medicinal, or economic benefits, purple loosestrife is documented as a strong competitor in local ecosystems. Since there were no native insects that feed on its foliage, *L. salicaria* spread rapidly and became a pernicious weed to local wetlands in Quebec (Balogh 1986). After the first documented record in Neepawa, Manitoba, in 1896, more and more purple loosestrife appeared in local ecosystems. In the 1950s, huge amounts of purple loosestrife were documented near the Assiniboine River and the Red River in Winnipeg, Canada (Lindgren 1999). Purple loosestrife has been introduced throughout the Erie Canal as well as the St. Lawrence River (Cygan 2003). It became established around areas adjacent to the Great Lakes in both Canada and the United States from 1940 to 1985 (Stuckey 1980).

L. salicaria can be found throughout all of the United States with the exceptions of Florida and Hawaii (due to temperature). For now, the most serious infestation is located in the northeastern United States. The plant invades wetland ecosystems by displacing other native plants, leading to negative impacts to wetland habitats. While purple loosestrife prefers disturbed wetlands, it can also be found invading and colonizing both tidal and inland habitats such as riverbanks, natural wetlands (which are undisturbed), wet meadows, bogs, swamps, roadsides, and reservoirs (Regional Lythrum salicaria Working Group 2004). Animal populations, such as waterfowl, start to decline due to the loss of food resources displaced by purple loosestrife. Considered one of the most highly distributed species in North America, this perennial plant grows with strong

vitality (Thomsen 2012). According to Jacobs (2008), purple loosestrife outcompeted over 20 wetland plant species under variable nutrient levels and seasonal flooding within a few years. It was observed by Mal et al. (1992) that an area with abundant *Typha angustifolia*, a common species of cattail, can be completely replaced by *L. salicaria* within four years. Species of waterfowl, such as the black tern, decreased in population due to habitat loss. For example, the disappearance of about 1,000 breeding pairs of black terns coincided with a large amount of purple loosestrife blooming in Montezuma National Wildlife Refuge, New York (Jacobs 2008). By occupying shallow water areas, fish species like northern pike cannot survive in areas where purple loosestrife is densely clustered. This can negatively affect other species that feed on fish who reside in shallow water, causing them to lose a source of food. Additionally, other species like reeds, sedges, rushes, cattails, muskrats, canvasback ducks, bog turtles, especially in Massachusetts and New York where the *L. salicaria* populations are high, are declining. The direct presence of purple loosestrife in a waterbody can lead to altered decomposition rates, nutrient cycling, and water chemistry (Warne 2016). The leaves of purple loosestrife decompose quicker than those of other aquatic plants and earlier in the season than endemic species, causing spikes in certain nutrients such as phosphorus. Therefore, the higher concentration of phosphorous in leaves of purple loosestrife can cause eutrophication in waterbodies and accelerate decomposition of plant material in the spring.

Invasive species have been attributed to a 42% decrease of endangered species in the United States (Stein and Flack 1997; Wilcove 1998). The losses caused by invasive species are too hard to calculate, let alone for one kind of notorious plant such as the purple loosestrife. A rough estimate of $45 million is spent annually to control *L. salicaria* and to repair forage losses up to 131,152 ha in the United States (Malecki et al. 1993; Blossey et al. 2001; Brown 2005) and annual costs of $2.6 million are lost indirectly due to the related degradation of agricultural activities (Malecki et al. 1991). Purple loosestrife can cause agricultural losses by reducing wetland pastures and hay meadows. Additionally, the plant is less desirable to livestock than native plants, and it can impede the flow of water in irrigation systems (Malecki et al. 1993). Invasions of purple loosestrife have also caused economic losses for wild rice and hay farmers, reducing the acreage of available pasture that can be used for growing crops (Thompson et al. 1987; Blossey and Schroeder 1992). In addition to economic costs associated with the agricultural industry, purple loosestrife can also affect municipalities by incurring costs due to decreased water flow and drainage (Mullin 1998). In studying the tidal Hudson River freshwater marsh system, McGlynn (2009) found that dominant plant species was less important in invasive plant–native plant interactions than species-specific characteristics and zonation.

MONITORING AND SIGHTINGS

The Massachusetts Department of Environmental Protection (MassDEP) is committed to protect, enhance, and restore the quality and value of the waters of the Commonwealth with guidance from the Federal Clean Water Act and works to secure the environmental, recreational, and public health benefits of clean water for the citizens of Massachusetts. This chapter summarizes monitoring result and sightings of purple loosestrife in Massachusetts' lake and river riparian areas. The data were compiled by MassDEP and Massachusetts Department of Conservation and Recreation (MA DCR) staff mostly in the summer field season (i.e., May to September). MassDEP compiled the data which represents a portion of the 2,290 Massachusetts waterbodies surveyed by MassDEP from 1994 to 2015. Data layers and maps of sites with purple loosestrife were created using ArcGIS® ArcMap 10.1 (ESRI, Redlands, California).

Purple loosestrife was found in many river and lake riparian areas in Massachusetts (Tables 19.1 and 19.2). Figure 19.3 depicts the distribution of purple loosestrife in Massachusetts which occurs mainly in the northern and eastern parts of the state. There are hitherto (as of April 12, 2017) 2,050 waterbodies that have been investigated and of these, 387 (18.8%) were found to have purple loosestrife during field visits. Of the 387 waterbodies (Figure 19.3), 58 are riverine (15.1%) and 329 are lakes (84.9%). Twenty-eight of the 33 major watersheds (84.8%) in Massachusetts contain *L. salicaria*. These numbers do not represent purple loosestrife in wetlands. Thus, the mapped sighting ratio (387 out of 2,050 waterbodies) likely underestimates the true presence of purple loosestrife in Massachusetts. Also, it is possible that early in the season, purple loosestrife could be hard to distinguish from native loosestrife, so there may be some discrepancies. Future mapping

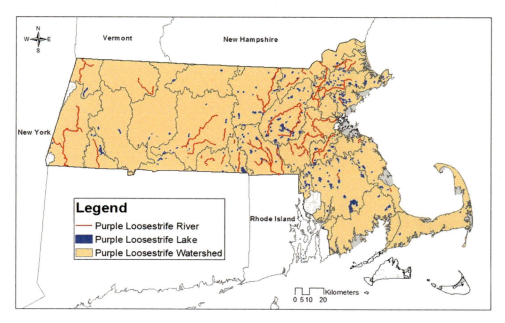

FIGURE 19.3 Distribution map depicting *Lythrum salicaria* in Massachusetts.

TABLE 19.1

List of Town Names and Waterbodies that Contain *Lythrum salicaria*

Town	Waterbody
Abington	Cleveland Pond
	Cushing Pond
	Island Grove Pond
	Cleveland Pond
	Cushing Pond
	Island Grove Pond
Andover	Bakers Meadow Pond
	Ballardvale Impoundment
	Brackett Pond
	Collins Pond
	Field Pond
	Fosters Pond
	Gravel Pit Pond
	Hussey Pond
	Pomps Pond
	Shawsheen River
	Unnamed Tributary
	Ballardvale Impoundment
	Gravel Pit Pond
	Pomps Pond
Ashburnham	Ward Pond
Ashland	Ashland Reservoir
Attleboro	Dodgeville Pond
	Farmers Pond
	Lake Como
	Luther Reservoir
	Manchester Pond Reservoir
	Mechanics Pond
	Orrs Pond
	Hebbronville Pond
Auburn	Dark Brook Reservoir (South Basin)
	Stoneville Pond
	Tinker Hill Pond
Avon	Brockton Reservoir
	Waldo Lake
Ayer	Flannagan Pond
	Grove Pond
	Plow Shop Pond
	Sandy Pond
Becket	Center Pond
	Shaw Pond
	Ward Pond
	Long Bow Lake
	Shaw Pond
Bedford	Fawn Lake
Bellingham	Silver Lake Spring Brook
Berlin	Gates Pond
	Wheeler Pond
Beverly	Wenham Lake

Billerica	Content Brook
	Nutting Lake (East Basin)
	Pond Street Pond
	Richardson Pond North
	Winning Pond
Blackstone	Mill River
Bolton	West Pond
Boston	Sprague Pond
Boxford	Baldpate Pond
	Howes Pond
	Lowe Pond
	Lower Four Mile Pond
	Stevens Pond
	Baldpate Pond
	Spofford Pond
	Stiles Pond
	Towne Pond
Boylston	Sewall Pond
Brewster	Flax Pond
Bridgewater	Lake Nippenicket
Brimfield	Unnamed Tributary
Brockton	Ellis-Brett Pond
	Upper Porter Pond
Burlington	Butterfield Pond
	Sandy Brook
	Vine Brook
	Mill Pond
Canton	Glen Echo Pond
	Reservoir Pond
	Ponkapoag Pond
Carlisle	Fiske Street Pond
	Meadow Pond
Carver	Federal Pond
Charlton	Cady Brook
	Pikes Pond
	Sibley Pond
Chelmsford	Elm Street Pond
	Heart Pond
	Newfield Pond
	Russell Millpond
Cheshire	Cheshire Reservoir, Middle Basin
	Cheshire Reservoir, North Basin
Chester	Littleville Lake
Chicopee	Chicoppee Reservoir
Clinton	Clamshell Pond
	Coachlace Pond
	Lancaster Millpond
	South Meadow Pond (East Basin)
	South Meadow Pond (West Basin)
Cohasset	Aaron River Reservoir
	Lily Pond
Concord	Goose Pond
	Batemans Pond

(Continued)

TABLE 19.1

(Continued)

Town	Waterbody
	Great Meadow Pond #3
Danvers	Putnamville Reservoir
Deerfield	Deerfield River
Dighton	Muddy Cove Brook Pond
Douglas	Mumford River
Dover	Charles River
Dracut	Beaver Brook
	Richardson Brook
Dudley	Easterbrook Pond
	Larner Pond
	Sylvestri Pond
	Wallis Pond
	French River
Dunstable	Massapoag Pond
Duxbury	Lower Chandler Pond
Easton	Longwater Pond
	New Pond
	Shovelshop Pond
	Longwater Pond
	New Pond
	Shovelshop Pond
	Ward Pond
Egremont	Prospect Lake
Erving	Northfield Mountain Reservoir
Essex	Chebacco Lake
Fall River	Cook Pond
Fitchburg	Sawmill Pond
Foxborough	Beaumont Pond
	Bird Pond
	Carpenter Pond
	Cocasset Lake
	Neponset Reservoir
Framingham	Farm Pond
	Framingham Reservoir #1
	Framingham Reservoir #2
	Framingham Reservoir #3
	Saxonville Pond
	Waushakum Pond
	Learned Pond
Franklin	Beaver Pond
Gardner	Bents Pond
Georgetown	Penn Brook
	Parker River
Gloucester	Babson Reservoir
	Fernwood Lake
	Goose Cove Reservoir
	Haskell Pond
	Upper Banjo Pond
	Wallace Pond
Grafton	Blackstone River
	Flint Pond

	Lake Ripple
	Quinsigamond River
Granby	Forge Pond
Great Barrington	Green River
Groton	Baddacook Pond
Groveland	Johnsons Pond
Halifax	Monponsett Pond
Hamilton	Beck Pond
	Gravelly Pond
	Round Pond
Hanover	Forge Pond
Hanson	Reservoir
	Riley Pond
	Wampatuck Pond
	Factory Pond
Harvard	Robbins Pond
Haverhill	Lake Saltonstall
Hingham	Weir River
Holden	Chaffin Pond
	Maple Spring Pond
	Quinapoxet Reservoir
	Dawson Pond
Holliston	Chicken Brook
Holyoke	Clear Pond
	Lake Bray
	Whiting Street Reservoir
	Wright Pond
Hopedale	Spindleville Pond
	Hopedale Pond
Hopkinton	North Pond
	Hopkinton Reservoir
	Whitehall Reservoir
Hudson	Assabet River
Ipswich	Hood Pond
Kingston	Foundry Pond
	Russell Pond
Lakeville	Assawompset Pond
	Long Pond
	Pocksha Pond
Lancaster	Nashua River
	North Nashua River
Lanesborough	Berkshire Pond
	Cheshire Reservoir, South Basin
Lawrence	Spicket River
Lee	Goose Pond Brook
	Housatonic River
Leicester	Moose Hill Reservoir
	Smiths Pond
	Town Meadow Brook
	Waite Pond
Lenox	Housatonic River

(Continued)

TABLE 19.1
(Continued)

Town	Waterbody
Leominster	Rockwell Pond
Lincoln	Farrar Pond
Littleton	Fort Pond
	Nagog Pond
	Spectacle Pond
Ludlow	Minechoag Pond
	Red Bridge Impoundment
Lunenburg	Lake Whalom
Lynn	Breeds Pond
	Flax Pond
	Floating Bridge Pond
	Walden Pond
Lynnfield	Hawkes Pond
	Pillings Pond
	Suntaug Lake
Mansfield	Sweets Pond
	Plain Street Pond
Marlborough	Fort Meadow Reservoir
	Hager Pond
	Milham Reservoir
Mashpee	Ashumet Pond
Medfield	Charles River
	Stop River
Mendon	Nipmuck Lake
Middleborough	Fuller Street Pond
Middleton	Creighton Pond
	Lower Boston Brook Pond
	Middleton Pond
	Unnamed Tributary
	Upper Boston Brook Pond
Millbury	Unnamed Tributary
Millis	Bogastow Brook
Millville	Blackstone River
Milton	Russell Pond
	Turners Pond
	Popes Pond
Monson	Chicopee Brook Pond
Natick	Fisk Pond
	Lake Cochituate
New Bedford	Sassaquin Pond
North Andover	Berry Pond
	Brook Street Pond
	Farnum Street Pond
	Salem Street Pond
	Stearns Pond
North Attleborough	Falls Pond (North Basin)

	Falls Pond (South Basin)
	Whiting Pond
North Reading	Ipswich River
	Martins Pond
	Bradford Pond
	Eisenhaures Pond
	Swan Pond
Northborough	Assabet River
	Bartlett Pond
	Smith Pond
	Solomon Pond
Northbridge	Blackstone River
	Whitins Pond
	Linwood Pond
Norton	Chartley Pond
	Meadow Brook Pond
	Norton Reservoir
	Winnecunnet Pond
Norwell	Jacobs Pond
Norwood	Ellis Pond
	Hawes Brook
	Neponset River
Otis	Benton Pond
	Otis Reservoir
	Royal Pond
Oxford	Hudson Pond
	Texas Pond
Palmer	Palmer Reservoir
	Quaboag River
	Thompson Lake
	Forest Lake
Paxton	Asnebumskit Pond
Peabody	Browns Pond
	Crystal Pond
	Devils Dishfull Pond
	Elginwood Pond
	Pierces Pond
	Spring Pond
	Winona Pond
Pembroke	Furnace Pond
	Oldham Pond
Pepperell	Heald Pond
	Nashua River
	Pepperell Pond
Petersham	Connors Pond
Plainville	Lake Mirimichi
	Turnpike Lake
	Wetherells Pond
	Chestnut Street Pond
	Lake Mirimichi
Plymouth	Billington Sea
	Russell Millpond
	Ship Pond
	Unnamed Tributary

(Continued)

TABLE 19.1
(Continued)

Town	Waterbody
	White Island Pond (East Basin)
	White Island Pond (West Basin)
Rayham	Kings Pond
Rockland	Studleys Pond
Rockport	Cape Pond
	Rum Rock Lake
Rowley	Mill River
Rutland	Long Pond
	Muschopauge Pond
Sandisfield	West Branch Farmington River
Saugus	Birch Pond Reservoir
	First Pond
	Saugus River
	Spring Pond
	Griswold Pond
Scituate	Musquashcut Pond
	Old Oaken Bucket Pond
Sharon	Gavins Pond
	Massapoag Brook
	Upper Leach Pond
Sheffield	Konkapot River
Shrewsbury	Flint Pond
Somerset	Somerset Reservoir
South Hadley	Lower Pond
Southborough	Sudbury Reservoir
Southbridge	Morse Pond
Spencer	Sevenmile River
Springfield	Lake Lorraine
	Noonan Cove
	Fivemile Pond
	Lake Lookout
	Loon Pond
	Mona Lake
	Porter Lake
	Porter Lake West
	Watershops Pond
	Long Pond
	Mill Pond
Sterling	West Waushacum Pond
Stoughton	Ames Long Pond
	Farrington Pond
	Pinewood Pond
	Woods Pond
Sturbridge	Quinebaug River
Sudbury	Grist Mill Pond
	Stearns Mill Pond
Sutton	Aldrich Pond
	Blackstone River
	Marble Pond
	Singletary Pond
Taunton	Bear Hole Pond (little)
	Johnson Pond
	Lake Rico

	Lake Sabbatia
	Middle Pond
	Prospect Hill Pond
	Watson Pond
	Whittenton Impoundment
	Middle Pond
Tewksbury	Ames Pond
	Long Pond
	Round Pond
	Shawsheen River
Tyngsborough	Althea Lake
	Flint Pond
Upton	West River
	Lake Wildwood
Uxbridge	Blackstone River
	Lackey Pond
Wakefield	Crystal Lake
	Lake Quannapowitt
Walpole	Cobbs Pond
	Ganawatte Farm Pond
	Neponset River
	Traphole Brook
	Willet Pond
Ware	Beaver Lake
	Ware River
Wareham	Glen Charlie Pond
Warren	Quaboag River
Wayland	Heard Pond
	Sudbury River
Webster	Webster Lake
Wellesley	Rosemary Brook
Wenham	Coy Pond
	Pleasant Pond
West Bridgewater	Hockomock River
	West Meadow Pond
Westborough	Assabet River
	Assabet River Reservoir
	Chauncy Lake
	Hocomonco Pond
Westford	Stony Brook
Westminster	Round Meadow Pond
Weston	Stony Brook
Whitman	Hobart Pond
Wilbraham	Nine Mile Pond
Williamstown	Green River
	Hoosic River
Wilmington	Lubber Pond East
	Lubber Pond West
	Lubbers Brook
	Maple Meadow Brook
Winchendon	Whitney Pond
Winchester	Winter Pond
Worcester	Coes Reservoir
	Patch Reservoir
	Salisbury Pond
Wrentham	Crocker Pond

TABLE 19.2

Geo-referenced Records of *Lythrum salicaria* in Massachusetts as Shown in Figure 19.3

Waterbody	Agency	Report Source	Longitude	Latitude	First Year Reported
Massapoag Brook	MassDEP	The Neponset River Watershed 1994	−71.1642662	42.1205358	1994
Neponset River	MassDEP	The Neponset River Watershed 1994 Resource Assessment Report	−71.2506977	42.1381175	1994
Neponset River	MassDEP	The Neponset River Watershed 1994 Resource Assessment Report	−71.2004073	42.1772345	1994
Hawes Brook	MassDEP	The Neponset River Watershed 1994 Resource Assessment Report	−71.2064474	42.1735363	1994
Traphole Brook	MassDEP	The Neponset River Watershed 1994 Resource Assessment Report	−71.2078652	42.1586282	1994
Ames Long Pond	MassDEP	Taunton River Watershed Water Quality Assessment Report 62-AC-1 CN 94.0, December 2005	−71.1187699	42.0875920	1996
Ames Long Pond	MassDEP	Appendix C—DWM 1996 and 2001 Lake Survey Data in the Taunton River Watershed	−71.1187699	42.0875920	1996
Brockton Reservoir	MassDEP	Appendix C—DWM 1996 and 2001 Lake Survey Data in the Taunton River Watershed	−71.0551783	42.1163914	1996
Brockton Reservoir	MassDEP	Taunton River Watershed Water Quality Assessment Report 62-AC-1 CN 94.0, December 2005	−71.0551783	42.1163914	1996
Cleveland Pond	MassDEP	Taunton River Watershed Water Quality Assessment Report 62-AC-1 CN 94.0, December 2005	−70.9828404	42.1199308	1996
Cleveland Pond	MassDEP	Appendix C—DWM 1996 and 2001 Lake Survey Data in the Taunton River Watershed	−70.9828404	42.1199308	1996
Crocker Pond	MassDEP	Taunton River Watershed Water Quality Assessment Report 62-AC-1 CN 94.0, December 2005	−71.2947890	42.0676913	1996
Crocker Pond	MassDEP	Appendix C—DWM 1996 and 2001 Lake Survey Data in the Taunton River Watershed	−71.2947890	42.0676913	1996
Cushing Pond	MassDEP	Taunton River Watershed Water Quality Assessment Report 62-AC-1 CN 94.0, December 2005	−70.9811913	42.1300566	1996
Cushing Pond	MassDEP	Appendix C—DWM 1996 and 2001 Lake Survey Data in the Taunton River Watershed	−70.9811913	42.1300566	1996
Gavins Pond	MassDEP	Taunton River Watershed Water Quality Assessment Report 62-AC-1 CN 94.0, December 2005	−71.2116854	42.0814356	1996
Gavins Pond	MassDEP	Appendix C—DWM 1996 and 2001 Lake Survey Data in the Taunton River Watershed	−71.2116854	42.0814356	1996
Hobart Pond	MassDEP	Taunton River Watershed Water Quality Assessment Report 62-AC-1 CN 94.0, December 2005	−70.9287075	42.0856020	1996
Hobart Pond	MassDEP	Appendix C—DWM 1996 and 2001 Lake Survey Data in the Taunton River Watershed	−70.9287075	42.0856020	1996
Island Grove Pond	MassDEP	Taunton River Watershed Water Quality Assessment Report 62-AC-1 CN 94.0, December 2005	−70.9399656	42.1110552	1996
Island Grove Pond	MassDEP	Appendix C—DWM 1996 and 2001 Lake Survey Data in the Taunton River Watershed	−70.9399656	42.1110552	1996
Johnson Pond	MassDEP	Taunton River Watershed Water Quality Assessment Report 62-AC-1 CN 94.0, December 2005	−71.0514379	41.9265631	1996
Johnson Pond	MassDEP	Appendix C—DWM 1996 and 2001 Lake Survey Data in the Taunton River Watershed	−71.0514379	41.9265631	1996
Long Pond	MassDEP	Appendix C—DWM 1996 and 2001 Lake Survey Data in the Taunton River Watershed	−70.9445534	41.8014573	1996
Longwater Pond	MassDEP	Taunton River Watershed Water Quality Assessment Report 62-AC-1 CN 94.0, December 2005	−71.0961075	42.0688526	1996

Longwater Pond	MassDEP	Appendix C—DWM 1996 and 2001 Lake Survey Data in the Taunton River Watershed	−71.0961075	42.0688526	1996
Middle Pond	MassDEP	Taunton River Watershed Water Quality Assessment Report 62-AC-1 CN 94.0, December 2005	−70.9900826	41.8677608	1996
Middle Pond	MassDEP	Appendix C—DWM 1996 and 2001 Lake Survey Data in the Taunton River Watershed	−70.9900826	41.8677608	1996
Lake Mirimichi	MassDEP	Taunton River Watershed Water Quality Assessment Report 62-AC-1 CN 94.0, December 2005	−71.2917133	42.0272167	1996
Lake Mirimichi	MassDEP	Appendix C—DWM 1996 and 2001 Lake Survey Data in the Taunton River Watershed	−71.2917133	42.0272167	1996
New Pond	MassDEP	Taunton River Watershed Water Quality Assessment Report 62-AC-1 CN 94.0, December 2005	−71.1375423	42.0260035	1996
New Pond	MassDEP	Appendix C—DWM 1996 and 2001 Lake Survey Data in the Taunton River Watershed	−1.1375423	42.0260035	1996
Norton Reservoir	MassDEP	Taunton River Watershed Water Quality Assessment Report 62-AC-1 CN 94.0, December 2005	−71.2007819	41.9846846	1996
Norton Reservoir	MassDEP	Appendix C—DWM 1996 and 2001 Lake Survey Data in the Taunton River Watershed	−71.2007819	41.9846846	1996
Lake Rico	MassDEP	Taunton River Watershed Water Quality Assessment Report 62-AC-1 CN 94.0, December 2005	−70.9969743	41.8747225	1996
Lake Rico	MassDEP	Appendix C—DWM 1996 and 2001 Lake Survey Data in the Taunton River Watershed	−70.9969743	41.8747225	1996
Lake Sabbatia	MassDEP	Taunton River Watershed Water Quality Assessment Report 62-AC-1 CN 94.0, December 2005	−71.1081674	41.9445610	1996
Lake Sabbatia	MassDEP	Appendix C—DWM 1996 and 2001 Lake Survey Data in the Taunton River Watershed	−71.1081674	41.9445610	1996
Shovelshop Pond	MassDEP	Taunton River Watershed Water Quality Assessment Report 62-AC-1 CN 94.0, December 2005	−71.1009649	42.0691099	1996
Shovelshop Pond	MassDEP	Appendix C—DWM 1996 and 2001 Lake Survey Data in the Taunton River Watershed	−71.1009649	42.0691099	1996
Sweets Pond	MassDEP	Taunton River Watershed Water Quality Assessment Report 62-AC-1 CN 94.0, December 2005	−71.2566549	41.9916367	1996
Sweets Pond	MassDEP	Appendix C—DWM 1996 and 2001 Lake Survey Data in the Taunton River Watershed	−71.2566549	41.9916367	1996
Turnpike Lake	MassDEP	Taunton River Watershed Water Quality Assessment Report 62-AC-1 CN 94.0, December 2005	−71.3124012	42.0155418	1996
Turnpike Lake	MassDEP	Appendix C—DWM 1996 and 2001 Lake Survey Data in the Taunton River Watershed	−71.3124012	42.0155418	1996
Upper Porter Pond	MassDEP	Taunton River Watershed Water Quality Assessment Report 62-AC-1 CN 94.0, December 2005	−71.0442045	42.1032222	1996
Upper Porter Pond	MassDEP	Appendix C—DWM 1996 and 2001 Lake Survey Data in the Taunton River Watershed	−71.0442045	42.1032222	1996
Waldo Lake	MassDEP	Taunton River Watershed Water Quality Assessment Report 62-AC-1 CN 94.0, December 2005	−71.0482845	42.1094618	1996
Waldo Lake	MassDEP	Appendix C—DWM 1996 and 2001 Lake Survey Data in the Taunton River Watershed	−71.0482845	42.1094618	1996
Watson Pond	MassDEP	Taunton River Watershed Water Quality Assessment Report 62-AC-1 CN 94.0, December 2005	−71.1186846	41.9507941	1996
Watson Pond	MassDEP	Appendix C—DWM 1996 and 2001 Lake Survey Data in the Taunton River Watershed	−71.1186846	41.9507941	1996
West Meadow Pond	MassDEP	Taunton River Watershed Water Quality Assessment Report 62-AC-1 CN 94.0, December 2005	−71.0383792	42.0405202	1996
West Meadow Pond	MassDEP	Appendix C—DWM 1996 and 2001 Lake Survey Data in the Taunton River Watershed	−71.0383792	42.0405202	1996
Winnecunnet Pond	MassDEP	Taunton River Watershed Water Quality Assessment Report 62-AC-1 CN 94.0, December 2005	−71.1312668	41.9709330	1996
Winnecunnet Pond	MassDEP	Appendix C—DWM 1996 and 2001 Lake Survey Data in the Taunton River Watershed	−71.1312668	41.9709330	1996

(Continued)

TABLE 19.2
(Continued)

Waterbody	Agency	Report Source	Longitude	Latitude	First Year Reported
Whittenton Impoundment	MassDEP	Taunton River Watershed Water Quality Assessment Report 62-AC-1 CN 94.0, December 2005	−71.1070860	41.9291042	1996
Whittenton Impoundment	MassDEP	Appendix C—DWM 1996 and 2001 Lake Survey Data in the Taunton River Watershed	−71.1070860	41.9291042	1996
Assawompset Pond	MassDEP	Appendix C—DWM 1996 and 2001 Lake Survey Data in the Taunton River Watershed	−70.9189452	41.8381170	1996
Beaumont Pond	MassDEP	Appendix C—DWM 1996 and 2001 Lake Survey Data in the Taunton River Watershed	−71.1989278	42.0522164	1996
Carpenter Pond	MassDEP	Appendix C—DWM 1996 and 2001 Lake Survey Data in the Taunton River Watershed	−71.2737942	42.0715149	1996
Chartley Pond	MassDEP	Appendix C—DWM 1996 and 2001 Lake Survey Data in the Taunton River Watershed	−71.2333304	41.9474469	1996
Cocasset Lake	MassDEP	Appendix C—DWM 1996 and 2001 Lake Survey Data in the Taunton River Watershed	−71.2597188	42.0588861	1996
Kings Pond	MassDEP	Appendix C—DWM 1996 and 2001 Lake Survey Data in the Taunton River Watershed	−71.0544926	41.9361331	1996
Meadow Brook Pond	MassDEP	Appendix C—DWM 1996 and 2001 Lake Survey Data in the Taunton River Watershed	−71.1641185	41.9375309	1996
Muddy Cove Brook Pond	MassDEP	Appendix C—DWM 1996 and 2001 Lake Survey Data in the Taunton River Watershed	−71.1312403	41.8142012	1996
Pocksha Pond	MassDEP	Appendix C—DWM 1996 and 2001 Lake Survey Data in the Taunton River Watershed	−70.8935724	41.8291440	1996
Prospect Hill Pond	MassDEP	Appendix C—DWM 1996 and 2001 Lake Survey Data in the Taunton River Watershed	−71.0931736	41.9343313	1996
Somerset Reservoir	MassDEP	Appendix C—DWM 1996 and 2001 Lake Survey Data in the Taunton River Watershed	−71.1384162	41.7824457	1996
Upper Leach Pond	MassDEP	Appendix C—DWM 1996 and 2001 Lake Survey Data in the Taunton River Watershed	−71.1513062	42.0702976	1996
Ward Pond	MassDEP	Appendix C—DWM 1996 and 2001 Lake Survey Data in the Taunton River Watershed	−71.1228875	42.0063244	1996
Ellis-Brett Pond	MassDEP	Appendix C—DWM 1996 and 2001 Lake Survey Data in the Taunton River Watershed	−71.0447806	42.0914975	1996
Mill River	MassDEP	Charles River Water Quality Assessment Report, 1997–2000	−71.5156703	42.0407739	1997
Charles River	MassDEP	Charles River Water Quality Assessment Report, 1997–2001	−71.3395731	42.1930188	1997
Berkshire Pond	MassDEP	Hudson River Basin 1997 Water Quality Assessment Report	−73.19612	42.504771	1997
Cheshire Reservoir, North Basin	MassDEP	Hudson River Basin 1997 Water Quality Assessment Report	−73.1739645	42.5473169	1997
Cheshire Reservoir, Middle Basin	MassDEP	Hudson River Basin 1997 Water Quality Assessment Report	−73.191871	42.531451	1997

Name	Agency	Source	Longitude	Latitude	Year
Cheshire Reservoir, South Basin	MassDEP	Hudson River Basin 1997 Water Quality Assessment Report	-73.1983974	42.5154111	1997
Lake Como	MassDEP	Ten Mile River Basin 1997 Water Quality Assessment Report	-71.3576397	41.9262715	1997
Falls Pond (South Basin)	MassDEP	Ten Mile River Basin 1997 Water Quality Assessment Report	-71.3242070	41.9602576	1997
Orrs Pond	MassDEP	Ten Mile River Basin 1997 Water Quality Assessment Report	-71.3326962	41.9286110	1997
Plain Street Pond	MassDEP	Ten Mile River Basin 1997 Water Quality Assessment Report	-71.2777427	42.0062287	1997
Fuller Street Pond	MassDEP	Ten Mile River Basin 1997 Water Quality Assessment Report	-70.8131810	41.9071626	1997
Chebacco Lake	MassDEP	North Coastal Watershed 1997/1998 Water Quality Assessment Report	-70.8111522	42.6119840	1997
Flax Pond	MassDEP	North Coastal Watershed 1997/1998 Water Quality Assessment Report	-70.9514364	42.4827035	1997
Griswold Pond	MassDEP	North Coastal Watershed 1997/1998 Water Quality Assessment Report	-71.0498670	42.4774936	1997
Lake Quannapowitt	MassDEP	North Coastal Watershed 1997/1998 Water Quality Assessment Report	-71.0785558	42.5149376	1997
Spring Pond	MassDEP	North Coastal Watershed 1997/1998 Water Quality Assessment Report	-71.0514912	42.4801724	1997
Lily Pond	MassDEP	North Coastal Watershed 1997/1998 Water Quality Assessment Report	-70.8159141	42.2246376	1997
Babson Reservoir	MassDEP	North Coastal Watershed 1997/1998 Water Quality Assessment Report	-70.6620453	42.6281521	1997
Beck Pond	MassDEP	North Coastal Watershed 1997/1998 Water Quality Assessment Report	-70.8228555	42.6040701	1997
Birch Pond Reservoir	MassDEP	North Coastal Watershed 1997/1998 Water Quality Assessment Report	-71.0014522	42.4786663	1997
Breeds Pond	MassDEP	North Coastal Watershed 1997/1998 Water Quality Assessment Report	-70.9786343	42.4797990	1997
Browns Pond	MassDEP	North Coastal Watershed 1997/1998 Water Quality Assessment Report	-70.9550201	42.4998821	1997
Cape Pond	MassDEP	North Coastal Watershed 1997/1998 Water Quality Assessment Report	-70.6308641	42.6404998	1997
Coy Pond	MassDEP	North Coastal Watershed 1997/1998 Water Quality Assessment Report	-70.8218642	42.5919722	1997
Falls Pond (North Basin)	MassDEP	Ten Mile River Basin 1997 Water Quality Assessment Report	-71.3233843	41.9689758	1997
Fernwood Lake	MassDEP	North Coastal Watershed 1997/1998 Water Quality Assessment Report	-70.6997665	42.6109725	1997
Floating Bridge Pond	MassDEP	North Coastal Watershed 1997/1998 Water Quality Assessment Report	-70.9398485	42.4855152	1997
Goose Cove Reservoir	MassDEP	North Coastal Watershed 1997/1998 Water Quality Assessment Report	-70.6639814	42.6439760	1997
Gravelly Pond	MassDEP	North Coastal Watershed 1997/1998 Water Quality Assessment Report	-70.8106273	42.5987887	1997

(Continued)

TABLE 19.2
(Continued)

Waterbody	Agency	Report Source	Longitude	Latitude	First Year Reported
Haskell Pond	MassDEP	North Coastal Watershed 1997/1998 Water Quality Assessment Report	−70.7384855	42.6113303	1997
Hawkes Pond	MassDEP	North Coastal Watershed 1997/1998 Water Quality Assessment Report	−71.0212112	42.5029144	1997
Pillings Pond	MassDEP	North Coastal Watershed 1997/1998 Water Quality Assessment Report	−71.0293596	42.5299953	1997
Rum Rock Lake	MassDEP	North Coastal Watershed 1997/1998 Water Quality Assessment Report	−70.6192624	42.6464618	1997
Saugus River	MassDEP	North Coastal Watershed 1997/1998 Water Quality Assessment Report	−70.9989210	42.4620059	1997
Upper Banjo Pond	MassDEP	North Coastal Watershed 1997/1998 Water Quality Assessment Report	−70.6943219	42.6122605	1997
Round Pond	MassDEP	North Coastal Watershed 1997/1998 Water Quality Assessment Report	−70.8179910	42.5983467	1997
Walden Pond	MassDEP	North Coastal Watershed 1997/1998 Water Quality Assessment Report	−71.0039981	42.4941163	1997
Wallace Pond	MassDEP	North Coastal Watershed 1997/1998 Water Quality Assessment Report	−70.7084723	42.6043935	1997
Whiting Pond	MassDEP	Ten Mile River Basin 1997 Water Quality Assessment Report	−71.3349764	41.9934412	1997
Rosemary Brook	MassDEP	Charles River Water Quality Assessment Report, 1997–2003	−71.2516891	42.3043036	1997
Stony Brook	MassDEP	Charles River Water Quality Assessment Report, 1997–2002	−71.2829360	42.3787810	1997
Crystal Lake	MassDEP	North Coastal Watershed 1997/1998 Water Quality Assessment Report	−71.0735447	42.4916140	1997
Dodgeville Pond	MassDEP	Ten Mile River Basin 1997 Water Quality Assessment Report	−71.2918061	41.9251093	1997
Farmers Pond	MassDEP	Ten Mile River Basin 1997 Water Quality Assessment Report	−71.3032357	41.9517620	1997
Hebbronville Pond	MassDEP	Ten Mile River Basin 1997 Water Quality Assessment Report	−71.3187011	41.9099740	1997
Luther Reservoir	MassDEP	Ten Mile River Basin 1997 Water Quality Assessment Report	−71.3388399	41.9383352	1997
Mechanics Pond	MassDEP	Ten Mile River Basin 1997 Water Quality Assessment Report	−71.2940074	41.9461536	1997
Wetherells Pond	MassDEP	Ten Mile River Basin 1997 Water Quality Assessment Report	−71.3375762	41.9999234	1997
Spring Pond	MassDEP	North Coastal Watershed 1997/1998 Water Quality Assessment Report	−70.9468041	42.4969367	1997
Chestnut Street Pond	MassDEP	Ten Mile River Basin 1997 Water Quality Assessment Report	−71.3597279	42.0216992	1997
Manchester Pond Reservoir	MassDEP	Ten Mile River Basin 1997 Water Quality Assessment Report	−71.3158334	41.9464202	1997
Prospect Lake	MassDEP	Chicopee River Basin 1998 Water Quality Assessment Report	−73.4522242	42.1949523	1998

Name	Agency	Report	Latitude	Longitude	Year
Lake Bray	MassDEP	Connecticut River Basin 1998 Water Quality Assessment Report	42.2666603	−72.6166715	1998
Forge Pond	MassDEP	Connecticut River Basin 1998 Water Quality Assessment Report	42.2715846	−72.4671967	1998
Porter Lake	MassDEP	Connecticut River Basin 1998 Water Quality Assessment Report	42.0735051	−72.5639194	1998
Whiting Street Reservoir	MassDEP	Connecticut River Basin 1998 Water Quality Assessment Report	42.2389377	−72.6377995	1998
Beaver Lake	MassDEP	Chicopee River Basin 1998 Water Quality Assessment Report 36-AC-2 47.0, April 2001	42.2603274	−72.3052244	1998
Fivemile Pond	MassDEP	Chicopee River Basin 1998 Water Quality Assessment Report	42.1416829	−72.5106631	1998
Forest Lake	MassDEP	Chicopee River Basin 1998 Water Quality Assessment Report	42.2072563	−72.3081122	1998
Long Pond	MassDEP	Chicopee River Basin 1998 Water Quality Assessment Report 36-AC-2 47.0, April 2001	42.3567719	−71.9921341	1998
Long Pond	MassDEP	Chicopee River Basin 1998 Water Quality Assessment Report	42.3567719	−71.9921341	1998
Coes Reservoir	MassDEP	Blackstone River Basin 1998 Water Quality Assessment Report	42.2544586	−71.8418253	1998
Dark Brook Reservoir (South Basin)	MassDEP	Blackstone River Basin 1998 Water Quality Assessment Report	42.1882716	−71.8658970	1998
Flint Pond	MassDEP	Blackstone River Basin 1998 Water Quality Assessment Report	42.2495864	−71.7360011	1998
Lake Ripple	MassDEP	09-G008-04	42.2141715	−71.6936628	1998
Lake Ripple	MassDEP	Blackstone River Basin 1998 Water Quality Assessment Report	42.2134635	−71.6963299	1998
Singletary Pond	MassDEP	Blackstone River Basin 1998 Water Quality Assessment Report	42.1600274	−71.7801300	1998
Tinker Hill Pond	MassDEP	Blackstone River Basin 1998 Water Quality Assessment Report	42.1846460	−71.8741419	1998
Whitins Pond	MassDEP	Blackstone River Basin 1998 Water Quality Assessment Report	42.1140136	−71.6901504	1998
Flint Pond	MassDEP	Blackstone River Basin 1998 Water Quality Assessment Report 51-AC-1 48.0, May 2011	42.2415305	−71.7306085	1998
Chaffin Pond	MassDEP	Appendix A—1998 DEP DWM Nashua River Basin QA/QC Report	42.3309848	−71.8391709	1998
Dawson Pond	MassDEP	Appendix A—1998 DEP DWM Nashua River Basin QA/QC Report	42.3352095	−71.8521791	1998
Flannagan Pond	MassDEP	Appendix A—1998 DEP DWM Nashua River Basin QA/QC Report	42.5579469	−71.5675232	1998
Grove Pond	MassDEP	Appendix A—1998 DEP DWM Nashua River Basin QA/QC Report	42.5525657	−71.5844890	1998
Plow Shop Pond	MassDEP	Appendix A—1998 DEP DWM Nashua River Basin QA/QC Report	42.5558002	−71.5923299	1998
Robbins Pond	MassDEP	Appendix A—1998 DEP DWM Nashua River Basin QA/QC Report	42.5375850	−71.6046534	1998
Sawmill Pond	MassDEP	Appendix A—1998 DEP DWM Nashua River Basin QA/QC Report	42.5445173	−71.8465410	1998
Lake Whalom	MassDEP	Appendix A—1998 DEP DWM Nashua River Basin QA/QC Report	42.5739661	−71.7404572	1998
Meadow Pond	MassDEP	Blackstone River Basin 1998 Water Quality Assessment Report	42.5534691	−71.3432969	1998

(Continued)

TABLE 19.2
(Continued)

Waterbody	Agency	Report Source	Longitude	Latitude	First Year Reported
Spectacle Pond	MassDEP	Chicopee River Basin 1998 Water Quality Assessment Report	−71.5168652	42.5633255	1998
Quaboag River	MassDEP	Chicopee River Basin 1998 Water Quality Assessment Report	−72.1796508	42.2233783	1998
Asnebumskit Pond	MassDEP	Appendix A—1998 DEP DWM Nashua River Basin QA/QC Report	−71.9090451	42.3231830	1998
Coachlace Pond	MassDEP	Appendix A—1998 DEP DWM Nashua River Basin QA/QC Report	−71.6974005	42.4131214	1998
East Waushacum Pond	MassDEP	Appendix A—1998 DEP DWM Nashua River Basin QA/QC Report	−71.7478236	42.4115697	1998
Heald Pond	MassDEP	Appendix A—1998 DEP DWM Nashua River Basin QA/QC Report	−71.6344636	42.6699219	1998
Lake Lookout	MassDEP	Connecticut River Basin 1998 Water Quality Assessment Report	−72.5382418	42.1147402	1998
Lancaster Millpond	MassDEP	Appendix A—1998 DEP DWM Nashua River Basin QA/QC Report	−71.6847860	42.4075777	1998
Loon Pond	MassDEP	Connecticut River Basin 1998 Water Quality Assessment Report	−72.4987251	42.1434933	1998
Maple Spring Pond	MassDEP	Appendix A—1998 DEP DWM Nashua River Basin QA/QC Report	−71.8818377	42.3766150	1998
Minechoag Pond	MassDEP	Chicopee River Basin 1998 Water Quality Assessment Report	−72.4587528	42.1633418	1998
Mona Lake	MassDEP	Chicopee River Basin 1998 Water Quality Assessment Report	−72.5206812	42.1426150	1998
Moose Hill Reservoir	MassDEP	Chicopee River Basin 1998 Water Quality Assessment Report	−71.9563492	42.2714098	1998
Muschopauge Pond	MassDEP	Appendix A—1998 DEP DWM Nashua River Basin QA/QC Report	−71.9213771	42.3838652	1998
Noonan Cove	MassDEP	Connecticut River Basin 1998 Water Quality Assessment Report	−72.5487494	42.1071250	1998
Northfield Mountain Reservoir	MassDEP	Connecticut River Basin 1998 Water Quality Assessment Report	−72.4405873	42.6134969	1998
Palmer Reservoir	MassDEP	Chicopee River Basin 1998 Water Quality Assessment Report	−72.3101546	42.1633650	1998
Porter Lake West	MassDEP	Connecticut River Basin 1998 Water Quality Assessment Report	−72.5711189	42.0736919	1998
Quinapoxet Reservoir	MassDEP	Appendix A—1998 DEP DWM Nashua River Basin QA/QC Report	−71.8832363	42.3962827	1998
Round Meadow Pond	MassDEP	Appendix A—1998 DEP DWM Nashua River Basin QA/QC Report	−71.8941192	42.5451980	1998
Sandy Pond	MassDEP	Appendix A—1998 DEP DWM Nashua River Basin QA/QC Report	−71.5556476	42.5618947	1998

Location	Agency	Source	Longitude	Latitude	Year
South Meadow Pond (East Basin)	MassDEP	Appendix A—1998 DEP DWM Nashua River Basin QA/QC Report	−71.7060627	42.4148445	1998
South Meadow Pond (West Basin)	MassDEP	Appendix A—1998 DEP DWM Nashua River Basin QA/QC Report	−71.7127434	42.4148406	1998
Thompson Lake	MassDEP	Chicopee River Basin 1998 Water Quality Assessment Report	−72.2990214	42.1749037	1998
Waite Pond	MassDEP	Blackstone River Basin 1998 Water Quality Assessment Report	−71.8911113	42.2491804	1998
Watershops Pond	MassDEP	Connecticut River Basin 1998 Water Quality Assessment Report	−72.5439428	42.1057506	1998
Long Pond	MassDEP	Chicopee River Basin 1998 Water Quality Assessment Report	−72.5081176	42.1542203	1998
Mill Pond	MassDEP	Connecticut River Basin 1998 Water Quality Assessment Report	−72.4951695	42.1117920	1998
Chicopee Brook Pond	MassDEP	Chicopee River Basin 1998 Water Quality Assessment Report	−72.3095784	42.1198885	1998
Connors Pond	MassDEP	Chicopee River Basin 1998 Water Quality Assessment Report	−72.1616303	42.4642839	1998
Hopedale Pond	MassDEP	Blackstone River Basin 1998 Water Quality Assessment Report	−71.5536066	42.1397139	1998
Lackey Pond	MassDEP	Blackstone River Basin 1998 Water Quality Assessment Report	−71.6921029	42.0931662	1998
Linwood Pond	MassDEP	Blackstone River Basin 1998 Water Quality Assessment Report	−71.6520998	42.1010953	1998
Lower Pond	MassDEP	Connecticut River Basin 1998 Water Quality Assessment Report	−72.5722162	42.2545682	1998
Patch Reservoir	MassDEP	Blackstone River Basin 1998 Water Quality Assessment Report	−71.8513623	42.2700807	1998
Red Bridge Impoundment	MassDEP	Chicopee River Basin 1998 Water Quality Assessment Report	−72.4020896	42.1802859	1998
Smiths Pond	MassDEP	Blackstone River Basin 1998 Water Quality Assessment Report	−71.8693452	42.2413182	1998
Stoneville Pond	MassDEP	Blackstone River Basin 1998 Water Quality Assessment Report	−71.8468045	42.2187351	1998
Rockwell Pond	MassDEP	Appendix A—1998 DEP DWM Nashua River Basin QA/QC Report	−71.7684382	42.5271890	1998
Salisbury Pond	MassDEP	Blackstone River Basin 1998 Water Quality Assessment Report Appendix B	−71.8058591	42.2771991	1998
Lake Wildwood	MassDEP	Blackstone River Basin 1998 Water Quality Assessment Report	−71.6412087	42.1763307	1998
Mill River	MassDEP	Parker River Watershed Water Quality Assessment Report Appendix C	−70.8990290	42.7396810	1999
Winter Pond	MassDEP	Boston Harbor 1999 Water Quality Assessment Report	−71.1530817	42.4563023	1999
Russell Pond	MassDEP	Boston Harbor 1999 Water Quality Assessment Report	−71.0734873	42.2362559	1999
Cobbs Pond	MassDEP	Boston Harbor 1999 Water Quality Assessment Report	−71.2445201	42.1595441	1999
Ellis Pond	MassDEP	Boston Harbor 1999 Water Quality Assessment Report	−71.2229887	42.1821424	1999
Neponset Reservoir	MassDEP	Boston Harbor 1999 Water Quality Assessment Report	−71.2445409	42.0846820	1999

(Continued)

TABLE 19.2
(Continued)

Waterbody	Agency	Report Source	Longitude	Latitude	First Year Reported
Pinewood Pond	MassDEP	Boston Harbor 1999 Water Quality Assessment Report	−71.1395441	42.1298261	1999
Farrington Pond	MassDEP	Boston Harbor 1999 Water Quality Assessment Report	−71.1155279	42.1101472	1999
Ponkapoag Pond	MassDEP	Boston Harbor 1999 Water Quality Assessment Report	−71.0951069	42.1923267	1999
Reservoir Pond	MassDEP	Boston Harbor 1999 Water Quality Assessment Report	−71.1231824	42.1688080	1999
Woods Pond	MassDEP	Boston Harbor 1999 Water Quality Assessment Report	−71.1127164	42.1167650	1999
Flint Pond	MassDEP	Merrimack River Basin 1999 Water Quality Assessment Report	−71.4318174	42.6729165	1999
Newfield Pond	MassDEP	Merrimack River Basin 1999 Water Quality Assessment Report	−71.3899233	42.6337843	1999
Massapoag Pond	MassDEP	Merrimack River Basin 1999 Water Quality Assessment Report	−71.4965387	42.6485085	1999
Ganawatte Farm Pond	MassDEP	Boston Harbor 1999 Water Quality Assessment Report	−71.2417500	42.1064274	1999
Memorial Pond	MassDEP	Boston Harbor 1999 Water Quality Assessment Report	−71.2474683	42.1454564	1999
Parker River	MassDEP	Parker River Watershed Water Quality Assessment Report Appendix C	−71.0122130	42.7302460	1999
Willet Pond	MassDEP	Boston Harbor 1999 Water Quality Assessment Report	−71.2372682	42.1793243	1999
Bird Pond	MassDEP	Boston Harbor 1999 Water Quality Assessment Report	−71.2231683	42.1613874	1999
Popes Pond	MassDEP	Boston Harbor 1999 Water Quality Assessment Report	−71.0955672	42.2487895	1999
Sprague Pond	MassDEP	Boston Harbor 1999 Water Quality Assessment Report	−71.1369968	42.2335018	1999
Beaver Brook	MassDEP	Merrimack River Basin 1999 Water Quality Assessment Report Appendix C	−71.3231210	42.6624310	1999
Spicket River	MassDEP	Merrimack River Basin 1999 Water Quality Assessment Report	−71.1836624	42.7191350	1999
Weir River	MassDEP	Boston Harbor 1999 Water Quality Assessment Report	−70.8677397	42.2372405	1999
Turners Pond	MassDEP	Boston Harbor 1999 Water Quality Assessment Report	−71.0773782	42.2608348	1999
Penn Brook	MassDEP	Parker River Watershed Water Quality Assessment Report Appendix C	−70.9871310	42.7297300	1999
Penn Brook	MassDEP	10-D020–04	−70.9841492	42.7187277	1999
Stony Brook	MassDEP	Merrimack River Basin 1999 Water Quality Assessment Report	−71.4096830	42.6102380	1999
Silver Lake	MassDEP	Ipswich River Watershed 2000 Water Quality Assessment Report	−71.4645777	42.0632423	2000
Fosters Pond	MassDEP	Shawsheen River Watershed 2000 Water Quality Assessment Report 83-AC-2 86.0, July 2003	−71.1369818	42.6067165	2000

Gravel Pit Pond	MassDEP	Shawsheen River Watershed 2000 Water Quality Assessment Report 83-AC-2 86.0, July 2003	-71.1604601	42.6722056	2000
Gravel Pit Pond	MassDEP	Shawsheen River Watershed 2000 Water Quality Assessment Report	-71.1604601	42.6722056	2000
Ballardvale Impoundment	MassDEP	Shawsheen River Watershed 2000 Water Quality Assessment Report 83-AC-2 86.0, July 2003	-71.1580528	42.6197553	2000
Ballardvale Impoundment	MassDEP	Shawsheen River Watershed 2000 Water Quality Assessment Report	-71.1580528	42.6197553	2000
Pomps Pond	MassDEP	Shawsheen River Watershed 2000 Water Quality Assessment Report 83-AC-2 86.0, July 2003	-71.1513484	42.6352457	2000
Pomps Pond	MassDEP	Shawsheen River Watershed 2000 Water Quality Assessment Report	-71.1513484	42.6352457	2000
Devils Dishfull Pond	MassDEP	Ipswich River Watershed 2000 Water Quality Assessment Report	-71.0068499	42.5406443	2000
Field Pond	MassDEP	Ipswich River Watershed 2000 Water Quality Assessment Report 92-AC-1 088.0, April 2004	-71.1087937	42.6083542	2000
Lower Boston Brook Pond	MassDEP	Ipswich River Watershed 2000 Water Quality Assessment Report 92-AC-1 088.0, April 2004	-71.0083967	42.6174828	2000
Lower Four Mile Pond	MassDEP	Ipswich River Watershed 2000 Water Quality Assessment Report 92-AC-1 088.0, April 2004	-71.0032797	42.6803774	2000
Lowe Pond	MassDEP	Ipswich River Watershed 2000 Water Quality Assessment Report 92-AC-1 088.0, April 2004	-70.9858738	42.6759112	2000
Lubber Pond East	MassDEP	Ipswich River Watershed 2000 Water Quality Assessment Report 92-AC-1 088.0, April 2004	-71.1974308	42.5562190	2000
Lubber Pond West	MassDEP	Ipswich River Watershed 2000 Water Quality Assessment Report 92-AC-1 088.0, April 2004	-71.2031072	42.5553548	2000
Martins Pond	MassDEP	Ipswich River Watershed 2000 Water Quality Assessment Report 92-AC-1 088.0, April 2004	-71.1252953	42.5955576	2000
Stevens Pond	MassDEP	Ipswich River Watershed 2000 Water Quality Assessment Report 92-AC-1 088.0, April 2004	-70.9973757	42.6836135	2000
Federal Pond	MassDEP	Buzzards Bay Watershed 2000 Water Quality Assessment Report 95-AC-2 085.0, November 2003	-70.7078164	41.8762623	2000
White Island Pond (East Basin)	MassDEP	07-T003-02	-70.6175309	41.8091834	2000
White Island Pond (East Basin)	MassDEP	07-T004-01	-70.6175309	41.8091834	2000
White Island Pond (East Basin)	MassDEP	08-F008-02	-70.6175309	41.8091834	2000
White Island Pond (East Basin)	MassDEP	11-B007-01	-70.6175309	41.8091834	2000

(Continued)

TABLE 19.2
(Continued)

Waterbody	Agency	Report Source	Longitude	Latitude	First Year Reported
White Island Pond (East Basin)	MassDEP	14-B007–01	−70.6175309	41.8091834	2000
White Island Pond (East Basin)	MassDEP	Buzzards Bay Watershed 2000 Water Quality Assessment Report 95-AC-2 085.0, November 2003	−70.6178754	41.8117381	2000
White Island Pond (East Basin)	MassDEP	Buzzards Bay Watershed 2000 Water Quality Assessment Report	−70.6178754	41.8117381	2000
White Island Pond (East Basin)	MassDEP	08-F008–01	−70.6275557	41.8099941	2000
White Island Pond (West Basin)	MassDEP	12-B003–02	−70.6275557	41.8099941	2000
White Island Pond (West Basin)	MassDEP	13-B007–02	−70.6275557	41.8099941	2000
White Island Pond (West Basin)	MassDEP	13-B009–03	−70.6275557	41.8099941	2000
White Island Pond (West Basin)	MassDEP	14-B005–02	−70.6275557	41.8099941	2000
White Island Pond (West Basin)	MassDEP	Buzzards Bay Watershed 2000 Water Quality Assessment Report 95-AC-2 085.0, November 2003	−70.6277405	41.8062552	2000
White Island Pond (West Basin)	MassDEP	Buzzards Bay Watershed 2000 Water Quality Assessment Report	−70.6277405	41.8062552	2000
Ames Pond	MassDEP	Shawsheen River Watershed 2000 Water Quality Assessment Report	−71.2253468	42.6382918	2000
Bakers Meadow Pond	MassDEP	Shawsheen River Watershed 2000 Water Quality Assessment Report	−71.1602603	42.6507736	2000
Bents Pond	MassDEP	Millers River Watershed 2000 Water Quality Assessment Report	−71.9846729	42.5580259	2000
Brackett Pond	MassDEP	Ipswich River Watershed 2000 Water Quality Assessment Report	−71.1095852	42.6134520	2000
Bradford Pond	MassDEP	Ipswich River Watershed 2000 Water Quality Assessment Report	−71.0789855	42.5999909	2000
Collins Pond	MassDEP	Ipswich River Watershed 2000 Water Quality Assessment Report	−71.1090134	42.6115557	2000

Creighton Pond	MassDEP	Ipswich River Watershed 2000 Water Quality Assessment Report	−71.0442255	42.6269243	2000
Crystal Pond	MassDEP	Ipswich River Watershed 2000 Water Quality Assessment Report	−70.9999708	42.5491935	2000
Eisenhaures Pond	MassDEP	Ipswich River Watershed 2000 Water Quality Assessment Report	−71.0896937	42.5860316	2000
Elginwood Pond	MassDEP	Ipswich River Watershed 2000 Water Quality Assessment Report	−71.0032608	42.5477493	2000
Farnum Street Pond	MassDEP	Ipswich River Watershed 2000 Water Quality Assessment Report	−71.0833779	42.6549048	2000
Fawn Lake	MassDEP	Shawsheen River Watershed 2000 Water Quality Assessment Report	−71.2734103	42.5137746	2000
Glen Charlie Pond	MassDEP	Buzzards Bay Watershed 2000 Water Quality Assessment Report	−70.6462652	41.7920967	2000
Hood Pond	MassDEP	Ipswich River Watershed 2000 Water Quality Assessment Report	−70.9548440	42.6699159	2000
Hussey Pond	MassDEP	Shawsheen River Watershed 2000 Water Quality Assessment Report	−71.1523517	42.6727427	2000
Middleton Pond	MassDEP	Ipswich River Watershed 2000 Water Quality Assessment Report	−71.0280532	42.5917317	2000
Pierces Pond	MassDEP	Ipswich River Watershed 2000 Water Quality Assessment Report	−70.9969165	42.5293390	2000
Pleasant Pond	MassDEP	Ipswich River Watershed 2000 Water Quality Assessment Report	−70.8952933	42.6140959	2000
Pond Street Pond	MassDEP	Shawsheen River Watershed 2000 Water Quality Assessment Report	−71.2556560	42.5869750	2000
Putnamville Reservoir	MassDEP	Ipswich River Watershed 2000 Water Quality Assessment Report	−70.9468334	42.6008152	2000
Richardson Pond North	MassDEP	Shawsheen River Watershed 2000 Water Quality Assessment Report	−71.2457842	42.5879255	2000
Round Pond	MassDEP	Shawsheen River Watershed 2000 Water Quality Assessment Report	−71.2319936	42.6029166	2000
Salem Street Pond	MassDEP	Ipswich River Watershed 2000 Water Quality Assessment Report	−71.0455120	42.6366929	2000
Spofford Pond	MassDEP	Ipswich River Watershed 2000 Water Quality Assessment Report	−71.0207771	42.6959440	2000
Stearns Pond	MassDEP	Ipswich River Watershed 2000 Water Quality Assessment Report	−71.0694217	42.6173500	2000
Stiles Pond	MassDEP	Ipswich River Watershed 2000 Water Quality Assessment Report	−71.0366642	42.6889085	2000
Suntaug Lake	MassDEP	Ipswich River Watershed 2000 Water Quality Assessment Report	−71.0048430	42.5217843	2000
Swan Pond	MassDEP	Ipswich River Watershed 2000 Water Quality Assessment Report	−71.0504704	42.5915270	2000
Towne Pond	MassDEP	Ipswich River Watershed 2000 Water Quality Assessment Report	−71.0287626	42.6546551	2000
Wenham Lake	MassDEP	Ipswich River Watershed 2000 Water Quality Assessment Report	−70.8907252	42.5904622	2000
Whitney Pond	MassDEP	Baseline Lake Survey 2003 Technical Memo	−72.0412151	42.6830822	2000
Whitney Pond	MassDEP	Millers River Watershed 2000 Water Quality Assessment Report	−72.0347660	42.6814818	2000
Winona Pond	MassDEP	Ipswich River Watershed 2000 Water Quality Assessment Report	−71.0099919	42.5344892	2000

(Continued)

TABLE 19.2
(Continued)

Waterbody	Agency	Report Source	Longitude	Latitude	First Year Reported
Butterfield Pond	MassDEP	Shawsheen River Watershed 2000 Water Quality Assessment Report	−71.2106699	42.4697954	2000
Howes Pond	MassDEP	Ipswich River Watershed 2000 Water Quality Assessment Report	−71.0021204	42.6545333	2000
Long Pond	MassDEP	Shawsheen River Watershed 2000 Water Quality Assessment Report	−71.2518748	42.5978382	2000
Mill Pond	MassDEP	Ipswich River Watershed 2000 Water Quality Assessment Report	−71.1737213	42.5129751	2000
Upper Boston Brook Pond	MassDEP	Ipswich River Watershed 2000 Water Quality Assessment Report	−71.0184112	42.6199362	2000
Bogastow Brook	MassDEP	Shawsheen River Watershed 2000 Water Quality Assessment Report	−71.3920350	42.1861990	2000
Brook Street Pond	MassDEP	Ipswich River Watershed 2000 Water Quality Assessment Report	−71.0877541	42.6491289	2000
Shaw Pond	MassDEP	06-E012−01	−73.1227752	42.252142	2001
Shaw Pond	MassDEP	Farmington River Watershed 2001 Biological Assessment	−73.1239117	42.2544760	2001
Sylvestri Pond	MassDEP	French and Quinebaug River Watershed 2001 Water Quality Assessment Report	−71.9847640	42.0673699	2001
Larner Pond	MassDEP	French and Quinebaug River Watershed 2001 Water Quality Assessment Report	−71.9060326	42.0597896	2001
Lake Nippenicket	MassDEP	Taunton River Watershed 2001 Water Quality Assessment Report Appendix C	−71.0416911	41.9696598	2001
Ashland Reservoir	MassDEP	Concord Watershed 2001 DWM Water Quality Monitoring Data	−71.4648923	42.2402611	2001
Assabet River Reservoir	MassDEP	Concord Watershed 2001 DWM Water Quality Monitoring Data	−71.6459850	42.2636753	2001
Bartlett Pond	MassDEP	Concord Watershed 2001 DWM Water Quality Monitoring Data	−71.6183754	42.3179137	2001
Batemans Pond	MassDEP	Concord Watershed 2001 DWM Water Quality Monitoring Data	−71.3664673	42.4939747	2001
Chauncy Lake	MassDEP	Concord Watershed 2001 DWM Water Quality Monitoring Data	−71.6123353	42.2943741	2001
Fisk Pond	MassDEP	Concord Watershed 2001 DWM Water Quality Monitoring Data	−71.3683511	42.2819821	2001
Fort Meadow Reservoir	MassDEP	Concord Watershed 2001 DWM Water Quality Monitoring Data	−71.5481107	42.3686485	2001
Framingham Reservoir #1	MassDEP	Concord Watershed 2001 DWM Water Quality Monitoring Data	−71.4493238	42.2896421	2001
Framingham Reservoir #3	MassDEP	Concord Watershed 2001 DWM Water Quality Monitoring Data	−71.4695991	42.3001460	2001

Great Meadow Pond #3	MassDEP	Concord Watershed 2001 DWM Water Quality Monitoring Data	−71.3349052	42.4756991	2001
Grist Mill Pond	MassDEP	Concord Watershed 2001 DWM Water Quality Monitoring Data	−71.4791471	42.3554728	2001
Hager Pond	MassDEP	Concord Watershed 2001 DWM Water Quality Monitoring Data	−71.4877578	42.3490361	2001
Heard Pond	MassDEP	Concord Watershed 2001 DWM Water Quality Monitoring Data	−71.3835387	42.3526192	2001
Hopkinton Reservoir	MassDEP	Concord Watershed 2001 DWM Water Quality Monitoring Data	−71.5210262	42.2547944	2001
Nutting Lake (East Basin)	MassDEP	Concord Watershed 2001 DWM Water Quality Monitoring Data	−71.2650031	42.5372640	2001
Russell Millpond	MassDEP	Concord Watershed 2001 DWM Water Quality Monitoring Data	−71.3329065	42.5703497	2001
Saxonville Pond	MassDEP	Concord Watershed 2001 DWM Water Quality Monitoring Data	−71.4136220	42.3201977	2001
Stearns Mill Pond	MassDEP	Concord Watershed 2001 DWM Water Quality Monitoring Data	−71.4530771	42.3858152	2001
Waushakum Pond	MassDEP	05-G006–02	−71.423861	42.2641737	2001
Waushakum Pond	MassDEP	Concord Watershed 2001 DWM Water Quality Monitoring Data	−71.4262402	42.2639502	2001
Whitehall Reservoir	MassDEP	Concord Watershed 2001 DWM Water Quality Monitoring Data	−71.5753633	42.2294215	2001
Winning Pond	MassDEP	Concord Watershed 2001 DWM Water Quality Monitoring Data	−71.3013147	42.5514593	2001
Lake Cochituate	MassDEP	Concord Watershed 2001 DWM Water Quality Monitoring Data	−71.3670553	42.2894878	2001
Forge Pond	MassDEP	South Shore Coastal Watersheds 2001 Water Quality Assessment Report 94-AC-2 CN 93.0, March 2006	−70.8800038	42.1062243	2001
Jacobs Pond	MassDEP	South Shore Coastal Watersheds 2001 Water Quality Assessment Report 94-AC-2 CN 93.0, March 2006	−70.8492691	42.1619900	2001
Lower Chandler Pond	MassDEP	South Shore Coastal Watersheds 2001 Water Quality Assessment Report 94-AC-2 CN 93.0, March 2006	−70.7624539	42.0312782	2001
Old Oaken Bucket Pond	MassDEP	South Shore Coastal Watersheds 2001 Water Quality Assessment Report 94-AC-2 CN 93.0, March 2006	−70.7501611	42.1791203	2001
Old Oaken Bucket Pond	MassDEP	South Coastal Watersheds DWM YEAR 2001 Water Quality Monitoring Data (PG. C2)	−70.7501611	42.1791203	2001
Oldham Pond	MassDEP	South Shore Coastal Watersheds 2001 Water Quality Assessment Report 94-AC-2 CN 93.0, March 2006	−70.8364494	42.0664190	2001

(Continued)

TABLE 19.2
(Continued)

Waterbody	Agency	Report Source	Longitude	Latitude	First Year Reported
Oldham Pond	MassDEP	South Shore Coastal Watersheds 2001 Water Quality Assessment Report, South Coastal Watersheds DWM Year 2001 Water Quality Monitoring Data (PG. C2)	−70.8364494	42.0664190	2001
French River	MassDEP	French and Quinebaug River Watershed 2001 Water Quality Assessment Report	−71.8841820	42.0246995	2001
Quinebaug River	MassDEP	French and Quinebaug River Watershed 2001 Water Quality Assessment Report	−72.1185692	42.1095615	2001
Russell Pond	MassDEP	South Shore Coastal Watersheds 2001 Water Quality Assessment Report, South Coastal Watersheds DWM Year 2001 Water Quality Monitoring Data (PG. C2)	−70.7470003	41.9778691	2001
Gates Pond	MassDEP	Concord Watershed 2001 DWM Water Quality Monitoring Data	−71.6108718	42.3782027	2001
Clear Pond	MassDEP	Westfield River Watershed DWM 2001 Water Quality Monitoring Data	−72.6595532	42.1855163	2001
Easterbrook Pond	MassDEP	French and Quinebaug River Watershed 2001 Water Quality Assessment Report	−71.9249238	42.0537809	2001
Easterbrook Pond	MassDEP	French and Quinebaug River Watershed 2004 Benthic Macroinvertebrate Bioassessment	−71.9249240	42.0537810	2001
Elm Street Pond	MassDEP	Concord Watershed 2001 DWM Water Quality Monitoring Data	−71.3736433	42.5605568	2001
Farrar Pond	MassDEP	Concord Watershed 2001 DWM Water Quality Monitoring Data	−71.3542989	42.4141745	2001
Fiske Street Pond	MassDEP	Concord Watershed 2001 DWM Water Quality Monitoring Data	−71.3754055	42.5534551	2001
Fort Pond	MassDEP	Concord Watershed 2001 DWM Water Quality Monitoring Data	−71.4663567	42.5071911	2001
Foundry Pond	MassDEP	South Shore Coastal Watersheds 2001 Water Quality Assessment Report, South Coastal Watersheds DWM Year 2001 Water Quality Monitoring Data (PG. C2)	−70.7091488	41.9842148	2001
Framingham Reservoir #2	MassDEP	Concord Watershed 2001 DWM Water Quality Monitoring Data	−71.4439073	42.2744049	2001
Furnace Pond	MassDEP	South Shore Coastal Watersheds 2001 Water Quality Assessment Report, South Coastal Watersheds DWM Year 2001 Water Quality Monitoring Data (PG. C2)	−70.8255009	42.0557280	2001
Furnace Pond	MassDEP	Baseline Lake Survey 2003 Technical Memo	−70.8255009	42.0557280	2001
Heart Pond	MassDEP	Concord Watershed 2001 DWM Water Quality Monitoring Data	−71.3877340	42.5657450	2001
Hocomonco Pond	MassDEP	Concord Watershed 2001 DWM Water Quality Monitoring Data	−71.6497506	42.2724040	2001
Hudson Pond	MassDEP	French and Quinebaug River Watershed 2001 Water Quality Assessment Report	−71.8343498	42.1433074	2001
Learned Pond	MassDEP	Concord Watershed 2001 DWM Water Quality Monitoring Data	−71.4188407	42.2873503	2001
Littleville Lake	MassDEP	Westfield River Watershed 2001 Water Quality Assessment Report	−72.8866628	42.2789380	2001

Long Bow Lake	MassDEP	Westfield River Watershed DWM 2001 Water Quality Monitoring Data	−73.0914248	42.2529731	2001
Milham Reservoir	MassDEP	Concord Watershed 2001 DWM Water Quality Monitoring Data	−71.6092840	42.3422667	2001
Morse Pond	MassDEP	French and Quinebaug River Watershed 2001 Water Quality Assessment Report	−72.0144229	42.0319671	2001
Musquashcut Pond	MassDEP	South Coastal Watersheds DWM Year 2001 Water Quality Monitoring Data (PG. C2)	−70.7583101	42.2275200	2001
Nagog Pond	MassDEP	Concord Watershed 2001 DWM Water Quality Monitoring Data	−71.4412077	42.5151087	2001
Pikes Pond	MassDEP	French and Quinebaug River Watershed 2001 Water Quality Assessment Report	−71.9404468	42.1617335	2001
Royal Pond	MassDEP	Farmington River Watershed 2001 Biological Assessment	−73.1609142	42.1804095	2001
Russell Millpond	MassDEP	South Shore Coastal Watersheds 2001 Water Quality Assessment Report, South Coastal Watersheds DWM Year 2001 Water Quality Monitoring Data (PG. C2)	−70.6335513	41.9169630	2001
Ship Pond	MassDEP	South Shore Coastal Watersheds 2001 Water Quality Assessment Report, South Coastal Watersheds DWM Year 2001 Water Quality Monitoring Data (PG. C2)	−70.5337638	41.8697059	2001
Sibley Pond	MassDEP	French and Quinebaug River Watershed 2001 Water Quality Assessment Report	−72.0107046	42.1515962	2001
Smith Pond	MassDEP	Concord Watershed 2001 DWM Water Quality Monitoring Data	−71.6603615	42.2898519	2001
Studleys Pond	MassDEP	South Shore Coastal Watersheds 2001 Water Quality Assessment Report	−70.9209521	42.1195205	2001
Sudbury Reservoir	MassDEP	Concord Watershed 2001 DWM Water Quality Monitoring Data	−71.5058924	42.3172382	2001
Ward Pond	MassDEP	Farmington River Watershed 2001 Biological Assessment	−73.1012513	42.2492359	2001
Wright Pond	MassDEP	Westfield River Watershed 2001 Water Quality Assessment Report	−72.6586467	42.1812338	2001
Factory Pond	MassDEP	South Shore Coastal Watersheds 2001 Water Quality Assessment Report, South Coastal Watersheds DWM Year 2001 Water Quality Monitoring Data (PG. C2)	−70.8748442	42.0899688	2001
Town Meadow Brook	MassDEP	French and Quinebaug River Watershed 2001 Water Quality Assessment Report	−71.9200884	42.2220138	2001
West Branch Farmington River	MassDEP	Farmington River Watershed 2001 Biological Assessment	−73.0737135	42.1526264	2001
Texas Pond	MassDEP	French and Quinebaug River Watershed 2001 Water Quality Assessment Report	−71.8928149	42.1717096	2001
Wallis Pond	MassDEP	French and Quinebaug River Watershed 2001 Water Quality Assessment Report	−71.9070449	42.0660729	2001
Housatonic River	MassDEP	Housatonic River Watershed 2002 Water Quality Assessment Report Appendix C	−73.2404010	42.3940380	2002
Green River	MassDEP	Housatonic River Watershed 2002 Water Quality Assessment Report Appendix C	−73.3781260	42.1785560	2002
Charles River	MassDEP	Charles River Watershed 2002 Biological Assessment	−71.2724073	42.2621874	2002
Goose Pond Brook	MassDEP	Housatonic River Watershed 2002 Water Quality Assessment Report Appendix C	−73.2271930	42.2943020	2002
Shaw Pond	DCR	DCR(2008) Data from Lakes and Ponds, 2002	−73.1239117	42.2544760	2002

(Continued)

**TABLE 19.2
(Continued)**

Waterbody	Agency	Report Source	Longitude	Latitude	First Year Reported
Beaver Pond	DCR	DCR(2008) Data from Lakes and Ponds, 2002	−71.4175437	42.0799485	2002
Baddacook Pond	DCR	DCR(2008) Data from Lakes and Ponds, 2002	−71.5304640	42.6189623	2002
Chicken Brook	MassDEP	Charles River Watershed 2002 Biological Assessment	−71.4422633	42.1785228	2002
Green River	MassDEP	Appendix D—Hudson Watershed 2002 Biological Assessment	−73.2312640	42.6764870	2002
Hoosic River	MassDEP	Appendix D—Hudson Watershed 2002 Biological Assessment	−73.2092170	42.7293160	2002
Stop River	MassDEP	Charles River Watershed 2002 Biological Assessment	−71.3059926	42.1539616	2002
Housatonic River	MassDEP	Housatonic River Watershed 2002 Water Quality Assessment Report Appendix C	−73.2408830	42.2833520	2002
Nipmuck Lake	DCR	DCR(2008) Data from Lakes and Ponds, 2002	−71.5699120	42.0964290	2002
Bear Hole Pond (little)	DCR	DCR(2008) Data from Lakes and Ponds, 2002	−70.9934554	41.8688229	2002
Bear Hole Pond (little)	DCR	DCR(2008) Data from Lakes and Ponds, 2007	−70.9934554	41.8688229	2002
Bear Hole Pond (little)	DCR	DCR(2008) Data from Lakes and Ponds, 2008	−70.9934554	41.8688229	2002
Konkapot River	MassDEP	Housatonic River Watershed 2002 Water Quality Assessment Report Appendix C	−73.311830	42.047410	2002
Center Pond	DCR	DCR(2008) Data from Lakes and Ponds, 2003	−73.0687461	42.2967178	2003
Long Pond	DCR	DCR(2008) Data from Lakes and Ponds, 2003	−71.9921341	42.3567719	2003
Long Pond	DCR	DCR(2008) Data from Lakes and Ponds, 2008	−71.9921341	42.3567719	2003
Watson Pond	DCR	DCR(2008) Data from Lakes and Ponds, 2003	−71.1186846	41.9507941	2003
Watson Pond	DCR	DCR(2008) Data from Lakes and Ponds, 2008	−71.1186846	41.9507941	2003
Winnecumnet Pond	DCR	DCR(2008) Data from Lakes and Ponds, 2003	−71.1312668	41.9709330	2003
Farm Pond	MassDEP	Baseline Lake Survey 2003 Technical Memo	−71.4267259	42.2815529	2003
Baldpate Pond	MassDEP	05-G004–02	−71.0014233	42.6988862	2003
Baldpate Pond	MassDEP	Baseline Lake Survey 2003 Technical Memo	−71.0023711	42.6982584	2003
Quaboag River	MassDEP	07-K004–05	−72.2635427	42.1814299	2003
Quaboag River	MassDEP	08-L006–05	−72.2635427	42.1814299	2003

Quaboag River	MassDEP	11-G004–05		−72.2635427	42.1814299	2003
Quaboag River	MassDEP	Baseline Lake Survey 2003 Technical Memo		−72.2630590	42.1686640	2003
Mumford River	MassDEP	Blackstone River Watershed 2003 Biological Assessment		−71.7293980	42.0832710	2003
Prospect Hill Pond	DCR	DCR(2008) Data from Lakes and Ponds, 2003		−71.0931736	41.9343313	2003
Spindleville Pond	MassDEP	Blackstone River Watershed 2003 Biological Assessment		−71.5312797	42.1161304	2003
Blackstone River	MassDEP	Blackstone River Watershed 2003 Biological Assessment		−71.5456216	42.0171721	2003
Pepperell Pond	MassDEP	Baseline Lake Survey 2003 Technical Memo		−71.5839579	42.6448131	2003
Nine Mile Pond	DCR	DCR (2008) Data from Lakes and Ponds, 2003		−72.4344914	42.1490848	2003
Turnpike Lake	DCR	DCR (2008) Data from Lakes and Ponds, 2004		−71.3124012	42.0155418	2004
Hood Pond	DCR	DCR(2008) Data from Lakes and Ponds, 2004		−70.9548440	42.6699159	2004
Richardson Brook	MassDEP	Merrimack River Watershed 2004 Benthic Macroinvertebrate Assessment		−71.2734717	42.6676737	2004
Unnamed Tributary	MassDEP	French and Quinebaug River Watershed 2004 Benthic Macroinvertebrate Bioassessment		−72.1988670	42.1223380	2004
Lake Saltonstall	DCR	DCR (2008) Data from Lakes and Ponds, 2004		−71.0658718	42.7827085	2004
Flax Pond	MassDEP	05-G012–01		−70.0260112	41.7659812	2005
Lake Lorraine	MassDEP	05-G013–02		−72.5129923	42.1450906	2005
Webster Lake	MassDEP	05-G014–01		−71.8487243	42.0543495	2005
Oldham Pond	DCR	DCR(2008) Data from Lakes and Ponds, 2005		−70.8364494	42.0664190	2005
Blackstone River	MassDEP	05-K004–04		−71.6520398	42.1538730	2005
Blackstone River	MassDEP	08-A028–03		−71.6800238	42.1748750	2005
Blackstone River	MassDEP	08-A035–05		−71.6223140	42.0987190	2005
Ashumet Pond	MassDEP	05-G011–02		−70.5318225	41.6347157	2005
Baddacook Pond	MassDEP	05-G001–01		−71.5304640	42.6189623	2005
Sudbury River	MassDEP	05-J003–02		−71.3973643	42.3254347	2005
Sudbury River	MassDEP	05-J004–02		−71.3973643	42.3254347	2005
Sudbury River	MassDEP	05-J005–02		−71.3973643	42.3254347	2005
Sudbury River	MassDEP	06-F003–02		−71.3973643	42.3254347	2005
Sudbury River	MassDEP	06-F004–02		−71.3973643	42.3254347	2005
Sudbury River	MassDEP	07-L003–02		−71.3973643	42.3254347	2005
Furnace Pond	DCR	DCR (2008) Data from Lakes and Ponds, 2005		−70.8255009	42.0557280	2005

(Continued)

TABLE 19.2
(Continued)

Waterbody	Agency	Report Source	Longitude	Latitude	First Year Reported
Hocomonco Pond	DCR	DCR (2008) Data from Lakes and Ponds, 2005	−71.6497506	42.2724040	2005
Chicoppee Reservoir	DCR	DCR (2008) Data from Lakes and Ponds, 2005	−72.5512770	42.1708408	2005
Chicoppee Reservoir	DCR	DCR (2008) Data from Lakes and Ponds, 2007	−72.5512770	42.1708408	2005
Goose Pond	DCR	DCR (2008) Data from Lakes and Ponds, 2005	−71.3312044	42.4430841	2005
Johnsons Pond	DCR	DCR (2008) Data from Lakes and Ponds, 2005	−71.0535516	42.7316414	2005
Deerfield River	MassDEP	05-F013–01	−72.6266247	42.5228109	2005
Shawsheen River	MassDEP	05-I006–13	−71.1508026	42.6474461	2005
Shawsheen River	MassDEP	05-I003–08	−71.2151746	42.5681537	2005
Shawsheen River	MassDEP	05-I004–08	−71.2151746	42.5681537	2005
Shawsheen River	MassDEP	05-I003–10	−71.2151746	42.5681537	2005
Shawsheen River	MassDEP	05-I004–10	−71.2151746	42.5681537	2005
Shawsheen River	MassDEP	05-I004–11	−71.1930607	42.5996629	2005
Unnamed Tributary	MassDEP	05-Q003–10	−71.0109215	42.5930941	2005
Unnamed Tributary	MassDEP	05-Q004–07	−71.0109215	42.5930941	2005
Unnamed Tributary	MassDEP	05-Q005–07	−71.0109215	42.5930941	2005
Ipswich River	MassDEP	05-Q004–03	−71.1106017	42.5611646	2005
Lubbers Brook	MassDEP	05-Q004–02	−71.1466976	42.5588828	2005
Maple Meadow Brook	MassDEP	05-Q004–01	−71.1500663	42.5519109	2005
Spring Brook	MassDEP	05-I003–03	−71.2578761	42.4948081	2005
Spring Brook	MassDEP	05-I004–03	−71.2578761	42.4948081	2005
Spring Brook	MassDEP	05-I006–03	−71.2578761	42.4948081	2005
Vine Brook	MassDEP	05-I004–04	−71.2242658	42.4964526	2005

			Longitude	Latitude	Year
North Nashua River	MassDEP	05-O004-01	−71.7219246	42.4951686	2005
Content Brook	MassDEP	05-I004-09	−71.2182168	42.5796709	2005
Content Brook	MassDEP	05-I006-09	−71.2182168	42.5796709	2005
Sandy Brook	MassDEP	05-I004-05	−71.2125419	42.4974268	2005
Unnamed Tributary	MassDEP	05-I004-12	−71.1994003	42.6288780	2005
West River	MassDEP	11-F004-04	−71.601117	42.1003609	2006
West River	MassDEP	06-G004-03	−71.601117	42.1003609	2006
West River	MassDEP	11-F004-04	−71.601117	42.1003609	2006
Glen Echo Pond	DCR	DCR (2008) Data from Lakes and Ponds, 2006	−71.0887317	42.1521704	2006
First Pond	DCR	DCR(2008) Data from Lakes and Ponds, 2006	−71.0511939	42.4754089	2006
Blackstone River	MassDEP	06-G004-02	−71.5682244	42.0219477	2006
Blackstone River	MassDEP	07-J005-02	−71.5705414	42.027186	2006
Wheeler Pond	DCR	DCR (2008) Data from Lakes and Ponds, 2006	−71.6304343	42.3572406	2006
Solomon Pond	DCR	DCR (2008) Data from Lakes and Ponds, 2006	−71.6260766	42.3397019	2006
West Pond	DCR	DCR (2008) Data from Lakes and Ponds, 2006	−71.5807312	42.4276008	2006
Unnamed Tributary	MassDEP	06-C001-03	−70.6135325	41.9260270	2006
Unnamed Tributary	MassDEP	06-C006-03	−70.6135325	41.9260270	2006
Assabet River	MassDEP	06-A008-03	−71.6284508	42.3048530	2006
Assabet River	MassDEP	06-A015-03	−71.6284508	42.3048530	2006
Assabet River	MassDEP	06-F003-01	−71.6284508	42.3048530	2006
Assabet River	MassDEP	06-F004-01	−71.6284508	42.3048530	2006
Assabet River	MassDEP	07-L004-01	−71.6284508	42.3048530	2006
Assabet River	MassDEP	07-L005-01	−71.6284508	42.3048530	2006
Assabet River	MassDEP	08-M003-01	−71.6284508	42.3048530	2006
Assabet River	MassDEP	08-M004-01	−71.6284508	42.3048530	2006
Assabet River	MassDEP	09-N004-01	−71.6284508	42.3048530	2006
Assabet River	MassDEP	06-A015-01	−71.6322420	42.2740602	2006
Assabet River	MassDEP	06-A008-08	−71.5861363	42.3803093	2006
Wampatuck Pond	DCR	DCR (2008) Data from Lakes and Ponds, 2007	−70.8668332	42.0611168	2007

(Continued)

TABLE 19.2
(Continued)

Waterbody	Agency	Report Source	Longitude	Latitude	First Year Reported
Wampatuck Pond	DCR	DCR (2008) Data from Lakes and Ponds, 2008	−70.8668332	42.0611168	2007
Cook Pond	DCR	DCR(2008) Data from Lakes and Ponds, 2007	−71.1748547	41.6752804	2007
Ward Pond	DCR	DCR (2008) Data from Lakes and Ponds, 2007	−71.8813490	42.6814691	2007
Cady Brook	MassDEP	07-M003–02	−72.0087044	42.1194732	2007
Quinsigamond River	MassDEP	09-L004–02	−71.7080924	42.2305379	2008
Quinsigamond River	MassDEP	08-K006–02	−71.7080924	42.2305379	2008
North Pond	DCR	DCR (2008) Data from Lakes and Ponds, 2008	−71.5558719	42.1965281	2008
Fisk Pond	DCR	DCR (2008) Data from Lakes and Ponds, 2008	−71.3683511	42.2819821	2008
Field Pond	DCR	DCR (2008) Data from Lakes and Ponds, 2008	−71.1087937	42.6083542	2008
Nashua River	MassDEP	08-Q005–04	−71.5677653	42.6696867	2008
Blackstone River	MassDEP	08-A028–02	−71.6881594	42.1795812	2008
Blackstone River	MassDEP	08-A028–01	−71.7306827	42.181217	2008
Ware River	MassDEP	08-L005–04	−72.2857550	42.2387258	2008
Billington Sea	MassDEP	08-F007–05	−70.6810010	41.9348301	2008
Collins Pond	DCR	DCR (2008) Data from Lakes and Ponds, 2008	−71.1090134	42.6115557	2008
Stearns Pond	DCR	DCR (2008) Data from Lakes and Ponds, 2008	−71.0694217	42.6173500	2008
Upper Leach Pond	DCR	DCR (2008) Data from Lakes and Ponds, 2008	−71.1513062	42.0702976	2008
Aaron River Reservoir	DCR	DCR (2008) Data from Lakes and Ponds Pioneer Infestation, 2008	−70.8270316	42.2076141	2008
Clamshell Pond	DCR	DCR (2008) Data from Lakes and Ponds, 2008	−71.6781971	42.3993398	2008
Althea Lake	DCR	DCR (2008) Data from Lakes and Ponds Pioneer Infestation, 2008	−71.3807537	42.6679916	2008
Berry Pond	DCR	DCR (2008) Data from Lakes and Ponds, 2008	−71.0868735	42.6200740	2008
Sewall Pond	DCR	DCR (2008) Data from Lakes and Ponds, 2008	−71.7461921	42.3230767	2008

Sevenmile River	MassDEP	08-L006–01	−72.0049161	42.2647794	2008
Nashua River	MassDEP	08-Q006–02	−71.6719879	42.4447381	2008
Unnamed Tributary	MassDEP	08-A032–06	−71.7877508	42.2054101	2008
Aldrich Pond	MassDEP	09-G008–06	−71.7402591	42.1633137	2009
Marble Pond	MassDEP	09-G008–05	−71.7422507	42.1654355	2009
Monponsett Pond	MassDEP	09-G006–01	−70.8452535	42.0045958	2009
Monponsett Pond	MassDEP	12-B004–01	−70.8452535	42.0045958	2009
Monponsett Pond	MassDEP	13-B008–01	−70.8452535	42.0045958	2009
Monponsett Pond	MassDEP	14-B006–01	−70.8452535	42.0045958	2009
Monponsett Pond	MassDEP	14-B008–01	−70.8452535	42.0045958	2009
Monponsett Pond	MassDEP	09-G005–04	−70.836781	42.0014672	2009
Monponsett Pond	MassDEP	09-G006–05	−70.836781	42.0014672	2009
Monponsett Pond	MassDEP	12-B002–04	−70.836781	42.0014672	2009
Monponsett Pond	MassDEP	12-B004–04	−70.836781	42.0014672	2009
Monponsett Pond	MassDEP	12-B006–05	−70.836781	42.0014672	2009
Monponsett Pond	MassDEP	12-B008–05	−70.836781	42.0014672	2009
Monponsett Pond	MassDEP	13-B006–05	−70.836781	42.0014672	2009
Monponsett Pond	MassDEP	13-B008–05	−70.836781	42.0014672	2009
Monponsett Pond	MassDEP	14-B004–05	−70.836781	42.0014672	2009
Monponsett Pond	MassDEP	14-B006–05	−70.836781	42.0014672	2009
Monponsett Pond	MassDEP	14-B008–05	−70.836781	42.0014672	2009
Monponsett Pond	MassDEP	15-B009–05	−70.836781	42.0014672	2009
Riley Pond	MassDEP	09-G010–04	−71.6704740	42.1044164	2009
Reservoir	MassDEP	10-B005–04	−70.8526283	42.0298844	2010
Reservoir	MassDEP	12-B004–06	−70.8526283	42.0298844	2010
Reservoir	MassDEP	13-B006–06	−70.8526283	42.0298844	2010
Saugus River	MassDEP	10-D020–01	−71.0364004	42.5095513	2010
Hockomock River	MassDEP	13-C040–01	−71.0354529	41.988368	2013
Benton Pond	DCR	DCR (2008) Data from Lakes and Ponds	−73.0473909	42.1841285	N/A

(Continued)

TABLE 19.2
(Continued)

Waterbody	Agency	Report Source	Longitude	Latitude	First Year Reported
Sassaquin Pond	MassDEP	Matfield River Watershed	−70.9487998	41.7347795	N/A
Otis Reservoir	DCR	DCR (2008) Data from ACT, GeoSyntec-ACT	−73.0383897	42.1507965	N/A
West Waushacum Pond	DCR	DCR (2008) Data from Lakes and Ponds	−71.7644996	42.4143945	N/A

needs to include information from wetlands and other infested areas.

MANAGING ACTIONS

The most efficient way to protect the environment from invasive species is integrated management, including education and prevention, early detection, control, ecosystem restoration, and environmental regulation and enforcement. Among these management options, prevention has been proved to be the most efficient and cost-effective approach in stopping or minimizing the spread of aquatic invasive species. Mapping sightings of purple loosestrife provides a spatial understanding of the status of this invasive species in Massachusetts' lake and river riparian areas. This can increase the public's awareness of this invasive species and ultimately help to minimize the human-mediated spread to other waterbodies. This type of proactive method should be promoted.

In the meantime, for areas with purple loosestrife, reactive actions should also be taken. Developing a long-term control plan is key, especially when time and resources are scarce (Warne 2016). When a new population of purple loosestrife is identified, landowners and managers should focus first on preventing immediate spread by targeting isolated plants or small mats which are susceptible to physical control methods. Critically important areas should receive priority, such as areas that are habitat for endangered or threated species. Follow-up monitoring is another crucial step to ensure that purple loosestrife populations that have been managed or are in the process of being managed do not reemerge.

CHEMICAL TECHNIQUES

Using herbicides from mid-summer to early fall to control purple loosestrife is possible. Herbicides can be ideal for treating smaller populations and require repeat treatments in order to get long-term results. Spraying the solution in mist form is not acceptable since mist can drift away and influence other vegetation. Similarly, rainy days should be avoided as the extra water could carry the solution further than intended (Minnesota Department of Natural Resources 2017). Triclopyr-formulated solutions (2, 4-D,

Rodeo and Pondmaster) for wetland areas are needed at 0.5% by volume in spray solution and 50 gallons spray solution is suggested as standard requirement (Jacobs 2008). When using backpack spraying, the solution can be adjusted to 1–1.5%. Multi-year herbicide treatment is needed to eradicate purple loosestrife (Jacobs 2008). Herbicides such as 2, 4-D for *L. salicaria* might be considered as a control method, but it may lead to chemical pollution of the local soil and ecosystem. Other chemicals such as glyphosate and imazapyr are also used for chemical treatment. More information about chemical control on purple loosestrife and other 18 invasive species can be found in Appendix III. Regardless of the herbicide chosen for a purple loosestrife management project, the directions on the label should be followed. Before chemical application, it is required to obtain an aquatic nuisance species control license (BRP WM04 Form) from Massachusetts Department of Environmental Protection.

MECHANICAL OR PHYSICAL TECHNIQUES

When treating small areas, or areas that have plants less than 2 years old (i.e., root system is not well-developed), digging, hand pulling, cutting, and mowing are the most effective albeit time-consuming eradication methods (Jacobs 2008; Warne 2016). Digging, hand pulling, and cutting is best performed throughout the summer before the plant begins to seed. When plotting the soil around purple loosestrife, it is better to minimize the disturbance to the soil system. After removing the plant stem above ground, roots need to be removed and put into bags to prevent resprouting; the spreading of seeds potentially in the soil needs to be prevented too. Grazing and tilling are not recommended because it would be considered more likely to spread the seeds. Notwithstanding the aerenchyma cells can tolerate low-oxygen conditions, flooding can still be an effective method to control it and will eliminate almost all of purple loosestrife within an area, but sometimes it might also eliminate other desirable species (Jacobs 2008). However, flooding at levels less than 30 cm will not kill seedlings and can actually allow mature purple loosestrife to thrive.

BIOLOGICAL TECHNIQUES

There are several biological treatments for controlling the spread of purple loosestrife. In 1992, the Manitoba Purple Loosestrife Project was set up as an introduction to use biological management methods to control purple loosestrife. This project used *Neogalerucella* spp. and resulted in a reduction in purple loosestrife populations in targeted areas (Warne 2016). Currently, there are four kinds of tested beetles that feed on purple loosestrife. *Galerucella calmariensis* and *Galerucella pusilla* are two beetles that feed on the leaves of purple loosestrife (Figure 19.4) and eat the terminal bud area; also, the larvae feed on the bottom side of leaves. Massachusetts Association of Wetland Managers and its partners did a study to raise *Galerucella* sp. beetles for release in wetlands and only ten beetles were released at two spots along the Concord River in 2004 (Edward Rainer, personal communication). Ten years later, there is hardly any purple loosestrife along the Concord River in Billerica; however, loosestrife did come back from seed even the beetles are still eating the new plants

(Figure 19.4). In Massachusetts, the leaf-eating *Galerucella* sp. is also used for purple loosestrife control in other areas such as along the Sudbury River (Flint 2014). Fifteen years ago, Massachusetts Office of Coastal Zone Management (MA CZM) Wetlands Restoration Program published a Guidance Document for the Purple Loosestrife Biocontrol briefly reviewing the use of biocontrol measures in Massachusetts to control purple loosestrife and providing information to people who may be interested in participating in the Project (MA CZM 2007). However, since 2010, new beetle release is not supported by MA Division of Ecological Restoration anymore. *Hylobius transversovittatus* is a weevil that feeds on the leaf from the edge toward the center while the larval form mines into the root of the plant depleting energy reserves. *Nanophyes marmoratus*, another beetle, forages on the seeds of purple loosestrife (Malecki et al. 1993; Blossey and Schroeder 1995; Hight et al. 1995). Two other species of beetle *Nanophyes brevis* and *Bayeriola salicariae* also eat seed capsules. Field testing of these two species has yet to be conducted and further

FIGURE 19.4 Purple loosestrife leaves eaten by *Galerucella* beetles along Concord River, Massachusetts. (Photos by Edward Rainer.)

research for these two subjects is required to prevent them from becoming the next new invasive species (Blossey 2002). More information about biological control on purple loosestrife is provided in the "Success Story" section.

There is no silver bullet to completely control purple loosestrife. For example, removing the plant during its pre-pollinating period would be considered an effective way of control, but it is costly and cannot be applied to a large area due to the requirement of manpower. Another option such as using flooding to extinguish the infested area is an undifferentiated method that may break the balance of local ecosystem. Therefore, integrated management practices are recommended in aquatic invasive species control and management (Wong and Gerstenberger 2015). Figure 19.5 provides an outline for choosing the best management techniques given the size of the area and the percent of the land that contains purple loosestrife. Following any management actions, restoration efforts should be conducted utilizing native plant species. Another example is that biocontrol has provided suppression of purple loosestrife at many sites in North America, it is not effective at or applicable to every infestation location (Blossey et al. 2015); so, the authors proposed that integrated purple loosestrife management is needed to control this invasive species with physical, cultural, and chemical control tools.

It is critical to not compost any plant material that may be live still as this can lead to the growth of new plants. Discarding plant waste in areas of critical importance should also be avoided. The ideal method of disposal for purple loosestrife is by burning. Composting of plant material can be conducted when brought to an upland waste management facility where the composting temperatures can reach high enough to ensure the death of any remaining live plant material. If the above methods of disposal cannot be achieved, another option is to seal the plant material into bags which are left in direct sunlight for at least three weeks.

SUCCESS STORY

The use of exotic biological control agents as a method of managing invasive pests is not widely accepted. Concerns exists about the safety of such treatment methods due to spillover effects on non-target species, as well as a lack of previous successful cases of introductions and a lack of monitoring of the impacts of biological controls on both target and non-target species (Malecki et al. 1993). However, it is proposed that biological control can be cost-effective, long-lasting, self-sustaining, and non-polluting given proper research and testing of the proposed control species (Malecki et al. 1993; Blossey et al. 1994). One initial study

Percent of Area Covered with Purple Loosestrife	Size of Area to be Managed			
	Isolated Plants	Small infestation <0.5 hectare / 1 acre	Medium <0.5 – 2 hectares / 1 - 4 acres	Large >2 hectares / > 4 acres
1-10%	• Mechanical • Chemical	• Mechanical • Chemical	• Biological	• Biological
10-25%	• Mechanical • Chemical	• Mechanical • Chemical	• Biological	• Biological
25-50%	• Mechanical • Chemical • Biological	• Mechanical • Chemical • Biological	• Biological	• Biological
>50%	• Biological	• Biological	• Biological	• Biological

FIGURE 19.5 Suggested control methods for purple loosestrife based on the size of the control area and the percentage of the area infested with purple loosestrife. (From Warne 2016.)

conducted jointly by the USDA Agricultural Research Service and the US Fish and Wildlife Service sought to find natural predators of *L. salicaria* by examining 120 species of insects native to habitats associated with that of purple loosestrife. Fourteen species were found to be host-specific, and three were further chosen as being the most promising control agents due to their substantial effect on the growth of *L. salicaria*: *Galerucella calmariensis*, *Galerucella pusilla*, and *Hylobius transversovittatus*. Fifty plants of similar taxonomy to purple loosestrife were also tested to determine selectivity of the proposed control agent. Only two species native to North America were identified: swamp loosestrife and winged loosestrife, both of the family Lythraceae (*Decodon verticillatus* and *Lythrum alatum*, respectively). However, it was observed that when given a choice between the North American plants and *L. salicaria*, the latter was predominantly chosen. Additionally, a separate study conducted by Blossey et al. (1994) tested the selectivity of *G. calmariensis* and *G. pusilla* on 48 non-target species; normal adult and larval feeding only occurred on the target plant with slight to moderate feeding occurring on several other species of *Lythrum*; normal oviposition only occurred on the target plant with moderate oviposition occurring only on winged loosestrife. In 1992, the three biological control species identified were approved for release into the United States (Malecki et al. 1993).

In 1994, populations of *G. calmariensis* and *G. pusilla* were released throughout central New York in four different release sizes (20, 60, 180, and 540) at 36 different sites (Grevstad 1999). Over the next three years, monitoring was conducted at each site to determine the remaining numbers of each species. Prior to the introductions, the size and density of each *L. salicaria* strand was measured to determine the effectiveness of the introduced control agents at managing purple loosestrife. Ten years later, Grevstad returned to examine the long-term effects of the introduction. He found that *G. pusilla* was present at 22 of the 36 original sites, and *G. calmariensis* was present at 10 sites (Grevstad 2006). Sites that had beetles during the final year of monitoring in 1997 were more likely to have beetles ten years after introduction, especially in

the case of *G. pusilla*. Sites with higher release sizes were also found to have higher population persistence than those with lower release sizes. In terms of their effectiveness as a control agent for purple loosestrife, stems were found to be on average shorter after ten years; however, overall plant density and size did not change. Grevstad (2006) suggested several factors may have led to the low overall impact on the target species. First, he proposed that ten years may not have been a long enough time to see beneficial impacts. Second, the initial densities of the introduced control agents may not have been high enough. And third, seasonal variations in beetle abundances may have allowed purple loosestrife to rebound on a year-to-year basis. He proposed, therefore, that large-scale benefits could be achieved by releasing more biological control agents into more locations, ideally with release sizes of at least 500 individuals.

Herbicides are also a potential solution to combat the spread of *L. salicaria*. One study looked at the effectiveness of 14 herbicides administered at four sites over a ten-year period (Knezevic et al. 2018). Overall, purple loosestrife was significantly impacted by herbicides across all locations. However, the age of the *L. salicaria* strand had an effect on the number of treatments required for adequate control to be achieved. Strands that were 3 years old required on average two to three years of sequential treatment. Strands that were 5 years old required two to five years of sequential treatment. Strands that were 10 years old required three to nine years of sequential treatment. Additionally, certain older strands showed regrowth after treatment had ceased due to no new growth being observed. Because of this, it is advised older strands of *L. salicaria* be observed for several years after a final treatment to ensure that growth has been ceased. While this study showed that herbicides can be an effective control method, non-target grassy and broadleaf plant species were negatively affected by treatment alongside *L. salicaria*.

Both studies which examined the long-term effects of the two different control methods (biological and chemical) showed promising results, but also came with several negative side effects. The best management technique then would integrate both management methods, coupled with active mapping of purple loosestrife locations statewide and long-term site monitoring. New or young strands that are identified can be treated with herbicides, which were shown to be most effective on younger *L. salicaria*. In areas that have larger or older strands of purple loosestrife, biological control methods can be applied, as they are long-lasting without the need for yearly reapplications.

REFERENCES

Balogh, G. 1986. Ecology, Distribution and Control of Purple Loosestrife (*Lythrum salicaria*) in Northwest Ohio. Thesis for Master of Science, Ohio State University, p. 107.

Blossey, B. 2002. Purple Loosestrife. In: Van Driesche R, Blossey B, Hoddle M, Lyon S, Reardon R (eds), Biological control of invasive plants in the Eastern United States. USDA Forest Service Publication FHTET-2002–04, Morgantown, West Virginia, p. 413.

Blossey, B., C.B. Randall, and M. Schwarzländer. 2015. *Biology and Biological Control of Purple Loosestrife* (2nd ed.). U.S. Department of Agriculture and U.S. Forest Service. 107 p.

Blossey, B., D. Schroeder. 1992. Biocontrol of *Lythrum salicaria* in the United States. Final Report. Cornell University, United States Fish and Wildlife Service, Washington State Dept. of Agriculture, and Washington State Dept. of Wildlife.

Blossey, B., D. Schroeder. 1995. Host specificity of three potential biological weed control agents attacking flowers and seeds of *Lythrum salicaria* (Purple Loosestrife). *Biological Control* 5: 47–53.

Blossey, B., D. Schroeder, S. Hight, and R. Malecki. 1994. Host specificity and environmental impact of two leaf beetles (*Galerucella calmariensis* and *G. pusilla*) for biological control of purple loosestrife (*Lythrum salicaria*). *Weed Science* 42: 134–140.

Blossey, B., L.C. Skinner, J. Taylor 2001. Impact and management of purple loosestrife (*Lythrum salicaria*) in North America. *Biodiversity and Conservation* 10: 1787–1807.

Brown, M.J. 2005. Purple loosestrife—*Lythrum salicaria* L. In: C.L. Duncan and J.K. Clark (eds). *Invasive Plants of Range and Wildlands and Their Environmental, Economic, and Societal Impacts*. Lawrence, KS: Weed Science Society of America, pp. 128–146.

Cao, L., J. Larson, R. Sturtevant. 2020. *Lythrum salicaria L.: U.S. Geological Survey, Nonindigenous Aquatic Species Database*. Gainesville, FL: U.S. Geological Survey.

Chandler, M.A., L.C. Skinner, L.C. Van Riper. n.d. *Biological Control of Invasive Plants in Minnesota*. Minnesota: Minnesota Department of Agriculture & Minnesota Department of Natural Resources. http://files.dnr.state.mn.us/natural_resources/invasives/biocontrolofplants.pdf (Accessed April 20, 2017).

Cygan, D. 2003. *Biological Control of Purple Loosestrife*. New Hampshire: New Hampshire Department of Agriculture, Markets & food, Plant Industry Division and New Hampshire Department of Transportation, Bureau of Environment.

Flint, S. 2014. *Loosestrife-eating Beetles Come to Local Backyard*. London: OARS For the Assabet Sudbury and Concord Rivers Newsletter. p. 1 and 6.

Grevstad, F. 2006. Ten-year impacts of the biological control agents *Galerucella pusilla* and *G. calmariensis* (Coleoptera: Chrysomelidae) on purple loosestrife (*Lythrum salicaria*) in Central New York State. *Biological Control* 39: 1–8.

Grevstad, F. 1999. Experimental invasions using biological control introductions: The influence of release size on the chance of population establishment. *Biological Invasions* 1: 313–323.

Henderson, R. 1987. Status and control of purple loosestrife in Wisconsin. *WDNR-Research Management Findings No. 4*, p. 4.

Hight, S.D., B. Blossey, J. Laing, and R. DeClerck-Floate. 1995 Establishment of insect biological control agents from Europe against *Lythrum salicaria* in North America. *Environmental Entomology* 44: 965–977.

Jacobs, J. 2008. Ecology and Management of Purple Loosestrife (*Lythrum salicaria* L.). *Invasive Species Technical Note* No. MT-21. United States Department of Agriculture, Natural Resources Conservation Service. www.nrcs.usda.gov/wps/portal/nrcs/detail/mt/home/?cid=nrcs144p2_056795 (Accessed April 10, 2017).

Katovich, E.J.S., R.L. Becker, and J.L. Byron. 2003. Winter survival of late emerging purple loosestrife (*Lythrum salicaria*) seedlings. *Weed Science* 51(4): 565–568.

Knezevic, S., O.A. Osipitan, M. Oliveira, and J. Scott. 2018. *Lythrum salicaria* (Purple Loosestrife) control with herbicides: Multiyear applications. *Invasive Plant Science and Management* 11: 143–154.

Lavoie, C. 2009. Should we care about purple loosestrife? The history of an invasive plant in North America. *Biological Invasions* 12(7): 1967–1999.

Lesica, P., M.T. Lavin, and P.F. Stickney. 2012. *Manual of Montana Vascular Plants*. Fort Worth, TX: BRIT Press. 771 p.

Lindgren, C.J. 1999. Performance of a biological control agent, *Galerucella calmariensis* L. (Coleoptera: Chrysomelidae) on purple loosestrife *Lythrum salicaria* L. in Southern Manitoba (1993–1998). *Proceedings of the X International Symposium on Biological Control of Weeds*, Stonewall, MB, pp. 367–382.

Lindgren, C.J., R.T. Clay. 1993. Fertility of 'morden pink' *Lythrum virgatum* L. transplanted into wild stands of *L. salicaria* L. in Manitoba. *Horticultural Science* 28(9): 954.

MA CZM (Massachusetts Office of Coastal Zone Management). 2007. *Guidance Document for the Purple Loosestrife Biocontrol Project*. Cambridge, MA: MA CZM, p. 34.

Mal, T.K., J. Lovett-Doust, L. Lovett-Doust, and G.A. Mulligan. 1992. The biology of Canadian weeds 100. *Lythrum salicaria. Canadian Journal of Plant Science* 72: 13051330.

Malecki, R.A., S. Hight, L. Kok, D. Schroeder, and J. Coulson. 1991. Information for the preparation of an environmental assessment. Host plant specificity testing of *Hylobius transversovittatus, Galerucella calmarienses and G. pusilla* for use in the biological control of *Lythrum salicaria* L. in North America. New York Cooperative Fish and Wildlife Research Unit, Department of Natural Resources, Fernow Hall, Cornell University, Ithaca, NY, p. 79.

Malecki, R.A., B. Blossey, S.D. Hight, D. Schroeder, L.T. Kok, and J.R. Coulson. 1993. Biological control of purple loosestrife; a case for using insects as control agents, after rigorous screening, and for integrating release strategies with research. *Biological Science* 43: 680–686.

McGlynn, C.A. 2008. Native and invasive plant interactions in wetlands and the minimal role of invasiveness. *Biological Invasions* 11: 1929–1939.

Minnesota Department of Natural Resources. 2017. Controlling purple loosestrife with herbicides. www.dnr.state.mn.us/invasives/aquaticplants/purpleloosestrife/control_herbicides.html (Accessed April 10, 2017).

Mullin, B.H. 1998. The biology and management of purple loosestrife (*Lythrum salicaria*). *Weed Technology* 12: 397–401.

Pellet, M. 1977. Purple loosestrife spreads down river. *The American Bee Journal.* 117: 214–215.

Regional *Lythrum salicaria* Working Group. 2004. *Purple Loosestrife (Lythrum salicaria) in the Chesapeake Bay Watershed: A Regional Management Plan*, p. 36.

Shamsi, S.R.A., F.H. Whitehead. 1974. Comparative eco-physiology of *Epilobium hirsutum L.*, and *Lythrum salicaria L.* I. General biology, distribution and germination. *Journal of Ecology* 62: 279–290.

Stein, B.A., S.R. Flack 1997. *Species report card: The State of U.S. plants and animals.* Arlington, VA: The Nature Conservancy, p. 26.

Stuckey, R.L. 1980. Distributional history of *Lythrum salicaria* (purple loosestrife) in North America. *Bartonia* 47: 3–20.

Thomsen, A. 2012. Perennial self-sowers. *Fine Gardening* 143: 32–37.

Thompson, D.Q., R.L. Stuckey, and E.B. Thompson. 1987. *Spread, impact and control of purple loosestrife (Lythrum salicaria) in North American wetlands. United States Fish and Wildlife Service, Fish and Wildlife Research No. 2.* Washington, DC: United States Department of Interior, p. 55.

Warne, A. 2016. *Purple Loosestrife (Lythrum salicaria) Best Management Practices in Ontario.* Peterborough, ON: Ontario Invasive Plant Council, p. 40.

Wilcove, D.S., D. Rothstein, J. Bulow, A.L. Phillips, and E. Losos. 1998. Quantifying threats to imperiled species in the United States. *Bioscience* 48: 607–615.

Wong, W.H., S.L. Gerstenberger (eds). 2015. *Biology and Management of Invasive Quagga and Zebra Mussels in the Western United States.* Boca Raton, FL: CRC Press, p. 566.

Appendix I

MassDEP Equipment Decontamination Protocol[1]

MassDEP

Massachusetts Department of Environmental Protection Division of Watershed Management

STANDARD OPERATING PROCEDURE

Field Equipment Decontamination to Prevent the Spread of Invasive Aquatic Organisms

CN 59.6
August, 2015

1.0 SCOPE AND APPLICATION

The following SOP has been developed under the assumption that Massachusetts' waterways are threatened by the spread of non-indigenous species. This SOP provides guidance to DWM field staff on the prevention and minimization of the spread of invasive aquatic organisms from one waterbody to another. It covers field and lab decontamination procedures for any equipment and gear used in streams, rivers, lakes, ponds, and impoundments throughout Massachusetts whenever any invasives have been observed or are reasonably suspected to be present and spread to non-contaminated areas by monitoring activity. A complete list of confirmed and suspected aquatic invasive species sightings by WPP and other agencies in Massachusetts can be found from the WPP Toolbox named Mass AIS Waterbody List. If a survey is going to be conducted in any of the listed waterbodies, the decontamination procedures will be implemented.

This SOP applies to controlling the spread of invasive organisms only. For additional guidance regarding equipment decontamination procedures to prevent and minimize cross contamination of samples, see CN 59.0.

2.0 SUMMARY

By following simple checking and cleaning steps, the inadvertent spread of invasive organisms (e.g., Eurasian milfoil, Didymosphenia geminata or Didymo, zebra mussel, asian clam, others) from infected to non-infected waterbodies by the actions of DWM field staff can be avoided or minimized. Special procedures are required because infestation can occur from very small amounts of invasive material (in some cases, microscopic), which can remain viable long after being removed from the water.

3.0 SAFETY AND LOGISTICAL CONSIDERATIONS

Care should be taken at all times during decontamination activities. Although these procedures do not involve the use of hazardous chemicals, gloves and safety glasses are recommended when using washing solutions. Clean tap water should be available at all times in case of splashes on skin or into eyes. For cleaning at DWM-Worcester, a shower/eyewash station is available, if necessary.

Employ a mitigation-hierarchy approach in implementing this SOP by practicing stepwise decision-making. First, seek to avoid contact with invasives such as Didymo and thus the need to decontaminate equipment. If avoidance is not possible, minimize contact with invasives and recognize where/when decontamination is required.

Avoidance of the need to decontaminate equipment in the field is preferred. This can be achieved, for example, by using equipment only once at known or potentially contaminated sites, then bringing all equipment back to the lab for cleaning prior to reuse. It is recognized, however, that this is not always possible for DWM field activities.

Because application of these procedures requires additional time for preparation, use in the field, and post-survey cleanup, adequate planning is required to ensure implementation. Specifically, plan to sample known/suspected invasive sites toward the end of the survey, if possible.

4.0 DECONTAMINATION EQUIPMENT AND MATERIALS

The equipment and materials needed to ensure adequate decontamination (DECON) for invasives vary depending on the type of contaminate and the location of decontamination activity. Soaking baths are generally more effective to render organisms non-viable. Spray washing is more useful for physical rinsing of organisms.

In general, the materials that may be needed for decontamination are as follows:

Field DECON "Kit": EACH CREW/VEHICLE

- Cleaning solution bath (cooler): Solution varies (see Appendix B for options; e.g., 5% hand-dishwashing soap solution); in tub with secure lid; minimum non-displaced solution depth of 6 inches; include energy dissipation to avoid excessive splashing, if possible
- Cleaning solution spray: In pressure sprayer#1
- Decon tubes: Containing cleaning solution for unattended deployment tubes and attended multiprobe sondes
- Spare cleaning solution (preferably non-P-containing, non-anti-microbial)
- Rinse water spray (tap water): In pressure sprayer#2 (use tap water spray if P-detergent or bleach is the cleaning solution)
- Clean rags and paper towels
- Trash bags
- PPE (gloves, safety glasses)
- Wristwatch
- Dedicated plastic sample container (for specimens)
- Bleach (1 gallon in reserve if needed)
- Portable eye wash (w/tap water)
- Quarters ($) for manual, pressure-spray car wash

Lab Decontamination

- Hot water (>60°C or >140°F)
- Cleaning solution: Varies (e.g., 5% hand-dishwashing soap solution)
- Tap water rinse
- Brushes, clean rags, and paper towels
- Trash container
- Bleach
- Portable pressure hose (boat/trailer washing), if available
- PPE (gloves, safety glasses)

5.0 GENERAL PROCEDURES

Checking and cleaning equipment (e.g., removing plant fragments) prior to moving to another waterbody is a recognized best management practice that should normally be employed whenever and wherever appropriate and practicable, regardless of whether invasives are suspected or confirmed to be present or not. This is especially true for macroscopic plant fragments and visible organisms. The DECON procedures outlined in this SOP represent additional measures to be used when necessary.

STEP 1: DETERMINE PRESENCE/ABSENCE POTENTIAL FOR INVASIVES

In the office and the field, perform background review and pre-survey field reconnaissance adequate to determine the likelihood of invasive organisms being present at each area to be visited. Agencies, watershed groups, etc. can assist in this determination. Based on the results, plan accordingly.

In the field upon arrival at each site, CHECK for the visual presence of invasive organisms in the area to be sampled. If invasives are found or reasonably suspected to be present and spread to non-contaminated areas by sampling activity (see the Mass AIS Waterbody List in the WPP Toolbox), all activities involving contact with water must employ "DECON" procedures as contained in this SOP. If invasives are encountered but were not expected, do not perform the activity without the proper decontamination equipment or restrict the sampling equipment to a single waterway, and then clean/dry later at the office/lab.

Example DECON approach involving "Didymo": As of January 2008, the only watershed in Massachusetts where Didymo has been found (albeit only out of state in upstream NH and VT tributaries) is the Connecticut River watershed. Therefore, the proposed Year-2 monitoring approach by DWM in 2008 will be to:

- Employ DECON procedures at all DWM stations visited in the Connecticut River watershed *and* watch for actual Didymo outbreaks in this watershed as much as possible.

- Watch for actual Didymo outbreaks as much as possible in other Year-2 watersheds. If observed in a watershed (in Massachusetts or upstream in other states), employ DECON procedures at all DWM stations visited in that watershed.

Any potential new invasive species observed in any site, whether it is a new site or a site has been infested by other invasive species, needs to be confirmed before it is reported to other agencies. A standard reporting protocol on invasive species sightings are described below (please also refer to Appendix K about field sampling protocol and verification for suspected invasives).

New Invasive Species Sightings Verification and Reporting Guidance: WPP field survey crew will take an extra sample of the suspicious invasive species. If possible, a photo should also be taken to record the habitat of the suspicious invasive species. If the suspicious species is an animal, the sample needs to be preserved with 50% ETOH in a container. For a plant, the sample should be collected as a whole or a segment; if plants have fruits and/or flowers, the plants need to be collected as a whole. Keep the plant sample in a bag or bucket of water at all times: don't leave them out of the water in the sun because they will wither quickly and become useless as specimens. The extra sample will be brought back to the lab and examined by at least one additional WPP biologist for verification. If the sample is not confirmed within WPP, external research experts will be consulted until the sample is finally confirmed. Upon the final confirmation, the sighting will be added to the list of Mass AIS Waterbody List in the WPP Toolbox. This new invasive sighting will be reported to other agencies and shared with the general public. If the sample is not identified as an invasive species, the waterbody will not be added to the Mass AIS Waterbody list. Contact David Wong for AIS confirmation along with Richard Chase and Art Johnson for AIS reporting. Preserve specimens for further analysis if it is needed. Destroy specimens prior to disposal as trash.

STEP 2: ASSEMBLE FIELD DECON "KIT"

At the office, prepare the field DECON kit and include it with survey equipment inside each DWM vehicle. The field DECON kit should be available at all times in case it is needed. The cleaning solution can also be made in the field as needed.

STEP 3: EMPLOY CHECK–CLEAN–DRY PROCEDURES (AND DECON AS NECESSARY)

If You Are Visiting Waterways Known or Reasonably Suspected of Containing Invasives, You Must Use Check, Clean, Dry Procedures. If Cleaning or Drying in the Field Is Not Practical, Restrict Equipment to a Single Waterway, and Then Clean/Dry Later at the Office/Lab

In general, do not wear felt soled boots/kickers or boots with porous surfaces for monitoring activity, especially where Didymo is known or suspected to be present. If it must be done for some reason, restrict felt soles (and other "porous" equipment) to a single waterway, and then thoroughly clean/dry later at the office/lab. For traction (in lieu of felt-soles), use attachable, washable cleats. If cleats are used, do not drive vehicles with them on, and make sure to decontaminate Velcro straps.

Ensure that all washwater drains to isolated ground or to the sanitary sewer system. Keep washwater away from storm drains to prevent the spread of invasives (in case decontamination is not 100% effective).

> **CHECK:** Before you arrive at and leave a river or lake site, thoroughly inspect all equipment for plant/algal fragments and debris, *and*
> **CLEAN:** Remove all plant fragments and organic debris and dispose as trash. Clean equipment using the most practical treatment for your situation (which will not adversely affect your gear). See Appendices E–J for specific DECON cleaning procedures, *and/or*
> **DRY:** Drying will kill many invasive organisms, but must be complete. Slightly moist organisms may survive for months. To ensure effective drying, the item must be completely dry to the touch, inside and out, then left dry for at least another 48 hours before use.

STEP 4: POST-SURVEY CLEANUP

Perform post-survey cleanup to ensure proper disposal of all spent cleaning solutions and to ensure clean equipment for the next users.

6.0   TRAINING

All field staff using field equipment shall be trained in this SOP. Trained staff may also train designated interns in the procedures. Training shall consist of instruction on office and field procedures, including field identification of invasives and decontamination steps. Each trainee will conduct a cleaning procedure under supervision of the trainer. Training in lab microscopy for microscopic invasives (e.g., Didymo and zebra mussel veligers) is also planned.

7.0   QUALITY CONTROL

Random interviews and audits can be performed by DWM QA staff and others to ensure compliance with this SOP.

For surveys involving sample collection, application of these decontamination procedures must not impact sample integrity.

8.0   MAINTENANCE

Survey planning, preparation, and execution with regard to decontamination procedures is the responsibility of the monitoring coordinators and survey crew leads. If problems arise with decontamination equipment or procedures, see DWM's QA Analyst.

9.0   CORRECTIVE ACTIONS

Known failures to follow this SOP will be documented and corrective actions (e.g., communication with applicable staff) taken. Adjustments to this SOP will be made as needed based on hands-on experience, improvements in procedures, or the nature and extent of invasives.

10.0   POLLUTION PREVENTION AND WASTE MINIMIZATION

- When possible, use prescribed tap water spray and drying procedures as an alternative to chemical use.
- Use the prescribed amounts of cleaning reagents (not more than necessary). Use non-P-containing detergent only.
- Ensure proper disposal of waste materials throughout the decontamination procedure.
- Return spent washwater solutions used in the field to the office/lab for disposal down the sanitary sewer drain.

NOTES

1   Modified from Chase and Wong (2015).

REFERENCES

Kilroy, C. 2005. Tests to determine the effectiveness of methods for decontaminating materials that have been in contact with *Didymosphenia geminata*; Prepared for Biosecurity New Zealand; NIWA Client Report: CHC2005–005, NIWA Project: MAF05501 National Institute of Water and Atmospheric Research Ltd; www.niwa.co.nz. 36 p.

Mass DCR and Mass DFG August 2009. Massachusetts Interim Zebra Mussel Action Plan. www.mass.gov/eea/docs/dcr/ watersupply/lakepond/downloads/zebra-mussel-interim-action-plan.pdf. (Accessed August 5, 2015).

Scholl. C. 2006. Aquatic invasive species: A guide for proactive and reactive management. Report to Wisconsin Department of Natural Resources. http://dnr.wi.gov/aid/documents/ais/aisguide06.pdf. (Accessed on August 4, 2015).

Spaulding, S., and L. Elwell. 2007. *Increase in Nuisance Blooms and Geographic Expansion of the Freshwater Diatom Didymosphenia geminata: Recommendations for Response.* Denver, CO: US EPA Region & Livingston MT: Federation of Fly Fishers, p. 33.

Special Session on *Didymosphenia geminata*, Western Division American Fisheries Society Meeting, May 15–16, 2006, Bozeman, Montana. post-meeting update.

Warrington P. 1994. Collecting and preserving aquatic plants. Environmental Protection Department, Ministry of Environment, Lands and Parks. Government of British Columbia. www.env.gov.bc.ca/wat/wq/plants/plantcollect.pdf. (Accessed August 4, 2015).

Zook, B., and S. Phillips. 2015. Recommended uniform minimum protocols and standards for watercraft interception programs for dreissenid mussels in the Western United States. In: W.H. Wong and S.L. Gerstenberger (eds). *Biology and Management of Invasive Quagga and Zebra Mussels in the Western United States*, pp. 175–204.

NOTE #1 CLEANING SOLUTIONS AND/OR ACTIONS TO PREVENT THE SPREAD OF INVASIVE AQUATIC ORGANISMS

Invasives of Concern	Disinfectant	Concentration	Contact Time	Reference
Dreissenid mussels (zebra and quagga mussels)	Vinegar	100%	20 minutes [A]	1,2,3,4, 5
	Chlorine/bleach	1 oz per gallon water	1 hour	
	Power wash with hot wash	>45°C	20 seconds	
	Steam/scalding hot wash	60°C	10 seconds	
	Freeze	<0 °C	4–24 hours	
	Salt bath	Saturated	30 minutes [A]	
	Ethanol	50%	10 minutes	
	Lysol	As sold	10 minutes	
	Drying	Dry to the touch	One week if kept dry; up to four weeks if subject to cool, wet weather	
Didymo (*Didymosphenia geminata*)	Dishwashing detergent[B] (non-absorbent)	5% (two large cups or 500 ml with water added to make 10 l)	>1 minute (soak or spray)	6
	Salt	5% v/v (e.g., 1 l dry salt in 19 l hot tap water)	>1 minute (soak or spray)	
	Bleach (non-absorbent)	2% (one small cup or 200 ml with water added to make 10 l)	>1 minute (soak or spray)	
	Hot water (non-absorbent)	Very hot water *kept above* 60°C (140°F; hotter than most tap water)	>1 minute (soak)	
		hot water *kept above* 45°C (uncomfortable to touch)	>20 minutes (soak)	
	Hot water (absorbent material)	hot water kept above 45°C	>60 minutes (soak)	
	Hot water plus detergent (absorbent material)	hot water kept above 45°C and containing 5% dishwashing detergent	>30 minutes (soak)	
	Freezing	<0 °C	12 hours	
	Drying	Dry to the touch	>2 days after dry	

(Continued)

NOTE #1 CLEANING SOLUTIONS AND/OR ACTIONS TO PREVENT THE SPREAD OF INVASIVE AQUATIC ORGANISMS
(Continued)

Invasives of Concern	Disinfectant	Concentration	Contact Time	Reference
	Other (TBD)	TBD	TBD	TBD
Other microscale invasives	BPJ similar to that for Didymo or mussels			
Aquatic macrophytes (following check–clean procedures)	Drying	Dry to the touch	>2 days after dry	
	Power wash with hot wash	≥60°C	2 minutes	7

[A] It is intended for zebra mussel veligers, not adults.

[B] If necessary, use unscented detergent to avoid nausea from prolonged breathing of perfumed air.

(1) DiVittorio, J., Grodowitz, M., Snow, J., Manross, T. (2012) Inspection and Cleaning Manual for Equipment and Vehicles to Prevent the Spread of Invasive Species. US Department of the Interior, Bureau of Reclamation, Technical Memorandum No. 86-68220-07-05.

(2) Massachusetts Department of Conservation and Recreation. 2010. Prevent the Spread of Zebra Mussels. Massachusetts DCR Lakes and Ponds Program. www.mass.gov/eea/docs/dcr/watersupply/lakepond/downloads/zebmussbro10.pdf. (Accessed on August 4, 2015).

(3) Comeau, S. Ianniello, R.S., Wong, W.H., Gerstenberger, S.L. 2015. Boat Decontamination with Hot Water Spray: Field Validation. In: Biology and Management of Invasive Quagga and Zebra Mussels in the Western United States (Eds Wong, W.H., Gerstenberger, S. L.). Pages 161–173.

(4) Wong, W.H., personal observation.

(5) Waller, D. L., Fisher, S.W., Dabrowska H. 1996. Prevention of Zebra Mussel Infestation and Dispersal during Aquaculture Operations. The Progressive Fish-Culturist. 58:77–84.

(6) New Zealand Ministry for Primary Industries. Didymo: *Didymosphenia geminate*. www.biosecurity.govt.nz/cleaning. (Accessed on August 4, 2015).

(7) Blumer, D.L., Newman, R.M., Gleason, F.K. 2009. Can Hot Water Be Used to Kill Eurasian Watermilfoil? *Journal of Aquatic Plant Management*. 47: 122–127.

NOTE #2 FIELD DECONTAMINATION EQUIPMENT CHECKLIST FOR MULTIPROBE DEPLOYMENT SURVEYS

- QuickGuide protocol for decon
- Non-porous (NP) waders *only*
- 6″-Diameter decon soaking tube (1) filled ¾ with decon solution 3″-diameter decon soaking tub with cable notch (1) filled ¾ with 5% decon solution
- Decon soaking can/cooler (1) filled 1/3–1/2 with 5% v/v decon solution (for misc. items)
- Decon sprayer (1 minimum; 2 each crew if possible) filled with 5% decon solution
- Extra decon ingredients
- Large bags for used/contaminated equipment
- Watch/timer
- Bungee cords sufficient to tie down decon tubes
- 2nd set of "new" bungee cords (if needed)
- 2nd set of "new" concrete blocks (if needed)
- 2nd set of "new" deploy tubes (if available/needed)

NOTE #3 GENERAL INSPECTION AND CLEANING PROTOCOL FOR DWM BOATING EQUIPMENT

Prior to departure from the DWM office/lab, thoroughly inspect and *check* all watercraft, motors, and trailers for the presence of plant/animal remnants and organic debris. Remove all plant fragments, algal clumps, and other organisms and dispose of as trash. If necessary, *clean*, rinse, and drain (as appropriate) boat motor, trailer, anchors, chains, bilge water, fish tanks, fishing gear, and other equipment using portable spray unit or tap water hose or, if a designated wash area is available,[1] use it to rinse all surfaces of boats and equipment. If available, use motor flusher or "muff" device with water hookup to rinse out engine cooling system for 2–3 minutes. If not available, run cold motor out of water for 3–5 seconds *max* to remove most of the water.

When removing boats/trailer from the water, repeat the procedure of checking/cleaning (at the lake just sampled or at a separate washing station prior to entering another waterbody): Check, clean, rinse with salt water, and drain surface areas of a boat, motor, trailer, anchors, chains, bilge water, fish tanks, fishing gear, and other equipment using portable spray unit or tap water hose. If a designated wash area is available, make sure that all water enters the sanitary sewer system. In general, avoid going from one lake (X) to another waterbody (Y) where the same boating equipment must be used *and* where invasives are known or reasonably suspected to be present and spread by sampling activity to non-contaminated areas. If this cannot be avoided and after extensive checking and cleaning, DECON by cleaning and rinsing all equipment and surfaces that came in contact with the water using a portable spray unit (containing either tap water or an appropriate non-phosphorus, non-antibacterial solution[2]), *or*, preferably if available, a designated hot water pressure wash area or manual hot water pressure-spray car wash facility. Drain bilge water on land away from storm drains. To remove cooling water remaining in the motor when exiting the lake, you can pull motor out of water while running and then turn off gas feed or turn motor off after 3–5 seconds. When reaching Lake Y, check for plant fragments again prior to launch.

Note: Avoid contaminating waterways with residual cleaning solution by using P-free, non-antimicrobial detergent. Although most liquid detergents for hand-washing dishes are now phosphorus-free, detergents for automatic dishwashers typically can have a high phosphorus content, as tripolyphosphate is a preferred water-softening agent. Use of soap without anti-bacterial chemicals (e.g., triclosan, triclocarban) also avoids potential pollution with these persistent biocides.

[1]As of August 2015, DWM does NOT have a designated pressure washing area at the DWM-Worcester location. Use a nearby pressure-spray wash facility and request reimbursement for $ spent.

[2]Ensure that cleaning solution used is appropriate for the invasive(s) of concern. Readily available cleaners include bleach, salt, and hand-dishwashing detergents. 5% solution (v/v) = 1 part cleaner to 20 parts water. For a 3 gallon spray unit (w/max fill line of 2 ¼ gallons), add about 1 pint of liquid dishwashing detergent or bleach and fill to max line).

NOTE #4 DECON PROTOCOL FOR FIELD CLEANING SAMPLING EQUIPMENT AND GEAR

Examples of DWM sampling gear that come into contact with ambient water and require checking and cleaning include, but are not limited to, boots and waders, felt-soled kickers, nets, gloves, ropes/tapes, bottle baskets, sediment samplers, Van Dorn samplers, integrated tube samplers, attended multi-probe sondes, deployed multi-probe sonde tubes, anchor/cable/lock assemblies for deployed multi-probe sonde tubes, Secchi disk, velocimeters, boats, motors, and trailers.

The following procedures apply to checking and cleaning equipment used for all types of DWM surveys in which more than one location is visited on the same day using the same equipment *and* where invasives are known to be present or suspected to be spread by sampling activity to non-contaminated areas[1]:

1. **Check** for and manually remove plant/animal remnants and organic debris from all equipment prior to and following sampling at each site. Dispose as trash at the site.
2. **Clean** gloves, boots, and equipment by immersing in a tub(s) containing an appropriate non-phosphorus, non-antibacterial solution.[2] Keep immersed for

the recommended time to kill the invasive of concern. When done, remove the items and rinse with tap water (optional). Cover the tub with the lid for reuse at the next site.

Soak solution.

3. For equipment that cannot be soaked in the tub, clean by spraying with an appropriate non-phosphorus, non-antibacterial solution for the recommended time using a portable spray device (away from waterbody). Make sure to cover all surfaces thoroughly. Rinse with second sprayer containing tap water (optional).

Soak solution.   Tap water rinse (optional).

4. **Dry** items to suppress invasives. Where cleaning at the lab is appropriate, these same standards apply, but also include drying (to ensure effective drying, the item must be *completely dry* to the touch, inside and out, then left dry for at least another 2–30 days, depending on the invasive of concern) (see Appendix C).

5. Although strongly discouraged, if you wear felt soled boots/kickers where Didymo and/or other microscopic invasives are known or suspected to be present, restrict equipment to a single waterway, and then clean/dry later at the office/lab.

6. Dispose of all used tub washwater back at the lab down the sink.

7. The cleaning solution can be made up at the lab prior to departure or in the field as needed.

8. For attended and deployed multi-probe sondes (and tubes), employ the following special procedures:

   (a) In general, pre-design the deployment surveys to avoid the need to redeploy multi-probe sondes from one location to another. This will avoid the need for field decontamination, but they will still require decontamination washing at the office/lab. It is recognized that this is not always possible.

   (b) For lab or field washing, scrub and rinse or soak entire apparatuses (sondes, deployment tubes, cables, L-brackets, locks, bungee cords, and anchor blocks) with appropriate P-free non-antibacterial solution for the recommended contact time. Do not use bleach on sonde units. For bungee cords (and other porous surfaces), soak for 30 minutes in hot water kept above 45°C containing 5% dishwashing detergent or other applicable solution (at the lab).

   (c) If *unattended* deployment tubes must be moved from one location to another, use field decontamination kit at the retrieval site to clean/rinse the entire deployment apparatus using P-free, non-antibacterial solution, followed by tap water rinse, *and* use lab-cleaned bungee cords only.

   (d) For *attended* multi-probe sondes used at multiple sites on the same survey run, clean/rinse sondes and coiled cable after each use with P-free non-antibacterial cleaning solution for the recommended contact time. Do not use bleach on sonde units. Do not clean the datalogger or its comm. ports.

[1] DECON procedures are also required whenever work is performed outside Massachusetts where Didymo is known or suspected to be present.

[2] Ensure that cleaning solution used is appropriate for the invasive(s) of concern. Readily available cleaners include bleach, vinegar, hand-dishwashing detergents, and salt.

NOTE #5 USE OF DECON SPRAYERS

Setup: To be performed by lead monitoring coordinator

1. Sprayers have been preset to produce a flat spray (no adjustment needed).
2. Fill sprayer unit(s) to max fill line with decon solution. Do not exceed max fill line.
3. Pressurize unit(s) and test pressure relief valve and spray (to make it ready for use). Approximately 10–20 pump cycles should be sufficient. Do not exceed max pressure limits.
4. Lock pump top and load sprayer(s) on cart or in vehicle. Protect sprayer wand(s) from damage by securing it inside the vehicle.
5. In case additional decon solution is needed in the field and to avoid having to make it up in the field, a second decon sprayer per crew should be set up and provided.
6. For maintenance issues, see operator's manual,.
7. For inflation precautions, see below.

Using the Sprayer: By Survey Staff

1. To save time, one person can be spraying boots, gloves, and equipment while the other performs other decontamination tasks.
2. If needed, unlock pump and add more pressure to get a better spray.
3. Rinse all "contaminated" items sufficiently for the solution to have the required contact time (not 1 minute of constant spraying), but be aware of how much spray solution is being used at each site (especially if you don't have a spare sprayer).
4. If solution runs out, use spare sprayer or make up new solution.
5. If you need to open pressurized tank, pull on relief valve first to release pressure before opening.
6. For other precautions, see below.

WARNING
AIR COMPRESSOR
INFLATION PROCEDURE

An air compressor can be used to pressurize Solo sprayers equipped with a built-in inflation valve. Compressors pressurize sprayers much more rapidly than hand pumping; for this reason extra caution must be exercised when using compressed air. **Sprayers must be functioning properly, unmodified and have all component parts.** The following procedures must be followed. Failure to do so could result in serious injury to the operator or others.

1) Before **each** use of a compressor with the sprayer be certain that:
   a) The umbrella valve (or valve cone on older models) is present, installed and functioning properly. The valve is located on the bottom of the cylinder (see owner's manual, page 2, "Pump Maintenance" item #7).
   b) The pump assembly is screwed snug to the sprayer tank.
   c) The pressure relief valve is functioning properly. Check by pulling up on the valve until the red stem shows. The valve stem should move freely and spring back to its original position when released. The O-ring and the valve stem must be greased (see page 2, "Maintenance" in the owner's manual).

If any of the above items or other components is malfunctioning, STOP!, do not use an air compressor to pressurize. Repairs must be made prior to use.

2) Additional Precautions:
   a) Fill the sprayer with liquid formula.
   b) Do not stand over the sprayer handle while pressurizing with an air compressor.
   c) Do not stand over the sprayer while releasing the handle from the locked position. Compressed air can cause the pump handle to pop upward if the valve cone or umbrella valve is worn, damaged or if the pressure relief valve is not functioning (see item 1c above).

3) When pressurizing, place pump handle in the unlocked position, resting on top of the pump assembly. Begin pressurizing slowly. If the pump handle rises...**STOP!, DISCONTINUE PRESSURIZING.** Pull up on the pressure release valve and hold until all the pressure is released; then remove the pump, check and replace the umbrella valve (or valve cone). Compressed air can cause the pump handle to pop upward if the valve cone or umbrella valve is worn, damaged or not installed completely. Do not pressurize until the worn or damaged part has been replaced, installed properly and the handle does not rise during pressurization. Should you have questions, please call Solo (800) 296-7656.

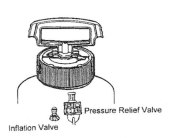

Figure 1 – Handle in the unlocked, resting position.

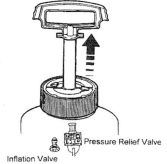

Figure 2 – Handle rising during pressurization; indicating a potentially unsafe condition.

Appendix II

A Guide for Using Hand Pulling and Benthic Barriers to Control Aquatic Invasive Species Pioneer Populations (Modified from MA DCR 2007)

CONTROL OF SUBMERGED AQUATIC INVASIVE PLANT SPECIES

HAND PULLING

Summary: This inexpensive techniques favored for controlling small pioneer infestations or where a large pool of volunteer labor is available. Although this technique is very species-specific and causes minimal damage to non-target species or other biota, many submerged non-native species spread by fragmentation, so extreme caution must be exercised when hand pulling to prevent additional spread. Although hand pulling is an inexpensive management technique, the use of SCUBA divers may increase the cost, and post-monitoring is essential.

Safety Considerations: It is essential to carry out this technique with a partner, rather than alone, and to take into consideration weather conditions, such as extreme heat or approaching storms. Life jackets are strongly recommended, and always follow the boating/water rules and regulations. If SCUBA divers are required, the divers will follow the safety guidelines outline by PADI, SCUBA, or the certifying company.

Materials: Plant bags and nets; spotter boat (if needed); scuba divers (if needed); plant guide, note pad, and markers (permanent ink); life vests; water, sunscreen, polarized glasses, hat with a brim.

Procedure:

1. If plants are in deep water, arrange for a spotter boat and two volunteers: one to drive and one to net any fragments that may float up. If the plants are confined to shallow waters, waders and view scopes (or mask/snorkel) may suffice.

2. Have someone on shore keep notes on the amount of plants removed, the time spent on removal, names of participants, etc. If using volunteer help, it is important to have them sign in and out so that they are accounted for.

3. Begin at the furthest boundary of the defined area and line up the participants along the boundary. Work toward the shore with the volunteers maintaining the line formation. Participants should go no more than chest deep, depending on the slope of the shore and type of bottom.

4. Remove target plants from the base, removing the entire root while disturbing the sediment as little as possible. Place the removed plants carefully in a collecting bag, and take care not to fragment the plants or leave any fragments in the water. The spotter boat should remain nearby, downwind/current, and have a volunteer with a net collect any fragments and place them in a storage bag. The boat operator needs to be very cautious of the volunteers in the water.

5. Repeat steps 3 and 4 until all the target plants have been removed. Depending

on the number of volunteers, type of sediment, depth of water, and other factors, expect this project to take a few days since the removal of the plants can create turbidity and make visibility very difficult.

6. Once the removal is complete, dispose of the contents of the storage bags far from the water so that they cannot cause a reinfestation.

7. Record the final details of the procedure and include date, time, site, town, volunteers involved, size of area pulled, approximate volume of plants removed, how they were disposed of, and other relevant notes.

8. To ensure effectiveness of treatment, monitor the site monthly during the growing season for reappearance of the non-native species. Remove any surviving plants.

BENTHIC BARRIERS

Summary: The installation of benthic barriers can be used to control the growth of aquatic submerged and floating-leaved plants. Benthic barriers are most suitable for small areas such as around docks and swim beaches. This technique can be repeated over a long period of time if the mats are maintained and cleaned.

Safety Considerations: The barrier needs to be securely anchored to the lake bottom or gases from decaying plants can build up beneath the barrier, causing it to rise off the bottom where it may create a hazard for boaters and swimmers. Caution must be used in selecting anchors so that they do not pose a hazard for swimmers, and will not roll as gases build up. Installation should be done on a day with appropriate weather conditions (calm and with no predicted storms). Use the buddy system when performing routine maintenance or inspection of the barrier.

Apparatus/materials: Anchors (sand bags, rocks, concrete blocks, flat weight discs); durable material (resistant to decay, limits light, and is less likely to float); area markers (anchored buoys/floating milk jugs) and rope; life vests; sunscreen, hat with brim, polarized glasses, water (optional); note pad and pen.

Procedure:

1. If possible, select a date in early spring for installation.

2. Purchase project materials, including benthic barriers, area markers, and weights. Note that there are many companies that both sell and install the benthic barriers.

3. If the barrier is not being professionally installed, identify an adequate number of volunteers who will install the barrier. Notify boaters, swimmers, and homeowners.

4. Define the area of barrier installation and mark area very clearly with visible, durable markers or buoys. Fishing lures and anchors can tear the fabric, so it is very important to forewarn boaters and fishermen of the barriers presence.

5. Record the details of the procedure and include date, time, site, town, number of volunteers (or name professional company involved), size of area covered, and other relevant notes.

6. One of the easier ways to install, maintain, and move a benthic barrier is to create a frame of marine plywood or rebar and PVC pipes for the barrier. Slits in the benthic material will allow gases to escape and sand bags can be used to secure the frame. Another option is to roll the material outward over the treatment area, anchoring as you proceed. Weights should then be placed along the edges, corners, and in the center of the barrier to anchor it down. Select weights that will not roll (such as cinder blocks).

7. Monitor the site monthly during the growing season to ensure that the mat remains well-anchored and free of sediments, and to assess effectiveness of treatment. Also monitor the area on a routine basis to check for reappearance of invasive species. Keep accurate records of the monitoring visits.

CONTROL OF FLOATING AQUATIC INVASIVE SPECIES

HAND PULLING

Summary: This inexpensive technique is favored for small- to medium-sized infestations or where a large pool of volunteer labors is available. Since the floating-leaved plants do not spread by fragmentation, this technique is very species-specific with minimal risk of additional spread. Although hand pulling is an inexpensive management technique, post-monitoring for regrowth and new plants is essential because many plants drop seeds that remain viable for years.

Safety Considerations: It is essential to carry out this technique with a partner and to take into consideration weather conditions. Life jackets are recommended and always follow the boating regulations.

Materials: Laundry baskets/leaf tip bags (or other containers to transport plants); kayak/canoe; pontoon boat (optional); plant guide, note pad, and pens; life vests; water, sunscreen, polarized lasses, hat with a brim; waders, mask/snorkel, view scope (optional); 100 pounds limit hanging scale (optional).

Procedure:

1. *Timing:* Mid-June is the best time to pull floating-leaved plants since they are visible at the water's surface, are still small, and have not produced seeds. Water chestnut must be pulled prior to August, before the nuts have formed. If hand pulling after August is inevitable, then take care to pull plants very carefully and to turn them upside down during retrieval to prevent the seeds from dropping into the water.

2. *Boats:* Kayaks and canoes are ideal vessels for hand pulling floating leaved species. Kayaks are easier to transport to remote locations, and are more maneuverable, especially in shallow water and dense plant growth. Since kayaks only require one person to paddle, their use maximizes the use of available volunteers. Canoes have the advantage of being able to carry a greater quantity of pulled plants, but generally require two volunteers to paddle and steer. Pontoon boats increase the efficiency of the hand pulling project in several ways. They can carry all the supplies, volunteers, and kayaks/canoes to the location and then once unloaded, can be used to store and transport the plants (e.g., an average pontoon boat can carry around 1,000 pounds of water chestnut plants). Kayakers and canoes can save time and energy by trading full laundry baskets for empty ones at the pontoon boat, instead of traveling to and from shore each time.

3. If plants are confined to shallow waters near shore, then volunteers with waders may suffice.

4. Plastic laundry baskets are ideal collection containers because they are inexpensive, lightweight, can be secured to the kayak with bungee cords, and drain excess water. Two handled plastic tubs also work well in canoes. For hand pulling projects near shore, leaf tip bags will suffice.

5. Prior to beginning a hand pulling event, give volunteers a brief history of the species, its impacts, and emphasize species identification.

6. Keep a list of the names of participants and have volunteers sign in and out so that they can be accounted. Remove target plants from the base, removing the entire root and stem. Place the removed plants carefully in the collecting basket/bag.

7. Once the basket is full, return to shore (or a stationary pontoon boat) and turn in the basket to be weighed (or plants counted). Assign a volunteer to keep notes on the weight or number of plants removed. This data is helpful for tracking population decreases year to year as a result of the hand pulling effort. If uniform bags or baskets are used, ten loads can be averaged to determine the approximate weight and plant count per load. This will eliminate the need

to weigh or count plants for every load collected. Plants should be disposed of on dry land, far above the high water mark. Some towns or local companies may be willing to haul away and compost or incinerate the removed plants.

8. Record the final details, including date, weather, beginning and end time, site, town, number of volunteers, size of area pulled, number (or weight) of plants removed, how they were disposed of, and other relevant notes.

9. Return to the site in one month and hand pull any new growth. Continue to monitor the site monthly (if possible) during the growing season.

CONTROL OF EMERGENT INVASIVE SPECIES

HAND PULLING

Summary: This inexpensive, labor-intensive technique is suitable for small pioneer infestations but is not feasible on a large scale. Although this technique is very species-specific and creates minimal damage to non-target species or other biota, it does require a long-term commitment. Both purple loosestrife and *Phragmites* have deep underground root systems and can regrow if any sections remain in the soil. Seeds already present in the soil rapidly germinate when the site has been disturbed, and therefore, continual monitoring and removal are necessary. These species need to be pulled prior to the formation of seeds; otherwise, the disturbance that occurs during hand pulling will only aid in their dispersal. Also, covering the area with black plastic or other light barrier may thwart new growth, and/or mowing the dry areas prevents returning plants from storing nutrients or developing seeds/flowers. Additional success may be achieved by planting native species in the disturbed area.

Safety Considerations: It is essential to carry out this technique with a partner or group and to take into consideration weather conditions, such as extreme heat or approaching storms. Dress appropriately by wearing long sleeves, long pants, and gloves. While

working, volunteers should keep a 10-ft distance from each other to avoid accidental injury from rakes and other tools.

Materials: Shovels, rakes, trowels, clippers; plant guide, note pad, and permanent markers; water, sunscreen, polarized glasses, hat with a brim; long sleeve shirts, pants, and gloves; waders (optional).

Procedure:

1. If plants are in deep water, a boat can be used; if they are confined to shallow water and the shoreline, chest waders are ideal.

2. Have a volunteer keep notes on the amount of plants removed, the time spent on removal, names of participants, etc. It is important to have volunteers sign in and out so that they can remain accounted.

3. Timing for the removal of emergent plants is species-specific; however, it is best to hand-pull the plants in the spring when they are smaller, less established, and have not produced seeds. Although it is easier to identify purple loosestrife when it is in bloom, the plants need to be removed before they have formed seeds (usually mid-July). Pulling plants when they are in seed will only aid in distributing the 1–2 million seeds per plant. However, if the plants must be pulled after they have bloomed, place a plastic bag carefully over the flowering portion of the plant and secure it around the stem. This will minimize seed dispersal.

4. Begin on shore and slowly work outward into the stand.

5. Remove target plants and try to extract the entire root. Place the removed plants carefully in a collecting bag, and take care not to fragment the plants or leave any fragments in the water.

6. Proper disposal of the removed plants is very important. Do not compost *Phragmites* or purple loosestrife. The plants should be incinerated or burned. If this is not feasible, the plants should remain in plastic bags and allowed to rot.

Appendix III

Literature Research of Chemical Control
Methods on 19 Invasive Species

Invasive Species	Chemicals*	Effective Minimal Concentration	Notes	References
Asian clam *Corbicula fluminea*	Ammonium nitrate	59.3 mg/l	Under hypoxic conditions, clam mortality rate began to steadily rise at 50 mg/l with effectiveness plateauing at around 200 mg/l. Under normoxic conditions, mortality rate did not begin to show signs of increase until around 125 mg/l. At 200 mg/l, normoxic testing began to show similar mortality levels to that of 75 mg/l under hypoxic conditions	Rosa et al. 2015; Rehman et al. 2021
	Potassium chloride	45 mg/l	Seemed to be most effective at increasing clam mortality rate from around 100 mg/l to 250 mg/l under hypoxic conditions. Under normoxic conditions, clam mortality rate saw a significant increase at around 150 mg/l to 900 mg/l with the highest mortality rate resulting from around 450 mg/l	Rosa et al. 2015; Densmore et al. 2018
	Niclosamide	0.08 mg/l	Required a significantly smaller concentration than other chemical agents to begin showing effectiveness at increasing Asian Clam mortality rates. Under hypoxic conditions, clam mortality rate increased to 10% with a 0.1 mg/l concentration of niclosamide, with effectiveness slowing at around 0.5 mg/l and 90% Asian clam mortality. Under normoxic conditions, niclosamide concentrations of around 0.2 mg/l began to slightly increase clam mortality. Clam mortality then reached about 90% at 1.2 mg/l under normoxic conditions and then began to plateau	Rosa et al. 2015
	Dimethoate	18 mg/l	Dimethoate concentration usage under hypoxic conditions began to increase clam mortality at around 45 mg/l. Effectiveness under hypoxic conditions then reached its maximum at around 100 mg/l and was near 100% clam mortality rate. Under normoxic conditions, clam mortality rate began to increase around 150 mg/l and reached its peak around 400 mg/l	Rosa et al. 2015; Van Scoy et al. 2016
	polyDADMAC	12.3 mg/l	Under hypoxic conditions, polyDADMAC effectiveness began to increase at approximately 10 mg/l, with mortality rate peaking at around 500 mg/l. Under normoxic conditions, polyDADMAC also began to increase at 10 mg/l and peaked at around 1200 mg/l. However, the clam mortality rate % difference between hypoxic and normoxic curves only varied slightly, with hypoxic conditions being more effective. Both hypoxic and normoxic curves also began to increase exponentially early on, but slowed quite quickly at around 200 mg/l	Rosa et al. 2015; Santos Ltd. 2021

(Continued)

Substance	Concentration	Notes	References
Chlorpyrifos	0.05 mg/l	Exposure to substance between 0.5 ppm and 1 ppm, caused a significant reduction in cholinesterase activity and a reduced capacity to burrow into the substrate. All measured concentrations were within 25% of the nominal concentrations with all initial concentrations, and all but two of the final concentrations, falling above the nominal target concentration. Freshwater bivalves often occur in close association with agricultural landscapes and this, along with their near continuous contact with waterborne contaminants during feeding, makes the potential for exposure to agricultural chemicals quite high	Cooper and Bidwell 2006; Rosa et al. 2015
Zebra mussel *Dreissena polymorpha*	100 mg/l	During the winter of 2006, KCl was used at a concentration of approximately 100 mg/l (ppm) to successfully eradicate zebra mussels from a 12-acre abandoned rock quarry in Virginia without harming other aquatic fauna. Whereas KCl is not an US Environmental Protection Agency (EPA) approved pesticide and requires special permitting for use, it is generally considered a reasonably safe alternative to other molluscicides because of its low potential to impact non-target species (Densmore Et al. 2018). The acute lethal concentrations of the KCL on juvenile salmonoid populations are not expected to be of concern. Concentrations used to eradicate the zebra mussel populations (around 100 mg/l) are not nearly high enough to be of great concern as the salmonoid populations survived for 96 hours in a controlled environment with up to 800 mg/l of KCl exposure. However, as previously stated, long-term effects on the species were not accounted for and the buildup of the chemical agent in aquatic ecosystems was not fully examined. The warmer summer months required a smaller concentration of the chemical agent to reach the 48 h LC50 value. In June and July, the concentration used was just over 150 mg/l, while in the autumn and winter the concentrations required to reach the 48 h LC50 shot up to 500 mg/l by January. By using the mussels' filtering behavior to concentrate BioBullets, the absolute quantity of active ingredient added to the water can be reduced substantially. Another study was performed on the Colorado River but used KCl levels that far exceeded those of previous studies and would be deemed to a high concentration for the survival of other non-target species according to most of the other papers found	Fisher et al. 1991; Aldridge et al. 2006; Costa et al. 2008; Davis et al. 2018; Moffitt et al. 2016; Densmore et al. 2018

Invasive Species	Chemicals*	Effective Minimal Concentration	Notes	References
	polyDADMAC	6.25 mg/l	Chemical was tested with seasonal variability to find most effective concentration and time frame for usage. The warmer summer months required a smaller concentration of the chemical agent to reach the 48 h LC50 value. The concentration in use remained quite consistent at around 10–20 mg/l between June and October, with concentration requirements to reach 48 h LC50 reaching near 100 mg/l by November	Costa et al. 2008; Rosa et al. 2015
	Bayluscide WP 70 (active ingredient is the niclosamide ethanolamine salt) (Bayer 73) (Clonitralid)	0.006 mg/l	Chemical was tested with seasonal variability to find most effective concentration and time frame for usage. The warmer summer months required a smaller concentration of the chemical agent to reach the 48 h LC50 value. From June to August, the concentration hovered around 0.07 mg/l used, and more than doubled to 0.15 mg/l between August and September	Dawson 2003; Costa et al. 2008
	Sodium chloride	5,000 mg/l	Adult mussels required a 4× longer exposure period to exhibit complete mortality when exposed to NaCl at 30,000 mg/l (24 h) compared to KCl (6 h). At 10,000 mg/l, NaCl took 8× longer (96 h) than KCl (12 h) to cause 100% mortality of adult mussels. Sodium chloride is less effective at causing mortality than KCl within the exposure periods tested	Davis et al. 2015b; Davis et al. 2018
	Distilled white vinegar	25%	All concentrations (25, 50, 75, and 100%) caused complete mortality within 4 hours. The use of such an acid solution may be problematic in certain applications where the materials that need to be disinfected could be harmed by the chemical treatment. It could be assumed that the time needed to kill veligers is shorter than needed to kill adults	Claudi et al. 2012; Davis et al. 2015a
Chinese mystery snail *Cipangopaludina chinensis*	Rotenone	25 mg/l	The use of chemicals such as rotenone has a limited effect on *B. chinensis* due to the impenetrability of the shell, large shell size, and operculum	Haak et al. 2014; Dalu et al. 2015
	Copper sulfate	1.25 mg/l	Within one month of treatment, more than 27,000 dead snails were removed from two ponds	Freeman 2010; Boone et al. 2012; Haak et al. 2014

Northern snakehead *Channa argus*	Rotenone	0.025 ppm	It has been found that a rotenone concentration of 0.075 mg/l is strong enough to kill all northern snakehead in an enclosed area after 24 hours, but it also includes the most devastating damage. The commercial products are most commonly applied at a concentration 1.0–2.0 ppm which equates to 0.025–0.100 ppm of actual rotenone; however, treatment rates can range from 0.5 to 10.0 ppm	Lazur et al. 2011; Turner et al. 2007; Abdel-Fattah 2011; Stinson 2018
	Deltamethrin	0.061 µg/l (asked Jonathan on 8/27/22)	Environmentally relevant concentrations (0.121, 0.242, 0.485, and 0.970 µg/l) inhibited the biochemicals, antioxidants, and immune responses and disease resistance of snakehead fish. Deltamethrin exposure induced tissues damage, oxidative damage, and immunotoxicity in fish. The LC50 at 96 hours was determined to be 1.94 µg/l	(Kong et al. 2021; Shrivastava et al. 2011
Hydrilla *Hydrilla verticillata*	Diquat dibromide	0.25 ppm	Herbicide-treated hydrilla plants were severely necrotic indicating mortality 14 and 21 days after application (but not different from each). Greater than 95% control was observed against all floating species tested when either formulation was applied at 4.7 or 7.0 l/ha. The performance of diquat on *Eichhornia crassipes* (water hyacinth) and some other species can be enhanced if copper is also applied together. Copper is the most common active ingredient in algaecides	EPA 1995; Langeland et al. 2002; Mattson et al. 2004
	Endothall	1.0 mg ae/l	Hydrilla required higher concentrations and longer exposure times than did milfoil to achieve acceptable levels of control. Although root biomass was greatly reduced by some of these treatments (especially at concentrations of 4.0 and 5.0 mg/l), the presence of apparently healthy regrowth suggested that complete recovery could occur. It is also found that endothall killed coontail, lagarosiphon, and hydrilla and some species of *Myriophyllum* and *Potamogeton* but not *Egeria* or species of *Chara* or *Nitella*. Potassium endothall is used extensively in Florida	Netherland et al. 1991; Hofstra and Clayton 2001; Mattson et al. 2004; Poovey and Getsinger 2010; Wisconsin DNR 2012; Giannotti et al. 2014; Ortiz et al. 2019
	Triclopyr	0.25 mg/l	Triclopyr produced epinastic shoots in all species, except the charophytes; however, these growth effects along with some loss of turgor and color change in stems were temporary	Hofstra and Clayton 2001

(Continued)

Invasive Species	Chemicals*	Effective Minimal Concentration	Notes	References
	Dichlobenil	0.5 mg/l	All plants treated with dichlobenil exhibited some loss of vigor when compared to untreated control plants, and some species had more pronounced shoot loss, browning of stems, and stem fragmentation; however, all symptoms were transient with plant recovery in 35–50 days after treatment. The onset of symptoms was related to dichlobenil concentration rather than exposure time, with all susceptible species exhibiting symptoms irrespective of exposure time, but for some species only at the higher concentration	Hofstra and Clayton 2001
	Bispyribac-sodium	Not available	Bispyribac was registered by the EPA for aquatic uses in 2011 and is sold under the trade name Tradewind. Bispyribac is a systemic herbicide that blocks an enzyme necessary for growth and development specific to plants. As the herbicide accumulates in plant tissue, the plant dies over a period of weeks to months. Tradewind herbicide may be applied to slow-moving or quiescent bodies of water where there is minimum or no outflow; it is not applicable in flowing systems. There is no post-application restriction against use of treated water for drinking or recreational purposes (e.g., swimming, fishing)	Giannotti et al. 2014
	Copper complexes	3 ppm	Copper is a fast-acting, broad-spectrum, contact herbicide that kills a wide range of aquatic plants and algae. Although copper is a micronutrient required by living plants and animals in small amounts, too much copper kills plants by interfering with plant enzymes, enzyme co-factors, and plant metabolism in general. Copper has long been used in natural and industrial waters for algae control, often applied directly to water as copper sulfate crystals. Chelated copper (i.e., copper ion combined with an organic molecule) is typically used for aquatic plant management as it remains active in the water column longer than copper sulfate. Copper is toxic to fish, and because copper is an element, it will accumulate in the sediments regardless of its bioavailability. Drinking water considerations are not typically the limiting factor for copper application. National Primary Drinking Water Regulations include an action level for copper at 1.3 mg/l, but copper applied to lakes and reservoirs tends to quickly bind with sediment and organic matter, settling out of the water column. Copper can be highly toxic to mollusks and fish at relatively low doses (approximately 1–5 ppm), thus some states (e.g., Florida and New York) restrict its use to prevent impacts to fisheries. Copper alone may only provide fair control of hydrilla at application rates of approximately 3 ppm. Copper can be added at lower rates in combination with other herbicides (e.g., diquat or endothall) for better control of hydrilla	Giannotti et al. 2014

Penoxsulam	Not available	Penoxsulam is currently sold under the trade name of Galleon. It is a systemic herbicide that is applied to plant foliage to control floating or immersed plants, or to the water column for submersed plant control. Penoxsulam works by inhibiting the plant enzyme acetolactate synthase, similar to imazamox. Treatment must maintain herbicide concentrations at sufficient levels for 90–120 days for optimum performance. Penoxsulam is broken down in water both microbially and through photolysis, and its half-life in water is approximately 25 days. Because of the long contact time, penoxsulam is not generally applied in areas of high water exchange. Penoxsulam needs a surfactant for foliar and exposed sediment applications to increase efficacy by binding the herbicide to the plant. There are no restrictions on consumption of treated water for potable use or by livestock, pets, or other animals. The label for penoxsulam-containing herbicides includes a number of restrictions on irrigation usage after treatment	Wright et al. 2018
Topramezone	30 ppb	Topramezone is currently sold under the trade name Oasis. It is a systemic herbicide that is applied to the water column for submersed plant control, directly to foliage of floating and emergent vegetation or to dewatered sites. Topramezone is the first herbicide belonging to new chemical class called pyrazolones. In sensitive plant species, topramezone inhibits the enzyme 4-hydroxyphenylpyruvate dioxygenase, leading to a disruption of the synthesis and function of chloroplasts. Consequently, chlorophyll is destroyed by oxidation resulting in bleaching of the growing shoot tissue (white or pink coloration) and subsequent death of the aboveground portion of the plant. Generally, topramezone is applied at 30–50 ppb and maintained at or near the initial concentration for a minimum of 60 days. Applications are made to actively growing plants early in the growing season before mature plants can build tubers. Topramezone is absorbed into the plant tissue and symptoms generally first appear in seven to ten days. Water with concentrations higher than 45 ppb are restricted for potable uses. There are no restrictions on consumption of treated water by livestock, pets, or other animals up to the maximum concentration of 50 ppb. There are no restrictions on use of treated water for recreational purposes, including swimming and fishing up to the maximum concentration of 50 ppb	Giannotti et al. 2014

(Continued)

Invasive Species	Chemicals*	Effective Minimal Concentration	Notes	References
	Flumioxazin	Not available	Flumioxazin is sold under the trade name Clipper, Schooner, Redeagle, or Flumigard. It is a contact herbicide that causes chlorosis (yellowing) and necrosis (browning) of exposed plant tissue. It moves within treated leaves, but does not translocate to other areas of the plant. Plants exposed to flumioxazin die because of the disruption of cell membranes. Once inside the plant cell, flumioxazin inhibits a key enzyme, protoporphyrinogen oxidase. Plant necrosis and death is rapid, taking a few days, up to two weeks. In general, at least four hours of contact time is required for good control. The primary breakdown pathway of flumioxazin in water is by hydrolysis, and it is highly dependent on water pH. Under high pH values (>9), flumioxazin half-life in water is 15–20 minutes. Under more neutral pH values (7–8), half-life in water is approximately 24 hours. Flumioxazin controls a wide variety of submersed and floating aquatic weeds and can be tank-mixed with other contact or systemic herbicides. There is no post-application restriction against use of treated water for drinking or recreational purposes (e.g., swimming, fishing). Treated water is restricted for use for irrigation for up to five days	Giannotti et al. 2014
	Imazamox	50 ppb	Imazamox was registered for aquatic use in 2007. Imazamox is a systemic herbicide that works by inhibiting the plant enzyme acetolactate synthase (ALS), which regulates the production of amino acids in plants. When ALS is inhibited, plants die. Animals do not produce these enzymes, so imazamox has low toxicity to animals. Enzyme inhibiting herbicides act very slowly. Imazamox is broken down in the water by photolysis and microbial degradation. Its half-life in water is 7–14 days. Imazamox is absorbed rapidly into plant tissues, and growth of susceptible plants is inhibited within a few hours after application, dying in approximately one to two weeks. In Florida, aquatic plant management programs include submersed applications for hydrilla control. At concentrations of 50–150 ppb, imazamox acts as a growth regulator for hydrilla, persisting for up to several months. At concentrations of 150–250 ppb, imazamox acts with herbicidal activity, killing hydrilla in a few weeks after application. Imazamox application rates are restricted to less than 50 ppb within ¼ mile of potable drinking water intakes, and irrigation is restricted at rates above 50 ppb	Giannotti et al. 2014

Fluridone	0.05 ppb	Low concentrations of fluridone (0.05 and 0.5 ppb) caused transient increases in the number of both subterranean and axillary turions by mature hydrilla, but higher concentrations (5 and 50 ppb) inhibited development of these tissues. Growth (shoot dry weight) of young plants treated with low concentrations of fluridone (0.05 and 0.5 ppb) was not affected. The 5.0 ppb-fluridone treatment did not affect the growth of young plants until after six weeks of exposure. The 50 ppb fluridone treatment prevented any significant change in young plant shoot dry weight over the 12-week study. There was no significant change in shoot dry weight of mature plants regardless of the treatment	Netherland et al. 1993b; Netherland and Getsinger 1995; Mattson et al. 2004
Curly-leaf pondweed *Potamogeton crispus*	Diquat	1 mg ai/l	Turion numbers and curly-leaf pondweed injury were observed. Although herbicide effects were more pronounced with the 21°C than the 16°C diquat treatments, plants showed signs of appreciable recovery by 6 week after treatment (WAT). Treatments conducted when water temperatures at 23°C were most effective in sustaining plant injury. All diquat treatments at the end of the study were effective in reducing shoot and root biomass. Endothall and diquat treatments in the early spring have resulted in e ffective biomass reduction and the inhibition of turion production	Poovey et al. 2002; Mattson Et al. 2004; Wisconsin DNR 2012; Barr and Ditomaso 2014
	2,4-D	10 µg/l	Low rates of endothall combined with 2,4-D or triclopyr can provide selective control of two invasive exotic species, Eurasian watermilfoil and curly-leaf pondweed, if applied in early spring when most native species are dormant	Poovey et al. 2002; Mattson et al. 2004; Belgers et al. 2007; Skogerboe and Netherland 2008)
	Endothall	1 mg ai/l	All endothall applications reduced shoot and root biomass; however, early and mid-spring treatments provided better control than late spring treatments. Both the 16°C and 21°C treatments reduced shoot biomass by 90% compared to the 23°C treatment, which reduced shoot biomass by 60%. Endothall and diquat treatments in the early spring have resulted in effective biomass reduction and the inhibition of turion production. Endothall effectively controlled Eurasian watermilfoil and curly-leaf pondweed at all of the application rates, and no significant regrowth was observed at 8 week after treatment (WAT)	Poovey et al. 2002; Skogerboe and Getsinger 2002; Mattson et al. 2004; Wisconsin DNR 2012); Barr and Ditomaso 2014

(Continued)

Invasive Species	Chemicals*	Effective Minimal Concentration	Notes	References
	Fluridone	1 mg ai/l	During the first and second years after treatment, it is assumed that fluridone was no longer having a direct effect on the vegetation because concentration of the herbicide reached a low level of 2.3 ppb at 91 days after treatment (DAT) on August 29, 2002 (Figure 1)	Poovey et al. 2002; Mattson et al. 2004; Valley et al. 2006; Wisconsin DNR 2012
Eurasian watermilfoil, *Myriophyllum spicatum*	Diquat	Not available	Eurasian watermilfoil was highly susceptible to diquat, with 85–100% biomass reductions for all diquat treatments. The dark exposure period did not increase diquat efficacy on Eurasian watermilfoil as all diquat treatments were similar. Reductions in Eurasian watermilfoil biomass of 97% to 100% was achieved using similar diquat concentrations at half-lives of 2.5 and 4.5 hours	Mattson et al. 2004; Wersal et al. 2010
	2,4-D	2 ppm ae (acid equivalent)	A plastic curtain suspended across the outlet of each treatment cove kept 94–98% of the herbicide from drifting into open water. Both control methods eliminated Eurasian watermilfoil from treatment sites within four to six weeks. The bottom fabrics eliminated all species of rooted, submerged aquatic plants. The 2,4-D treatments initially reduced the mean standing crop of coontail (*Ceratophyllum demersum* L.), elodea (*Elodea canadensis* Michaux), variable-leaf watermilfoil (*Myriophyllum heterophyllum* Michaux), and wild celery (*Vallisneria americana* Michaux) by 14–85%. These native plants, however, recovered about 80–120% of their standing crops within 10–12 weeks after herbicide treatments	Mattson et al. 2004; Hesel et al 1996; Thum et al. 2017
	Endothall	0.5 mg ae/l	Severe Eurasian watermilfoil injury (>85% biomass reduction) occurred when exposed to 0.5 mg ae/l for 48 hours, 1.0 mg ae/l for 36 hours, 3.0 mg ae/l for 18 hours, and 5.0 mg ae/l for 12 hours. Milfoil control increased with increasing endothall concentrations and/or exposure times. All treatments resulted in a significant reduction of shoot and root biomass levels compared to the untreated reference aquaria at four weeks posttreatment, except 1 mg/l for 2 hours	Netherland et al. 1991; Mattson et al. 2004
	Fluridone	10 µg/l	Fluridone concentrations ranging 10–100 µg/l were used on Eurasian watermilfoil treatment and reduction in root biomass ranged from 53% at 10 µg/l to 79% at 20 µg/l while shoot biomass saw a reduction of 68% at 10 µg/l and 84% at 90 µg/l	Hall et al. 1984; Netherland et al. 1993a; Mattson et al. 2004

Carfentrazone-ethyl	0.1 mg ai/l	It is reported that carfentrazone-ethyl resulted in complete control (100% biomass reduction of this species; conversely, carfentrazone-ethyl was not efficacious against Eurasian watermilfoil because biomass reductions were only 25% and 37% for the 0.20 mg ai/l with light and dark treatments	Gray et al. 2007; Wersal et al. 2010; Wisconsin DNR 2012
Triclopyr	1.00 mg/l	This chemical can be applied in either river or lake systems. The proposed potable water tolerance level is 0.5 mg/l	Getsinger et al. 1997; Thum et al. 2017
Bensulfuron methyl	25 μg/l	Eurasian milfoil biomass averaged 50% less than that of untreated references approximately after six weeks of exposure to all treatments. By 12 weeks, the biomass was reduced to 96–98% compared to untreated plots for all chemical application rates	Getsinger et al. 1994
Variable milfoil *Myriophyllum heterophyllum*			
Diquat	Not available	Contained infestation in the short term by reducing footprint and density of patch, but it did not eradicate the variable milfoil	Mattson et al. 2004; Carrol County 2008
2,4-D	500 μg ai/l	"In greenhouse studies, 2,4-D ester at 500 and 1500 μg ai L^{-1} exposed for 3, 8, and 24 hours provided 98 to 100% control of variable-leaf milfoil (). Bugbee et al. (2003) also reported that 227 kg ha-1 2,4-D ester as Navigate controlled nearly all the variable-leaf milfoil in treated field sites" (Glomski and Netherland 2008)	Mattson et al. 2004; Netherland and Glomski 2007; Skogerboe and Netherland 2008; Glomski and Netherland 2008a
4-Amino-3-chloro-6-(4-chloro-2-fluoro-3-methoxyphenyl)-5-fluoro-pyridine-2-benzyl ester (SX-1552 or XDE-848 BE; proposed ISO common name in review)	0.3 μg/l	Variable watermilfoil is one of the most sensitive species evaluated. Symptomology occurred within one week after treatment (WAT)	Richardson et al. 2016
Endothall	0.5 mg ae/l	Variable milfoil can be partially controlled by endothall	Mattson et al. 2004
Fluridone	8 ppb	Variable milfoil also tends to be quite susceptible to fluridone and the ALS inhibitors at rates in the range of 8–20 ppb. Studies suggest the phenology of variable milfoil will require that these treatments be applied early in the season prior to or just as the plants start to come out of winter dormancy	Mattson et al. 2004; Netherland 2007; Glomski and Netherland 2008b; Wisconsin DNR 2012

(Continued)

Invasive Species	Chemicals*	Effective Minimal Concentration	Notes	References
	Carfentrazone-ethyl	100 μg ai/l	Carfentrazone at 100 μg ai/l for 6–30 hours was reported to provide 61–81% control of variable-leaf milfoil. Doubling the rate of carfentrazone did not improve efficacy. While there is no published literature regarding field applications of carfentrazone to control variable-leaf milfoil, recent field trials in North Carolina have demonstrated good control	Glomski and Netherland 2008a; Wisconsin DNR 2012
Parrot-feather *Myriophyllum aquaticum*	Diquat	0.19 mg ai/l	In all treatments, diquat at each concentration and application time significantly reduced biomass of parrot-feather by 52–82% across diquat concentrations. Allowing for a dark exposure after herbicide application did not result in increased efficacy of diquat on parrot-feather at the concentrations tested. Although parrot-feather biomass was reduced with respect to untreated plants, regrowth was evident, and plants would have recovered given sufficient time. Plant recovery was occurring through regrowth via root crowns and the formation of new shoots from the nodes of surviving plants	EPA 1995; Langeland et al. 2002; Mattson et al. 2004; Hofstra et al. 2006; Wersal et al. 2010
	Endothall	2.5 mg ae/l	Endothall at 5.0 mg ae L21 and fluridone at 0.02 mg ai/l (0.02 ppmv) did reduce parrot-feather biomass; however, reductions were only 30% and 26% of untreated reference plants, respectively, at 6 WAT, Endothall was promising in reducing parrots-feather cover and biomass in the first year contained trial. It is also reported that reduction of parrot-feather biomass was not impressive (i.e., only 30% or 26% of untreated plants)	Hofstra et al. 2006; Wersal and Madsen 2010
	2,4-D	2.0 mg ae/l	It is one of the most effective herbicides for parrot-feather control. The foliar application of 2,4-D resulted in 85% control of parrot-feather at six week after treatment (WAT)	Wersal and Madsen 2010
	Imazamox	281 g ai/ha	At four week after treatment (WAT), imazamox at 561 and 281 g ai/ha controlled 63.3% and 78.3% of parrot-feather, respectively. However, by 6 WAT regrowth of plant tissue in the imazamox treatments was observed and control decreased to 63.3% and 56.7% for the 561 and 281 g ai/ha rates, respectively	Wersal and Madsen 2007

Imazapyr	584 g ai/ha	Imazapyr provided excellent control of parrot-feather. At four weeks after treatment (WAT), imazapyr at 1,123 and 584 g ai/l rates reduced 73.6% and 80.0%, respectively. Additionally, *M. aquaticum* treated with imazapyr at 1,123 and 584 g ai/ha was controlled 90.0% six WAT with an increase to 100.0% for both treatments by eight WAT. It was observed that beyond five WAT, *M. aquaticum* control with any imazamox rate was significantly less than all imazapyr treatments except imazapyr at 281 g ai/ha	Patten 2003; Wersal and Madsen 2007
Fluridone	0.01 mg ai/l	Fluridone applications were not efficacious on parrot-feather from one to six weeks after treatment (WAT). The fluridone-treated plants had symptomatic development of pink and chlorotic shoots within the first two WAT, although this did not appear to impact on percent cover and did not result in reduced biomass	Hofstra et al. 2006; Wersal and Madsen 2010
Glyphosate	1.5 l/ha	The best effectiveness of control of the glyphosate alone was 85% at the dose of 7.5 l/ha, increasing to 100% when associated with Aterbane® and Veget'oil®. The control reached 100% for all glyphosate doses associated with Dash®. Moreover, glyphosate at the dose of 7.5 l/ha associated with Assist® provided a 98% control, while glyphosate doses of 3.5, 5.5, and 7.5 l/ha associated with Agral® provided a 100% control. Glyphosate at doses of 5.5 and 7.5 l/ha associated with Dash® and Agral® was more effective in reducing the regrowth and dry biomass (100%). Thus, glyphosate + Dash® and Agral® promoted the highest control (above 95%), the lowest regrowth, and the highest reduction in the dry biomass of *M. aquaticum*	Cerveira et al. 2020
Dichlobenil	6.75 kg ai/ha	Dichlobenil (granular form of Prefix®-D) was scattered onto tanks at a rate of either 6.75 or 20.25 kg ai/ha. It was promising in reducing parrot-feather cover and biomass in the first year contained trial. Dichlobenil was seen as less likely to be an acceptable choice of product for parrots-feather control compared with triclopyr, due to the longer persistence of dichlobenil	Hofstra et al. 2006

(Continued)

Invasive Species	Chemicals*	Effective Minimal Concentration	Notes	References
	Clopyralid	1.5 kg ai/ha	Plants treated with clopyralid showed some initial loss of vigor, with reduced cover (and shoot number) for approximately eight weeks after treatment (WAT). By 17 WAT, the percent plant cover in the clopyralid tubs was within 20% of that observed in the control tubs, but the biomass did not differ significantly from control plants. Clopyralid was selected in this study because it provides selective control of a range of dicot species with little damage to monocots. The advantage of this for a drain-weed herbicide is that ditch bank grasses will not be damaged and continue to provide or maintain bank stability in the absence of other vegetation	Hofstra et al. 2006
	Triclopyr	2.0 kg ai/ha	Results indicate that triclopyr is effective at controlling parrot-feather under contained experimental conditions, reducing biomass to zero (or near), with little or no plant recovery in contrast to the results for glyphosate. Under field conditions, triclopyr has successfully reduced the cover and presence of parrot-feather. All triclopyr treatments had significantly lower percent cover than glyphosate tubs from 15 week after treatment (WAT). This trend continued for the duration of the study. Therefore, triclopyr provided substantially better control of parrot-feather than glyphosate	Hofstra et al. 2006
	Carfentrazone-ethyl	0.10 mg ai/l	Carfentrazone-ethyl has shown variable control of parrot-feather where ratings ranged from 29% to 70%. The use of carfentrazone-ethyl was also effective at reducing parrot-feather at both concentrations and application times. A 64% and 65% reduction in parrot-feather biomass was obtained when carfentrazone-ethyl was applied at 0.20 mg ai/l during a dark and light exposure period, respectively. In another study, similar biomass reductions of parrot-feather (63%) using carfentrazone-ethyl at 0.20 mg ai/l	Gray et al. 2007; Wersal et al. 2010
Fanwort *Cabomba caroliniana*	Fluridone	5 g/l	Treatments included static exposures of 0, 5, 7.5, 10, 15, 20, and 30 g fluridone per liter. For fanwort, phytoene levels increased 82% and β-carotene decreased 88% when exposed to 5 g fluridone per liter for 14 days. Effects on these two pigments persisted through 84 days after treatment (DAT) for fanwort. Leaf chlorophyll decreased with increasing fluridone concentration in fanwort, whereas in water marigold, decreased chlorophyll was observed on plants treated with rates of 7.5 g/l and higher. Despite these observed differences in pigment response, all fluridone treatments significantly reduced shoot dry weight biomass	Nelson et al. 2002

Species	Herbicide	Concentration	Description	References
	Carfentrazone-ethyl	100 ug ai/l	At 2 mg ai/l, there was a 60–90% biomass reduction in mesocosms with likely eradication of a field site. Another study involving carfentrazone indicates that control of fanwort was not as effective as endothall amine salt was, but its performance may have been inhibited by the heavy shading employed to reduce algal growth in the study	Hunt et al. 2015; Hofstra et al. 2021
	Endothall	1 mg ai/l	Endothall (dipotassium salt) reduced fanwort biomass; however, with viable plant material remained, it indicates the potential for rapid regrowth after treatment, and a high degree of uncertainty of outcome where the herbicides are to be used for the management of field populations. Symptoms within three to seven days were less effective than endothall amine	Mattson et al. 2004; Willey 2012; (); Hunt et al. 2015; Hofstra et al. 2021
	Flumioxazin	0.1 mg ai/l	Flumioxazin reduced fanwort biomass; however, with viable plant material remained, it indicates the potential for rapid regrowth after treatment, and a high degree of uncertainty of outcome where the herbicides are to be used for the management of field populations	Hofstra et al. 2021
	Endothall amine salt	0.5 mg ae/l	It could be an effective tool for fanwort control, even at low concentration (0.5 mg ae/l) provided contact times can be maintained for at least 50 hours. Field verification is now required to confirm the findings of this research	Hunt et al. 2015
	Triclopyr	0.5 mg ai/l	Triclopyr reduced fanwort biomass; however, with viable plant material remained, it indicates the potential for rapid regrowth after treatment, and a high degree of uncertainty of outcome where the herbicides are to be used for the management of field populations	Hofstra et al. 2021
European naiad *Najas minor*	Diquat	Not available	In July of 2010, 3.5 acres of the pond was sprayed with the herbicide diquat to eradicate the newly discovered European spiny naiad without more information on this project	Mattson et al. 2004; Desmerais 2016
	2,4-D	Not available	Limited 2,4-D treatment information in the Final Generic Environmental Report	Mattson et al. 2004
	Endothall	Not available	Limited information on endothall treatment on European naiad in the Final Generic Environmental Report	Mattson Et al. 2004
	Imazamox	Not available	It is a systemic herbicide, so contact exposure time for submersed vegetation such as the European naiad may be lengthy	Richardson 2008

(Continued)

Invasive Species	Chemicals*	Effective Minimal Concentration	Notes	References
	Fluridone	Not available	Spiny naiad (*Najas marina* L.) is a native species susceptible to fluridone that was originally present in low abundances but its abundance decreased significantly following treatment and was undetected in transect surveys for at least three years	Mattson et al. 2004; Wagner et al. 2007
South American waterweed *Egeria densa*	Diquat	1.8 kg ai/ha	Control of this invasive waterweed and effects on native species of diquat (bottom injection) were assessed by the point intercept method. One year after treatment, the waterweed was absent from all points except one. After another diquat treatment, no invasive waterweed was found. The native plant community was resilient with an increase in species richness from 11 pretreatment to 18 two-year posttreatment	Mattson et al. 2004; Bugbee et al. 2020
	Oxytetracycline	5 µg/l + 20 µg/l oxytetracycline amended with nitrogen and phosphorus	In a microcosm study, increased susceptibility to oxytetracycline exposure was noted in some paired plantings (e.g., *E. densa* root development), relative to individual plants in these treatments; however, no clear explanation for this response is available	Hanson et al. 2006
	Fluridone	Not available	It can be used between March 1 and November 30). Fluridone is a slow-acting, selective, systemic herbicide used to control submersed aquatic weeds. It has an average half-life of 20 days in pond water and three months in pond hydrosol for fluridone formulations. Fluridone is applied at rates of 5–30 ppb depending on plant susceptibility. Because of the slow-acting nature of fluridone, 8–16 weeks of exposure is necessary. This usually requires monitoring and additional applications to maintain the prescribed dose	Mattson et al. 2004; Caudill et al. 2021
Swollen bladderwort *Utricularia inflata*	Diquat	Not available	Diquat is listed as "E" for "Excellent Control" for bladderwort	Miller and Lewis 1980
	Endothall	Not available	Endothall is listed as "G" for "Good Control" for bladderwort	Miller and Lewis 1980
	Fluridone	Not available	Fluridone as a 480 g/l aqueous suspension (4 pounds/gal AS) or 2.5% granule at rates 0.1–1.0 ppm ai total water treatment were applied over the water surface, directly under the water surface, or as a layered treatment to the bottom of the body of water provided control of many submersed and emersed aquatic plant species	Parka and Arnold 1979

Species	Herbicide	Rate	Description	References
Water chestnut *Trapa natans*	2,4-D	0.5 mg ai/l	All application rates of 2,4-D significantly controlled water chestnut by 60–65% compared to the untreated reference. The Oswego County Soil and Water Conservation District turned to chemical control in 2005 when mechanical control methods were slowly losing effectiveness at fighting invasive water chestnut and 2,4-D is one of the two chemicals on large patches with erratic at best result	Mattson et al. 2004; Poovey and Getsinger et al. 2007; OARS 2017
	Fluridone	Not available	Partial control for water chestnut but control can sometimes be achieved with a higher dose or other conditions	Mattson et al. 2004; Mustafa et al. 2016
	Triclopyr	0.5 mg ai/l	Control of water chestnut with triclopyr significantly increased from 33% for plants treated with 0.5 mg ai/l to 66% for plants treated with 2.0 mg ai/l. Application rates of 2,4-D and triclopyr significantly reduced water chestnut shoot growth; Technology LLC claims to have success with the 2, 4-D and not the triclopyr, when removing the water chestnut. Although triclopyr may not be as effective as 2,4-D, it can be used in public waters where 2,4-D use is not allowed or may be allowed with special conditions	Poovey and Getsinger 2007; Jacobs 2008; Mustafa et al. 2016; OARS 2017
	Imazapyr	Not available	It is one of the nine herbicides approved for control of water chestnut in Massachusetts	OARS 2017
	Imazamox	Not available	Since it is approved for control of water chestnut, imazamox has been used with initial good results	OARS 2017
	Glyphosate	71 ha/l at 53.8% ai	Its application may have a significant impact on the surviving plants' ability to produce viable fruit. The seeds may either be non-viable or of smaller mass and lack vigor	Rector et al. 2015; OARS 2017
Water hyacinth *Eichhornia crassipes*	Diquat	0.25 ppm	To protect aquatic organisms, EPA is requiring labeling that limits application of diquat dibromide to one-third or one-half of the dense weed areas in a waterbody, and prohibits subsequent applications for two weeks. Diquat dibromide's primary route of environmental dissipation is strong adsorption to soil particles. Diquat does not hydrolyze or photodegrade and is resistant to microbial degradation under aerobic and anaerobic conditions. When used as an aquatic herbicide, diquat dibromide is removed from the water column by adsorption to soil sediments, aquatic vegetation, and organic matter. Adsorbed diquat dibromide is persistent and immobile, and is not expected to be a groundwater contaminant. It can be used with copper, which also enhances the performance of diquat on *Eichhornia crassipes* (water hyacinth) and some other species	EPA 1995; Langeland et al. 2002; Mattson et al. 2004

(Continued)

Invasive Species	Chemicals*	Effective Minimal Concentration	Notes	References
	2,4-D	1.5 kg ai/ha	Maintenance of infestations at the "lowest feasible level" is generally accomplished by repeated applications of herbicides, most often 2,4-D. 1999). 2,4-D is sold in liquid or granular forms as sodium and potassium salts, as ammonia or amine salts, and as an ester. Doses of 50–150 pounds per acre are usual for submersed weeds. Herbicide 2,4-D application for plant suppression is excellent	Center Et al. 1999; Mattson et al. 2004; Fawad et al. 2015
	Glyphosate	2.0 kg ai/ha	A study showed that the spraying of the glyphosate freshwater mixture using a helicopter to control invasive aquatic plants does not dramatically reduce water quality conditions. It also found that the spraying did not fully repair the DO levels of the wetlands that were put into hypoxic conditions by the water hyacinth	Jacobs 2008; Fawad et al. 2015; Waltham and Fixler 2017
European water clover, *Marsilea quadrifolia*	Diquat	Not available	It is regarded as capable of only partial control for pepperwort. Control can sometimes be achieved, but may require a higher dose or specific natural conditions that are difficult to control	Mattson et al. 2004
	Glyphosate	10 mg/l	Glyphosate is regarded as capable of only partial control. Control can sometimes be achieved, but may require a higher dose of other conditions	Mattson et al. 2004
	Aryloxyphenoxypropionate (Clincher)	0.5 mg/l	Approximately 60–70% survival rate at the lowest of diluted concentrations. The toxicological tests showed that *M. quadrifolia* is sensitive to a wide range of herbicides that have different mechanisms of action, such as the interruption of fatty acid	Bruni et al. 2013
	Sulfonamide + triazolopyrimidine (Viper)	20 ml/l	Approximately 80% survival rate at the lowest of diluted concentrations. The inhibition of enzymes involved in amino acid biosynthesis (Viper)	Bruni et al. 2013
	Cyclohexanone (Aura)	10 mg/l	Under 50% survival rate at the lowest of diluted concentrations	Bruni et al. 2013
	Penoxsulam	Not available	Penoxsulam is the only herbicide tested for its effectiveness in controlling this invasive species that is approved for aquatic use in Michigan. It was shown to be effective in the laboratory. This may contribute to the reduced weed density detected in rice fields	Hackett et al. 2018
Yellow floating heart *Nymphoides peltata*	Diquat	Not available	Limited scholarly literature information on direct usage of diquat on yellow floating heart but it is listed as one of the herbicides capable of consistent control over the species	Mattson et al. 2004

2,4-D	Not available	Regulation of the use of herbicides in or near water is stringent in Canada and there are no reports of use to control *N. peltata*. Countryman reported that "some success" was achieved in Lake Champlain, VT, using "various formulations" of 2,4-D	Mattson et al. 2004; Darbyshire and Francis 2008
Glyphosate	Not available	To completely eradicate the plant, all rhizome fragments in the sediment must be removed or killed with a herbicide like glyphosate. In some trials, the application to floating leaves between July and September resulted in about 40–50% control (for one season), but in others regrowth occurred before the end of the season	Darbyshire and Francis 2008; Warren 2009; Willey 2012
Dichlobenil	Not available	In the UK, dichlobenil has been effective in small localized areas within a larger waterbody, although in Canada conditions under which this agent can be used are currently limited. In the UK, it was recommended that no more than 20% of a waterbody be treated at any one time and this herbicide was not recommended for use in water flowing at more than 90 m/hour	EPA 1995; Darbyshire and Francis 2008
Florpyrauxifen-benzyl**	3 (PDU)/ac-f	55-acre infestation of YFH at a rate of 3 Prescription Dose Units (PDU)/ ac-f. Total surface coverage of yellow floating heart in the reservoir was reduced by more than 90% within 15 days after the treatment, and to less than 3 acres within 50 days after the treatment. No blooming flowers were observed after treatment and the surface coverage was close to 0% within 17 days after treatment in the cove. The effect of the herbicide treatment also appeared to carry over into the following growing season as the total surface coverage of yellow floating heart in the reservoir was less than 8 acres one year after the treatment in July 2020	Lamb et al. 2021
Endothall	Not available	A recent study on herbicide efficacy on *N. cristata* in south Florida showed endothall (dipotassium salt) was most effective, giving 98–100% control at 1.5–2.5 ppm. The longer term impacts of these treatments were not reported	Willey 2012

(Continued)

Invasive Species	Chemicals*	Effective Minimal Concentration	Notes	References
Common reed grass *Phragmites australis*	Glyphosate	25% strength rodeo (dosage not specified) + 1 l glyphosate + 2 l of water	Glyphosate and imazapyr are two broad-spectrum herbicides commercially available and known to control *Phragmites*. These herbicides are best applied in late summer/early fall after the plant has flowered either as a cut stump treatment or as a foliar spray. Together, glyphosate and imazapyr were found to be the only herbicides that resulted in greater than 90% biomass reduction of *Phragmites australis* in controlled mesocosm studies. However, it must be noted that these chemicals are non-selective and will impact native plants if they come in contact with the herbicides. Glyphosate should be applied after plants are in full bloom in late summer. Glyphosate is not as effective as imazapyr; however, it costs less and has good results with follow-up treatment or where water level management is available	Startford 1995; Avers et al. 2007; Hazelton et al. 2014
	Imazapyr	25% strength arsenal (dosage not specified)	Imazapyr should be applied to actively growing green foliage after full leaf elongation. If stand has substantial amount of old stem tissue, allow new growth to reach approximately 5 ft tall before treatment. In a long-term study, it was shown that imazapyr was 15% more successful than glyphosate in removing *Phragmites australis* as measured by the number of live stems, mean height, and percent cover remaining. No significant differences were observed between either herbicide treatments in regard to plant recolonization over the two-year time course	Startford 1995; Avers et al. 2007; Mozdzer et al. 2008
	Imazamox	280 g ai/ha	Compared to glyphosate and imazapyr, imazamox provided the lowest level of control	Knezevic et al. 2013
	Fluazifop-butyl	1 l fluazifop-butyl + 2 l of water	Increasing rates of glyphosate and fluazifop-butyl volumes of application increase the efficiency of the herbicides in the control of common reed. Rope-Wick method would better control compared with application of herbicide with sprayer. The cost of management of common reed has been reduced and glyphosate gave a good control on common reed compared to fluazifop-butyl	Al-Wagaa et al. 2019

Purple loosestrife *Lythrum salicaria*	Triclopyr	50 gallons spray solution per acre at 0.5 % solution (two quarts per 100 gallon solution)	Triclopyr formulated for use in wetlands applied as a foliar broadcast application at six to eight quarts per acre is most effective when purple loosestrife is in the bud to mid-bloom stage. This is when plants are actively growing, the toxic chemical is moved into roots, and when roots reserves needed to regenerate from defoliation are at their lowest level. The second optimum timing is in September when starches in the roots are being converted to sucrose important for cold tolerance and surviving freezing temperatures over winter. Annual application may be needed to target plants regrowing from root stocks and the seed bank. This herbicide is more selective and will not harm monocot species such as cattails	Jacobs 2008
	Glyphosate	2 quarts per acre as a 1% spray solution	Glyphosate herbicides are very effective for killing purple loosestrife. Glyphosate should be applied to actively growing plants and best results can be expected when it is applied during bloom. If applied at late bloom, plants may still produce seeds which can be clipped and bagged to prevent spread. One study found glyphosate most effective when applied in mid-August (late-bloom). Selective spot treatment can be used to avoid injuring neighboring non-target plants that can fill in the areas opened by removing the purple loose strife	Jacobs 2008
	2,4-D	1 quarter per acre	Broadcast applications of 2,4-D at one to two quarts per acre or a 0.5–1% solution applied to the bud or early bloom stage will kill top growth. Repeated applications will be needed to target plants regenerating from roots. An aquatic surfactant can be used to improve effectiveness	Jacobs 2008
	Imazapyr	1 pint per acre	A non-ionic surfactant with an aquatic label is needed for imazapyr to be effective	Jacobs 2008

Source: * Not all chemicals in this table are permitted in Massachusetts. More information about many chemicals' modes of actions, as well as their advantages and disadvantages, can be found in Table 1.6.

** Florpyrauxifen-benzyl (i.e., ProcellaCOR) is a relatively new chemical that has been used widely in selective control of milfoils and other aquatic invasive plants such as hydrilla, water hyacinth, and parrot-feather. However, there is very limited scholarly research.

REFERENCES

Abdel-Fattah, S. 2011. Aquatic Invasive Species Early Detection and Rapid Response- Assessment of Chemical Response Tools. International Joint Commission, Great Lakes Regional Office. 104 pp.

Aldridge, D.C., P. Elliott, and G.D. Moggridge. 2006. Microencapsulated BioBullets for the control of biofouling zebra mussels. *Environmental Science & Technology* 40: 975–979.

Al-Wagaa, A.H., I.A.H. Al-Obadui, H.A.K. Alfarttoosi, and O.A. AL-Gburi. 2019. Evaluating the performance of rope-wick herbicide applicator to control common reed. *IOP Conference Series: Earth and Environmental Science* 388: 012003.

Avers, B., R. Fahlsing, E. Kafcas, J. Schafer, T. Collin, L. Esman, E. Finnell, A. Lounds, R. Terry, J. Hazelman, J. Hudgins, K. Getsinger, and D. Schuen. 2007. A Guide to the Control and Management of Invasive Phragmites. Michigan Department of Environmental Quality, Lansing.

Barr, T.C.III., and J.M. Ditomaso. 2014. Curly leaf pondweed (*Potamogeton crispus*) turion control with acetic acid and benthic barriers. *Journal of Aquatic Plant Management* 52: 31–38.

Belgers, J.D.M., R.J. Van Lieverloo, L.J.T. Van Der Pas, and P.J. Van Der Brink. 2007. Effects of the herbicide 2,4-D on the growth of nine aquatic macrophytes. *Aquatic Botany* 86: 260–268.

Boone, C., C. Bond, K. Buhl, and D. Stone. 2012. Copper Sulfate. Oregon, USA: National Pesticide Information Center, Oregon State University. http://npic.orst.edu/factsheets/cuso4gen.html. (Accessed March 21, 2017).

Bruni, I., R. Gentili, F. De Mattia, P. Cortis, G. Rossi, and M. Labra. 2013. A multi-level analysis to evaluate the extinction risk of and conservation strategy for the aquatic fern *Marsilea quadrifolia* L. in Europe. *Aquatic Botany* 111: 35–42.

Bugbee, G.J., C.S. Robb, and S.E. Stebbins. 2020. Efficacy of diquat treatments on Brazilian waterweed, effects on native macrophytes and water quality: A case study. *Journal of Aquatic Plant Management* 58: 83–91.

Caudill, J., J.D. Madsen, and W. Pratt. 2021. Operational aquatic weed management in the California Sacramento–San Joaquin River Delta. *Journal of Aquatic Plant Management* 59: 112–122.

Center, T., F. Dray, Jr., G. Jubinsky, and M.J. Grodowitz. 1999. Biological Control of Water Hyacinth Under Conditions of Maintenance Management: Can Herbicides and Insects Be Integrated? *Environmental Management* 23: 241–256.

Cerveira, W.R., A.F. Silva, J.H.C. Cervoni, C. Cruz, and R.A. Pitelli. 2020. The addition of adjuvants on glyphosate enhances the control of aquatic plant *Myriophyllum aquaticum* (Vell.). *Planta Daninha* 38. DOI:10.1590/s0100–83582020380100090.

Claudi, R., A. Graves, A.C. Taraborelli, R.J. Prescott, and S. Mastitsky. 2012. Impact of pH on survival and settlement of dreissenid mussels. *Aquatic Invasions* 7: 21–28.

Cooper, N.L., and J.R. Bidwell. 2006. Cholinesterase inhibition and impacts on behavior of the Asian clam, *Corbicula fluminea*, after exposure to an organophosphate insecticide. *Aquatic Toxicology* 76: 258–267.

Costa, R., D.C. Aldridge, and G.D. Moggridge. 2008. Seasonal variation of zebra mussel susceptibility to molluscicidal agents. *Journal of Applied Ecology* 45: 1712–1721.

Dalu, T., R.J. Wasserman, M. Jordaan, W.P. Froneman, and O.L.F. Weyl. 2015. An assessment of the effect of rotenone on selected non-target aquatic fauna. *PLoS One* 10(11): e0142140.

Darbyshire, S.J., and A. Francis 2008. The biology of invasive alien plants in Canada. 10. Nymphoides peltata (S. G. Gmel.) Kuntze. *Canadian Journal of Plant Science* 88: 811–829.

Davis, E.A., W.H. Wong, and W.N. Harman. 2015a. Distilled white vinegar (5% acetic acid) as a potential decontamination method for adult zebra mussels. *Management of Biological Invasions* 6(4): 423–428.

Davis, E.A., W.H. Wong, and W.N. Harman. 2015b. Comparison of three sodium chloride chemical treatments for adult zebra mussel decontamination. *Journal of Shellfish Research* 34(3): 1029–1036.

Davis, E.A., W.H. Wong, and W.N. Harman. 2018. Toxicity of potassium chloride compared to sodium chloride for zebra mussel decontamination. *Journal of Aquatic Animal Health* 30: 3–12.

Dawson, V.K. 2003. Environmental fate and effects of the lampricide Bayluscide: A review, *Journal of Great Lakes Research* 29(Supplement 1): 475–492.

Densmore, C.L., L.R. Iwanowicz, A.P. Henderson, V.S. Blazer, B.M. Reed-Grimmett, and L.R. Sanders. 2018. An Evaluation of the Toxicity of Potassium Chloride, Active Compound in the Molluscicide Potash, on Salmonid Fish and Their Forage Base. *USGS Science for a Changing World* 1–46.

EPA R.E.D. 1995. *Diquat Dibromide*. The United States Environmental Protection Agency. https://archive.epa.gov/pesticides/reregistration/web/pdf/0288fact.pdf.

Fawad, M., H. Khan, B. Gul, M.A. Khan, and K.B. Marwat. 2015. Comparative effect of herbicidal and non-chemical control methods against water hyacinth. *Pakistan Journal of Weed Science Research* 21(4).

Fisher, S.W., P. Stromberg, K.A. Bruner, and L.D. Boulet. 1991. Molluscicidal activity of potassium to the zebra mussel, *Dreissena polymorpha*: Toxicity and mode of action. *Aquatic Toxicology* 20(4): 219–234.

Freeman M. 2010. Invasive snails pose threat if not eradicated. www.mailtribune.com/apps/pbcs.dll/article?AID=/20101030/NEWS/10300304. (Accessed March 21, 2017).

Getsinger, K.D., G.O. Dick, R.M. Crouch, and L.S. Nelson. 1994. Mesocosm evaluation of bensulfuron methyl activity on Eurasian watermilfoil, vallisneria, and American pondweed. *Journal of Aquatic Plant Management* 32: 1–6.

Getsinger, K.D., L.S. Nelson, L.M. Glomski, E. Kafcas, J. Schafer, S. Kogge, and M. Nurse. 2007. *Control of Phragmites in a Michigan Great Lakes Marsh*. Vicksburg, MS: U.S. Army Engineer Research and Development Center, p. 120.

Getsinger, K.D., E.G. Turner, J.D. Madsen, and M.D. Netherland. 1997. Restoring native vegetation in a Eurasian water-milfoil dominated plant community using the herbicide triclopyr. *Regulated Rivers: Research & Management: An International Journal Devoted to River Research and Management* 13(4): 357–375.

Giannotti, A.L., T.J. Egan, M.D. Netherland, M.L. Williams, and A.K. Knecht. 2014. Hydrilla Shows Increased Tolerance to Fluridone And Endothall In the Winter Park Chain of Lakes: Considerations for Resistance Management and Treatment Options. https://conference.ifas.ufl.edu/aw14/Presentations/Grand/Thursday/Session%209A/0850%20Giannotti.pdf

Glomski, L.M., and M.D. Netherland. 2008a. Effect of water temperature on 2, 4-D ester and carfentrazone-ethyl applications for control of variable-leaf milfoil. *Journal of Aquatic Plant Management* 46: 119–121.

Glomski, L.M., and M.D. Netherland. 2008b. Efficacy of fluridone, penoxsulam, and bispyribac-sodium on variable-leaf milfoil. *Journal of Aquatic Plant Management* 46(1): 193–196.

Gray, C.J., J.D. Madsen, R.M. Wersal, and K.D. Getsinger. 2007. Eurasian watermilfoil and parrotfeather control using carfentrazone-ethylethyl. *Journal of Aquatic Plant Management* 45: 43–46.

Haak, D.M., B.J. Stephen, R.A. Kill, N.A. Smeenk, and C.R. Allen. 2014. Toxicity of copper sulfate and rotenone to Chinese mystery snail (*Bellamya chinensis*). *Management of Biological Invasions* 5: 371–375.

Hackett, R.A., B.C. Cahill, and A.K. Monfils. 2018. *2018 Status and Strategy for European Water-clover (Marsilea quadrifolia L.) Management*. Lansing, MI: Michigan Department of Environmental Quality, p. 30.

Hall, J.F., H.E. Westerdahl, and T.J. Stewart. 1984. Growth response of *Myriophyllum spicatum* and *Hydrilla verticillata* when exposed to continuous low concentrations of fluridone. US Army Engineers Waterways Experimental Station, Vicksburg, MS, USA. Technical Report A-84–1.

Hanson, M.L., C.W. Knapp, and D.W. Graham. 2006. Field assessment of oxytetracycline exposure to the freshwater macrophytes *Egeria densa* Planch. and *Ceratophyllum demersum* L. *Environmental Pollution* 141: 434–442.

Hazelton, E.L.G., T.J. Mozdzer, D. Burdick, K.M. Kettenring, and D. Whigham D. 2014. Phragmites australis management in the United States: 40 years of methods and outcomes. *Annals of Botany Plants* 6.

Hesel, D.R., D.T. Gerber, and S. Engel. 1996. Comparing spring treatments of 2,4-D with bottom fabrics to control a new infestation of Eurasian watermilfoil. *Journal of Aquatic Plant Management* 34: 68–71.

Hofstra, D., P.D. Champion, and T.M. Dugdale. 2006. Herbicide trials for the control of parrotsfeather. *Journal of Aquatic Plant Management* 44: 13–18.

Hofstra, D.E., D. Clements, D.M. Rendle, and P.D. Champion. 2021. Response of fanwort (*Cabomba caroliniana*) to selected aquatic herbicides in New Zealand. *Journal of Aquatic Plant Management* 59(1): 35–39.

Hofstra, D.E., and J.S. Clayton. 2001. Evaluation of selected herbicides for the control ofexotic submerged weeds in New Zealand: I The use of endothall, triclopyr and dichlobenil. *Journal of Aquatic Plant Management* 39: 20–24.

Hunt, T.D., T.D. Dugdale, D. Clements, and M. Fridman. 2015. Concentration–exposure time relationships for controlling fanwort (*Cabomba caroliniana*) with endothall amine salt and carfentrazone. *Journal of Aquatic Plant Management* 53: 144–149.

Jacobs, J. 2008. Ecology and Management of Purple Loosestrife (*Lythrum salicaria* L.). Invasive Species Technical Note No. MT-21. United States Department of Agriculture, Natural Resources Conservation Service. www.nrcs.usda.gov/wps/portal/nrcs/detail/mt/home/?cid=nrcs144p2_056795

Knezevic, S.Z., R.E. Rapp, A. Datta, and S. Irmak. 2013. Common reed (*Phragmites australis*) control is influenced by the timing of herbicide application. *International Journal of Pest Management* 59(3): 224–228.

Kong, Y., M. Li, X. Shan, G.n. Wang, and G. Han. 2021. Effects of deltamethrin subacute exposure in snakehead fish, Channa argus: Biochemicals, antioxidants and immune responses. *Ecotoxicology and Environmental Safety* 209: 111821.

Lamb, B, T.A.A. McCrea, S.H. Stoodley, and A.R. Dzialowski. 2021. Monitoring and water quality impacts of an herbicide treatment on an aquatic invasive plant in a drinking water reservoir. *Journal of Environmental Management* 288: 112444.

Langeland, K.A., O.N. Hill, T.J. Koshnick, and W.T. Haller. 2002. Evaluation of a New Formulation of Reward Landscape and Aquatic Herbicide for Control of Duckweed, Waterhyacinth, Waterlettuce, and Hydrilla. *Journal of Aquatic Plant Management* 40: 51–53.

Lazur, A., S. Early, and J.M. Jacobs. 2011. Acute toxicity of 5% rotenone to northern snakeheads. *North American Journal of Fisheries Management* 26(3): 628–630.

Mattson, M.D., P.J. Godfrey, R.A. Barletta, and A. Aiello. 2004. Eutrophication and aquatic plant management in Massachusetts. Final Generic Environmental Report. Edited by Kenneth J. Wagner. Department of Environmental Protection and Department of Conservation and Recreation, EOEA Commonwealth of Massachusetts.

Miller, J. F, and G.W. Lewis. 1980. Duckweeds water hyacinth coontail bladderwort emersed aquatic weeds. In: *Identification and Control of Weeds in Southern Ponds*. The University of Georgia Cooperative Extension Service, College of Agriculture, p. 26.

Moffitt, C.M., K.A. Stockton-Fiti, and R. Claudi. 2016. Toxicity of potassium chloride to veliger and byssal stage dreissenid mussels related to water quality. *Management of Biological Invasions* 7(3): 257–268.

Mozdzer, T.J., C.J. Hutto, P.A. Clarke, and D.P. Field. 2008. Efficacy of imazapyr and glyphosate in the control of non-native phragmites australis. *Restoration Ecology* 16(2): 221–224.

Mustafa, J., R. Smolenski, and R. Sullivan. 2016. The Future of Coes Reservoir: An Investigation of Removal Methods for Water Chestnuts. IQP-DR2-WS15. Worcester Polytechnic Institute, p. 121.

Nelson, L.S., A.B. Stewart, and K.D. Getsinger. 2002. Fluridone effects on fanwort and water marigold. *Journal of Aquatic Plant Management* 40: 58–63.

Netherland, M.D. 2007. Evaluation of registered and EUP herbicides for control of variable milfoil. In *Proceedings of the Sixty-First Annual Meeting—Northeastern Weed Science Society*, p. 123.

Netherland, M.D., and K.D. Getsinger. 1995. Laboratory evaluation of threshold fluridone concentrations under static conditions for controlling hydrilla and Eurasian watermilfoil. *Journal of Aquatic Plant Management* 33: 33–36.

Netherland, M.D., K.D. Getsinger, and E.G. Turner. 1993a. Fluridone concentration and exposure time requirements for control of Eurasian watermilfoil and hydrilla. *Journal of Aquatic Plant Management* 31: 189–189.

Netherland, M.D., and L.M. Glomski. 2007. Evaluation of Aquatic Herbicides for Selective Control of Variable Milfoil (*Myriophyllum heterophyllum* Michx). Final Report to the New Hampshire Dept. of Environ. Services, p. 96.

Netherland, M.D., W.R. Green, and K.D. Getsinger. 1991. Endothall concentration and exposure time relationships for the control of Eurasian watermilfoil and hydrilla. *Journal of Aquatic Plant Management* 29: 61–67.

Netherland, M.D., W.R. Green, and K.D. Getsinger. 1993b. Effects of fluridone on hydrilla growth and reproduction. *Journal of Aquatic Plant Management* 29: 189–194.

OARS. 2017. *Water chestnut management guidance & five-year management plan for the Sudbury, Assabet, and Concord River watersheds*. Concord, MA: OARS "Planning and Guidance for Water Chestnut Management in the SuAsCo Watershed".

Ortiz, M.F., S.J. Nissen, and C.J. Gray. 2019. Endothall behavior in *Myriophyllum spicatum* and *Hydrilla verticillate*. *Pest Management Science* 75(11): 2942–2947.

Parka, S., and W. Arnold. 1979. Fluridone, a new herbicide for aquatic plant management. *Journal of Aquatic Plant Management* 17: 27–30.

Patten, K. 2003. Evaluating Imazapyr in Aquatic environments. *Agricultural and Environmental News* 205.7 pp.

Poovey, A.G., and K.D. Getsinger. 2007. Subsurface applications of triclopyr and 2,4-D amine for control of water chestnut (*Trapa natans* L.). *Journal of Aquatic Plant Management* 45: 63–66.

Poovey, A.G., and K.D. Getsinger. 2010. Comparative response of monoecious and dioecious hydrilla to endothall. *Journal of Aquatic Plant Management* 48: 15–20.

Poovey, A.G., J.G. Skogerboe, and C.S. Owens. 2002. Spring treatments of diquat and endothall for curlyleaf pondweed control. *Journal of Aquatic Plant Management* 40: 63–67.

Rector, P.R., P.J. Nitzsche, and S.S. Mangiafico. 2015. Temperature and herbicide impacts on germination of water chestnut seeds. *Journal of Aquatic Plant Management* 53: 105–112.

Richardson, R.J. 2008. Aquatic plant management and the impact of emerging herbicide resistance issues. *Weed Technology* 22(1): 8–15.

Richardson, R.J., E.J. Haug, and M.D. Netherland. 2016. Response of seven aquatic plants to a new arylpicolinate herbicide. *Journal of Aquatic Plant Management* 54: 26–31.

Rosa, I., R. Garrido, A. Ré, J. Gomes, J. Peirera, F. Gonçalves, and R. Costa. 2015. Sensitivity of the invasive bivalve *Corbicula fluminea* to candidate control chemicals: The role of dissolved oxygen conditions. *Science of the Total Environment* 536: 825–830.

Santos Ltd. 2021. (rep.). *Qualitative Tier 2 Assessment* (PolyDADMAC). EHS Support.

Shrivastava, B., A. Shrivastava, A. Kumar, J.L. Bhatt, S.P. Bajpai, S.S. Parihar, and V. Bhatnaga. 2011. Impact of deltamethrin on environment, use as an insecticide and its bacterial degradation—A preliminary study. *International Journal of Environmental Sciences* 1(5): 977–985.

Skogerboe, J.G., and K.D. Getsinger. 2002. Endothall species selectivity evaluation: Northern latitude aquatic plant community. *Journal of Aquatic Plant Management* 40: 1–5.

Skogerboe, J.G., and M.D. Netherland. 2008. Draft Report Following April 2008 Aquatic Herbicide Treatments of Three Bays on Lake Minnetonka. October. Report to the Minnesota Department of Natural Resources.

Stinson, H. 2018. The Northern Snakehead, Channa Argus, as an Invasive Species. *Eukaryon*, Lake Forest College News.

Thum, R.A., S. Parks, J.N. Mcnair, P. Tyning, P. Hausler, L. Chadderton, A. Tucker, and A. Monfils. 2017. Survival and vegetative regrowth of Eurasian and hybrid watermilfoil following operational treatment with auxinic herbicides in Gun Lake, Michigan. *Journal of Aquatic Plant Management* 55: 103–107.

Turner, L., S. Jacobson, and L. Shoemaker. 2007. Risk Assessment for Piscicidal Formulations of Rotenone. Compliance Services International, p. 104.

Valley, R.D., W. Crowell, C.H. Welling, and N. Proulx. 2006. Effects of a low-dose fluridone treatment on submersed aquatic vegetation in a eutrophic Minnesota lake dominated by Eurasian watermilfoil and coontail. *Journal of Aquatic Plant Management* 44(1): 19–25.

Van Scoy, A, A. Pennell, X. Zhang. 2016. Environmental fate and toxicology of dimethoate. *Reviews of Environmental Contamination and Toxicology* 237: 53–70.

Wagner, K.I., J. Hauxwell, P.W. Rasmussen, F. Koshere, P. Toshner, K. Aron, D.R. Helsel, S. Toshner, S. Provost, M. Gansberg, J. Masterson, and S. Warwick. 2007. Whole-lake herbicide treatments for Eurasian watermilfoil in four Wisconsin lakes: effects on vegetation and water clarity. *Lake and Reservoir Management* 23(1): 83–94.

Waltham, N.J., and S. Fixler. 2017. Aerial herbicide spray to control invasive water hyacinth (*Eichhornia crassipes*): Water quality concerns fronting fish occupying a tropical floodplain wetland. *Tropical Conservation Science* 10: 1–10.

Warren, D. 2009. Invasive Species Management Plan for Oswego Lake. Planning & Building Services, City of Oswego, OR. 17 pp.

Wersal, R.M., and J.D. Madsen. 2007. Comparison of imazapyr and imazamox for control of parrotfeather (*Myriophyllum aquaticum* (Vell.) Verdc.). *Journal of Aquatic Plant Management* 45: 132–136.

Wersal, R.M., and J.D. Madsen. 2010. Comparison of subsurface and foliar herbicide applications for control of parrotfeather (*Myriophyllum aquaticum*). *Invasive Plant Science and Management* 3: 262–267.

Wersal, R.M., J.D. Madsen, J.H. Massey, W. Robles, and J.C. Cheshier. 2010. Comparison of Daytime and Night-time Applications of Diquat and Carfentrazoneethyl for Control of Parrotfeather and Eurasian Watermilfoil. *Journal of Aquatic Plant Management* 48: 56–58.

Willey, L.N. 2012. Biology and control of the invasive aquatic plant crested floating heart (*Nymphoides cristata*) (Thesis, University of Florida).

Wisconsin Department of Natural Resources (DNR). 2012a. *Endothall Chemical Fact Sheet*. Madison, WI. https://dnr.wi.gov/lakes/plants/factsheets/FluridoneFactsheet.pdf

Wisconsin Department of Natural Resources (DNR). 2012b. *Fluridone Chemical Fact Sheet*. Madison, WI. https://dnr.wi.gov/lakes/plants/factsheets/FluridoneFactsheet.pdf

Wright, B., H. Landis, and T. Nelson. 2018. Chemical Management of Hydrilla for Drinking Water Utilities. Project #4747, Water Research Foundation, Denver, Colorado, p. 120.

Appendix IV

Laws and Statutes Pertaining to Aquatic Invasive/Nuisance Species in the New England States and New York

This document serves as an overview of the major laws/statutes pertaining to aquatic species, wetlands, and invasive species, within the six New England States and New York. The laws are cited in the following tables by title, chapter, and section. The section titles are shown, as well as a summary and/or the full or abbreviated version of the section referenced.

CONNECTICUT

INVASIVE SPECIES

Title/Chapter/ Section	Section Title	Summary	Penalty
Title 15. Ch. 268 § 15–180	Transporting vessel or trailer without inspecting for and properly removing and disposing of vegetation and aquatic invasive species. Penalty	• No person shall transport a vessel, as defined in section 15–127, or any trailer used to transport such vessel, in the state without first inspecting such vessel for the presence of vegetation and aquatic invasive species, as determined by the commissioner, and properly removing and disposing of any such vegetation and aquatic invasive species that are visible and identifiable without optical magnification from such vessel or trailer	Any person who violates the provisions of this section shall be fined a maximum $100 for each violation
Title 22a. Ch. 446i § 22a–381a	Duties and recommendations of the Invasive Plants Council	• To developed programs to educate the public on the issues of invasive plants; make recommendations, to control the spread of invasive plants and to make information available, publish updates	
Title 22a. Ch. 446i § 22a–381c	Prohibition on purchase of invasive or potentially invasive plants by state agencies	• No state agency, department or institution shall purchase any plant listed as invasive or potentially invasive	
Title 22a. Ch. 446i § 22a–381d	Prohibited actions on certain invasive plants. Exceptions. Municipal ordinances prohibited. Penalty	• List of 67 invasive species as well as 19 other invasive plants that no person shall import, move, sell, purchase, transplant, cultivate or distribute	Any person who violates the provision of this section will be find a maximum of $100 per plant

WETLANDS

Title/Chapter/ Section	Section	Summary	Penalty
Title 22a. Ch. 446i § 22a–381d	Prohibited actions re certain invasive plants. Exceptions. Municipal ordinances prohibited. Penalty	• No person shall import, move, sell, purchase, transplant, cultivate or distribute any of the following invasive plants: Curly leaved Pondweed (*Potamogeton crispus*); (2) fanwort (*Cabomba caroliniana*); (3) eurasian watermilfoil (*Myriophyllum spicatum*); (4) variable watermilfoil (*Myriophyllum heterophyllum*); (5) water chestnut (*Trapa natans*); (6) egeria (*Egeria densa*); (7) hydrilla (*Hydrilla verticillata*); (8) common barberry (*Berberis vulgaris*); (9) autumn olive (*Elaeagnus umbellata*) etc.	Any person who violates the provisions of this section shall be fined a maximum of $100 per plant

RELEVANT ACTS

Act	Summary	Year
Public Act 10–20: To authorize conservation officers to enforce certain prohibitions concerning invasive plants	• Any conservation officer, special conservation officer or patrolman may, within the boundaries of the state, may examine the contents of any boat, ship, automobile or other vehicle, box, locker, basket, creel, crate, game bag or game coat or other package in which he has probable cause to believe that any fish, crustacean, bird, or quadruped is being kept, in violation of any said statutory provisions or any regulation • Note: Section 1. Section 26–6 of the general statutes is repealed, and the following is substituted in lieu thereof (*Effective October 1, 2010*)	2010
Public Act 12–167: An act requiring the inspection of vessels and vessel trailers for aquatic invasive species	• Section 1. Section 15–180 of the general statutes is repealed, and the following is substituted in lieu thereof (Effective July 1, 2012) • No person shall transport a vessel, as defined in section 15–127, or any trailer used to transport such vessel, in the state without first inspecting such vessel for the presence of vegetation and aquatic invasive species, as determined by the commissioner, and properly removing and disposing of any such vegetation and aquatic invasive species that are visible and identifiable without optical magnification from such vessel or trailer	2012
Public Act No. 21–12 An act concerning aquatic invasive species effects on lakes and related funding, certain group fishing events and eliminating the season limitation for the taking or trout	• Section 1. Section 22a-339h of the general statutes is repealed and the following is substituted in lieu thereof (Effective October 1, 2021): • Section 2. Not later than October 1, 2021, the Commissioner of Energy and Environmental Protection shall submit a report, in accordance with section 11–4a of the general statutes having cognizance of matters relating to the environment on the amount of revenue collected from the Aquatic Invasive Species fees authorized pursuant to section 14–21bb of the general statutes and the number, value and purpose of any grants to municipalities and not-for-profit organizations awarded from the Connecticut Lakes, Rivers and Ponds Preservation account in accordance with section 14–21aa of the general statutes	2021

MAINE

Many of Maine's revised statutes have been repealed. As a result, as of August 2021, this section represents a summary of available information; some of the relevant laws that have been repealed as of the access date, August 5, 2021 are also noted.

INVASIVE SPECIES

Title/ Chapter/ Section	Section Title	Summary	Penalty
Title 38. Ch. 3 § 410-N	Aquatic Nuisance Species Control	• Defines terms such as Aquatic Plant, and Invasive Aquatic Plant • Department shall prepare educational materials that inform the public about problems associated with invasive aquatic plants, how to identify invasive aquatic plants, why it is important to prevent the transportation of aquatic plants and the prohibitions relating to aquatic plants contained in section 419-C • The department shall investigate and document the occurrence of invasive aquatic plants in state waters and may undertake activities to control invasive aquatic plant populations	
Title 38. Ch. 3 § 419-C	Prevention of the spread of invasive aquatic plants	No person may: • Transport any aquatic plant or parts of any aquatic plant on the outside of a vehicle, boat, personal watercraft, boat trailer or other equipment on a public road • Possess, import, cultivate, transport, or distribute any invasive aquatic plant or parts of any invasive aquatic plant in a manner that could cause the plant to get into any state water • Sell or offer for sale in this State any invasive aquatic plant • Failure to remove any aquatic plant or parts of any aquatic plant, from the outside of a vehicle, boat, personal watercraft, boat trailer or other equipment on a public road	A person who violates this section commits a civil violation for which a forfeiture not to exceed $500 may be adjudged for the first violation and a forfeiture not to exceed $2,500 may be adjudged for a subsequent violation
Title 38. Ch 20-A § 1862	Program to prevent infestation of and to control invasive aquatic plants	• Program: The commissioner and commissioner of Inland Fisheries and Wildlife jointly shall implement to inspect watercraft, watercraft trailers and outboard motors at or near the border of the State and at boat launching sites for the presence of invasive aquatic plants and to provide educational materials to the public and to watercraft owners regarding invasive aquatic plants • The program will establish inspection stations at boat launching sites • The program will provide informational material	
Title 38. Ch 20-B §1872	Action plan to protect State's inland waters	• The task force shall also recommend to the department an action plan to protect the State's inland waters from invasive aquatic plants and nuisance species. This includes the identification of inland waters known to be infested, vulnerability assessment, lake monitoring program, response program, training, and public information material	
Title 38. Ch. 3 § 423	Discharge of waste from watercraft	• Discharge from watercraft prohibited. A person, firm, corporation, or other legal entity may not discharge, spill, or permit to be discharged sewage, septic fluids, garbage, or other pollutants from watercraft: Into inland waters of the State; On the ice of inland waters and on the banks of inland waters of the State	Penalty. Notwithstanding section 349, subsection 2, a person who is charged with a civil violation of this section is subject to a civil penalty, payable to the State, of not less than $500 and not more than $10,000 for each day of that violation

WETLANDS

Title/Chapter/Section	Section Title	Summary
Title 5. Ch. 357 § 6401	Water Resources Planning Committee	• Established in the department of agriculture, Conservation and Forestry who will plan for sustainable use of water resources focused on water withdrawal activities, watersheds at risk climate change, and review state of policy with regards to conserving water resources, effects of surface water quality improvements on water withdraws opportunities
Title 12. Ch. 200 § 401	Maine's rivers	• Rivers and streams are a natural resource, As the state has 32,000 miles of rivers and streams • Importance's of increasing the value and use of river and streams • Implementing policies that recognizes the importance of rivers and streams
Title 12. Part 5. Ch. 421 Wetlands		Repealed
Title 38. Ch. 3 § 401-L	Lake Assessment and Protection Program. Established	• Established within the department to monitor and protect the health and integrity of the State's Lakes
Title 38. Ch. 3 § 410-M	Lake Assessment and Protection	• The Lake Assessment and Protection Program (LAPP) will provide education and technical assistance relating to lake function and values, watershed planning and management to implement applicable laws and rule • The LAPP will conduct resource monitoring and research and promote the natural resources protections laws/protection of lakes laws

RELEVANT ACTS

Title/Chapter/ Section	Section Title	Summary	Year
Title 38. Ch. 3 § 480. Natural Resource Protection Act	A finding; purpose; short title	• The State's rivers and streams, great ponds, fragile mountain areas, freshwater wetlands, significant wildlife habitat, coastal wetlands and coastal sand dunes systems are resources of state significance • The Legislature further finds and declares that there is a need to facilitate research, develop management programs and establish sound environmental standards that will prevent the degradation of and encourage the enhancement of these resources • The Legislature further finds and declares that the cumulative effect of frequent minor alterations and occasional major alterations of these resources poses a substantial threat to the environment and economy of the State and its quality of life	1987

MASSACHUSETTS

INVASIVE SPECIES

Title/Chapter/ Section	Section Title	Summary	Penalty
Title 2. Ch. 21 § 37B	Aquatic Nuisance Control Program (ANCP)	• The Department of Environmental Management will establish and maintain the ANCP and work with the department of fish and game to promulgate rules and regulation to suppress, eradicate/control Aquatic nuisances. Study and promote methods of suppressing/controlling/ reducing the spread of aquatic nuisances • The ANCP shall receive and respond to complaints, work with municipalities, local interest organizations and agencies to develop log-range program to control aquatic nuisances • (f) Except as otherwise authorized by the department, no person shall knowingly or intentionally place, or cause to be placed, an aquatic nuisance in or upon inland waters. Mass. Ch. 21 §37B • (g) Except as otherwise authorized by the department, no person shall place in or upon inland waters any vessel that has an aquatic nuisance attached to the vessel unless it has been cleaned/decontaminated or treated to kill/remove the Aquatic nuisance in accordance with the department's regulations	• (h) Whoever: (1) violates subsection (f); (2) knowingly and willfully resists or obstructs the department/ authorized employees/ agents from suppressing or eradicating aquatic nuisances shall be subject to a civil assessment of not more than $5,000 for each violation; each day that such violation occurs or continues shall be deemed a separate violation • (3) Violate subsection (g) or knowingly violates any rule, regulation, order, or quarantine issued by the commissioner pursuant to this section shall be subject to the fines and sanctions set forth in section 5D of chapter 90B (Mass. Ch. 21 §37B)
Title 26. Ch. 111 § 5E	Application of chemicals to control aquatic nuisances; licenses; rules and regulations; violations; penalties; applicability of section	• No one may use chemicals in a waterbody to control aquatic nuisances without obtaining a license from DEP. After a public hearing, DEP may establish rules for licenses and the application of chemicals to control aquatic nuisances	• The regulations may include penalties not exceeding $500 per offense. Any unlicensed person who applies chemicals to a waterbody to control aquatic nuisances shall be fined a minimum of $25 and a maximum of $500
Title 19. Ch. 128 § 20A	Spread or growth of plants of water chestnuts (*Trapa natans*)	• Spread or growth of plants of water chestnuts (*Trapa natans*): no one many knowingly transplant, transport, plant, or traffic in water chestnuts or its seed or nuts or cause the water chestnuts' growth or spread	• Violations: Anyone who knowingly transplants, transports, plants, or traffics in water chestnut, or its seeds or nuts, or causes the water chestnuts' growth or spread, shall be fined a minimum of $10 and a maximum of $100 (Mass. Gen. Laws Ch. 128, § 29)

Title/Chapter/ Section	Section Title	Summary	Penalty
Title 19. Ch. 131 § 40	Removal, fill, dredging or altering of land bordering waters	• No person shall remove, fill, dredge or alter any bank, riverfront area, freshwater wetland, coastal wetland, beach, dune, flat, marsh, meadow or swamp bordering on the ocean or on any estuary, creek, river, stream, pond, or lake, or any land under said waters or any land subject to tidal action, coastal storm flowage, or flooding, other than while maintaining, repairing, or replacing	• Whoever violates any provision of this section, (a) shall be punished by a fine of not more than twenty-five thousand dollars or by imprisonment for not more than two years, or both such fine and imprisonment; or (b), shall be subject to a civil penalty not to exceed twenty-five thousand dollars for each violation
Title 19. Ch 132 § 11	Suppression of moths, caterpillars, worms, and beetles and any invasive plant or animal species rules and regulation contracts: studies for control of public nuisances: annual recommendations	• The chief superintended, with the approval of the commissioner makes rules and regulation that governs all operation in cities, towns, or person for the purpose of suppressing moths, caterpillars, worms and beetles and any invasive plant or animal species • The chief superintended may make contracts on behalf of the commonwealth. Study and promote improved methods of suppression such as public nuisances • The chief superintended may conduct investigation, gather, and distributed information concerning public nuisances	

Fish and Mammals and Other

Title/Chapter/ Section	Section Title	Summary	Penalty
Title 19. Ch. 131 § 26	Importation of fish, birds, mammals, reptiles, or amphibians; shipments; tags; violations; seizure and disposition	• Any animal that is lawfully taken/lawfully propagated without the commonwealth may be purchased by any dealer licensed under section 23. the Export and the sale must be lawful in the state, province/country in which the animal was taken from. All shipments need to bear the name of the consignee, consignor or to the carton, package, box/ crate. A mark of identification is required by state/province/country. • The dealer purchasing/dealing any bird/ mammal before offering the same for sale, required a number tag as discussed in section 25. Any tags required hereunder for fish shall be furnished by the director at cost	• Any fish, bird, mammal, reptile, or amphibian possessed, shipped, transported, or delivered in violation of this section or of any rule or regulation made under the authority thereof may be seized and shall be disposed of by the director of law enforcement for the best interest of the commonwealth.
Title 19. Ch. 131 § 19	License to put fish or spawn into inland waters; permit to import live fish or viable eggs, certification, inspection for disease, and disposition	• No one may put fish or spawn thereof into inland waters without a license issued under Ch. 131 section 23, or with the director's written approval • No one may bring, or cause to be brought into Massachusetts any live fish or viable fish eggs without a permit issued by the director	• Anyone who violates Ch. 131, § 19 shall be fined a minimum of $100 or a maximum of $500, or be imprisoned for a maximum of six months, or both fined and imprisoned (Mass. Gen. Laws Ch. 131, § 90)

WETLANDS

Title/Chapter/Section	Section Title	Summary	Penalty
Title 19. Ch. 131 § 40	Removal fill dredging or altering of land boarding waters	No person shall remove, fill, dredge or alter any bank, riverfront area, freshwater wetland, coastal wetland, beach, dunce, flat, march, meadow or swam, bordering on the ocean or on any estuary, creek, river, stream pond, lake, or any land under said waters or any land subject to tidal action, coastal storm flowage, or flooding	

RELEVANT ACTS

Title/Chapter/Section	Summary	Penalty	Year
Title 19. Ch. 131 § 40. The Wetland Protection Act	• This act sets forth a public review and decision-making process by which activities affecting areas subject to protection under M.G.L. c. 131 section 40 are to be regulated to protect and control: public and private water supply, ground water supply, fisheries and wildlife habitat, control flooding, and prevent storm damage, pollution and land containing shellfish	• Penalties can be found for Ch. 131 in Section 90	2017

NEW HAMPSHIRE STATUTES

INVASIVE SPECIES

Title/Chapter/Section	Section Title	Summary	Penalty
Title 50. Ch. 487 § 43	Aquatic invasive species decal	• No person shall operate a powerboat registered in another state on New Hampshire public waters without displaying a New Hampshire aquatic invasive species decal from the department of environmental services	Any person who violates this section shall be guilty of a violation punishable by a fine of $50 for a first offense, $100 for a second offense, and $250 for any subsequent offense
Title 50. Ch. 487 § 16-a	Exotic aquatic weed prohibition	• No exotic aquatic weeds shall be offered for sale, distributed, sold, imported, purchased, propagated, transported, or introduced in the state. The commissioner may exempt any exotic aquatic weed from any of the prohibitions of this section consistent with the purpose of this subdivision	It shall be unlawful to offer for sale, distribute, sell, import, purchase, propagate, negligently transport, or introduce exotic aquatic weeds into New Hampshire waterbodies. Notwithstanding RSA 487:7, any person engaging in such an activity shall be guilty of a violation (RSA 487: 16-b)
Title 50. Ch. 487 § 16-c	Transport of aquatic plants or exotic aquatic weeds on outside of boats, vehicles, and equipment	• No person shall negligently transport any aquatic plants or plant parts, or exotic aquatic weed or weed parts to or from any New Hampshire waters on the outside of a vehicle, boat, ski craft as defined in RSA 270:73, trailer, or other equipment	Any person who violates RSA 487:16-c through 487:16-e shall be guilty of a violation punishable by a fine of $50 for a first offense, $100 for a second offense, and $250 for any subsequent offense (RSA 487: 16-f)

Title/Chapter/Section	Section Title	Summary	Penalty
Title 50. Ch. 487 § 16-d	Draining of water conveyances	• When leaving waters of the state, a person shall drain their boat and other water-related equipment that holds water, including live wells and bilges • Devices used to control the draining of water from ballast tanks, bilges, and live wells shall be removed or opened while transporting boats and other water-related equipment • Commercial enterprises shall drain all water-related equipment holding water and live wells and bilges prior to transporting the equipment to another waterbody	Any person who violates RSA 487:16-c through 487:16-e shall be guilty of a violation punishable by a fine of $50 for a first offense, $100 for a second offense, and $250 for any subsequent offense (RSA 487: 16-f)
Title 50. Ch 487 § 29	Milfoil and Other Exotic Aquatic Plants Prevention; Grant Fund Report and Budget	• The department of environmental services shall submit an annual report shall include, but not be limited to, a description of prevention and research projects funded by the milfoil and other exotic aquatic plants prevention program and the extent of aid to municipalities or subdivisions of the state, non-profit corporations, and research institutions	
Title 50. Ch 487 § 30	Exotic Aquatic Weeds and Species Committee	• It is a committee to study exotic aquatic weeds and exotic aquatic species of wildlife in the state of NH • The committee will also aid the department of environmental services in the control and eradication milfoil	

FISH AND MAMMALS AND OTHER

Title/Chapter/Section	Section Title	Summary	Penalty
Title 18. Ch. 207 § 22-c	Wildlife damage control program; Administration	• The general wildlife damage mitigation program will shall address conflicts between wildlife and human populations by disseminating educational and technical information and providing assistance	
Title 18. Ch. 207 § 14	Import, possession, or release of wildlife	• No person shall import, possess, sell, exhibit, or release any live marine species or wildlife, or the eggs or progeny thereof, without first obtaining a permit from the executive director except as permitted under title XVIII	Any person who violates this section or any rule adopted under this section shall be guilty of a violation and guilty of an additional violation for each marine species or wildlife possessed contrary to the provisions of this section

WETLANDS

Title/Chapter/Section	Section Title	Summary
Title 50. Ch. 483 § 1	Statement of policy	• New Hampshire rivers are valued ecologic, economic, public health and safety, and social assets for the benefit of present and future generations • The state shall encourage and assist in the development of river corridor management plans and regulate the quantity and quality of instream flow along certain protected rivers or segments of rivers to conserve and protect outstanding characteristics including recreational, fisheries, wildlife, environmental, hydropower, cultural, historical, archaeological, scientific, ecological, aesthetic, community significance, agricultural, and public water supply so that these valued characteristics shall endure as part of the river uses to be enjoyed by New Hampshire people • If conflicts arise in the attempt to protect all valued characteristics within a river or stream, priority shall be given to those characteristics that are necessary to meet state water quality standards
Title 50. Ch. 483 § 3	Program established: Intent	• It is the intent of the legislature that the New Hampshire lakes management and protection program shall complement and reinforce existing state and federal water quality laws • The scenic beauty and recreational potential of lakes shall be maintained or enhanced, that wildlife habitat shall be protected, that opportunity for public enjoyment of lake uses be ensured, and that littoral interests shall be respected
Title 50.485 § 1	Declaration of purpose	• The purpose of this chapter is to protect water supplies, to prevent pollution in the surface and groundwaters of the state and to prevent nuisances and potential health hazards. In exercising all powers conferred upon the department of environmental services under this chapter, the department shall be governed solely by criteria relevant to the declaration of purpose set forth in this section

RELEVANT ACTS

Chapter and Section	Section Title	Summary	Year
Title 50 Ch. 482 § 1. Fill and Dredge in wetlands or Wetland Act	Finding public's purpose	• This act protects and preserves its submerged lands under tidal and fresh waters and its wetlands (both saltwater and freshwater), from despoliation and unregulated alteration as it will adversely affect the value of such areas as sources of nutrients for finfish, crustacea, shellfish and wildlife of significant value, will damage or destroy habitats and reproduction areas for plants, fish and wildlife of importance, will eliminate, depreciate, or obstruct the commerce, recreation and aesthetic enjoyment of the public, will be detrimental to adequate groundwater levels, will adversely affect stream channels and their ability to handle the runoff of waters, will disturb and reduce the natural ability of wetlands to absorb flood waters and silt, thus increasing general flood damage and the silting of open water channels, and will otherwise adversely affect the interest of the general public	1990

NEW YORK

INVASIVE SPECIES

Chapter/Article/Section	Section Title	Summary
Ch. 37. Article 3 § 33-C	Regulating disposal of sewage; littering of waterways	Regulating disposal of sewage; littering of waterways: • No person, whether engaged in commerce or otherwise, shall place, throw, deposit, or discharge, or cause to be placed, thrown, deposited, or discharged into the waters of this state, from any watercraft, marina or mooring, any sewage, or other liquid or solid materials which render the water unsightly, noxious, or otherwise unwholesome to be detrimental to the public health or welfare or to the enjoyment of the water for recreational purposes
Ch. 37. Article 3 § 35-D	Aquatic invasive species signs at public boat launches	• The department of environmental conservation shall design and establish universal signage which may be posted at any access point to the navigable waters of the state relating to the threat and mitigation of aquatic invasive species • Commencing within one year of the effective date of this section, owners of each public boat launch shall conspicuously post the universal sign of not less than eighteen inches by twenty-four inches
Title 6, Ch. 5, Subchapter C., Part 575.1	Purpose, scope, and applicability	• Establish procedures to identify and classify invasive species and to establish a permit system to restrict the sale, purchase, possession, propagation, introduction, importation, and transport of invasive species in New York
Title 6, Ch. 5, Subchapter C., Part 575.4	Regulated Invasive Species	• No person shall knowingly introduce into a free-living state or introduce by a means that one knew or should have known would lead to the introduction into a free-living state any regulated invasive species, although such species shall be legal to possess, sell, buy, propagate, and transport
Title 6, Ch. 5, Subchapter C., Part 575.9	Invasive Species Permits	• A person may possess, with the intent to sell, import, purchase, transport or introduce, a prohibited invasive species if the person has been issued a permit by the department for research, education, or other approved activity, under this Part • A person may only introduce a regulated invasive species into a free-living state if the person is issued a permit under this section expressly authorizing such introduction. Written application are required, noted criteria must be met
Title 6, Ch. 5, Subchapter C., Part 575.10	Penalties	• Any person who violates this Part or any license or permit or order issued by the department, or Department of Agriculture and Markets, pursuant to section 9–1709 of the Environmental Conservation Law or pursuant to the provisions of this Part shall be liable for all penalties and other remedies provided for in the Environmental Conservation Law including section 71–0703(9). With respect to Eurasian boars, any person who violates this Part shall also be liable for all penalties provided for in the Environmental Conservation Law, including section 71–0925. Such penalties and remedies may be in addition to any other penalty or remedy available under any other law, including but not limited to, permit revocation

Fish and Mammal and Other

Chapter/Article/Title/Section	Section Title	Summary
Ch. 43-B. Article 11 Title 14 § 1703	Importation, possession, and sale of fish without license of permit; prohibitions	• No person shall possess or transport into the state any fish except eels caught in that part of Missisquoi Bay in Lake Champlain lying in the Province of Quebec or in the Richelieu River

Wetlands

Chapter/Article/Title/Section	Section Title	Summary
Ch. 43-B. Article 15. Title 1 § 0105	Declaration of policy	It is public policy of the state of New York that: • The regulation and control of the water resources of the state of New York be exercised only pursuant to the laws of this state • The waters of the state be conserved and developed for all public beneficial uses
Ch. 43-B. Article 17. Title 5 § 0501	General prohibition against pollution	• It shall be unlawful for any person, directly or indirectly, to throw, drain, run, or otherwise discharge into such water's organic or inorganic matter that shall cause or contribute to a condition in contravention of the standards adopted by the department pursuant to section 17–0301
Ch. 43-B. Article 24 Title 1 § 0103	Declaration of policy	• It is declared to be the public policy of the state to preserve, protect and conserve freshwater wetlands and the benefits derived therefrom, to prevent the despoliation and destruction of freshwater wetlands, and to regulate use and development of such wetlands to secure the natural benefits of freshwater wetlands, consistent with the general welfare and beneficial economic, social, and agricultural development of the state
Ch. 43-B. Article 24. Title 7 § 0701	Permits	• Activities subject to regulation shall include any form of draining, dredging, excavation, removal of soil, mud, sand, shells, gravel, or other aggregate from any freshwater wetland, either directly or indirectly; and any form of dumping, filling, or depositing of any soil, stones, sand, gravel, mud, rubbish or fill of any kind, either directly or indirectly, erecting any structures, roads, the driving of pilings, or placing of any other obstructions whether changing flow of the water; any form of pollution and any other activity which substantially impairs any of the several functions served by freshwater wetlands

RELEVANT BILLS

Information	Title	Summary
Senate Bill S6858 (2011–2012 Legislative Session)	Requires Publication of Laws Relating to the Control of Invasive Species	*Purpose:* • "The purpose of this bill is to provide sportspersons with information on legislation intended to reduce the effects of invasive species and to prevent further invasions." *Summary of specific provisions:* • "This bill would require the Department of Environmental Conservation (DEC) to include in their existing publications information on legislation affecting the eradication, suppression, reduction, or management of invasive species."
Assembly Bill A7692 (2019–2020 Legislative Session)	Extends provisions relating to the Department of Environmental Conservation's Management of Aquatic Invasive Species, Spread Prevention and Penalties	• Extends provisions relating to the Department of Environmental Conservation's Management of Aquatic Invasive Species, Spread Prevention and Penalties
Senate Bill S7010C (2012–2022 Legislative Session	Authorizes the establishment of aquatic invasive species inspection stations in Adirondack Park to inspect motorized watercraft for the presence of organisms or organic material that may harbor invasive species	• Authorizes self-issuing certification; directs education and outreach; makes permanent certain provisions of law relating to aquatic invasive species spread prevention • To make permanent the requirement that boaters take reasonable precautions to clean, drain and dry their watercraft before launching in New York State Waters, and to grant the Department of Environmental Conservation with the authority to conduct inspections and decontaminations of watercraft for the prevention of invasive species within the Adirondack Park or a 10-mile radius of its boundary

RHODE ISLAND

INVASIVE SPECIES

Title/Chapter/ Section	Section Title	Summary	Penalty
Title 20. Ch. 1 § 26	Freshwater invasive aquatic plants—prohibition on importation and possession	• No person shall import, transport, disperse, distribute, introduce, sell, purchase, or possess in the state any species of non-native (exotic) freshwater invasive aquatic plants, as defined by the director. The director shall promulgate rules and regulations governing the prohibition and its applicability	Violations of this section shall be punishable by a fine of not more than five hundred dollars ($500)

Title/Chapter/ Section	Section Title	Summary	Penalty
Title 20. Ch. 1 § 27	Exotic invasive freshwater fish and invertebrate species—prohibition on liberation into freshwaters	• No person shall release or liberate, by design or accident, any species of non-native (exotic) fish or invertebrate, as defined by the director, into the freshwater lakes, ponds, rivers, streams, or wetlands of the state. The director shall promulgate rules and regulations governing the prohibition and its applicability	Violations of this section shall be punishable by a fine of not more than five hundred dollars ($500)
Title 20. Ch. 1 § 1–28	Regulation of exotic baitfish—prohibition on importation, sale, and transport of exotic baitfish species	• No person shall import, transport, culture, or sell exotic species of baitfish, as specified by the director, in Rhode Island. The director shall promulgate rules and regulations governing the prohibition and its applicability. The regulations shall include a list of approved native baitfish and a list of prohibited non-native or exotic species	Violations of this section shall be punishable by a fine of not more than five hundred dollars ($500)

FISH AND MAMMALS AND OTHER

Title/Chapter/ Section	Section Title	Summary	Penalty
Title 20. Ch. 18 § 1.	Acquisition or control of land for protecting wildlife	• The director may acquire by gift, lease, purchase, or easement, land within the state for the purpose of protecting, conserving, cultivating, or propagating any species of wildlife, plant, or animal	

WETLANDS

Title and Chapter	Section	Summary	Penalty
Title 46. Ch. 12 § 37	Waste from seagoing vessels	• Pertains to the owners and/or agents of any seagoing vessel entering the waters or waterways of this state which intends to transfer or discharge any type of waste or bilge at a certified shore facility or terminal pursuant to the Resource and Conservation Recovery Act (RCRA), 42 U.S.C. § 6901 et seq.	The first file a performance bond or other evidence of financial responsibility with the director in the amount of a minimum of $50,000 payable to the state of Rhode Island and Providence Plantations. If vessel causes damage to the environment, the bond shall be forfeited to the extent of the costs incurred by the state to rectify and clean up the damages. To the extent of any fine levied for violations of water pollution abatement laws. Any seagoing vessel transferring any type of waste or bilge without coverage shall be fined minimum of $5,000

Title and Chapter	Section	Summary	Penalty
Title 46. Ch. 15 § 46–15–10	Sec. 10. Public nuisances— abatement	• Any person, municipality, municipal water department or agency, special water district, or private water company, committing a violation shall be liable for the costs of abatement of any pollution and any public nuisance caused by the violation. The superior court is hereby given jurisdiction over actions to recover the costs of the abatement • Any activity or condition declared by this chapter to be a nuisance, or which is otherwise in violation of this chapter, shall be abatable in the manner provided by law or equity for the abatement of public nuisances. In addition, the water resources board may proceed in equity to abate nuisances or to restrain or prevent any violation of this chapter	
Title 46. Ch. 17.1 § 1–1	Transporting of waste and dredge materials over state waterways prohibited	• Except as permitted under § 46–23–18.1 or otherwise, no person, either as principal, agent, or servant, shall transport or cause to be transported, dump, or cause to be dumped within the territorial waters of the state, any waste or dredge material	Every person who shall violate any provision of this section shall, upon conviction, be fined not more than one thousand dollars ($1,000) or be imprisoned not more than one year, or both, for each offense
Title 46. Ch. 17.3 § 1	Legislative findings	• Non-indigenous aquatic species introduced into waters of the United States from ballast water have caused tremendous environmental and ecological damage, displacing native species, degrading water quality, and interfering with commerce, including the zebra mussel brought into the Great Lakes and the upper Susquehanna River • Ballast water practices of commercial shipping vessels may cause irreversible damage to waters of the state; and protective measures against the introduction of non-indigenous aquatic species from ballast water are necessary to minimize exposure of the waters of the state to unwanted contamination and damage caused by these species	Sec 11. Penalties and remedies. Any person who willfully or negligently violates any provision of this chapter, or any rule or regulation or other order promulgated by the water resources board, or any condition of any permit issued pursuant to the chapter, is guilty of a misdemeanor and, upon conviction, shall be subject to a fine of not more than five hundred ($500) dollars for each separate offense or to imprisonment for a period of not more than one year, or both (RI. Gen. Laws. Ch. 17.3 § 11)

Relevant Acts

Act	Section Title	Summary	Year
Title 46. Ch. 23 § 2–3. The Comprehensive Watershed and Marine Monitoring Act of 2004	Purpose	• The purpose of this chapter is to establish a comprehensive watershed and marine monitoring system that provides consistent and useful data to resource managers, decision-makers and the public concerning the health of the marine environment of the state. The Rhode Island environmental monitoring collaborative is hereby established with the purposes of organizing, coordinating, maintaining, and supporting the watershed and marine monitoring system of the state	2004
Title 20 Chapter 20 § 3.2–2 and 3.2.3. Rhode Island Freedom to Fish and Marine Conservation Act	Findings Freedom to Fish	• The rights and interests of people to engage in commercial and recreational fishing in Rhode Island's marine waters need to be recognized and protected. Various management measures, including the closure of marine waters or portions thereof to fishing, can be utilized to manage marine animals, and resources, but such measures must be developed to specific conservation or restoration needs and be based on scientific information; and emanate from an open management and regulatory process, incorporating full input from all affected stakeholders, conducted pursuant to the general laws of the state of Rhode Island • Closer of recreational or commercial fishing will on occur if it is deemed necessary to protect, manage or restore marine fish, shellfish, crustaceans, and associated marine habitats or other marine resources, protect public health or safety, or address some other public purpose	2003
Title 20. Ch. 20–1 § 24. Wildlife Damage Act	General provisions	• For purposes of this section, "wildlife population" includes, but is not limited to deer, indigenous Canadian geese, mute swans, cormorants, wild turkeys, crows, coyotes, and furbearers • The department of environmental management shall establish a program of financial assistance to farmers, when state or federal funds become available, for the purpose of establishing preventive practices to protect damage to crops by wildlife. The director of the department is authorized and empowered to establish rules and regulations to enforce the provisions of this section	2002

VERMONT

Invasive Species

Title/Chapter	Section	Summary	Penalty
Title 6. Ch. 083 § 981	Adoption of compact	Party states: • An annual loss of approximately 137 billion dollars from the depredation of pest is certain to continue/increase • Varying climatic, geographic, and economic factors may be affected differently by pest, and may pose serious damages to all states • The migratory character of pest infestation makes it necessary for states adjacent to and distant from one another to complement each other's activities when faced with conditions of infestation and reinfestation • Pest can cause damage to crop and products. in the interest to all states established an operation of an Insurance Fund	

Title/Chapter	Section	Summary	Penalty
Title 6. Ch. 084 § 1033	Detection and abundance surveys; eradication and suppression	• The Secretary may conduct detection and abundance surveys for plant pests of an injurious nature that may be present in the State to determine the necessity for establishing control practices • When the Secretary determines that a new injurious plant pest exists within the State or that an established pest requires control and the nature of the pest dictates immediate action, he or she may proceed with a plan of eradication or suppression	
Title 10. Ch. 050 § 1453	Aquatic Nuisance Control Program	They shall: • Receive/respond to aquatic nuisance complaints • Work with municipalities, local interest organizations, private individuals, and agencies of the state to develop long-range programs regarding aquatic nuisance controls • Work with federal, state, and local governments to obtain funding for aquatic nuisance control programs. • Implement an aquatic species rapid response program • Administer a grant-in-aid program under section 1458 of this title • place a sign that states that the water is infected with an aquatic nuisance and that the sign is easily visible from a ramp used to launch vessels at any fish and wildlife access area on a body of water infected with an aquatic nuisance • Provide the Commissioner of Fish and Wildlife and the Commissioner of Motor Vehicles with written educational information about aquatic nuisances	Regarding the signage: in violation of section 1454 of this title may be subject to a penalty of up to $1,000.00 pursuant to 23 V.S.A. § 3317
Title 10. Ch. 050 § 1454	Transport of aquatic plants and aquatic nuisance species	• A person shall not transport an aquatic plant, aquatic plant part, or aquatic nuisance species to or from any Vermont water • The inspection of vessels entering or leaving water • No-cost boat wash; aquatic nuisance species inspection station is open for vessels • Requires for the draining of vessel when leaving the state and profit to transport away from the area where the vessel left the water (including equipment used, live wells, ballast tanks, bilge areas	Violations. Pursuant to 4 V.S.A. § 1102, a violation of this section may be brought in the Judicial Bureau by any law enforcement officer, as that term is defined in 23 V.S.A. § 3302(2), or, pursuant to section 8007 or 8008 of this title, a violation of this section may be brought in the Environmental Division of the Superior Court
Title 10. Ch. 050 § 1457	Entrance upon lands to prevent the introduction and spread of new aquatic species	In taking steps to prevent the introduction/spread of new aquatic species/invasive the Secretary may, after first obtaining the permission of the landowner or lessee, enter upon lands for the following purposes: • To survey for, inspect, or investigate conditions relating to new aquatic species that may become invasive. • To collect information to issue coverage under rapid response general permits under section 1456 of this title • To conduct or use control techniques that are available under or authorized by a rapid response general permit issued under section 1456 of this title • To determine whether the rules of the agency adopted or issued under this chapter are being complied with	

Fish, Mammals, and Others

Title/Chapter/Section	Section Title	Summary	Penalty
Title 10. Ch. 111 § 4601	Taking fish; possession	• A person shall not take fish, except in accordance with this part and regulations of the Board, or possess a fish taken in violation of this part or regulations of the Board (Added 1961, No. 119, § 1, eff. May 9, 1961)	
Title 10. Ch. 001 § 18	Governing the importation and possession of wild animals, excluding fish	• It is the purpose of this regulation to carry out the mandate of the Vermont General Assembly to control through a permit program the importation and possession • It is the purpose of this statute and its regulations to protect the health, safety, and welfare of animals, both wild and domestic, to prevent damage to agriculture and livestock, and to protect the health, safety, and welfare of human inhabitants of the State of Vermont	Any wild animal that is taken, imported, or possessed in violation of this regulation, or is kept in violation of any permit issued may be confiscated and disposed of in accordance with 10 V.S.A. § 4513. Permit violations and violations of Part 4 of Title 10 may result in the revocation of the permit

Wetlands

Title/Chapter/Section	Section Title	Summary	Penalty
Title 10. Ch. 049 § 1421	Policy	• It is declared to be in the public interest to make studies, establish policies, make plans, make rules, encourage, and promote buffers adjacent to lakes, ponds, reservoirs, rivers, and streams of the State, • The purposes of the rules shall be to further the maintenance of safe and healthful conditions; prevent and control water pollution; protect spawning grounds, fish, and aquatic life; control building sites, placement of structures, and land uses; reduce fluvial erosion hazards; reduce property loss and damage; preserve shore cover, natural beauty, and natural stability; and provide for multiple use of the waters in a manner to provide for the best interests of the citizens of the State	Any person who violates a rule adopted under this chapter shall be subject to the civil penalty provision of 23 V.S.A. § 3317(b) (10 V.S.A. § 1426)

Index

Note: Page numbers in *italics* indicate figures; page numbers in **bold** indicate tables.

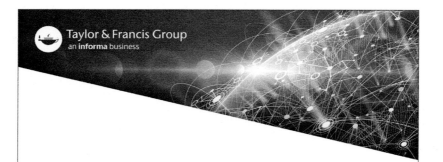